Introduction to Modern Virology

Introduction to Modern Virology

N. J. Dimmock
A. J. Easton
K. N. Leppard

Department of Biological Sciences
University of Warwick
Coventry

SIXTH EDITION

Blackwell
Publishing

© 1974, 1980, 1987, 1994, 2001, 2007 by Blackwell Publishing Ltd

BLACKWELL PUBLISHING
350 Main Street, Malden, MA 02148-5020, USA
9600 Garsington Road, Oxford OX4 2DQ, UK
550 Swanston Street, Carlton, Victoria 3053, Australia

First edition published 1974
Second edition published 1980
Third edition published 1987
Fourth edition published 1994
Fifth edition published 2001
Sixth edition published 2007 by Blackwell Publishing Ltd

1   2007

Library of Congress Cataloging-in-Publication Data

Dimmock, N. J.
    Introduction to modern virology/N. J. Dimmock, A. J. Easton, K. N. Leppard. – 6th ed.
        p. ; cm.
    Includes bibliographical references and index.
    ISBN-13: 978-1-4051-3645-7 (pbk. : alk. paper)
    ISBN-10: 1-4051-3645-6 (pbk. : alk. paper)   1. Virology.   2. Virus diseases.   I. Easton, A. J. (Andrew J.)   II. Leppard, Keith.   III. Title.
    [DNLM: 1. Viruses.   2. Virus Diseases.   QW 160 D582i 2007]

QR360.D56 2007
616.9'101–dc22

                                                                2006009426

A catalogue record for this title is available from the British Library.

Set in 10/12.5pt Meridien
by Graphicraft Limited, Hong Kong
Printed and bound in Singapore
by Markono Print Media Pte Ltd

The publisher's policy is to use permanent paper from mills that operate a sustainable forestry policy, and which has been manufactured from pulp processed using acid-free and elementary chlorine-free practices. Furthermore, the publisher ensures that the text paper and cover board used have met acceptable environmental accreditation standards.

For further information on
Blackwell Publishing, visit our website:
www.blackwellpublishing.com

Contents

Preface

This book, now in its sixth edition, provides a rounded introduction to viruses and the infections that they cause, and is aimed at undergraduate students at all levels and postgraduates wishing to learn about virology for the first time. It approaches the subject on a concept by concept basis, rather than considering each virus in turn. In this way, the important parallels and contrasts between different viruses and their infections are emphasized. Previous editions have underpinned our own teaching of virology at the University of Warwick for many years and have been widely adopted for undergraduate courses elsewhere. Our aim in writing this new edition has been to cover the breadth of this fascinating and important subject while keeping the text concise and approachable. It is thus suitable for students who may be studying virology as just one among many facets of biology or medicine, as well as for those who intend to focus on the subject in depth. A basic knowledge of cell and molecular biology is assumed, but other topics are introduced progressively in the text and explained as needed. An introduction to immunology is provided as a separate chapter because of its crucial relevance to the understanding of viral disease.

The pace at which information in the field of virology is accumulating has shown no signs of abating since the last revision of this volume. When incorporating these advances into the book, we have aimed to maintain a broad coverage of virology while emphasizing human and animal virus systems, although inevitably this has constrained our consideration of other viruses. Despite the relentless quest for knowledge, much still remains to be learned. In particular, the intimate interaction of viruses with their hosts at the molecular level is poorly understood. We have tried to indicate where such gaps in knowledge exist, and where future research is likely to be focused.

The public perception of viruses as significant threats to humans and animals, already heightened by the ongoing epidemic of HIV infection, has been brought into sharp relief with recent concerns over emerging viruses, such as the avian influenza viruses that have the potential to become pandemic strains of human influenza. It has never been more important than now to understand viruses and to spread that understanding as widely as possible. Although the discipline of virology is often considered to be highly specialized, we hope that readers will see the tremendous range of systems and technologies that virologists bring to bear in order to elucidate their subject and thus pick up some of the excitement of working in this field. Virology is a vibrant area and its study, far from being constraining, opens up a vista in which virus infections can be understood in the context of the biology of their hosts.

NEW TO THIS EDITION

This edition contains a number of important changes and innovations. The text has been reorganized to create four thematic sections on the fundamental nature of viruses, their growth in cells, their interactions with the host organism, and their role as agents of human disease. This clearer organization makes information more immediately available. We have added a new chapter on viral disease, and thoroughly revised and updated material in other chapters, adding sections on viruses as gene therapy vectors and emerging virus infections such as Ebola and SARS.

The presentation too has been comprehensively reorganized. The book is now illustrated in full color throughout and three types of text boxes have been included. Text features now comprise:

- Highlight boxes – draw attention to important points (pink, unnumbered boxes).
- Evidence boxes – provide experimental evidence for certain key facts and give additional detail (yellow, numbered boxes).
- Detail boxes – for in-depth study (green, numbered boxes).
- Integrated questions at the end of chapters – prompt students to digest and synthesize the information they have been reading about.
- Summaries at the end of chapters – review the key messages from the chapter.
- Additional readings – include suggestions for more information for the interested student or for research projects.

We hope all of these will be of use to students and teachers alike.

SUPPLEMENTS AVAILABLE

Website – With this edition, for the first time, we provide a website for instructors and students that includes:

- Artwork in high resolution for download.
- Animations that will illustrate some of the key processes in virology.

Artwork CD – Artwork from the book is available to instructors at www.blackwellpublishing.com/dimmock and by request on CD-ROM from this email address: artworkcd@bos.blackwellpublishing.com

We are grateful to the staff at Blackwell Publishing for their support for this new edition, and for their extensive input to it. We also acknowledge the contribution of the reviewers, Margo A. Brinton (Georgia State University), Julian A. Hiscox (Leeds University), Judy Kandel (California State University, Fullerton), Brian Martin (University of Birmingham), Nancy McQueen (California State University, Los Angeles), Andrew J. Morgan (University of Bristol), Jay Louise Nadeau (McGill University), Michael Roner (University of Texas at Arlington), A. C. R. Samson (University of Newcastle upon Tyne), and Juliet V. Spencer (University of San Francisco), who showed us many ways in which to improve our text.

Nigel Dimmock, Andrew Easton, and Keith Leppard
University of Warwick, July 2006

Advance Praise

"I have consistently used this book as a teaching resource. Concepts are explained clearly and background information is provided without excess detail. This book also contains some excellent figures."

Margo Brinton, Georgia State University

"The text is written in a style that undergraduate and graduate students alike will find appealing. The case examples and evidence boxes represent the applied side of virology, and should help to keep students interested in the material."

Michael Roner, University of Texas, Arlington

Part I

What is a virus?

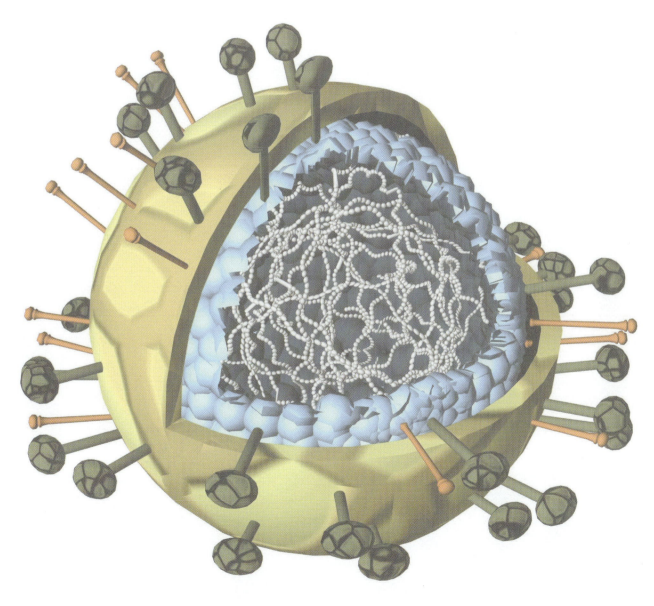

Towards a definition of a virus

Viruses occur universally, but they can only be detected indirectly. Although they are well known for causing disease, most viruses coexist peacefully with their hosts.

Chapter 1 Outline

It may come as a surprise to learn that every animal, plant and protist species on this planet is infected with viruses. Of course this is a generalization as not every species has been examined – far from it, as new species are being discovered almost every day, but those that have been tested all yield up new virus isolates. Further, not only do viruses occur universally but each species has its own specific range of viruses that, by and large, infects only that species. Thus, one can take the number of known human viruses (humans being the best studied host species) and multiply by the number of species in the world to obtain an estimate of the total number of extant virus genomes. These notions immediately inspire questions as what the viruses are doing there, and what selective advantage, if any, they afford to the species that hosts them. The answer to the first is the same as to the question as to what a lion is doing there – just existing in a particular environment, except the environment for a virus is another species. The answer to whether or not any benefit accrues for hosting a virus is not known – more is known about the downside of virus infections. However, it is clear that the viruses have not made their hosts extinct. At the moment all that is possible is to list some of the ways that viruses impact upon their host species (Box 1.1).

To understand the nature of viruses it is informative to consider the general aspects of their multiplication process and general properties.

Box 1.1

Some ways in which viruses impact upon their host species

- Some viruses impact on the health of their hosts, although probably most have no impact, or very little impact.
- There is a view that viruses only kill a large proportion of the hosts they infect when this is a new relationship; eventually this evolves into peaceful coexistence.
- A new virus–host relationship arises when a virus moves from its normal host to a new species; this is thought to be a rare event.
- It is axiomatic that the survival of a virus depends on the survival of its host species.
- At the organism level, different viruses have different lifestyles ranging from hit-and-run infections that make the host ill for a short period of time (days to a few weeks) to infections where there are no adverse signs. During the latter infections the virus may actively multiply but cause no disease, or for long periods may sit in a cell and do nothing.
- The impact on a host species can be adversely affected by external factors (e.g. nutritional status). Other factors like infection at a young age can exacerbate or ameliorate infection, depending on the virus.
- Virus infection of some plants, notably tulips, changes the color of their flowers.
- Viruses can make bacteria virulent, either by harboring a prophage (phage DNA which has integrated with the host's DNA) that encodes a toxin (e.g. *Corynebacterium diphtheriae* and the diphtheria toxin, *Vibrio cholerae* and cholera toxin) or by harboring "swarms" of prophages that incrementally contribute to bacterial virulence (e.g. *Salmonella enterica* serovar Typhimurium).

1.1 DISCOVERY OF VIRUSES

Although much is known about viruses (Box 1.2), it is instructive and interesting to consider how this knowledge came about. It was only just over 100 years ago at the end of the nineteenth century that the germ theory of disease was formulated, and pathologists were then confident that a causative microorganism would be found for each infectious disease. Further they believed that these agents of disease could be seen with the aid of a microscope, could be cultivated on a nutrient medium, and could be retained by filters. There were, admittedly, a few organisms which were so fastidious that they could not be cultivated *in vitro* (literally, in glass, meaning in the test tube), but the other two criteria were satisfied. However, a few years later, in 1892, Dmitri Iwanowski was able to show that the causal agent of a mosaic disease of tobacco plants, manifesting as a discoloration of the leaf, passed through a bacteria-proof filter, and could not be seen or cultivated. Iwanowski was unimpressed by his discovery, but Beijerinck repeated the experiments in 1898, and became

Box 1.2

Properties common to all viruses

- Viruses have a nucleic acid genome of either DNA or RNA.
- Compared with a cell genome, viral genomes are small, but genomes of different viruses range in size by over 100-fold (*c.* 3000 nt to 1,200,000 bp)
- Small genomes make small particles – again with a 100-fold size range.
- Viral genomes are associated with protein that at its simplest forms the virus particle, but in some viruses this nucleoprotein is surrounded by further protein or a lipid bilayer.
- Viruses can only reproduce in living cells.
- The outermost proteins of the virus particle allow the virus to recognize the correct host cell and gain entry into its cytoplasm.

convinced this represented a new form of infectious agent which he termed *contagium vivum fluidum*, what we now know as a virus. In the same year Loeffler and Frosch came to the same conclusion regarding the cause of foot-and-mouth disease. Furthermore, because foot-and-mouth disease could be passed from animal to animal, with great dilution at each passage, the causative agent had to be reproducing and thus could not be a bacterial toxin. Viruses of other animals were soon discovered. Ellerman and Bang reported the cell-free transmission of chicken leukemia in 1908, and in 1911 Rous discovered that solid tumors of chickens could be transmitted by cell-free filtrates. These were the first indications that some viruses can cause cancer.

Finally bacterial viruses were discovered. In 1915, Twort published an account of a glassy transformation of micrococci. He had been trying to culture the smallpox agent on agar plates but the only growth obtained was that of some contaminating micrococci. Upon prolonged incubation, some of the colonies took on a glassy appearance and, once this occurred, no bacteria could be subcultured from the affected colonies. If some of the glassy material was added to normal colonies, they too took on a similar appearance, even if the glassy material was first passed through very fine filters. Among the suggestions that Twort put forward to explain the phenomenon was the existence of a bacterial virus or the secretion by the bacteria of an enzyme which could lyse the producing cells. This idea of self-destruction by secreted enzymes was to prove a controversial topic over the next decade. In 1917 d'Hérelle observed a similar phenomenon in dysentery bacilli. He observed clear spots on lawns of such cells, and resolved to find an explanation for them. Upon noting the lysis of broth cultures of pure dysentery bacilli by filtered emulsions of feces, he immediately realized he was dealing with a bacterial virus. Since this virus was incapable of multiplying except at the expense of living bacteria, he called his virus a *bacteriophage* (literally a bacterium eater) or *phage* for short.

Thus the first definition of these new agents, the viruses, was presented entirely in negative terms: they could not be seen, could not be cultivated in the absence of cells and, most important of all, were not retained by bacteria-proof filters.

1.2 DEVELOPMENT OF VIRUS ASSAYS

Much of the early analytical virus work was carried out with bacterial viruses. Virologists of the time would much rather have worked with agents that caused disease in humans, animals, or crop plants, but the technology was not sufficiently advanced. It is simply not possible to analyze the details of virus growth in whole animals or plants, although viruses could be assayed in whole organisms (see below). Animal cell culture was not a practicable proposition until the 1950s when antibiotics became available for inhibiting bacterial contamination; plant cell culture is still technically difficult. This left bacterial viruses which infect cells that grow easily, in suspension culture, and quickly – experiments with bacterial viruses are measured in minutes, rather than the hours or days needed for animal viruses.

The observations of d'Hérelle in the early part of the twentieth century led to the introduction of two important techniques. The first of these was the preparation of stocks of bacterial viruses by lysis of bacteria in liquid cultures. This has proved invaluable in modern virus research, since bacteria can be grown in defined media to which radioactive precursors can be added to "label" selected viral components. Many animal viruses can be similarly grown in cultures of the appropriate animal cell. Secondly, d'Hérelle's observations provided the means of assaying these invisible agents. One method is to grow a large number of identical cultures of a susceptible bacterium species and to inoculate these with dilutions of the virus-containing sample. With more concentrated samples all the cultures lyse, but if the sample is diluted too far, none of the cultures lyse. However, in the intermediate range of dilutions not all of the cultures lyse, since not all receive a virus particle, and quantitation of virus is based on this. For example, in 10 test cultures inoculated with a dilution of virus corresponding to 10^{-11} ml, only three lyse. Thus, three cultures receive one or more viable phage particles while the remaining seven receive none, and it can be concluded that the sample contained between 10^{10} and 10^{11} viable phages per ml. It is possible to apply statistical methods to *end-point dilution* assays of this sort and obtain more precise estimates of virus concentration, normally termed the virus titer. The other method suggested was the *plaque assay*, which is now the more widely used and more useful. d'Hérelle observed that the number of clear spots or

Fig. 1.1 Plaques of viruses. (a) Plaques of a bacteriophage on a lawn of *Escherichia coli*. (b) Local lesions on a leaf of *Nicotiana* caused by tobacco mosaic virus. (c) Plaques of influenza virus on a monolayer culture of chick embryo fibroblast cells.

plaques formed on a lawn of bacteria (Fig. 1.1a) was inversely proportional to the dilution of bacteriophage lysate added. Thus the titer of a virus-containing solution can be readily determined in terms of *plaque-forming units* (PFU) per ml. If each virus particle in the preparation gives rise to a plaque, then the *efficiency of plating* is unity, however for many viruses preparations have particle to PFU ratios considerably greater than 1.

Both these methods were later applied to the more difficult task of assaying plant and animal viruses. However, because of the labor, time, cost, and ethical considerations, end-point dilution assays using animals are avoided where possible. For the assay of plant viruses, a variation of the plaque assay, the *local lesion assay* was developed by Holmes in 1929. He observed that countable necrotic lesions were produced on leaves of the tobacco plant, particularly *Nicotiana glutinosa*, inoculated with tobacco mosaic virus and that the number of local lesions depended on the amount of virus in the inoculum. Unfortunately, individual plants, and even individual leaves of the same plant, produce different numbers of lesions with the same inoculum. However, the opposite halves of the same leaf give almost identical numbers of lesions so two virus-containing samples can be compared by inoculating them on the opposite halves of the same leaf (Fig. 1.1b).

A major advance in animal virology came in 1952, when Dulbecco devised a plaque assay for animal viruses. In this case a suspension of susceptible cells, prepared by trypsinization of a suitable tissue, is placed in Petri dishes or other culture vessel. The cells attach to the surface and divide until a *monolayer* of cells (one cell in depth) is formed. The nutrient medium bathing the cells is then removed and a suitable dilution of the virus added. After a short period of incubation to allow virus particles to attach to the cells, nutrient agar is placed over the cells. After a further period of incubation of usually around 3 days, (but ranging from 24 hours to 24 days depending on the type of virus), a dye is added to differentiate living cells from the unstained circular areas that form the plaques (Fig. 1.1c). These days plaque assays are conducted using cell lines that can be maintained for many generations in the laboratory, rather than generating them from fresh tissue every time. Some viruses are not cytopathic (i.e. do not kill cells), but infected cells can always be recognized by the presence of virus protein or nucleic acids that they produce, providing that the appropriate specific detection reagents are available. An alternative for those tumor viruses that cause morphological transformation of cells (Chapter 20), is a focus-forming assay in which a single infectious particle leads to the formation of a discrete colony of cells; colonies can be counted as a measure of the input virus.

1.3 MULTIPLICATION OF VIRUSES

Although methods of assaying viruses had been developed, there were still considerable doubts as to the nature of viruses. d'Hérelle believed that the infecting phage particle multiplied within the bacterium and that its progeny were liberated upon lysis of the host cell, whereas others believed that phage-induced dissolution of bacterial cultures was merely the consequence of a stimulation of lytic enzymes endogenous to the bacteria. Yet another school of thought was that phages could pass freely in and out of bacterial cells and that lysis of bacteria was a secondary phenomenon not necessarily concerned with the growth of a phage. It was Delbruck who ended the controversy by pointing out that two phenomena were involved, lysis from within and lysis from without. The type of lysis observed was dependent on the ratio of infecting phages to bacteria (*multiplicity of infection*). At a low multiplicity of infection (with the ratio of phages to bacteria no greater than 2 : 1), then the phages infect the cells, multiply, and lyse the cells from within. When the multiplicity of infection is high, i.e. many hundreds of phages per bacterium, the cells are lysed directly, and rather than an increase in phage titer there is a decrease. Lysis is due to weakening of the cell wall when large numbers of phages are attached.

Convincing support for d'Hérelle's hypothesis was provided by the one-step growth experiment of Ellis and Delbruck in 1939. A phage preparation such as bacteriophage λ (lambda) is mixed with a suspension of the bacterium *Escherichia coli* at a multiplicity of infection of 10 PFU per cell, ensuring that virtually all cells are infected. Then after allowing 5 minutes for the phage to attach, the culture is centrifuged to pellet the cells and attached phage. Medium containing unattached phage is discarded. The cells are then resuspended in fresh medium. Samples of medium are withdrawn at regular intervals, cells removed and assayed for infectious phage. The results obtained are shown in Fig. 1.2. After a latent period of 17 minutes in which no phage increase is detected in cell-free medium, there is a sudden rise in PFU in the medium. This "burst" size represents the average of many different bursts from individual cells, and can be calculated from the total virus yield/number of cell infected. The entire growth cycle here takes around 30 minutes, although this will vary with different viruses and cells. The amount of cell-associated virus is determined by taking the cells pelleted from the medium, disrupting them, and assaying for virus infectivity as before. The fact that virus appears inside the cells before it appears in the medium demonstrates the intracellular nature of phage replication. It can be seen also that the kinetics of appearance of intracellular phage particles are *linear*, not exponential. This is consistent with particles being produced by assembly from component parts, rather than by binary fission.

1.4 THE VIRUS MULTIPLICATION CYCLE

We now know a great deal about the processes which occur during the multiplication of viruses within single cells. The precise details vary for individual viruses but have in common a series of events marking specific phases in the multiplication cycle. These phases are summarized in Fig. 1.3 and are considered in detail in section II of this book. The first stage is that of *attachment* when the virus attaches to the potential host cell. The interaction is specific, with the virus attachment protein(s) binding to target receptor molecules on the surface of the cell. The initial contact between a virus and host cell is dynamic and reversible, and often involves weak electrostatic interactions. However, the contacts quickly become much stronger with more stable interactions which in some cases are essentially

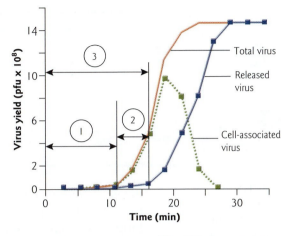

Fig. 1.2 A one-step growth curve of bacteriophage λ following infection of susceptible bacteria (*Escherichia coli*). During the *eclipse phase* (1), the infectivity of the cell-associated, infecting virus is lost as it uncoats; during the *maturation phase* (2) infectious virus is assembled inside cells (cell-associated virus), but not yet released; and the *latent phase* (3) measures the period before infectious virus is released from cells into the medium. Total virus is the sum of cell-associated virus + released virus. Cell-associated virus decreases as cells are lysed. This classic experiment shows that phages develop intracellularly.

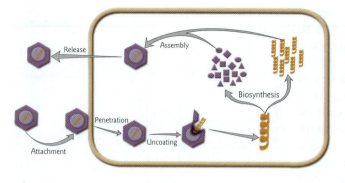

Fig. 1.3 A diagrammatic representation of the six phases common to all virus multiplication cycles. See text for details.

irreversible. The attachment phase determines the specificity of the virus for a particular type of cell or host species. Having attached to the surface of the cell, the virus must effect entry to be able to replicate in a process called *penetration* or *entry*. Once inside the cell the genome of the virus must become available. This is achieved by the loss of many, or all, of the proteins that make up the particle in a process referred to as *uncoating*. For some viruses the entry and uncoating phases are combined in a single process. Typically these first three phases do not require the expenditure of energy in the form of ATP hydrolysis. Having made the virus genome available it is now used in the *biosynthesis* phase when genome replication, transcription of mRNA, and translation of the mRNA into protein occur. The process of translation uses ribosomes provided by the host cell and it is this requirement for the translation machinery, as well as the need for molecules for biosynthesis, that makes viruses obligate intracellular parasites. The newly synthesized genomes may then be used as templates for further rounds of replication and as templates for transcription of more virus mRNA in an amplification process which increases the yield of virus from the infected cells. When the new genomes are produced they come together with the newly synthesized virus proteins to form progeny virus particles in a process called *assembly*. Finally, the particles must leave the cell in a *release* phase after which they seek out new potential host cells to begin the process again. The particles produced within the cell may require further processing to become infectious and this *maturation* phase may occur before or after release.

Combining the consideration of the steps which make up a virus multiplication cycle with the information in the graph of the results of a single step growth curve it can be seen that during the eclipse phase the virus is undergoing the processes of attachment, entry, uncoating, and biosynthesis. At this time the cells contain all of the elements necessary to produce viruses but the original infecting virus has been dismantled and no new infectious particles have yet been produced. It is only after the assembly step that we see virus particles inside the cell before they are released and appear in the medium.

1.5 VIRUSES CAN BE DEFINED IN CHEMICAL TERMS

The first virus was purified in 1933 by Schlessinger using differential centrifugation. Chemical analysis of the purified bacteriophage showed that it consisted of approximately equal proportions of protein and deoxyribonucleic acid (DNA). A few years later, in 1935, Stanley isolated tobacco mosaic virus in paracrystalline form, and this crystallization of a biological material thought to be alive raised many philosophical questions about the nature of life. In 1937, Bawden and Pirie extensively purified tobacco mosaic virus and showed it to be nucleoprotein containing ribonucleic acid (RNA). Thus virus particles may contain either DNA or RNA. However, at this time it was not known that nucleic acid constituted genetic material.

The importance of viral nucleic acid

In 1949, Markham and Smith found that preparations of turnip yellow mosaic virus comprised two types of identically sized spherical particles, only one of which contained nucleic acid. Significantly, only the particles containing nucleic acid were infectious. A few years later, in 1952, Hershey and Chase demonstrated the independent functions of viral protein and nucleic acid using the head–tail virus, bacteriophage T2 (Box 1.3).

In another classic experiment, Fraenkel-Conrat and Singer (1957) were able to confirm by a different means the hereditary role of viral RNA. Their experiment was based on the earlier discovery that particles of tobacco mosaic virus can be dissociated into their protein and RNA components, and then reassembled to give particles which are morphologically mature and fully infectious (see Chapter 11). When particles of two different strains

Box 1.3

Evidence that DNA is the genetic material of bacteriophage T2: the Hershey–Chase experiment

Bacteriophage T2 was grown in *E. coli* in the presence of ^{35}S (as sulfate) to label the protein moiety, or ^{32}P (as phosphate) to mainly label the nucleic acid. Purified, labelled phages were allowed to attach to sensitive host cells and then given time for the infection to commence. The phages, still on the outside of the cell, were then subjected to the shearing forces of a Waring blender. Such treatment removes any phage components attached to the outside of the cell but does not affect cell viability. Moreover, the cells are still able to produce infectious progeny virus. When the cells were separated from the medium, it was observed that 75% of the ^{35}S (i.e. phage protein) had been removed from the cells by blending but only 15% of the ^{32}P (i.e. phage nucleic acid) had been removed. Thus, after infection, the bulk of the phage protein appeared to have no further function and this suggested (but does not prove – that had to await more rigorous experiments with purified nucleic acid genomes) that the nucleic acid is the carrier of viral heredity. The transfer of the phage nucleic acid from its protein coat to the bacterial cell upon infection also accounts for the existence of the eclipse period during the early stages of intracellular virus development, since the nucleic acid on its own cannot normally infect a cell (Fig. 1.4).

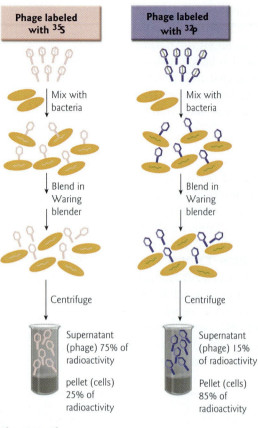

Phage labeled with ^{35}S

Mix with bacteria

Blend in Waring blender

Centrifuge

Supernatant (phage) 75% of radioactivity

pellet (cells) 25% of radioactivity

Phage labeled with ^{32}P

Mix with bacteria

Blend in Waring blender

Centrifuge

Supernatant (phage) 15% of radioactivity

Pellet (cells) 85% of radioactivity

Fig. 1.4 The Hershey–Chase experiment proving that DNA (labelled with ^{32}P) is the genetic material of bacteriophage T2.

(differing in the symptoms produced in the host plant) were each disassociated and the RNA of one reassociated with the protein of the other, and vice versa, the properties of the virus which was propagated when the resulting "hybrid" particles were used to infect host plants were always those of the parent virus from which the RNA was derived (Fig. 1.5).

The ultimate proof that viral nucleic acid is the genetic material comes from numerous observations that under special circumstances purified viral nucleic acid is capable of initiating infection, albeit with a reduced efficiency. For example, in 1956 Gierer and Schramm, and Fraenkel-Conrat independently showed that the purified RNA of tobacco mosaic virus can be infectious, provided precautions are taken to protect it from inactivation by ribonuclease. An extreme example is the causative agent of potato spindle tuber disease which lacks any protein component and consists solely of RNA. Because such agents have no protein coat, they cannot be called viruses and are referred to as *viroids*.

Synthesis of macromolecules in infected cells

Knowing that nucleic acid is the carrier of genetic information, and that only the nucleic acid of bacteriophages enters the cell, it is pertinent to review the events occurring inside the cell. The discovery in 1953, by Wyatt and Cohen, that the DNA of the T-even bacteriophages T2, T4, and T6 contains hydroxymethylcytosine (HMC) instead of cytosine made it possible for Hershey, Dixon, and Chase to examine infected bacteria for the presence of phage-specific DNA at various stages of intracellular growth. DNA was extracted from T2-infected *E. coli* at different times after the onset of phage growth, and analyzed for its content of HMC. This provided an estimate of the number of phage equivalents of HMC-containing DNA present at any time, based on the total nucleic acid and relative HMC content of the intact T2 phage particle. The results showed that, with T2, synthesis of phage DNA commences about 6 minutes after infection and the amount present then rises sharply, so that by the time the first infectious particles begin to appear 6 minutes later there are 50–80 phage equivalents of HMC. Thereafter, the numbers of phage equivalents of DNA and of infectious particles increase linearly and at the same rate up until lysis, even if lysis is delayed beyond the normal burst time.

Hershey and his co-workers also studied the synthesis of phage protein, which can be distinguished from bacterial protein by its interaction with specific antibodies. During infection of *E. coli* by T2 phage, protein can be detected about 9 minutes after the onset of the latent period, i.e. after DNA synthesis begins, and before infectious particles appear. A few minutes later there are approximately 30–40 phages inside the cell. Whereas the synthesis of viral protein starts about 9 minutes after the onset of the latent period, it was shown by means of pulse–chase experiments that the uptake of ^{35}S into intracellular protein is constant from the start of infection. A small quantity (a pulse) of ^{35}S (as sulfate) was added to the medium at different times after infection and was followed shortly by a vast excess of unlabelled sulfate (chase) to stop any further incorporation of label. When the pulse was made from the ninth minute onward, the label could be chased into material identifiable by its reaction with antibody (i.e. serologically) as phage coat protein. However, if the pulse was made early in infection, it could be chased into protein but, although this was nonbacterial, it did not react with antibodies to phage structural proteins. This early protein comprises mainly virus-specified enzymes that are concerned with phage replication but are not incorporated into phage particles. The concept of early and late, nonstructural and structural viral proteins is discussed in Part II.

These classical experiments are typical only of head–tail phages infecting *E. coli* under optimum growth conditions. *E. coli* is normally found in the anaerobic environment of the intestinal tract, and it is doubtful that it grows with its optimal doubling time of 20 minutes under natural conditions. Other bacterial cells grow more slowly than *E. coli* and their viruses have longer multiplication cycles.

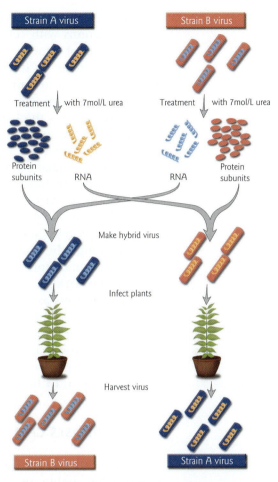

Fig. 1.5 The experiment of Fraenkel-Conrat and Singer which proved that RNA is the genetic material of tobacco mosaic virus.

1.6 MULTIPLICATION OF BACTERIAL AND ANIMAL VIRUSES IS FUNDAMENTALLY SIMILAR

The growth curves and other experiments described above have been repeated with many animal viruses with essentially similar results.

Bacterial and animal viruses both attach to their target cell through specific interactions with cell surface molecules. Like the T4 bacteriophage, the genomes of some animal viruses (e.g. HIV-1) enter the cell and leave their coat proteins on the outside. However, with most animal viruses, some viral protein, usually from inside the particle, enters the cell in association with the viral genome. In fact it is now known that some phage protein enters the bacterial cells with the phage genome. Such proteins are essential for genome replication. Many other animal viruses behave slightly differently, and after attachment are engulfed by the cell membrane, and taken into the cell inside a vesicle. However, strictly speaking this virus has not yet entered the cell cytoplasm, and is still outside the cell. The virus genome gains entry to the cytoplasm through the wall of the vesicle, when the particle is stimulated to uncoat. Again, the outer virion proteins stay in the vesicle – i.e. outside the cell. Animal viruses go through the same stages of eclipse, and virus assembly from constituent viral components with linear kinetics, as bacterial viruses. Release of progeny virions may happen by cell lysis (although this is not an enzymatic process as it is with some bacterial viruses), but frequently virus is released without major cell damage. The cell may die later, but death of the cell does not necessarily accompany the multiplication of all animal viruses. One major difference in the multiplication of bacterial and animal virus is that of time scale – animal virus growth cycles take in the region of 5–15 hours for completion.

1.7 VIRUSES CAN BE MANIPULATED GENETICALLY

One of the easiest ways to understand the steps involved in a particular reaction within an organism is to isolate mutants which are unable to carry out that reaction. Like all other organisms, viruses sport mutants in the course of their growth, and these mutations can affect all properties including the type of plaque formed, the range of hosts which the virus can infect, and the physicochemical properties of the virus. One obvious caveat, however, is that many mutations will be lethal to the virus and remain undetected. This problem was overcome in 1963 by Epstein and Edgar and their collaborators with the discovery of *conditional lethal mutants*. One class of these mutants, the *temperature-sensitive mutants*, was able to grow at a lower temperature than normal, the *permissive temperature*, but not at a higher, *restrictive* temperature at which normal virus could grow. Another class of conditional lethal mutants was the *amber* mutant. In these mutants a DNA lesion converts a codon within transcribed RNA into a triplet which terminates protein synthesis. They can only grow on a *permissive* host cell, which has an amber-suppressor transfer RNA (tRNA) that can insert an amino acid at the mutation site during translation.

The drawback to conditional lethal mutants is that mutation is random, but the advent of *recombinant DNA technology* has facilitated controlled mutagenesis, known as reverse genetics, at least for those viruses for which infectious particles can be reconstituted from cloned genomic DNA or cDNA (DNA that has been transcribed from RNA). What happens is that a piece of a cloned viral DNA or cDNA genome containing the target sequence is excised from the plasmid using two different restriction enzymes, so that it forms a unique restriction fragment which can eventually be re-inserted in the correct orientation. The fragment is then modified by *oligonucleotide-* or *site-directed mutagenesis* via the polymerase chain reaction (PCR) using appropriate mutagenic primers, is then inserted back into the original sequence in a plasmid, and then used to form a mutated virus particle. The PCR reaction is explained in Section 2.3.

1.8 PROPERTIES OF VIRUSES

With the assumption that the features of virus growth just described for particular viruses are true of all viruses, it is possible to compare and contrast the properties of viruses with those of their host cells. Whereas host cells contain both types of nucleic acid, viruses only contain one type. However, just like their host cells, viruses have their genetic information encoded in nucleic acid. Another difference is that the virus is reproduced solely from its genetic material, whereas the host cell is reproduced from the integrated sum of its components. Thus, the virus never arises directly from a pre-existing virus, whereas the cell always arises by division from a pre-existing cell. The experiments of Hershey and his collaborators showed quite clearly that the components of a virus are synthesized independently and then assembled into many virus particles. By contrast, the host cell increases its constituent parts, during which the individuality of the cell is continuously maintained, and then divides and forms two cells. Finally, viruses are incapable of synthesizing ribosomes, and depend on pre-existing host cell ribosomes for synthesis of viral proteins. These features clearly separate viruses from all other organisms, even chlamydia, which for many years were considered to be intermediate between bacteria and viruses.

1.9 ORIGIN OF VIRUSES

The question of the origin of viruses is a fascinating topic but as so often happens when hard evidence is scarce, discussion can generate more heat than light. There are two popular theories: viruses are either degenerate cells or vagrant genes. Just as fleas are descended from flies by loss of wings, viruses may be derived from pro- or eukaryotic cells that have dispensed with many of their cellular functions (*degeneracy*). Altern-atively, some nucleic acid might have been transferred accidentally into a cell of a different species (e.g. through a wound or by sexual contact) and, instead of being degraded, as would normally be the case, might have survived and replicated (*escape*). Although half a century has elapsed since these two theories were first proposed, we still do not have any firm indications if either, or both, are correct. Rapid sequencing of viral and cellular genomes is now providing data for computer analysis that is giving an ever better understanding of the relatedness of different viruses. However, while such analyses may identify the progenitors of a virus, they cannot decide between degeneracy and escape.

It is unlikely that all currently known viruses have evolved from a single progenitor. Rather, viruses have probably arisen numerous times in the past by one or both of the mechanisms outlined above. Once formed, viruses are subject to evolutionary pressures, just as are all other organisms, and this has led to the extraordinary diversity of viruses that exist today. Two processes that contribute significantly to virus evolution are recombination and mutation. Recombination takes place infrequently between the single molecule genomes of two related DNA or RNA viruses that are present in the same cell and generates a novel combination of genes. Of far greater significance is the potential for genetic exchange between related viruses with segmented genomes. Here, whole functional genes are exchanged, and this type of recombination is called *reassortment*. The only restriction is the compatibility between the various individual segments making up the functional genome. Fortunately, this seems to be a real barrier to the unlimited creation of new viruses, although it is not invincible, since pandemic influenza A viruses can be created in this way (see Section 18.7). Mutation is of particular significance to the evolution of RNA genomes as, in contrast to DNA synthesis, there is no molecular proof-reading mechanism during RNA synthesis. Mutations accumulated at a rate of approximately 3×10^{-4} per nucleotide per cycle of replication, whereas with DNA this figure is 10^{-9} to 10^{-10} per nucleotide per cycle. In other words, an RNA virus can achieve in one generation the degree of genetic variation which would take an equivalent DNA genome between 300,000 and 3000,000 generations to achieve. Once formed by reassortment, an influenza A virus evolves so rapidly that it takes only 4 years on average to mutate sufficiently to escape recognition by host defences and to reinfect that same individual.

KEY POINTS

- How viruses were discovered.
- Viruses multiply by assembling many progeny particles from a pool of virus-specified components, whereas cells multiply by binary fission.

- Viruses are evolutionarily unstable, and recombination, reassortment, and mutation of viral genomes provide a platform for natural selection.
- In animals the immune system provides powerful selection, and viruses evolve to evade its influence in a dynamic interaction.
- Viruses with RNA genomes mutate around one million times faster than DNA viruses, as there is no proof-reading mechanism for RNA replication.
- Viruses have probably originated independently many times.
- It is likely that every living organism on this planet is infected by a species-specific range of viruses.

FURTHER READING

Cann, A. 2001. *Principles of Molecular Virology*, 3rd edn. Academic Press, London.

Flint, S. J., Enquist, L. W., Krug, R. M., Racaniello, V. R., Skalka, A. M. 2003. *Principles of Virology: molecular biology, pathogenesis, and control*, 2nd edn. ASM Press, Harnden, VA.

Granoff, A., Webster, R. G. 1999. *Encyclopedia of Virology*, 2nd edn, vol 1. Academic Press, New York.

Hull, R. 2002. *Matthew's Plant Virology*, 4th edn. Academic Press, New York.

Knipe, D. M., Howley, P. M., Griffin, D. E., *et al.* 2001. *Field's Virology*, 4th edn. Lippincott Williams & Wilkins, Philadelphia.

Mahy, B. W. J. 2001. *A Dictionary of Virology*, 3rd edn. Academic Press, San Diego, CA.

Murphy, F. A., Gibbs, E. P. J., Horzinek, M. C., Studdert, M. J. 1999. *Veterinary Virology*, 3rd edn. Academic Press, New York.

Old, R. W., Primrose, S. B. 1994. *Principles of Genetic Manipulation*, 5th edn. Blackwell Scientific Publications, Oxford.

White, D. O., Fenner, F. J. 1994. *Medical Virology*, 4th edn. Academic Press, New York.

Zuckerman, A. J., Banatvala, J., Pattison, J. R. 1999. *Principles and Practice of Clinical Virology*, 4th edn. John Wiley & Sons, Chichester.

Also check Appendix 7 for references specific to each family of viruses.

2

Some methods for studying animal viruses

Viruses are usually detected indirectly – by the pathological effects which result from their multiplication, by interaction with antibody, or by identification of their nucleic acid genomes. Advances in knowledge are driven by technical developments.

Viruses are too small to be seen except by electron microscopy (EM) and this requires concentrations in excess of 10^{11} particles per ml, or even higher if a virus has no distinctive morphology, some fancy equipment, and a highly skilled operator. Thus viruses are usually detected by indirect methods. These fall into three categories: (i) *multiplication* in a suitable culture system and detection of the virus by the effects it causes; (ii) *serology*, which makes use of the interaction between a virus and antibody directed specifically against it; and (iii) detection of viral *nucleic acid*. However these days the *polymerase chain reaction* (PCR) is more likely to be employed as it is much quicker provided that the appropriate oligonucleotide primers are available (Section 2.3). Many viruses are uncultivatable, particularly those occurring in the gut, but some of these occur in such high concentration that they were actually discovered by EM. This chapter is not intended to be a technical manual, but to illustrate the principles governing the study of animal viruses.

2.1 SELECTION OF A CULTURE SYSTEM

The culture system for growing a virus always consists of living cells, and the choice is outlined in Box 2.1. Which culture system is used depends on the aims of the experiment, for example isolation of viruses, biochemistry of multiplication, structural studies, and study of natural infections.

Often a virus is first noticed because it is suspected of causing disease. By definition, disease can only be studied in the whole organism, preferably the natural host. However, this may be ruled

Box 2.1

Choosing a culture system for animal viruses

Culture system	Advantages	Limitations
Animal	Natural infection	Upkeep is expensive. Variation between individuals, even if inbred, means that large numbers needed. Ethical considerations
Organ culture, e.g. pieces of brain, gut, trachea	Natural infection Differentiated cell types present Fewer animals needed Less variation since one animal gives many organ cultures	Unnatural since cultures are no longer subject to homeostatic responses such as the immune system
Cell	Can be cloned therefore variation between individuals is minimal Good for biochemical studies as the environment can be controlled exactly and quickly	There are three types of cell cultures: primary cells, cell lines, and permanent cell lines. Primary cells are derived from an organ or tissue, remain differentiated, but survive for only a few passages. Cell lines are dedifferentiated but diploid and survive a larger number (about 50) of passages before they die. Continuous cell lines are also dedifferentiated, but immortal

out for humans on ethical or safety grounds. Alternatively, organ cultures and cells can be used. Logically, these should be from the natural host and obtained from those sites where the virus multiplies in the whole animal. However, it may be that cells from unrelated animals are susceptible, e.g. human influenza viruses were first cultivated by inoculating a ferret intranasally and found to grow best in embryonated chicken eggs. Usually, viruses grow poorly on initial isolation but adapt, due to selection of mutants, on being passed from culture to culture. Then there is the problem of knowing how similar the adapted virus is to the original primary isolate. PCR gets over this difficulty as it uses the original nucleic acid as template.

The usual way of detecting the presence of virus in an infected cell is by the pathology that it causes. This is known as the *cytopathic effect* or CPE. Often a virus or group of related viruses changes the morphology of the cell in a characteristic way, and this can be recognized by inspecting the cell culture through a microscope at low magnification. During the isolation of an unknown virus, such CPE gives an excellent clue as to

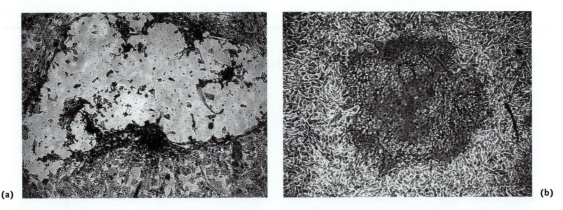

(a)   (b)

Fig. 2.1 Cytopathic effects caused by an influenza A virus and human respiratory syncytial virus (HRSV) in confluent cell monolayers (a layer of cells with a depth of just one cell). (a) Chick embryo cells infected by influenza A virus. In the clear central area infected cells have lysed. Some cell debris remains, and cells in the process of rounding up can be seen on the edge of the lesion. There are healthy cells around the periphery. (b) A monkey cell line infected with HRSV. HRSV does not lyse cells, but causes them to fuse together to form syncytia. A collection of syncytia forms the dark area in the center. Individual cells are magnified to approximately 3 mm in length, and are packed close together. Note the difference in morphology between the monkey cells and the chick cells: the monkey cells are slimmer and are more regularly packed together.

which further, more specific, diagnostic tests to employ. In the research laboratory, CPE provides a quick and easy check on the progress of the infection. An example of CPE is shown in Fig. 2.1.

Biochemical studies of virus infections require a cell system in which nearly every cell is infected. To achieve this, large numbers of infectious particles, and hence a system which will produce them, are required. Often, cells which are suitable for production of virus are different from those used for the study of virus multiplication. There is little logic in choosing a cell system, only pragmatism. Cells differ greatly and different properties make one cell the choice for a particular study and unsuitable for another. The ability to control the cell's environment is desirable, especially for labelling with radioisotopes, since a chemically defined medium must be prepared that lacks the nonradioactive isotope. Otherwise, the specific activity of the radioisotope would be reduced to an unusable level.

Whole organisms

The investigation of natural infections and disease is best done in the natural host. However, these are frequently unsuitable and the nearest approximation is usually a purpose-bred animal which has a similar range of defence mechanisms and can be maintained in the laboratory. The mouse has been extensively studied, its genetics are well understood, and inbred strains reduce genetic variability. Although the use of animals for studying virus diseases has been criticized by organizations concerned with animal rights, the student of virology will be aware, after reading Parts III and IV, that there is, as yet, no alternative for studying the complex interactions of viruses with the responses of the host. Although analysis of the processes involved would be so much easier if there were a test-tube system, none seem likely to be available in the foreseeable future.

Organ cultures

Organ cultures have the advantage of maintaining the differentiated state of the target cell. However, there are technical difficulties in their large-scale use, and as a result they have not been widely employed.

A commonly used organ culture system is that derived from the trachea, which has been used to grow a variety of respiratory viruses. Figure 2.2

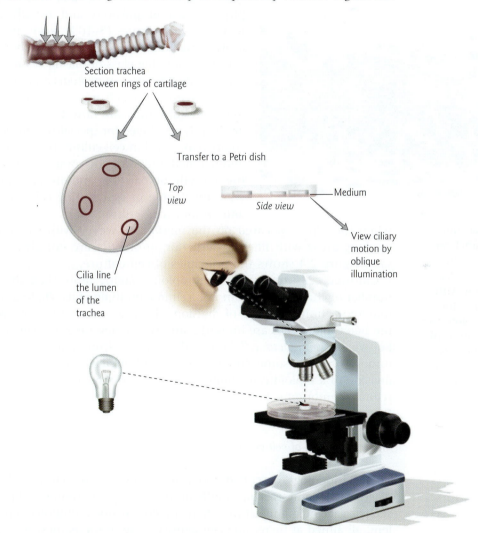

Fig. 2.2 Preparation of tracheal organ cultures.

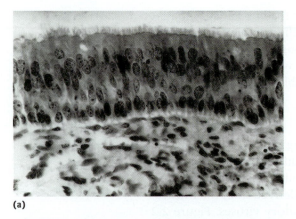

(a)

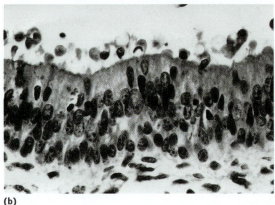

(b)

Fig. 2.3 Sections through tracheal organ cultures: (a) uninfected; (b) infected with a rhinovirus for 36 hours. Note the disorganization of the ciliated cells (uppermost layer) after infection. (Courtesy of Bertil Hoorn.)

shows the procedure used to prepare these cultures. Ciliated cells lining the trachea continue to beat in coordinated waves while the tissue remains healthy. Multiplication of some viruses causes the synchrony to be lost and eventually the ciliated cells to detach (Fig. 2.3). Virus is also released into fluids surrounding the tissue and can be measured if appropriate assays are available.

Cell cultures

Cells in culture are kept in an isotonic solution, consisting of a mixture of salts in their normal physiological proportions and usually supplemented with serum (5–10% v/v). In such a growth medium most cells rapidly adhere to the surface of suitable glass or plastic vessels. Serum is a complex mixture of proteins and other compounds without which mitosis does not occur. Synthetic substitutes are now available but these are mainly employed for specialized purposes. All components used in cell culture have to be sterile and handled under aseptic conditions to prevent the growth of bacteria and fungi. Antibiotics were invaluable in establishing cells in culture, and routine cell culture dates from the 1950s when they first appeared on the market. However with the advent of working areas with filtered sterile air, antibiotics are not always necessary. Figure 2.4 shows the principles of cell culture.

Cultured cells are usually heteroploid (having more than the diploid number of chromosomes but not a simple multiple of it). Diploid cell lines undergo a finite number of divisions, from around 10 to 100, whereas the heteroploid cells are immortal and will divide for ever. The latter are known as *continuous cell lines* and originate from naturally occurring tumors or from some spontaneous event which alters the control of division of a diploid cell. Diploid cell lines are most easily obtained from reducing embryonic kidney or whole body to a suspension of single cells. Frequently mouse or chicken embryos are used.

Modern methods of cell culture

The methodology described above is suited for research and clinical or diagnostic laboratories but is difficult to scale up for commercial purposes, such as vaccine manufacture. There are now various solutions to the problem, all aimed at increasing cell density. One of the earliest was to grow cells in suspension, and this has been refined to grow hybridoma cells (immortalized antibody-synthesizing or B cells) which produce *monoclonal antibodies* (MAbs). However, many cells only grow when anchored to a solid surface, so the technology has sought to increase the surface area available by, for example, providing spiral inserts to fit into conventional culture bottles (Fig. 2.5a). Another method is to grow cells on "microcarriers," tiny particles (about 200 μm diameter) on which cells attach and divide. The surface area afforded by 1 kg of microcarriers is about 2.5 m² and the space taken up (a prime consideration in commercial practice) is economical. This method combines the ease of handling cell suspensions with a solid matrix for the cell to grow on (Fig. 2.5b,c).

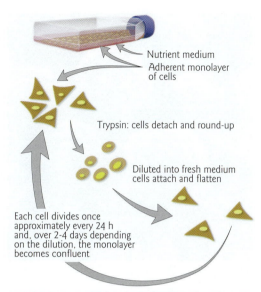

Nutrient medium
Adherent monolayer of cells

Trypsin: cells detach and round-up

Diluted into fresh medium cells attach and flatten

Each cell divides once approximately every 24 h and, over 2-4 days depending on the dilution, the monolayer becomes confluent

Fig. 2.4 Cell culture.

2.2 IDENTIFICATION OF VIRUSES USING ANTIBODIES (SEROLOGY)

Antibodies are proteins produced by the immune system of higher vertebrates in response to foreign materials (antigens) which those cells encounter. Such antibodies have a region that recognizes and binds specifically to that same antigen. Antibodies are secreted into the body fluids and are most easily obtained from blood. Blood is allowed to clot and antibodies remain in the fluid part (serum) which remains after clotting has removed cells and clotting proteins. This is then known as an *antiserum*.

The principle of identifying infectious virus by using an antibody of known specificity is shown in Fig. 2.6. If the antibody recognizes and binds to the virus, virus infectivity will be inhibited (Fig. 2.6, top line). Infectivity is only one of several virus properties that can be affected by antibody binding, and hence can be monitored in this type of assay. Another is inhibition of the agglutination of red blood cells by virus. This is a property of some viruses, like the influenza viruses, that attach to molecules on the surface of the red blood cell (RBC). At a certain virus to cell ratio the RBCs are linked together by virus and the cells are agglutinated (clumped together). This has nothing to do with infectivity, and when infectivity of a virus has been deliberately inactivated, that virus can still agglutinate RBCs efficiently, providing that its surface properties are unimpaired. A quantitative hemagglutination test can be devised by making dilutions of virus in a suitable tray and then adding a standard amount of RBCs to each well (Fig. 2.7). The amount of virus present is estimated as the dilution at which the virus causes 50% agglutination. This test has the advantage of speed – it takes just 30 minutes, compared with an average of 3 days for a plaque assay. However, it is insensitive;

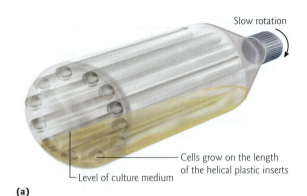

Slow rotation

Cells grow on the length
of the helical plastic inserts

Level of culture medium

(a)

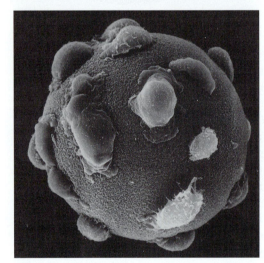

(b)

(c)

Fig. 2.5 (a) One way to increase cell density is by increasing the surface area to which cells can attach; this is a view from the end of a bottle lined with spiral plastic coils. The bottle is rotated slowly, at about 5 rev/h, so that a small volume of culture fluid can be used. Cells tolerate being out of the culture fluid for short periods. (b,c) Cells growing on microcarrier beads. (b) Scanning electron micrograph of pig kidney cells. (Courtesy of G. Charlier.) (c) Removal of cells from a microcarrier bead by incubation with trypsin. The cells are rounding up and many have already detached. Each bead is about 200 μm in diameter. The microcarriers shown are Cytodex (Pharmacia Ltd) (reproduced by permission).

e.g. approximately 10^6 plaque-forming units (PFUs) of influenza virus are needed to cause detectable agglutination.

In the hemagglutination-inhibition test a small amount of virus is added to serial dilutions of antibody before the addition of RBCs (Fig. 2.8). Blocking of agglutination indicates that antibody has bound to a virus particle, and hence identifies it. It can be used with a known antibody to identify an unknown virus, or *vice versa* using a known virus to identify the presence of virus-specific antibody in a serum sample. Alternatively virus that has been aggregated by reaction with specific antibody can be directly visualized by electron microscopy.

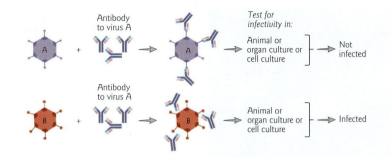

Fig. 2.6 A neutralization test. Virus A loses its infectivity after combining with A-specific antibody (it is neutralized). A-specific antibody does not bind to virus B, so infectivity of virus B is unaffected. The complete test requires the reciprocal reactions.

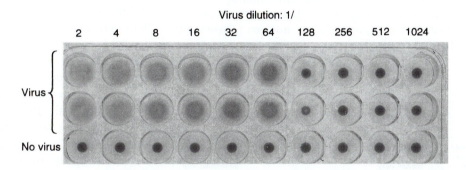

Fig. 2.7 Hemagglutination titration. Here an influenza virus is serially diluted from left to right in wells in a plastic plate. Red blood cells (RBCs) are then added to 0.5% v/v and mixed with each dilution of virus. Where there is little or no virus, RBCs settle to a button (from 1/128) indistinguishable from RBCs to which no virus was added (row 3). Where sufficient virus is present (up to 1/64), cells agglutinate and settle in a diffuse pattern. (Photograph by Andy Carver.)

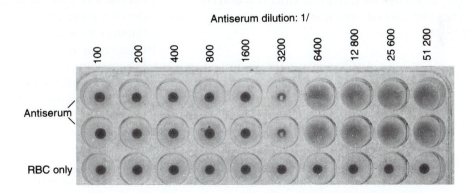
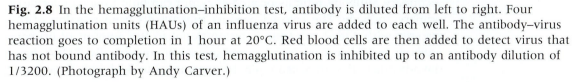

Fig. 2.8 In the hemagglutination–inhibition test, antibody is diluted from left to right. Four hemagglutination units (HAUs) of an influenza virus are added to each well. The antibody–virus reaction goes to completion in 1 hour at 20°C. Red blood cells are then added to detect virus that has not bound antibody. In this test, hemagglutination is inhibited up to an antibody dilution of 1/3200. (Photograph by Andy Carver.)

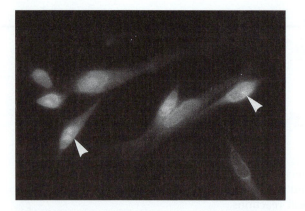

Fig. 2.9 Fluorescent antibody staining. An antibody covalently bound to a fluorescent dye has been used to detect an antigen present mainly in the nucleus (arrows) of influenza virus-infected cells.

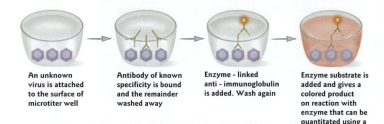

An unknown virus is attached to the surface of microtiter well

Antibody of known specificity is bound and the remainder washed away

Enzyme - linked anti - immunoglobulin is added. Wash again

Enzyme substrate is added and gives a colored product on reaction with enzyme that can be quantitated using a spectrophotometer

Fig. 2.10 Identification of an unknown virus by ELISA with a specific antiserum. The unknown virus would be used in parallel with a control virus preparation that is known to react with the antibody. The unknown virus is positively identified when the antibody reaction is identical to that using the control virus.

Antibody can also be employed to detect viral antigens inside the infected cell. When the cell is alive, antibodies cannot cross the plasma membrane and will therefore react only with antigens exposed on the surface of the cell. This permeability barrier is destroyed by "fixing" the cell in organic solvents such as acetone or methanol, which permeabilize the plasma membrane and enable antibody to enter the cytoplasm and nucleus and attach to antigens. Antibodies are "tagged" before use with a marker substance and hence can be detected *in situ*. Tags such as fluorescent dyes can be detected microscopically when illuminated with ultraviolet (UV) light (Fig. 2.9); enzymes (e.g. peroxidase, phosphatase) which leave a colored deposit on reaction with substrate can be seen by light microscopy; radioactive substances can be detected by deposition of silver grains from a photographic emulsion; and electron-dense molecules (e.g. colloidal gold particles or ferritin, an iron-containing protein) can be visualized by electron microscopy.

Antibodies covalently linked to a marker enzyme are now commonly used in a quantitative assay called the enzyme-linked immunosorbent assay (ELISA) (Fig. 2.10). In the example shown, a panel of antibodies is being used to identify an unknown virus that is attached to the surface of the plastic well. Here two antibodies are used: the primary antibody is specific for virus and is unlabelled; the secondary antibody is specific for conserved epitopes on the primary antibody and is covalently linked to an enzyme. This makes the ELISA more sensitive as several molecules of secondary antibody (and associated enzyme) can bind to one molecule of primary antibody. The colored product that results from reaction of the enzyme with added substrate is proportional to the amount of primary antibody bound and can be measured spectrophotometrically. ELISAs have the advantage that they can easily be automated to deal with large numbers of routine samples. An ELISA can also be used to measure antibody concentration by titrating the antibody and adding serial dilutions to a constant, low concentration of virus that has been bound to the tray.

2.3 DETECTION, IDENTIFICATION, AND CLONING OF VIRUS GENOMES USING PCR AND RT-PCR

As discussed above, all techniques have their advantages and limitations. For example, serological methods of virus detection are effective and quick but tell us nothing about the virus genome. Neutralization tests are simple, but are confined to viruses that can be cultivated, and are slow to give a result. This depends on the time a virus takes to kill a detectable number of cells, and this can range from several days to several weeks. Such a situation is far from ideal, and the problem was solved by the discovery of a technique which makes many, many copies of a chosen part of the virus genome. This is the polymerase chain reaction (PCR) which synthesizes DNA from a DNA template, and was devised in 1985 by Kari Mullis. If the virus of interest has an RNA genome, the region of interest has first to be converted into DNA using a primer (see below) and the retrovirus enzyme, reverse transcriptase (Section 8.3). If a unique sequence is chosen and there is a positive result, the virus present is immediately identified. The system is highly sensitive and can detect one copy of a DNA genome or around 1000 copies of an RNA genome. Thus it is a detection and identification system all in one. Normally a region of 100 bp (basepairs) or so is amplified, but with care whole genomes up to 15,000 bp can be copied. PCR has the added advantage of detecting virus in primary tissue, so that mutations associated with adaptation to cell culture are avoided. It is no more expensive than a neutralization assay.

The only prerequisite for PCR is knowing the sequence of the regions flanking the portion of the genome to be detected, so that oligonucleotide primers that are complementary to a sequence on each strand of DNA can be made. PCR requires two primers, each of around 20–30 nucleotides in length, and these are chemically synthesized. The DNA is denatured by heating to around 90°C and the primers added in high molar excess together with deoxyribonucleotide triphosphates (dNTPs; Fig. 2.11). On

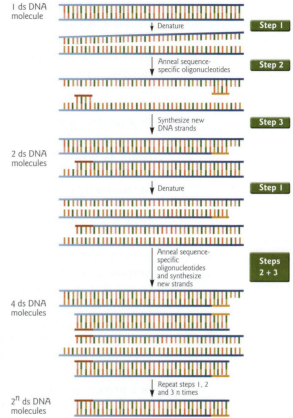

Fig. 2.11 An outline of the polymerase chain reaction (PCR). Step 1: denaturation; step 2: annealing of oligonucleotide primers; step 3: synthesis of new DNA by added polymerase. This is repeated n times. Note that end-product, the amplified DNA fragment or amplicon, is not formed until after the third annealing process.

cooling, the primers anneal to their respective template strands and the template is copied by the enzyme. It is convenient to use a polymerase that is not inactivated at high temperatures, such as the *Taq* polymerase from *Thermophilus **aquaticus***, a bacterium that lives in natural hot springs; otherwise fresh polymerase would have to be added after each denaturation step. The mix is again denatured and cooled so that further primers can anneal, and the next round of DNA synthesis take place. The defined PCR product is now present, but around 30 rounds of synthesis are required before there is sufficient product to be analyzed. To determine if the PCR result is positive, DNA is electrophoresed to determine if an amplified fragment (amplicon) of the expected size has been synthesized. Size is measured by comparison with DNA size markers electrophoresed in a parallel track. Confirmation of the result can be obtained by isolating the DNA fragment from the gel and sequencing it.

This technique rapidly gained acceptance and has very many applications. It is as widely used for diagnostic clinical virology as for research purposes. For diagnostic purposes PCR may be carried out in two phases. In the first, primers are chosen that amplify regions of the genome that are common to a whole group of viruses known to occur, say in the gut. On finding a positive, primers to a region that is unique to each virus type can then be used for exact identification.

PCR can amplify complete genes or even small genomes, and these can be cloned into an appropriate vector, which can express more genomes or a protein product when transfected into animal cells in culture. In this way vaccines can be made from noncultivable viruses. Cloned DNA that expresses a virus protein can also be injected directly into animals as an experimental vaccine – this is called a DNA vaccine.

KEY POINTS

- Viruses are usually detected only indirectly.
- The main methods for studying animal viruses center on growth in living cells, reaction with specific antibodies, and the polymerase chain reaction.
- It is difficult to study a virus that does not grow to high titer in cultured cells or in a convenient animal model.
- All techniques have their strengths and limitations, and should be chosen with care for the intended purpose, with those strengths and limitations in mind.
- For routine diagnostics, speed, automation, reliability, and cost are key factors.

FURTHER READING

Adolph, K. W. 1994. Molecular virology techniques. In *Methods in Molecular Genetics*, vol. 4. Academic Press, New York.

Bendinelli, M., Friedman, H. 1998. *Rapid detection of Infectious Agents*. Plenum, New York.

Clementi, N. 2000. Quantitative molecular analysis of virus expression and replication. *Journal of Clinical Microbiology* **38**, 2030–2036.

Crowther, J. R. 2000. The ELISA guide. Humana, Totowa, NJ.

Desselberger, U., Flewett, T. H. 1993. Clinical and public health virology: a continuous task of changing pattern. *Progress in Medical Virology* **40**, 48–81.

Harlow, E., Lane, D. 1988. *Antibodies: A Laboratory Manual*. Cold Spring Harbor Press, NY.

Liddell, J. E., Weeks, I. 1995. *Antibody Technology*. Bios Scientific Publishers, Oxford.

Mahy, B. W. J., Kangro, H. O. 1996. *Virology Methods Manual*. Academic Press, London.

Masters, J. R. W. 1999. *Animal cell culture*, 3rd edn. Oxford University Press, Oxford.

Payment, P., Trudel, M. 1993. *Methods and Techniques in Virology*. Marcel Dekker, New York.

Price, C., Newman, D. 1996. *Principles and Practice of Immunoassay*, 2nd edn. Stockton, London.

Taubenberger, J. K., Layne, S. P. 2001. Diagnosis of influenza virus: coming to grips with the molecular era. *Molecular Diagnosis* **6**, 291–305.

Wiedbrauk, D. L., Farkas, D. H. 1995. *Molecular Methods for Virus Detection*. Academic Press, New York.

Wild, D. 2001. *The Immunoassay Handbook*, 2nd edn. Nature Publishing Group, London.

Also check Appendix 7 for references specific to each family of viruses.

3

The structure of virus particles

All virus genomes are surrounded by proteins which:

- *Protect nucleic acids from nuclease degradation and shearing.*
- *Contain identification elements that ensure a virus recognizes an appropriate target cell (but plant viruses do not, and enter the cell directly by injection or injury).*
- *Contain a genome-release system that ensures that the virus genome is released from a particle only at the appropriate time and location.*
- *Include enzymes that are essential for the infectivity of many, but not all, viruses.*
- *Are called* structural proteins, *as they are part of the virus particle.*

All viruses contain protein and nucleic acid with at least 50%, and in some cases up to 90%, of their mass being protein. At first sight it would appear that there are many ways in which proteins could be arranged round the nucleic acid. However, viruses use only a limited number of designs. The limitation on the range of structures is due to restrictions imposed by considerations of efficiency and stability.

3.1 VIRUS PARTICLES ARE CONSTRUCTED FROM SUBUNITS

While proteins may have regular secondary structure elements in the form of α helix and β structure, the tertiary structure of the protein is not symmetrical. This is a consequence of hydrogen bonding, disulfide bridges, and the intrusion of proline in the secondary structure. Although it may be naïvely thought that the nucleic acid could be covered by a single, large protein molecule, this cannot be the case since proteins are irregular in shape, whereas most virus particles have a regular morphology (Fig. 3.1). However, that viruses must contain more than a single protein can also be deduced solely from considerations of the coding potential of nucleic acid molecules. A coding triplet has an M_r of approximately 1000 but specifies a single amino acid with an average M_r of about 100. Thus a nucleic acid can at best only specify one-tenth of its mass of protein. Since viruses frequently contain more than 50% protein by mass, it is apparent that more than one protein must be present.

Obviously, less genetic material is required if the single protein molecule specified is to be used as a repeated subunit. However, it is not essential that the coat be constructed from identical subunits, provided the combined molecular weights of the different subunits is sufficiently small in relation to the nucleic acid molecule which they protect. There is a further advantage in constructing a virus from subunits, since any misfolding of protein (occurring commonly at a frequency of 1 in 1000 and for which there is no repair mechanism) affects only a small part of a structural unit. Thus, provided that faulty subunits are not included in the virus particle during assembly, an error-free structure can be constructed with the minimum of wastage.

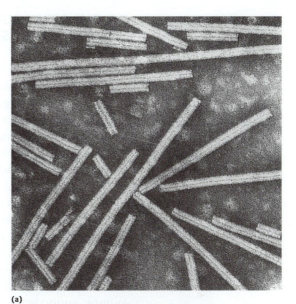

(a)

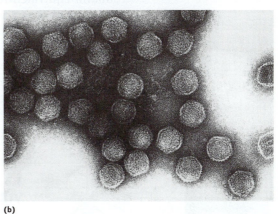

(b)

Fig. 3.1 Electron micrographs of virus particles showing their regular shape: rods and spheres. (a) Tobacco mosaic virus. (b) Bacteriophage Si1.

The necessary physical condition for the stability of any structure is that it be in a state of minimum free energy, so it can be assumed that the maximum number of interactions is formed between the subunits of a virus particle. Since the subunits themselves are nonsymmetrical, the maximum number of interactions can be formed only if they are arranged symmetrically, and there are a limited number of ways this can be done. Shortly after their seminal work on the structure of DNA, Watson and Crick predicted on theoretical grounds that the only two ways in which asymmetrical subunits could be assembled to form virus particles would generate structures with either cubic or helical symmetry. (However an important rider to the energy status of virus particles is that at least some are suspected of being metastable, and that their true minimum energy state is reached only after they have undergone uncoating during the process of infecting a cell (see Section 3.7).)

Fig. 3.2 Arrangement of identical asymmetrical components around the circumference of a circle to yield a symmetrical arrangement.

3.2 THE STRUCTURE OF FILAMENTOUS VIRUSES AND NUCLEOPROTEINS

One of the simplest ways of symmetrically arranging nonsymmetrical components is to place them round the circumference of a circle to form discs (Fig. 3.2). This gives a two-dimensional structure. If a large number of discs is stacked on top of one another, the result is a "stacked-disc" structure. Thus a symmetrical three-dimensional structure can be generated from a nonsymmetrical component such as protein and still leave room for nucleic acid. Examination of published electron micrographs of viruses reveals that some of them have a tubular structure. One such virus is the tobamovirus, tobacco mosaic virus (TMV) (Fig. 3.1). However, closer examination reveals that the TMV subunits are not arranged cylindrically, i.e. in rings, but helically. There is an obvious explanation for this. A helical nucleic acid could not be equivalently bonded in a stacked-disc structure. However, by arranging the subunits helically, the maximum number of bonds can still be formed and each subunit equivalently bonded, except, of course, for those at either end. All filamentous viruses so far examined are helical rather than cylindrical and the insertion of the nucleic acid may be the factor governing this arrangement. A helical arrangement offers considerable stability because of the subunits. This is greater than would be found with a cylinder which has no linking along the long axis (see Section 11.3). Many nucleoprotein structures inside enveloped viruses (Fig. 3.24) are constructed in the same way.

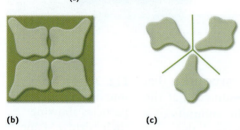

(a)

(b)   (c)

Fig. 3.3 Symmetrical arrangement of identical asymmetrical subunits by placing them on the faces of objects with cubic symmetry. (a) Asymmetrical subunits located at vertices of each triangular facet. (b) Asymmetrical subunits placed at vertices of each square facet. (c) Arrangement of asymmetrical subunits placed at each corner of a cube with faces as represented in (b).

3.3 THE STRUCTURE OF ISOMETRIC VIRUS PARTICLES

A second way of constructing a symmetrical particle would be to arrange the smallest number of subunits possible around the vertices or faces of an object with cubic symmetry, e.g. tetrahedron, cube, octahedron, dodecahedron (constructed from 12 regular pentagons), or icosahedron (constructed from 20 equilateral triangles). Figure 3.3 shows possible arrangements for objects with triangular and square faces. Multiplying the minimum number of subunits per face by the number of faces gives the smallest number of subunits which can be arranged around such an object. The minimum number of subunits is determined by the symmetry element of the face, i.e. a square face will have four subunits, a triangular face will have three subunits, etc.

For a tetrahedron the smallest number of subunits is 12, for a cube or octahedron it is 24 subunits, and for a dodecahedron or icosahedron it is 60 subunits. Although it may not be immediately apparent, these represent the few ways in which an asymmetrical object (such as a protein molecule) can be placed symmetrically on the surface of an object resembling a sphere. (This can be checked by using a ball and sticking on bits of paper of the shape shown in Fig. 3.3.) Examination of electron micrographs reveals that many viruses appear spherical in outline, but actually have icosahedral symmetry rather than octahedral, tetrahedral, or cuboidal symmetry. There are two possible reasons for the selection of icosahedral symmetry over the others. Firstly, since it requires a greater number of subunits to provide a sphere of the same volume, the size of the repeating subunits can be smaller, thus economizing on genetic information. Secondly, there appear to be physical restraints which prevent the tight packing of subunits required by tetrahedral and octahedral symmetry.

Symmetry of an icosahedron

An icosahedron is made up of 20 triangular faces, five at the top, five at the bottom and 10 around the middle, with 12 vertices (Fig. 3.4c). Each triangle is symmetrical and it can be inserted in any orientation (Fig. 3.4a). An icosahedron has three axes of symmetry: fivefold, threefold, and twofold (Fig. 3.4b). However, many viruses are icosahedral and yet achieve a much greater size than can apparently be accomplished with this simple arrangement (Box 3.1).

The triangulation of spheres – or how to make bigger virus particles

It is possible to enumerate all the ways in which this subdivision can be carried out. This can be demonstrated with a simple example. Starting with an icosahedron, arranging the subunits around the vertices will generate 12 groups of five subunits (Fig. 3.6a). Subdividing each triangular face into four smaller and identical equilateral triangular facets and incorporation of subunits at the vertices of those smaller triangles gives a structure containing a total of 240 subunits (Fig. 3.6b). At the

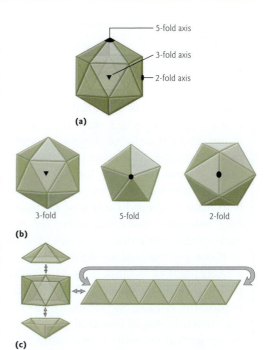

Fig. 3.4 (a) Properties of a regular icosahedron. Each triangular face is equilateral and has the same orientation whichever way it is inserted. Axes of symmetry intersect in the middle of the icosahedron. There are 12 vertices, which have fivefold symmetry, meaning that rotation of the icosahedron by one-fifth of a revolution achieves a position such that it is indistinguishable from its starting orientation; each of the 20 faces has a threefold axis of symmetry and each of the 30 edges has a twofold axis of symmetry (b). The icosahedron is built up of five triangles at the top, five at the bottom, and a strip of 10 around the middle (c). (Copyright 1991, from *Introduction to Protein Structure* by C. Branden & J. Tooze. Reproduced by permission of Routledge Inc., part of The Taylor & Francis Group.)

Box 3.1

The construction of more complex icosahedral virus particles

The combination of an icosahedral structure linked with evolutionary pressure to use small repeating subunits to form virus particles imposes a restriction on the achievable size of the virion. Since the size of the particle defines the maximum size of the nucleic acid that can be packed within it, it appears that viruses should not be able to package large genomes. However many viruses have very long genomes and contain more than 60 subunits. This apparent paradox is solved as follows: if $60n$ subunits are put on the surface of a sphere, one solution is to arrange them in n sets of 60 units, but the members of one set would not be equivalently related to those in another set. If in Fig. 3.5a, all the subunits, represented by light and dark spheres, were identical, then those represented by dark spheres are related equivalently to those represented by light spheres, but light sphere units do not have the same spatial arrangement of neighbors as dark sphere units and so cannot be equivalently related. (The problem could be solved if the structure was built out of n different subunits, but this would require the virus to encode many more structural proteins.) Thus accepting that we must build the structure out of identical subunits, we find that we can regularly arrange more than 60 asymmetrical subunits in the same way that Buckminster Fuller did on his geodesic domes (Fig. 3.5b). He subdivided the surface of a sphere into triangular facets, and arranged them with icosahedral symmetry. The device of triangulating the sphere represents the optimum design for a closed shell built of regularly bonded identical subunits. No other subdivision of a closed surface can give a comparable degree of equivalence. This is a minimum-energy structure and is the probable reason for the preponderance of icosahedral viruses.

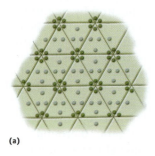

(a)

(b)

Fig 3.5 (a) Spatial arrangement of two identical sets of subunits. Note that any member of the set represented by dark spheres does not have the same neighbors as a member of the other set represented by light spheres. (b) An example of a geodesic dome – the United States Pavilion at Expo '67 in Montreal. (Courtesy of the US Information Service.)

Box 3.2

Calculating the number of subunits in a virus particle

The way in which each triangular face of the icosahedron can be subdivided into smaller, identical equilateral triangles is governed by the law of solid geometry. This can be calculated from the expression:

$$T = Pf^2$$

where T, the *triangulation number*, is the number of smaller, identical equilateral triangles, P is given by the expression $h^2 + hk + k^2$. In this expression, h and k are any pair of integers without common factors, i.e. h and k cannot be multiplied or divided by any number to give the same values, $f = 1, 2, 3, 4$, etc.

For viruses so far examined, the values of P are 1 ($h = 1, k = 0$), 3 ($h = 1, k = 1$), and 7 ($h = 1, k = 2$). Representative values of T are shown in Table 3.1. Once the number of triangular subdivisions is known, the total number of subunits can easily be determined since it is equal to $60T$.

vertices of each of the original icosahedron faces there will be rings of five subunits, called *pentamers* (dark spheres). However, at all the other (new) vertices generated by the triangular facets will be rings of six subunits, called *hexamers* (light spheres). Since some of the subunits are arranged as pentamers and others as hexamers, it should be apparent that they cannot be equivalently related; hence they are called *quasi-equivalent*, but this still represents the minimum-energy shape. Thus with the subunit being kept a constant size, greater subdivision allows the formation of larger virus particles.

T = 1: the smallest virus particle – satellite tobacco necrosis virus

In its simplest form one *subunit* used in the construction of a virus particle subunit is one protein. However no independently replicating virus is known to consist of only 60 protein subunits, but satellite viruses do (Table 3.1). These encode one coat protein but depend upon coinfection with a helper virus to provide missing replicative functions (Appendix 5). The single-stranded RNA genome of satellite tobacco necrosis virus is about 1000 nt. Presumably the volume of a 60-subunit structure is too small to accommodate the larger genome that is needed by an independent virus. The virion is only 17 nm in diameter, compared to the 30 nm of small independent viruses.

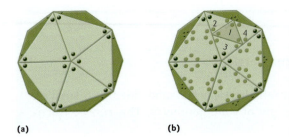

(a)   **(b)**

Fig. 3.6 Arrangement of 60n identical subunits on the surface of an icosahedron. (a) $n = 1$ and the 60 subunits are distributed such that there is one subunit at the vertices of each of the 20 triangular faces. Note that each subunit has the same arrangement of neighbors and so all the subunits are equivalently related. (b) $n = 4$. Each triangular face is divided into four smaller, but identical, equilateral triangular facets and a subunit is again located at each vertex. In total, there are 240 subunits. Note that, in contrast to the arrangements discussed in Box 3.1, each subunit, whether represented by a dark or light sphere, has the identical arrangement of neighbors: see the face in which triangles 1–4 have been drawn. However, since some subunits are arranged in pentamers and others in hexamers, the members of each set are "quasi-equivalently" related.

Detailed determination of the structure of a virus depends primarily on being able to grow crystals of purified virus, although valuable but lower resolution information can be obtained from cryo-electron microscopic examination of virus frozen in vitreous ice. The conditions required for crystallization of virus particles or proteins are not fully understood, and many will not form crystals at all. Large stable crystals are required that are then bombarded with X-rays. These are diffracted by atoms within the virion, and the image captured. Knowledge of the amino acid sequence of the proteins which comprise the particle makes it possible to determine the three-dimensional crystal structure. X-ray analysis gives resolution to around 0.3 nm and cryoelectron microscopy to around 15 nm. Both processes require high-powered computers which are used to make the necessary calculations and for image reconstruction.

The morphological units seen by electron microscopy are called capsomers and *the number of these need not be the same as the number of protein subunits*. The numbers of morphological units seen will depend on the size and physical packing of the subunits and on the resolution of electron micrographs. A repeating subunit may consist of a *complex of several proteins*, such as the four structural proteins of poliovirus (see below and Section 11.3), or a *fraction of a protein*, such as the adenovirus hexon protein, half of which is considered to be a single repeating subunit.

Table 3.1

Values of capsid parameters in a number of icosahedral viruses. The value of T was obtained from examination of electron micrographs, thus enabling the values of P and f to be calculated.

P	f	$T (= Pf^2)$	No. of subunits (60T)	Example
1	1	1	60	Tobacco necrosis satellite virus
3	1	3	180	Tomato bushy stunt virus, picornaviruses*
1	2	4	240	Sindbis virus
1	4	16	960	Herpesviruses
1	5	25	1500	Adenoviruses[†]

*In fact picornaviruses have a pseudo $T = 3$ structure (see below).
[†]See text.

T = 3: the molecular basis for quasi-equivalent packing of chemically identical polypeptides – tomato bushy stunt virus

Some plant virus particles achieve the *T* = 3, 180-subunit structure (Fig. 3.7) while encoding only a single virion polypeptide. They compensate for the physical asymmetry of quasi-equivalence by each polypeptide adopting one of three subtly different conformations. The virion polypeptide of tomato bushy stunt virus has three domains P, S, and R (Fig. 3.8a): this is folded so that P and S are external and hinged to each other, while R is inside the virion and has a disordered structure. An arm (a) connects S to R, h connects S and P (Fig. 3.8b).

Each triangular face is made of three identical polypeptides, but these are in different conformations to accommodate the quasi-equivalent packing. For example, the C subunit has the S and P domains orientated differently from the A and B subunits (Fig. 3.8c), while the arm (a) is ordered in C and disordered in A and B (not shown). The S domains form the viral shell with tight interactions, while the P domains (total = 180) interact across the twofold axes of symmetry to form 90 dimeric protrusions. This virion is 33 nm in diameter and can accommodate a single-stranded RNA genome about fourfold larger than that of the satellite viruses. Thus a larger particle can be achieved without any more genetic cost.

T = 3: with icosahedra constructed of four different polypeptides – picornaviruses

Picornaviruses are made of 60 copies of each of four polypeptides: VP1, VP2, VP3, and VP4. VP4 is entirely internal. The repeating subunit of picornaviruses is the complex of VP1, VP2, and VP3. This should generate a *T* = 1 particle. However despite the significant differences in amino acid sequence, the proteins adopt very similar conformations and, in geometric terms, appear as separate repeating subunits. For this reason the assembled picornavirus particles appear to have a *T* = 3 structure; more accurately this is a pseudo *T* = 3 (compare Figs 3.10 and 3.7).

The pentamers contain 15 polypeptides, with five molecules of VP1 forming a central vertex. These pentamers are the building blocks in the cell from which the virion is assembled. Use of three polypeptides gives a chemically more diverse structure and may be an adaptation to cope with the immune system of animal hosts.

The cell receptor attachment site of picornavirus particles

Many virus proteins including the VP1, VP2, and VP3 of picornaviruses, have the same type of structure – an antiparallel β barrel also known as a "jelly-roll" (Fig. 3.9). Crystallographic, biochemical, and immunological data have together identified a depression within the β barrel of VP1, which

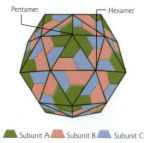

Pentamer — Hexamer

Subunit A   Subunit B   Subunit C

Fig. 3.7 Schematic diagram of a *T* = 3, 180-subunit virus. Each triangle is composed of three subunits, A, B, and C, which are asymmetrical by virtue of their relationship to other subunits (pentamers or hexamers). (Copyright 1991, from *Introduction to Protein Structure* by C. Branden & J. Tooze. Reproduced by permission of Routledge Inc., part of The Taylor & Francis Group.)

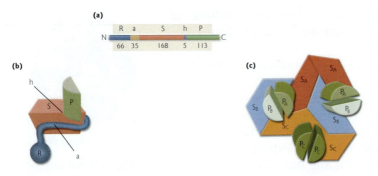

Fig. 3.8 (a) The linear arrangement of domains present in the single virion polypeptide of tomato bushy stunt virus. (b) Conformation of the polypeptide. The S domain forms the shell of the virion, while P points outwards and R is internal. (c) This shows a triangular face, composed of subunits A, B, and C, and the interaction of the P domains to form dimeric projections (see text). (Copyright 1991, from *Introduction to Protein Structure* by C. Branden & J. Tooze. Reproduced by permission of Routledge Inc., part of The Taylor & Francis Group.)

is thought to be the attachment site of picornaviruses. There are 60 attachment sites per virion. Apart from its intrinsic interest, the structure of the attachment site is important as the prime target for antiviral drugs (e.g. pleconaril – see Section 21.8) which can stop attachment of virus to the host cell. The arrangement of the β strands of VP1 is such that an annulus is formed around each fivefold axis of symmetry (Fig. 3.10). In the rhino (common cold) viruses, this is particularly deep and is called a "canyon." The canyon lies within the structure of the β barrel. Amino acids within the canyon are invariant, as expected from their requirement if they have to interact with the cell receptor, while amino acids on the rim of the canyon are variable. Only the latter interact with antibody. It is thought that the floor of the canyon has evolved so that it physically cannot interact with antibody. This avoids immunological pressure to accumulate mutations in order to escape from reaction with antibody, since these would at the same time render the attachment site nonfunctional, and hence be lethal to the virus.

An unknown structure – virus particles with 180 + 1 subunits and no jelly-roll β barrel – RNA bacteriophages

The leviviruses are 24 nm icosahedral RNA bacteriophages, and include MS2, R17, and Qβ. They encode two coat proteins. There are 180 subunits of one of these arranged with $T = 3$, but only a single copy of the second "A" protein in each particle. This is the attachment protein. It is not known how the single subunit is incorporated into the particle. The main coat protein does not form a jelly-roll β barrel like the others described above, but instead has five antiparallel β strands arranged like the vertical elements of battlements. Two subunits interact to form a sheet consisting of 10 antiparallel β strands.

$T = 25$: more complex animal virus particles – adenoviruses

Careful examination of electron micrographs of adenoviruses shows that $T = 25$. There are 240 hexamers and 12 pentamers, and a fiber projects from each vertex of the virus (Fig. 3.11a,b). The fibers, the pentamers,

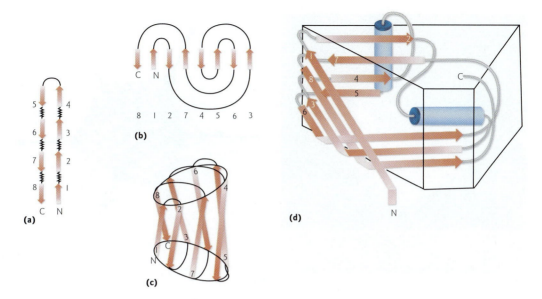

Fig. 3.9 The jelly-roll or antiparallel β barrel: a common structure of plant and animal virion proteins. The formation of an antiparallel β barrel from a linear polypeptide can be visualized to occur in three stages. First, a hairpin structure forms where β strands are hydrogen-bonded to each other: 1 with 8, 2 with 7, 3 with 6, and 4 with 5, creating antiparallel β strand pairs separated by loop regions of variable length (a). Second, these pairs become arranged side by side, so that further hydrogen bonds can be formed by newly adjacent strands, e.g. 7 with 4 (b) Third, the strand pairs wrap around an imaginary barrel forming a three-dimensional structure. The eight β strands are arranged in two sheets, each composed of four strands: strands 1, 8, 3, and 6 form one sheet and strands 2, 7, 4, and 5 form the second sheet (c). The dimensions of the barrel are such that each protein forms a wedge, and these wedges are subunits that are assembled into virus particles (d). ((a–c) Copyright 1991, from *Introduction to Protein Structure* by C. Branden & J. Tooze; reproduced by permission of Routledge Inc., part of The Taylor & Francis Group. (d) From Hogle *et al.*, 1985 *Science* **229**, 1358.)

and the hexamers are all constructed from different proteins. We are thus faced with the problem of arranging not one, but three different proteins in a regular fashion, while adhering to the design principles outlined above. This can be achieved by arranging the pentamers and the fibers at the vertices of the icosahedron and the hexamers on the faces of the icosahedron (Fig. 3.11c). However the formula 60*T* gives the number of subunits as 1500 (Table 3.1). How is this difference resolved?

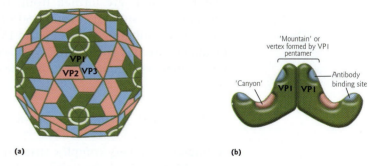

Fig. 3.10 A picornavirus particle (a) and the attachment site formed by VP1 shown as an annulus around the fivefold axis of symmetry. (b) A vertical section through a VP1 pentamer where the cross-section of the annulus is referred to as the "canyon." ((a) From Smith *et al.*, 1993 *Journal of Virology* **67**, 1148. (b) From Luo *et al.*, 1987 *Science* **235**, 182.)

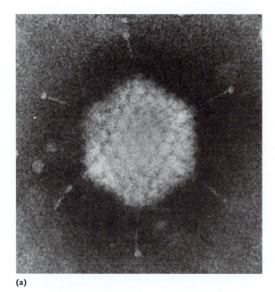

(a)

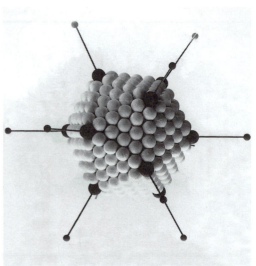

(b)

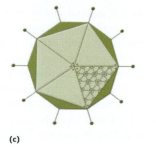

(c)

Fig. 3.11 The structure of adenoviruses. (a) A negatively stained electron micrograph of an adenovirus. (b) Model of an adenovirus to show arrangements of the capsomers. (c) Schematic diagram to show the arrangements of the subunits on one face of the icosahedron. Note the subdivision of the face into 25 smaller equilateral triangles. (Photographs courtesy of Nicholas Wrigley.)

The 240 hexons are composed of three identical polypeptides and each functions as *two* repeating subunits. Thus there are 1440 hexon subunits. The 12 pentons are formed from five identical polypeptides and each functions as a subunit, making 60 of these. Thus 1440 hexon subunits + 60 penton subunits makes up the 1500 predicted subunits. Actually the hexamers are not all spatially equivalent, as those surrounding the vertex pentamers contact five other hexons, while the others contact six hexons.

Double-shelled particles: a capsid within a capsid – reoviruses

A different and very complex structural arrangement is found in another class of isometric viruses, the reoviruses, which are composed of a capsid within a capsid. The diameter of the inner and the outer capsid being 51 nm and 73 nm respectively. Reovirus synthesizes 11 polypeptides. Of these eight are located in the virion, three forming the outer capsid and five forming the inner capsid (Table 3.2). Both capsids have icosahedral

Table 3.2

Reovirus polypeptides and their location in the outer or inner capsid of the virion, and some of their properties.

Location in virion or nonstructural	Protein	Encoding RNA segment	No. of polypeptides per virion	Function
Outer	μ1	M2	600	Main structural element; role in entry
Outer	σ3	S4	600	Main structural element
Outer, vertex	σ1	S1	36–48	Attachment protein; serotype determinant
Inner	λ1	L3	120	Main structural element
Inner	σ2	S2	150	Stabilizes λ1
Inner, vertex	λ2	L2	60	Forms turret; capping and export of mRNA
Inner	μ2	M1	12	Not known
Inner	λ3	L1	12	RNA-dependent RNA polymerase
Nonstructural	μNS	M3	0	Binds ssRNA; role in secondary transcription
Nonstructural	σNS	S3	0	Binds ssRNA
Nonstructural	σ1s	S1	0	Not known

symmetry. (Fig. 3.12) and the outer has a $T = 13$ symmetry and 780 (13 × 60) subunits. However it is not certain how the 600 molecules of protein μ1 and the 600 molecules of σ3 are arranged in the outer capsid to form the subunits. The third outer capsid protein, σ1, is the attachment protein which is located at each of the 12 vertices (Fig. 3.12). It is supported by the five molecules of protein λ2 vertices, which are part of the inner capsid (although not long ago attributed to the outer capsid). The structure of the inner capsid was recently solved to 0.36 nm. Three of the five inner capsid proteins (λ1, σ2, and λ2) are symmetrically arranged. The main skeleton is formed by protein λ1 built into 12 decamers, each of which surrounds a vertex. This is clamped into position by protein σ2. Protein λ2 forms a turret at each vertex that surround the outer core protein σ1, as mentioned. The turret is only seen as such after the outer core is stripped away. It has an important role in capping of viral mRNAs that are all made within the particle, and acts as a conduit for the export of mRNA from the particle (see Section 10.3). The 10 double-stranded RNAs are tightly coiled like the DNA inside phage heads. Each molecule is thought to be associated with a transcriptase complex and tethered near a vertex.

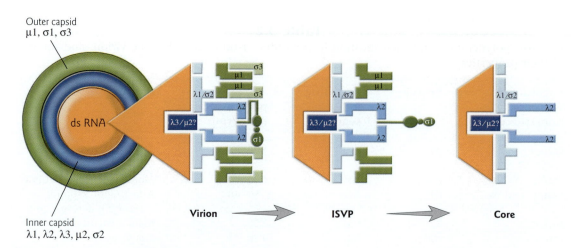

Fig. 3.12 The double capsid structure of reovirus showing the location of polypeptides in the virion. Here a section of the virion has been taken through the fivefold axis of symmetry of the vertex. In order to demonstrate protein function, two intermediates that occur during the virus uncoating process are also shown: the intermediate subviral particle (ISVP) and the viral core. In the ISVP, σ1 has achieved an extended conformation and σ3 has been lost from the outer capsid. Some molecules of σ1 may also be extended in the virion, but this is seen more frequently in the intermediate. The core is formed by the loss of the entire outer capsid and the change in conformation of the turret protein λ2 to form the channel through which mRNA synthesized in the particle will escape into the cell cytoplasm. (From M. L. Nibert *et al.* in B. N. Fields, D. M. Knipe and P. M. Howley eds., 1996 *Virology*, Lippincourt-Raven, p. 1562.)

3.4 ENVELOPED (MEMBRANE-BOUND) VIRUS PARTICLES

Although they appear complex, these viruses have a conventional isometric or helical structure that is surrounded by a membrane – a 4-nm-thick lipid bilayer containing proteins. Examples include many of the larger animal viruses, but only a few plant and bacterial viruses. Traditionally these viruses were distinguished from nucleocapsid viruses by treatment with detergents or organic solvents, which disrupts the membrane and destroys infectivity. Thus they were sometimes referred to as "ether-sensitive viruses." The envelope, which is derived from host cell membranes, is obtained by the virus *budding* from cell membranes, but most contain no cell proteins. How cell proteins are excluded and why retroviruses, the exception, do not exclude cell proteins from their virions are not understood (see Section 11.6).

An isometric core surrounded by an isometric envelope: Sindbis virus particles

Sindbis virus (a togavirus) has an icosahedral nucleocapsid that comprises a single protein, surrounded by an envelope from which viral spike proteins protrude. The core has $T = 3$ and 180 subunits, exactly like tomato bushy stunt virus described above. Surprisingly, the envelope also has icosahedral symmetry, but to everyone's surprise this is $T = 4$ and has 240 subunits. This apparent paradox was resolved when it was found that the two structures are complementary, so that the internal ends of the spike proteins fit exactly into depressions between the subunits of the nucleocapsid (Fig. 3.13). So far, this and its near relations are the only enveloped viruses that are known to have a geometrically symmetrical envelope.

A helical core surrounded by an approximately spherical envelope: the influenza virus particle

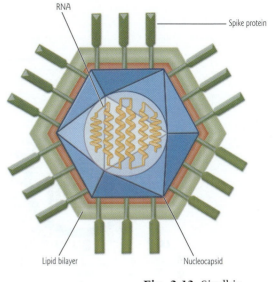

RNA

Spike protein

Lipid bilayer

Nucleocapsid

Fig. 3.13 Sindbis virus: an enveloped icosahedron. The core is $T = 3$, but the envelope is $T = 4$. See text for the explanation. (Courtesy of S. D. Fuller.)

One of the best studied groups of enveloped viruses is the influenza viruses. The helical core is composed of the matrix (M1) protein surrounding a ribonucleoprotein, itself composed of a flexible rod of RNA and nucleoprotein (NP). This is constructed as described in Section 3.2, and arranged in a twisted hairpin structure. The genome is segmented RNA and there are eight separate core structures. Each is associated with a transcriptase complex. The core is contained within a lipid envelope that is only roughly spherical, and hence often described as pleomorphic (Fig. 3.13). M1 protein also lines the inner membrane surface.

In electron micrographs (Fig. 3.14), a large number of protein spikes, projecting about 13.5 nm from the viral envelope, can be observed. These spikes, which have an overall length of 17.5 nm, are transmembrane glycoproteins like many of those in cell membranes. The spike layer consists solely of virus-specified glycoproteins, and comprises about 800 hemagglutinin (HA) and 200 neuraminidase (NA) proteins. HA proteins are trimers and NA proteins are tetramers. NA spikes are arranged nonrandomly in clusters on the virion surface. The HA functions in attachment and fusion-entry, and the NA releases infecting virus from receptors that do not lead to entry and infection. The NA also releases progeny virions which reattach to the host cell in which they were formed (see Section 5.1). Receptor analogs (e.g. Tamiflu) are antivirals that prevent virus release (see Section 21.8). The spikes are morphologically distinct. Figure 3.15 shows the structure of the influenza virus HA. Also in the membrane are a few molecules of an ion channel protein called M2. This allows the passage of protons into the core and is necessary for secondary uncoating.

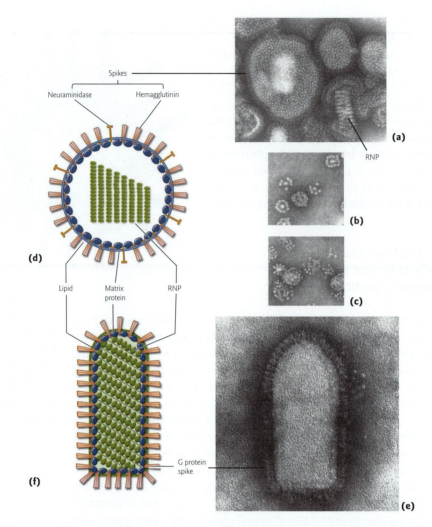

Fig. 3.14 Influenza A virus (an orthomyxovirus) and vesicular stomatitis virus (a rhabdovirus): viruses with enveloped helical structures. Although their morphology is different, these viruses are constructed in the same way. (a) Negatively stained electron micrograph of influenza A virus showing the internal helical ribonucleoprotein (RNP) core and the surface spikes. (b) Aggregates of purified neuraminidase. (c) Aggregates of purified hemagglutinin. Note the triangular shape of the spikes when viewed end-on. (d) Schematic representation of the structure of influenza virus. (e) Negatively stained electron micrograph of vesicular stomatitis virus. (f) Schematic representation of the structure of vesicular stomatitis virus. (Electron micrographs courtesy of Nicholas Wrigley and Chris Smale.)

A helical core surrounded by a nonspherical envelope: rhabdovirus particles

This group of viruses has a helical core consisting of just nucleoprotein, and an envelope like the influenza viruses. What is distinctive about these

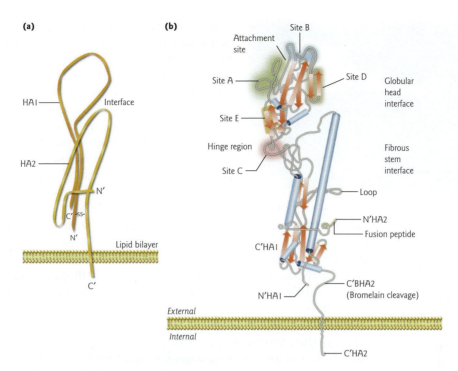

Fig. 3.15 The influenza virus hemagglutinin (HA). This is a homotrimer but only a monomer is shown here. The HA is synthesized as a single polypeptide which is proteolytically cleaved into the membrane-bound HA2 and the distal HA1. (a) An outline structure showing that HA1 and HA2 are both hairpin structures. (b) The crystal structure. The globular head of HA1 bears all the neutralization sites (A–E; shaded) and is made of a distorted jelly-roll β barrel like most of the icosahedral viruses. (From Wiley *et al.* (1981) *Nature (London)*, **289**, 373.)

viruses is that they are not isometric, but either bullet-shaped or bacilliform (rounded at both ends). These are unique morphologies. There are both plant and animal rhabdoviruses, and the latter are all bullet-shaped. The envelope contains a dense layer of spikes comprising just one protein, the viral attachment protein, G. The matrix protein underlies the membrane. Apart from their strikingly different overall shape, rhabdoviruses and influenza viruses have fundamentally a very similar structure (Fig. 3.14). The length of the bullet appears to be controlled by the size of the RNA genome, as defective-interfering rhabdoviruses, that have a massively deleted genome, form small bullet-shaped particles. Nothing is known of the structural geometry involved in the formation of rhabdovirus particles.

3.5 VIRUS PARTICLES WITH HEAD–TAIL MORPHOLOGY

While the head–tail architectural principle is unique to bacterial viruses (Fig. 3.16), many bacterial viruses have other morphologies (Table 3.3

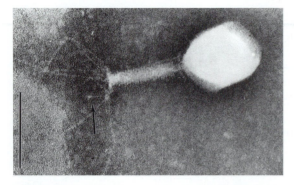

Fig. 3.16 Electron micrograph of bacteriophage T2 with six long tail fibers. Tail pins are not evident but a short fiber (indicated by the arrow) can be seen. The bar is 100 nm. (Courtesy of L. Simon.)

and Appendix 4). There is a large variation on the head–tail structural theme and bacteriophages can be subdivided into those with short tails, long noncontractile tails, and complex contractile tails (see Appendix 4). A number of other structures, such as base plates, collars, etc., may be present. Despite their complex structure, the design principles involved in head–tail phages are identical to those outlined earlier for the viruses of simpler architecture. Heads usually possess icosahedral symmetry, whereas tails usually have helical symmetry. All other structures, base plates, collars, etc., also possess a defined symmetry. The evolution of this elaborate structure may be connected with the way in which these bacterial viruses infect susceptible cells (see Section 5.4). In brief, the phage attaches to a bacterium via its tail, enzymatically lyses a hole in the cell wall and inserts its DNA, which is tightly packed into the phage head, into the cell using the tail as a conduit.

Some of the larger viruses fit none of these structures, and the rules governing their formation have not yet been elucidated. For example, the poxviruses of animals have a complex enveloped structure enclosing two lateral bodies and a biconcave core that includes all the enzymes required for viral mRNA synthesis, and the giant mimivirus, that infects protozoa, has a 400-nm nonenveloped, spherical particle that is surrounded by an icosahedral capsid and fibrils (see Appendixes 1 and 3).

Table 3.3

Distribution of the various types of virus particle structure among families of animal, plant, and bacterial viruses.

Type of particle structure	Animal viruses	Plant viruses	Bacterial viruses
Nonenveloped icosahedral	Common	Common	Uncommon: cortico-, levivi-, micro-, and tectiviruses
Nonenveloped helical	Not known	Common	Rare: ino- and rudiviruses
Enveloped icosahedral	Uncommon: arteri-, asfar-, borna-, flavi-, and togaviruses	Not known	Rare: cystoviruses
Enveloped helical	Common	Uncommon: bunya- and rhabdoviruses	Rare: lipothrixviruses
Head–tail	Not known	Not known	Common

3.6 FREQUENCY OF OCCURRENCE OF DIFFERENT VIRUS PARTICLE MORPHOLOGIES

The different virus morphologies discussed above do not occur with equal frequency among animal, plant and bacterial viruses (Table 3.3). There are relatively few purely icosahedral viruses in bacteria (see Appendix 4); nonenveloped helical viruses are common and occur almost exclusively in plants; enveloped icosahedral viruses and enveloped helical viruses are common in animals and rare in plants and bacteria. Finally head–tail virus morphology, in which an isometric head and a helical tail are joined together, is found only in bacteria. Unfortunately there is no real explanation as to why there should be these restrictions. There also exist some very large, very complex viruses (e.g. poxviruses of animals and mimivirus of amoebae: see Appendix) whose morphogenesis is beyond our current comprehension.

3.7 PRINCIPLES OF DISASSEMBLY: VIRUS PARTICLES ARE METASTABLE

It is important to remember that all virus particles not only have to be constructed to protect the genome, but they also have to disassemble to permit the genome to enter a new target cell. This is supremely important to the virus particle as it has only the one chance to do this successfully and hence propagate its genome. The notion is developing that the particle is metastable, i.e. it can spontaneously descend to a lower energy level and, in doing so, releases its genome. Not surprisingly there are a number of fail-safe devices that tell the virus when it is safe to let go the genome. One of the simplest systems is used by enveloped animal viruses like HIV-1. This undergoes a succession of interactions between cell receptors and virus envelope protein binding sites, the passwords needed to gain entry to a high security establishment. If everything is in order, the metastable envelope protein then undergoes profound rearrangements that allow a hidden hydrophobic segment to insert into the cell membrane. This initiates fusion of the lipid bilayer of the virus with that of the cell plasma membrane, and the virus genome automatically enters the cell cytoplasm. However, if the sequence of passwords proves incorrect, the virus detaches from the cell and the process can be repeated until the correct cell is found. Mechanisms of entry are discussed in Chapter 5.

KEY POINTS

- Virus particles consist mainly of nucleic acid and protein, arranged to form a regular geometric structure.

- Some viruses have particles that are surrounded by a membrane.
- Virus proteins protect the viral genome, identify the appropriate target cell, and get the genome into the target cell.
- Some viruses contain proteins with enzymatic functions that are needed for genome replication, and these enter the cell with the genome.
- Elongated virus particles have their protein subunits arranged in a helix, and are said to have helical symmetry.
- Isometric virus particles are usually icosahedrons – structures with 20 faces, with icosahedral symmetry.
- Membrane-bound viruses are formed by budding from a cell membrane.
- Membrane-bound viruses have an internal nucleoprotein structure with helical symmetry or with isometric symmetry.

FURTHER READING

Branden, C., Tooze, J. 1999. *Introduction to Protein Structure*, 2nd edn. Garland Publishing, New York.

Butler, P. J. G. 1984. The current picture of the structure and assembly of tobacco mosaic virus. *Journal of General Virology* **65**, 253–279.

Caspar, D. L. D., Klug, A. 1962. Physical principles in the construction of regular viruses. *Cold Spring Harbor Symposium of Quantitative Biology* **27**, 1–24.

Eiserling, F. A. 1979. Bacteriophage structure. In *Comprehensive Virology* (H. Fraenkel-Conrat, and R. R. Wagner, eds), Vol. 13, pp. 543–580. Plenum Press, New York.

Hogle, J. M., ed. 1990. Virus structure. *Seminars in Virology* **1**, 385–487 (several articles).

Johnson, J. E. 2003. Virus particle dynamics. *Advances in Protein Chemistry* **64**, 197–218.

Rixon, F. J., Chiu, W. 2003. Studying large viruses. *Advances in Protein Chemistry* **64**, 379–408.

Smith, T. J., Baker, T. 1999. Picornaviruses: epitopes, canyons, and pockets. *Advances in Virus Research* **53**, 1–23.

Stuart, D. 1993. Virus structures. *Current Opinion in Structural Biology* **3**, 167–174.

Tyler, K. L., Oldstone, M. B. A. 1998. Reoviruses I. Structure, proteins and genetics. *Current Topics in Microbiology and Immunology* **233**, 1–213.

Wiley, D. C., Skehel, J. J. 1987. The structure and function of the hemagglutinin membrane glycoprotein of influenza virus. *Annual Review of Biochemistry* **56**, 365–394.

Zhou, Z. H., Chiu, W. 2003. Determination of icosahedral virus structures by electron cryomicroscopy at sub-nanometer resolution. *Advances in Protein Chemistry* **64**, 93–124.

For images of virus particles see the Virus Particle Explorer (VIPER) website: http://viperdb.scripps.edu

Also check Appendix 7 for references specific to each family of viruses.

4

Classification of viruses

Viruses are found throughout the world and infect all known organisms. Viruses cause a range of different diseases and display a diversity of host range, morphology, and genetic makeup. Bringing order to this huge diversity requires the designation of classification groups to permit study of representative viruses that can inform us about their less well studied relatives.

Viruses represent one of the most successful types of parasite in the world and have been isolated from representatives of every known group of organisms from the smallest single-celled bacterium to the largest mammal. While in most cases the virus is specific for the host species in which it has been identified, some viruses are able to infect species from different phyla and even different kingdoms. The number of known viruses now reaches over 5000 with new viruses being discovered all the time. This very large number contains a diverse array of viruses which at first sight is very bewildering. To make easier the study of viruses and bring order to this apparent diversity, over the years a number of different systems has been proposed to generate classification schemes which will allow us to study representative viruses rather than each individually. All of the proposed classification schemes have different strengths and weaknesses but there is now general consensus. The International Committee on Taxonomy of Viruses has responsibility for assignment of new viruses to specific groupings.

4.1 CLASSIFICATION ON THE BASIS OF DISEASE

The first, and most common, experience of viruses is as agents of disease and it is possible to group viruses according to the nature of the disease with which they are associated. Thus, one can discuss hepatitis viruses or viruses causing the common cold. This is attractively simple. However, this method of grouping viruses, though reflecting an important characteristic, suffers from serious deficiencies. First, this approach is very anthropomorphic, focusing as it does on diseases that we recognize because they affect humans or our domestic livestock. This ignores the fact that most viruses either do not cause disease or cause a disease that we do not recognize because of a lack of understanding of the host; for example we understand little of the diseases caused by viruses of fish or amphibians. Similarly, it is possible for a single virus to cause more than one type of disease; a good example of this is varicella zoster virus which causes chickenpox in a first infection but when reactivated later in life causes shingles. This problem is compounded when considering viruses which infect more than one host as it is common to find that it causes either no disease in one host while dramatically affecting the other or that it may cause different diseases in different hosts. A classification based on disease, while it may be helpful in some settings, also fails in the important feature of being able to use the groupings to predict common fundamental features of the viruses in question. In the case of agents of hepatitis and the common cold, many different viruses with very different molecular makeups are involved and studying just one of these tells us little, if anything, about the others.

4.2 CLASSIFICATION ON THE BASIS OF HOST ORGANISM

An alternative approach has been to group viruses according to the host that they infect. This has the attraction that it emphasizes the parasitic nature of the virus–host interaction. However, there are several difficulties with this approach. This form of classification implies a fixed, unchanging, link between the virus and host in question. Some viruses are very restricted in their host range, infecting only one species, such as hepatitis B virus infecting humans, and so a designation based on the host is appropriate. However, others may infect a small range of hosts, such as poliovirus which can infect various primates, and the designation here must reflect this rather than name a single species. The most serious difficulty arises with viruses which infect and replicate within very different species. This can be seen with certain viruses which can infect and replicate within both plants and insects. Designation of a virus by the host it infects is therefore not always straightforward. Overriding all of these difficulties is the problem that even if a number of viruses infect a single species, this characteristic does not imply any other similarities in terms of disease or genetic makeup of the various viruses.

A different level of sophistication in terms of defining the host has been used for some viruses, notably the herpesviruses. Having shown that herpesviruses are similar in a number of ways, a classification which defined them in terms of the nature of the host cell they infected, for example with gammaherpesviruses infecting lymphoid cells in the host animal, was described. With the discovery of new herpesviruses which infect lymphoid cells but share characteristics with the other, nongammaherpesviruses, this definition is no longer sustainable. Consequently, studying a single virus which infects a single species or group of species, or indeed a virus which infects a particular cell type, tells us nothing about the fundamental nature of the potentially many other viruses which also infect that host.

4.3 CLASSIFICATION ON THE BASIS OF VIRUS PARTICLE MORPHOLOGY

The structural features of virus particles and the principles which underlie these structures have been described in Chapter 3. When viruses were first visualized in the electron microscope, defining classification groups on the basis of the observed particle shape or morphology was relatively simple. A key structural feature is whether or not the virus particle has a lipid envelope and this alone can be used as a designated feature, giving enveloped and nonenveloped viruses (see Section 3.4). If the virion is nonenveloped three morphological categories are defined, isometric, filamentous, and complex. Isometric viruses (see Section 3.3) appear approximately spherical but are actually icosahedrons or icosadeltahedron. Filamentous viruses (see Section 3.2) have a simple, helical, morphology. The complex viruses are those which do not neatly fit within the other two categories. Complex shapes for virus particles may be made up of a combination of isometric and filamentous components, such as is seen with bacteriophage T2 (Fig. 3.16), or they may have a structure which does not conform to the simple geometrical rules of the majority and appear to our eye to be irregular in shape. If the virion is enveloped a further level of classification is possible by describing the morphology of the nucleocapsid found within the membrane. Thus, there are isometric and helical nucleocapsids.

While a classification scheme based on morphology is simple and describes an unchanging feature of the virus, it suffers from several drawbacks. Primary amongst these is that knowing the shape of a virus particle does not allow us to predict anything about the biology, pathology, or molecular biology of similarly shaped viruses. Thus, two viruses with very similar morphologies may differ in all of their other fundamental characteristics. This drawback is also true even for viruses which appear to share a number of other features. For example, the polyomaviruses and the papillomaviruses were originally classified together on the basis of their very similar morphology and the similarity extended to other, deeper, features of their structures including the nature and organization of their genomes. A better understanding of these viruses at the molecular level has shown that they differ in several critical areas and they are now recognized as quite different entities. Despite these problems a preliminary description of a virus in terms of morphology is still common.

4.4 CLASSIFICATION ON THE BASIS OF VIRAL NUCLEIC ACIDS

The nucleic acid of a virus contains all the information needed to produce new virus particles. Some of this information is used directly to make virion components and some to make accessory proteins or to provide signals which allow the virus to subvert the biosynthetic machinery of a cell and redirect it towards the production of virus. Whereas the standard form of genetic material in living systems is double-stranded DNA, viruses contain a diverse array of nucleic acid forms and compositions. The nucleic acid content of a virus has been used as a basis for classifying viruses. The key aspect of this classification scheme is that it considers the nature of the virus genome in terms of the mechanisms used to replicate the nucleic acid and transcribe mRNA encoding proteins. A detailed consideration of the nature of virus nucleic acids and the mechanisms by which they are replicated and transcribed are to be found in Chapters 6–10. Here we will consider only how such features can be used to generate a classification scheme.

The nature of a particular nucleic acid sample is assessed by determining its base composition, sensitivity to DNase or RNase, buoyant density, etc. Single-stranded nucleic acids are distinguished from double-stranded by the absence of a sharp increase in absorbance of ultraviolet light upon heating and the nonequivalence of the molar proportions of adenine (A) and thymine (T) (or uracil (U)) or guanine (G) and cytosine (C). From these types of analysis, it appears that viruses utilize four possible kinds of viral nucleic acid: single-stranded DNA, single-stranded RNA, double-stranded DNA, and double-stranded RNA. Each kind of genome is found in many virus families, which between them contain members that infect a diverse array of animals, plants, and bacteria.

Classifying viruses – the Baltimore Scheme

As considered above, viruses exhibit great diversity in terms of morphology, genome structure, mode of infection, host range, tissue tropism, disease (pathology), etc. While, as we have seen, each of these properties can be used to place viruses into groups, classifying viruses solely on the basis of one or even two of these parameters does not lead to a system where studying one virus in a particular group can be used to draw inferences about other members of the same group. Also, classification on these grounds does not give a good basis for unifying discussions of virus replication processes. To circumvent these problems Nobel laureate David Baltimore proposed a classification scheme which encompasses all viruses, based on the nature of their genomes, and their modes of replication and gene expression. This system provides an opportunity to make inferences and predictions about the fundamental nature of all viruses within each defined group.

The original Baltimore classification scheme was based on the fundamental importance of messenger RNA (mRNA) in the replication cycle of viruses. Viruses do not contain the molecules necessary to translate mRNA and rely on the host cell to provide these. They must therefore synthesize mRNAs which are recognized by the host cell ribosomes. In the Baltimore scheme, viruses are grouped according to the mechanism of mRNA synthesis which they employ (Fig. 4.1). By convention, all mRNA is designated as positive (or "plus") sense RNA. Strands of viral DNA and RNA which are complementary to the mRNA are designated as negative (or "minus") sense and those that have the same sequence are termed positive sense. Using this terminology, coupled with some additional information about the replication process, a modified classification scheme based on the original proposed by Baltimore defines seven groups of viruses, with each commonly being referred to by the nature of the virus genomes it includes:

Class 1 contains all viruses that have double-stranded (ds) DNA genomes. In this class, the designation of positive and negative sense is not meaningful since mRNAs may come from either strand. Transcription can occur using a process similar to that found in the host cells.

Class 2 contains viruses that have single-stranded (ss) DNA genomes. The DNA can be of positive or negative sense, depending on the virus being studied. For viruses in class 2 the DNA must be converted to a double-stranded form before the synthesis of mRNA can proceed.

Class 3 contains viruses that have dsRNA genomes. All known viruses of this type have segmented genomes and mRNA is only synthesized from one template strand of each segment. The process of transcription from a dsRNA genome can be envisioned as occurring using a mechanism similar to that for transcription from a dsDNA genome. However, the enzymes necessary to carry out such a process do not exist in normal, uninfected, cells. Consequently, these enzymes must be encoded by the virus genome and must be carried into the cell by the virus to initiate the infectious process.

Class 4 contains viruses with ssRNA genomes of the same (positive) sense as mRNA and which can be translated. Synthesis of a complementary strand, generating a dsRNA intermediate, precedes synthesis of mRNA. As with the class 3 viruses the RNA synthesis must be carried out using virus-encoded enzymes, although these are not carried in the virus particle.

Class 5 contains viruses that have ssRNA genomes which are complementary in base sequence to the mRNA (negative-strand RNA viruses). Synthesis of mRNA requires novel virus-encoded enzymes, and generation of new virus genomes requires the synthesis of a dsRNA intermediate, the positive sense strand of which is used as a template for replication. Viral

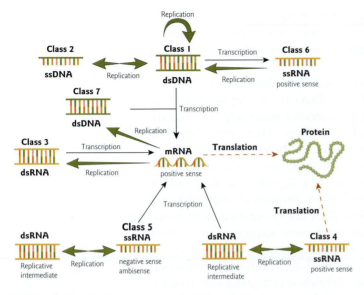

Fig. 4.1 The Baltimore classification scheme (see text for details). mRNA is designated as positive sense. Translation of protein from mRNA and positive sense RNA virus genomes is indicated with red arrows. The path of production of mRNA from double-stranded templates is shown with black arrows. Green arrows show the steps in the replication of the various types of genome with double-headed arrows, indicating production of a double-stranded intermediate from which single-stranded genomes are produced.

RNA-synthesizing enzymes are carried in the virion. Some class 5 viruses use the newly synthesized "antigenome" RNA strand as a template for the production of an mRNA and are referred to as "ambisense" viruses.

Class 6 contains viruses that have ssRNA genomes and which generate a dsDNA intermediate as a prelude to replication, using an enzyme carried in the virion.

Class 7 More recently, it has been suggested that some viruses, termed reversiviruses, should be transferred from class 1 into a new class 7. This is based on their replication from dsDNA via a positive sense ssRNA intermediate back to dsDNA. This represents the inverse of the class 6 replication strategy, with which class 7 has many similarities.

The Baltimore scheme has both strengths and weaknesses as a tool for understanding virus properties. A particular strength is that assignment to a class is based on fundamental, unchanging, characteristics of a virus. Once assigned to a class, certain predictions about the molecular processes of nucleic acid synthesis can be made, such as the requirement for novel virus-encoded enzymes. A weakness is that, whilst it brings together viruses with similarities of replication mechanism, the scheme takes no account of their biological properties. For example, bacteriophage T2 and variola virus (the cause of smallpox) are classified together in class 1 although they are totally dissimilar in both structure and biology. Similarly, the identification of a positive sense RNA genome is not sufficient to classify the virus unambiguously since viruses of classes 4 and 6 have similar genome nucleic acids.

4.5 CLASSIFICATION ON THE BASIS OF TAXONOMY

The International Committee on Taxonomy of Viruses (ICTV), first founded in the late 1960s, has established a taxonomic classification scheme for viruses. This uses the familiar systematic taxonomy scheme of *Order, Family, Subfamily,* and *Genus* (no Kingdoms, Phyla, or Class of viruses have been described within this scheme). However, the concept of a species in the classification of viruses is complex and in many cases remains a point of ongoing debate. For viruses with RNA genomes the concept of a species is made difficult by the absence of a proofreading function during genome replication. The result of this is that a virus exists as a member of a population where each member has a genome sequence which may be different to the others but which belongs to a collection of sequences which will combine to form a consensus for that virus. The virus is said to be a *quasi-species*, and there is no defined "correct" genome sequence.

In assigning a virus to a taxonomic group the ICTV considers a range of characteristics. These include host range (eukaryote or prokaryote, animal, plant, etc.), morphological features of the virion (enveloped, shape of capsid or nucleocapsid, etc.), and nature of the genome nucleic acid (DNA or RNA, single stranded or double stranded, positive or negative sense, etc.). Within these parameters additional features are considered. These include such things as the length of the tail of a phage or the presence or absence of specific genes in the genomes of similar viruses, and these aspects allow allocation of subdivisions in the taxonomic designation. With the advent of rapid nucleotide sequencing of virus genomes it is now possible to derive phylogenetic trees based on the degree of conservation of specific gene sequences and this further informs the designation of phylogenetic groupings (clades). Phylogeny is particularly useful when considering newly discovered viruses for which little other information may be available. For each of the individual genera defined by the ICTV a single virus has been designated as the type member. The type member is used essentially as the reference for the genus. For an example of the classification process see Box 4.1.

It can be seen from the details used in the classification of measles virus that the system utilizes aspects of all of the other classification schemes in an attempt to clearly identify each virus group. The classification of viruses is an ongoing project as new viruses are discovered. As our understanding of specific features grows, particularly in the area of phylogenetic analysis, the taxonomic groupings are constantly reassessed.

4.6 SATELLITES, VIROIDS, AND PRIONS

Satellites and viroids

Satellites and viroids have features which differentiate them from the better understood "conventional" viruses. However, they have features which are common to several viruses and they are often referred to as subviral agents. The similarities with conventional viruses have permitted a preliminary classification of viroids into families and genera.

Box 4.1

Example of phylogenetic assignment of measles virus using the ICTV criteria

This describes the key elements for classification. Other more detailed criteria are also used.
Order: *Mononegavirales*. Enveloped viruses containing one single-stranded negative sense RNA molecule. The genome RNA contains a number of defined features; the regulatory sequences at the 3′ and 5′ ends are complementary and flank the core, conserved, genes which are present in a fixed order in the genome. The nucleocapsids are helical. The ribonucleoprotein of the virus core is the "infectious unit." The virion contains prominent spike-like structures on the surface.
Family: *Paramyxoviridae*. The virions are usually spherical and contain spikes consisting of two or three glycoproteins arranged in homopolymers. A nonglycosylated matrix protein is associated with the inner surface of the lipid membrane. The genome RNA is between approximately 13,000 nt and 15,500 nt in length. Genome RNA does not contain a 5′ cap structure or a 3′ polyadenylate tail.
Subfamily: *Paramyxovirinae*. The genome contains between five and seven transcriptional regions. Phylogenetic sequence relatedness exists between the proteins, with the nucleocapsid protein, the matrix protein, and the polymerase protein generally the most highly conserved. Nucleocapsids have a diameter of 18 nm and a pitch of 5.5 nm. Surface spikes are approximately 8 nm in length. The genome length is a multiple of six nucleotides. All members have hemagglutinin activity associated with the virion.
Genus: *Morbillivirus*. Narrow host range which also distinguishes the members from each other. Members display greater sequence relatedness in phylogenetic trees within the genus than with members of the other genera in the subfamily. All members show the same pattern of transcription and RNA editing. All members generate intracytoplasmic and intranuclear inclusion bodies which are assemblies of virus nucleocapsids. Members show cross-reactivity in some, but not all, antibody neutralization and cross-protection tests.
Type member: measles virus.

Satellites are categorized by two forms, satellite viruses and satellite nucleic acid. Neither satellite viruses nor satellite nucleic acids encode the enzymes required for their replication and require coinfection with a conventional, helper, virus to provide the replicative enzymes in the same way as DI viruses (see Section 7.1). However, unlike DI viruses, the sequence of satellite genomes is significantly different to that of the helper virus. Satellite viruses encode the structural protein that encapsidate them but the satellite nucleic acids either encode only nonstructural proteins or no proteins at all. The satellite nucleic acids derive their structural proteins from the coinfecting helper virus. In some cases the sequences of the immediate termini of satellite genomes are similar to those of the helper virus, suggesting that these regions are involved in replication as is frequently the case for the helper virus genomes.

The presence of a satellite or satellite virus may affect the replication of the helper virus and may also increase or decrease the severity of the disease(s) caused by the helper virus. Satellite

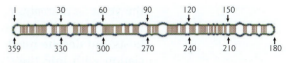

Fig. 4.2 Diagram of the circular single-stranded 359-nucleotide RNA of potato spindle tuber viroid, showing the maximized base-paired structure.

viruses and satellite nucleic acids with either DNA or RNA genomes have been identified. The RNA genomes range in size from approximately 350 nt to 1500 nt and the DNA genomes, which are either single stranded or double stranded, from range from 500 nt to 1800 nt. The RNA satellite genomes have been found in linear ssRNA, circular ssRNA, or dsRNA forms (see Section 22.5).

A satellite that is important because of its association with disease in humans is hepatitis delta virus (HDV). HDV is only found in association with hepatitis B virus (HBV) and is associated with enhanced pathogenicity of HBV. The HDV genome consists of a circular ssRNA molecule which is extensively base-paired and appears as a rod-like structure. HDV is unable to replicate without a helper virus (HBV) which provides the structural proteins which encapsidate the genome and allow HDV to be spread. The HDV genome is approximately 1700 nt in length and encodes two proteins. Both proteins are translated from a mRNA which is of opposite sense to the HDV genome RNA, analogous to the method of gene expression used by the class 5 viruses (see Chapter 10).

Viroids are novel agents of disease in plants; their infectious material consists of a single circular ssRNA molecule with no protein component. Viroids, with genomes ranging in size from 246 to 400 nucleotides, are the smallest self-replicating pathogens known. Up to 70% of the nucleotides in the genome RNAs are base-paired, and when genomes are examined in the electron microscope they appear as rod-shaped or dumb-bell-shaped molecules (Fig. 4.2), similar to the genome of HDV. Nucleotide sequence analysis of viroid RNA has shown that no proteins are encoded by either the genome or antigenome sense RNA. It is thought that the diseases associated with viroids result from the RNA interfering with essential host cell mechanisms.

Viroids and HDV appear to be replicated by the cellular DNA-dependent RNA polymerase II which normally recognizes a DNA template. Replication of viroid RNA is described in Section 7.5.

Prions

Prions were initially identified as the agents of a number of diseases characterized by slow progressive neurological degeneration which was fatal. The diseases are associated with a spongy appearance of the brain that is seen *post mortem*. Together the diseases are termed spongiform encephalopathies and include scrapie in sheep, bovine spongiform encephalopathy (BSE; "mad cow" disease) in cattle, and the human infections of kuru and variant Creutzfeldt–Jakob disease. Prions have also be identified in the yeast *Saccharomyces cerevissiae* and the fungus *Podospora anserina*. Prions were initially thought to be viruses which replicated slowly within their hosts, but despite much work no nucleic acid has yet been found in association with infectious material. Analysis of prions has shown that they are aberrant forms of normal cellular proteins which can induce changes in the shape of their normal homologs with catastrophic consequences for the host. The nature of these unusual agents and the diseases they cause is considered in Chapter 22.

KEY POINTS

- Viruses infect all known species.
- Viruses can be classified by a number of different methods including disease caused, host range, morphology and nature of the genome in association with the method of replication and transcription.
- Virus genomes consist of either DNA or RNA and may be single stranded or double stranded. Some viruses convert their genomes from RNA to DNA, or vice versa, for replication and transcription.
- The Baltimore system has the potential to predict general features of the replication and transcription of viruses from a knowledge of the nature of the genome.
- The best and most comprehensive classification scheme uses elements of all possible schemes, in association with phylogenetic analysis of nucleic acid sequences to generate a taxonomic system which accurately assigns viruses to a genus within a hierarchical structure.
- Satellites, satellite viruses, and viroids differ from viruses in fundamental ways but satellites and satellite viruses rely on the presence of specific viruses to replicate. Satellites, satellite viruses, and viroids are associated with serious disease in their hosts.

FURTHER READING

Baltimore, D. 1971. Expression of animal virus genomes. *Bacteriological Reviews* **35**, 235–241.

Fauquet, C. M., Mayo, M. A., Maniloff, J., Desselberger, U., Ball, L. A., eds. 2005. *Virus Taxonomy: VIIIth report of the International Committee on Taxonomy of Viruses*, 8th edn. Elsevier Press, Amsterdam.

Flores, R., Di Serio, F., Hernández, C. 1997. Viroids: the noncoding genomes. *Seminars in Virology* **8**, 65–73.

Karayiannis, P. 1998. Hepatitis D virus. *Reviews in Medical Virology* **8**, 13–24.

Symons, R. H., ed. 1990. Viroids and related pathogenic RNAs. *Seminars in Virology*, **1**(2).

The website of the International Committee on Taxonomy of Viruses (ICTV) http://www.ncbi.nlm.nih.gov/ICTVdb/index.htm

The GenBank sequence database (includes virus genome sequences) http://www.ncbi.nlm.nih.gov/Entrez/index.html

Part II

Virus growth in cells

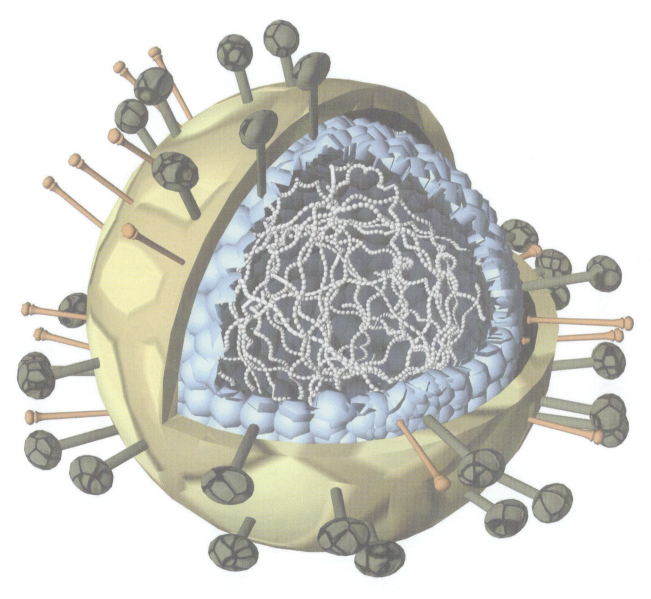

5

The process of infection: I. Attachment of viruses and the entry of their genomes into the target cell

Virus receptors are normal cell molecules that also serve as docking and entry molecules for a virus. They tell the virus that they have arrived at a suitable cell, and prepare the virus to release its genome. This is often a complex multistage process that the virus must perform correctly – if it gets it wrong, the virus dies.

Chapter 5 Outline

5.1 Infection of animal cells – attachment to the cell
5.2 Infection of animal cells – entry into the cell
5.3 Infection of plants
5.4 Infection of bacteria
5.5 Prevention of the early stages of infection

The process of infection begins with the coming together of a virus particle with its target cell, but this union occurs by different means with each of bacteriophages, plant viruses, and animal viruses. The initial interaction of animal viruses with animal cells occurs by simple diffusion, since particles the size of a virus are in constant Brownian motion when suspended in liquid. Diffusion of bacteriophage is probably also the force influencing their union with bacterial cells. By contrast, plant viruses are, in most cases, injected directly into the cell cytoplasm by the activities of virus-carrying pathogens, or else plants become infected following mechanical damage, very often as a result of wind action. Consequently, the way in which the union of virus and cell occurs is not so important in plant systems. However, viruses can also be transmitted from infected tissue to a noninfected plant from grafting, and diffusion of virus through the vascular system is most likely responsible for infection in this situation. Plants are a special case as all cells directly communicate with their neighbors, and functionally a plant behaves as a single cell.

Virtually all data for the attachment and entry of animal viruses comes from experiments with cells cultured *in vitro*, and most of these are dedifferentiated compared with their state *in vivo*. Thus infection of differentiated cells, which may also be part of a multicellular tissue structure *in vivo*, may not take place by exactly the same process. The efficiency of infection, for example, may be lower.

Terminology relating to attachment can be confusing. Consequently, the terms *virus attachment protein* (of the virus) and *cell receptor* (of the cell) are used here to describe the interacting components. In some situations it is useful to refer to cell receptor sites, which may be multivalent and consist of several cell receptors.

5.1 INFECTION OF ANIMAL CELLS – ATTACHMENT TO THE CELL

Animal cells are bounded by a lipid bilayer (the plasma membrane) into which is inserted a variety of proteins with which the cell communicates with its environment. Viruses have evolved to hijack such proteins in order to gain entry into that cell, and an animal cell cannot be infected unless it expresses the molecule which serves as a receptor for that particular virus on its surface.

Viruses attach to animal cells via receptor molecules found on the cell surface. These are usually proteins, but carbohydrates and, very occasionally, lipids are also used (Table 5.1). The virus–receptor interaction is highly specific, but a family of viruses may use the same receptor. One notable exception is the sugar, *N*-acetylneuraminic acid, which often forms the terminal moiety of a carbohydrate group of a glycoprotein or glycolipid, and is used as a receptor by members of several different families of viruses (Table 5.1). Some viruses can use more than one type of molecule as

---

Box 5.1

Evidence required to demonstrate receptor activity in a cell-surface molecule

- Candidate receptor molecules can be identified as cell components that bind to intact virus or to the known attachment protein(s) of the virus.
- Presence of a candidate receptor on cells susceptible to infection and absence from other cells is supportive circumstantial evidence.
- Proof that a candidate receptor can actually mediate infection requires either that a monoclonal antibody to the candidate receptor can block infection of cells by the virus or that a complementary DNA (cDNA) clone for the candidate receptor introduced into nonpermissive cells causes those cells to become infectable by the virus.
- A molecule may fail these tests and still be a genuine receptor for the virus, since not all antibodies to a receptor will necessarily block infection and there may be factors blocking infection in a nonpermissive cell other than the lack of the receptor.

Table 5.1

Some cell surface molecules used by animal viruses as receptors.

Molecule	Normal function	Virus	Type of receptor
Protein			
ICAM-1	Adhesion to other cells via CD54	Rhinoviruses (most but not all)	Primary
CAR (Coxsackie-adenovirus receptor)	A unique protein of unknown function	Many adenoviruses, Coxsackie B viruses	Primary
$\alpha_v\beta_x$ integrin	Adhesion to other cells via vitronectin	Adenoviruses	Coreceptor
$\alpha_v\beta_6$ integrin	Adhesion to other cells via vitronectin	Foot-and-mouth disease virus	Primary
CD4	Ligand for MHC II on T helper cells	HIV-1, HIV-2, SIV	Primary
CCR5	Proteins with 7-transmembrane domains that bind C–C chemokines	HIV-1, HIV-2, SIV	Coreceptor
CXCR4	Proteins with 7-transmembrane domains that bind C–X–C chemokines	HIV-1, HIV-2, SIV	Coreceptor
DC-SIGN	C-type lectin on dendritic cells that binds carbohydrate moieties	HIV-1, SIV	Primary, but unusual as bound virus is transferred from the dendritic cell to a T cell
CDW150 (SLAM)	Lymphocyte activation molecule	Measles virus	Primary
CD155	Ligand for vitronectin, nectin-3, and DNAM-1	Poliovirus	Primary
MHC I	Ligand for CD8; presents peptides to T cells	Human cytomegalovirus	Primary
MHC II	Ligand for CD4; presents peptides to T cells	Lactate dehydrogenase-elevating virus	Primary
CR2	Receptor for complement component C3d	Epstein–Barr virus	Primary
IgA receptor	Binds IgA for transport across the cell	Hepatitis B virus	Primary
Phosphate transporter	Transports phosphate	Some retroviruses	Primary
Virus-specific IgG bound to cells by Fc receptors (see text)	Binds to virus	Dengue virus *in vivo* and many others *in vitro*	Primary
β-adrenergic receptor	Binds the hormone adrenaline (epinephrin)	Reoviruses	Primary

Table 5.1 (*Cont'd*)

Molecule	Normal function	Virus	Type of receptor
Acetylcholine receptor	Binds the acetylcholine neurotransmitter	Rabies virus	Primary
Carbohydrate			
N-acetylneuraminic acid (only when terminal in the carbohydrate moiety)	Part of the carbohydrate moiety of glycoproteins and glycolipids. Gives cells much of their negative charge	Influenza virus A, B, C, paramyxoviruses, polyomavirus, encephalomyocarditis virus, reoviruses	Primary
Heparan sulbate	Extracellular matrix glycosaminoglycan	HIV-1, herpes simplex virus, dengue virus, Sindbis virus, cytomegalovirus, adeno-associated virus, respiratory syncytial virus, foot-and mouth disease virus	Coreceptor
Lipid			
Phosphatidylserine	Constituent of lipid bilayer	Vesicular stomatitis virus	Primary

ICAM, intercellular adhesion molecule; MHC, major histocompatibility complex; CR, complement receptor; IgA, immunoglobulin A; HIV, human immunodeficiency virus; SIV, simian immunodeficiency virus.

a primary receptor (e.g. reoviruses bind to the β-adrenergic receptor or *N*-acetylneuraminic acid). The unequivocal demonstration that a molecule serves as a receptor for virus infection requires some stringent tests (Box 5.1).

Viruses bind up to three different types of receptor molecule on the cell surface in succession. These are low affinity receptors, primary receptors and co- or secondary receptors (Table 5.1). In principle, receptors serve to overcome any repulsive forces that may exist between the virus and the cell, to allow the virus particle to have intimate contact with the lipid bilayer of the cell membrane, and to trigger the release of the viral genome into the cell. The first type of receptor is a high abundance molecule that has a low specificity, low affinity interaction with the virus. This serves to get the virus out of the fluid bathing the cell and in direct contact with molecules on the cell surface. Several viruses, including HIV-1, use heparans while others use the sugar, *N*-acetyl neuraminic acid (NANA), as their first receptor (Table 5.1). This binding is usually followed by interactions with further cell-surface molecules that lead to

Box 5.2

Attachment interactions of HIV-1 with its target cell

HIV-1 infects $CD4^+$ T cells or $CD4^+$ macrophages. Initially the virus uses a cellular protein (cyclophilin A) that is an essential component of the HIV-1 particle to bind a low affinity receptor, heparan. This interaction allows the virus to make initial contact with the cell. It then has the opportunity of contacting the primary receptor, CD4, which is a less abundant molecule. If the virus does not find a primary receptor molecule, it will dissociate from the cell completely and the process begins again. The search process consists of the virus rolling along the cell surface, dissociating from and reassociating with the low affinity receptor, until it comes in contact with CD4 molecules. The CD4–gp120 binding is a high affinity interaction. In turn, binding to CD4 causes gp120 to undergo conformational changes that expose another site that is specific for the coreceptors CCR-5 or CXCR4. However, binding of the viral gp120 with cellular CD4 keeps the virus and cell bilayers too far apart for fusion to take place. This is remedied by binding to the coreceptor molecule as this comprises seven transmembrane regions with only small loops on the extracellular surface. Thus binding to the coreceptor draws the bilayers closer together. It seems that this series of receptor interactions progressively ratchets the viral and cell bilayers into closer proximity so that fusion of the viral and cell lipid bilayers takes place, and the viral genome gains entry to the cell cytoplasm (see below). This succession of dependent binding steps is akin to a fail-safe system, telling the virus that it has bound to the right type of cell in which it will be able to replicate its genome, and leading to a position which favors virus–cell fusion.

infection. The binding of HIV-1 to its target cell, for example, is complex and it uses three cell receptors and three virus attachment sites (Box 5.2). However, NANA is the sole receptor for the influenza viruses.

Some viruses can actually use non-neutralizing, virus-specific antibody as an additional receptor (Table 5.1). The antibody binds via its constant regions to Fc receptors that certain cells have in their plasma membranes and this leads to infection. Such cells do not carry a regular virus receptor, and would not normally be infected. This process is known as *antibody-dependent enhancement* (ADE) of infectivity. While ADE can be readily demonstrated in cells in culture, it is rare *in vivo*. The classic example occurs with dengue fever virus (a flavivirus). This virus normally causes a mild subclinical or febrile illness in humans but in the presence of antibody is able to infect macrophages and to cause life-threatening infections (dengue hemorrhagic fever and dengue shock syndrome).

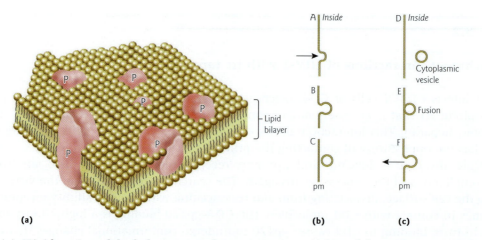

Fig. 5.1 (a) "Lipid sea" model of plasma membrane structure. (Adapted from Singer & Nicholson 1972 *Science* **175**, 723.) Proteins (P) may span the lipid bilayer or may not, and are free to move laterally like icebergs in the sea. (b) Pinocytosis by a plasma membrane (pm) inwards (A,B,C) and (c) exocytosis outwards (D,E,F).

5.2 INFECTION OF ANIMAL CELLS – ENTRY INTO THE CELL

To gain access to the interior of the cell, a virus attached at the cell surface must cross a membrane. Viruses with lipid envelopes achieve this by membrane fusion, at the surface or after being taken into the cell in a vesicle, while nonenveloped cells are internalized and then breach the integrity of the vesicle that contains them.

The plasma membrane that surrounds the cell is a very mobile and active structure. It comprises a lipid bilayer into which are inserted a variety of proteins. The membrane has been compared to a sea and the proteins as icebergs that can move laterally within the membrane (Fig. 5.1a). Cells are constantly taking samples of their immediate environment by endocytosis, during which the membrane invaginates and a vesicle is pinched off into the cytoplasm (Fig. 5.1b), or carrying out the reverse process to export substances from the cell such as enzymes, hormones, or neurotransmitters (Fig. 5.1c).

A virus, attached to cells as described above, may need to recruit further receptors before it is able to enter the cell and/or uncoat. Because the receptors in the lipid membrane are mobile, both the attached virus and free receptors can move laterally and find each other by random collision. This recruitment of a finite number of receptors is the final signal for the entry/uncoating stage to begin, for in addition to allowing the virus to find the correct target cell, interaction with receptors prepares the virus particle for uncoating and the entry of the viral genome into the cell cytoplasm. Uptake by the cell happens in two ways: enveloped viruses enter by *fusion* of the lipid bilayer of the virus with one of the cell membranes, while nonenveloped viruses are taken up into a vesicle by *receptor-mediated endocytosis*, and from there the virus genome escapes into the cell cytoplasm.

Entry of the genome of an enveloped virus into the cell by fusion of the lipid bilayers of the virus and the cell at neutral pH or at low pH

Membrane fusion is a universal biological phenomenon that occurs in a myriad of processes from fertilization to the trafficking of membrane vesicles within the cell. However lipid bilayers are immensely stable structures, composed of two monolayers arranged with the long hydrophobic chains of the constituent molecules inside and the polar, negatively charged head groups exposed on their inner and outer surfaces. As a result fusion is far from a spontaneous process. For fusion to occur, the viral and cell bilayers must first disrupt at a common point and so expose their hydrophobic interiors in an incompatible environment, and then combine and reform as one bilayer. It is still not completely understood how this occurs, but there is most information about HIV-1. Direct fusion of the HIV-1 lipid bilayer with the plasma membrane (Fig. 5.2a) is initiated by the succession of virus–receptor inter-

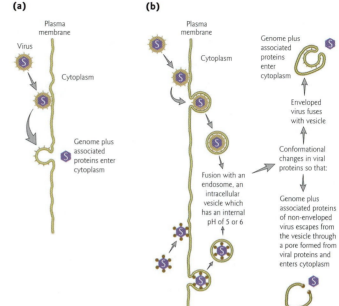

Fig. 5.2 Entry of animal virus genomes into cells. All viruses start by attaching to specific receptors on cells. (a) Entry by fusion of the lipid bilayers of an enveloped virus and the plasma membrane at neutral pH. (b) Entry by endocytosis is followed by a fusion of the vesicle with an endosome and a decrease in the pH of the endosome. This promotes conformational changes in viral proteins. For enveloped viruses (upper panel) this leads to fusion of the lipid bilayers of the virus and the endosome. For nucleocapsid virus particles (lower panel), the low pH causes conformational changes in viral proteins. This results in the insertion of newly exposed hydrophobic regions of the virion into the lipid bilayer of the vesicle and the escape of the viral genome and associated proteins into the cytoplasm.

actions (Box 5.2). This triggers conformational changes in the metastable envelope protein that result in the exposure of the hydrophobic terminal segments of the gp41 envelope protein molecules from their normally concealed position close to the virion membrane. (Gp41 is the membrane anchoring part of the envelope trimer (see Chapter 19).) Fusion has three essential steps. Firstly, some of the N-terminal hydrophobic regions of gp41 insert into the lipid bilayer of the cell membrane; secondly, a small number of gp41 envelope proteins combine together to form a hydrophobic channel between the two membranes, disrupting both bilayers and allowing the disturbed lipid to flow through the hydrophobic channel and the two bilayers to come together and form a fusion pore; lastly, the pore

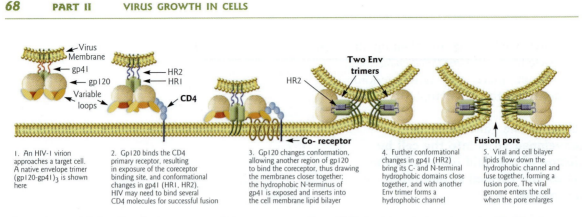

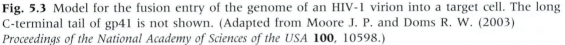

1. An HIV-1 virion approaches a target cell. A native envelope trimer (gp120-gp41)$_3$ is shown here

2. Gp120 binds the CD4 primary receptor, resulting in exposure of the coreceptor binding site, and conformational changes in gp41 (HR1, HR2). HIV may need to bind several CD4 molecules for successful fusion

3. Gp120 changes conformation, allowing another region of gp120 to bind the coreceptor, thus drawing the membranes closer together; the hydrophobic N-terminus of gp41 is exposed and inserts into the cell membrane lipid bilayer

4. Further conformational changes in gp41 (HR2) bring its C- and N-terminal hydrophobic domains close together, and with another Env trimer forms a hydrophobic channel

5. Viral and cell bilayer lipids flow down the hydrophobic channel and fuse together, forming a fusion pore. The viral genome enters the cell when the pore enlarges

Fig. 5.3 Model for the fusion entry of the genome of an HIV-1 virion into a target cell. The long C-terminal tail of gp41 is not shown. (Adapted from Moore J. P. and Doms R. W. (2003) *Proceedings of the National Academy of Sciences of the USA* **100**, 10598.)

enlarges so that the genome and associated proteins can pass through it and enter the cytoplasm, and the viral and cell bilayers become united. Figure 5.3 shows a more detailed version of the HIV-1 fusion reaction.

The fusion of other enveloped viruses, such as influenza A virus, is not triggered solely by protein–protein interactions but also requires exposure to an acid pH (Fig. 5.2b). The virus particle attaches to a cell and is then engulfed into a vesicle by receptor-mediated endocytosis. Strictly speaking, the virus in the vesicle is still on the *outside* the cell and not in the cytoplasm, as it is contained in what was plasma membrane (Fig. 5.2b). The release of the genome into the cytoplasm is dependent upon the internal environment of the endocytic vesicle becoming acidic (pH 5–6). This is achieved by fusion of the endocytic vesicle with an intracellular vesicle called an endosome. A membrane-bound endosomal protein then pumps protons into the lumen of the vesicle. The low pH initiates conformational changes in the viral envelope proteins (the influenza virus hemagglutinin) which release the hidden hydrophobic N-terminus of the membrane anchoring part of the membrane (HA2). Fusion of the viral lipid bilayer with the bilayer of the vesicle then proceeds exactly as described above for HIV-1.

Entry of the genome of a nonenveloped virus into the cell

Following attachment, all nonenveloped animal viruses are taken up by the cell by receptor-mediated endocytosis into an endocytic vesicle, like the enveloped influenza virus (above). Release of the genome into the cytoplasm is dependent upon conformational changes in virion proteins that may be initiated by binding to receptors in the vesicle and/or a decrease in the pH of the internal environment of the vesicle. The conformational changes release hidden hydrophobic regions that can insert in the bilayer of the vesicle, and form a channel, through which the genome and associated proteins enter the cytoplasm (Fig. 5.2b). More details of the entry of the poliovirus genome into the cells is shown in Fig. 5.4. Infection by this virus is low pH-independent.

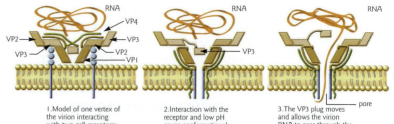

1. Model of one vertex of the virion interacting with two cell receptors; Part of VP3 plugs the fivefold axis, as shown

2. Interaction with the receptor and low pH cause conformational changes so that the N-terminal regions of VPI and the myristylated N-termini of VP4 interact with the cell membrane

3. The VP3 plug moves and allows the virion RNA to pass through the pore into the cytoplasm

The complex structure of uncoated genomes and secondary uncoating

Fig. 5.4 A model for the translocation of poliovirus RNA across the cell membrane. (Adapted from Belnap D. M. *et al.* (2000) *Journal of Virology* **74**, 1342.)

It should be pointed out that regardless of the nature of the virus particles, what enters the cell is a nucleoprotein structure, often with residual capsid components still associated, and not naked nucleic acid. In addition, either the genome that enters the cell is complexed as viral nucleoprotein, or it rapidly associates with cellular structures such as polymerases or ribosomes for the next phase of multiplication. The viral proteins associated with the genome include enzymes for nucleic acid synthesis, and internal structural virus proteins that form the core of some of the more complex viruses. There may also be *secondary uncoating* of the viral nucleocapsid inside the cell during which some viral proteins are removed or rearranged. Equally it is a common misconception that viral nucleic acids are stretched out (as illustrators frequently draw them). Single-stranded nucleic acid genomes form double-stranded secondary structures wherever regions of base complementarity can be brought together, and all types of nucleic acid will probably be tightly coiled. Thus nucleic acids and their associated proteins form complex structures.

The inefficiency of the infectious process

One of the striking aspects of infection of animal cells by viruses is its inefficiency. In poliovirus infections, for example, the majority of the RNA of the infecting virus is degraded after virus interacts with cells. This has several causes, including failure of the virus–cell interaction to complete the entry process or subsequent processes described above. Thus instead of changes in the virion capsid structure leading to release of the RNA genome into the cytoplasm and its complexing with ribosomes and later with replicase proteins, it is hydrolyzed by marauding ribonucleases. Such failure to initiate infection accounts, at least in part, for high physical particle to infectious particle ratios of 1000 : 1 or so (meaning that of 1000 particles only one initiates an infection). Neither electron microscopy nor biochemical studies can distinguish between these noninfectious and infectious particles. Also, most of the noninfectious particles contain an intact virus genome. Thus, it is likely that most of these "noninfectious" particles are in fact potentially infectious. It seems that by chance the majority of particles fails to complete the infectious process. The presence of an excess of noninfectious particles makes it difficult to define the functional infectious entry pathway for a virus.

5.3 INFECTION OF PLANTS

The infection of plant cells by viruses differs considerably from the infection of animal cells. Plants have rigid cell walls composed of cellulose, and consequently viruses must be introduced into the host cytoplasm by some traumatic process.

The cell wall that surrounds the plasma membrane of all plant cells prevents viruses from attaching and entering them in the ways described for animal cells (see Sections 5.1 and 5.2). Viruses can only reach the cytoplasm of plant cells when the tissue is damaged. Thus plants are infected either with the help of *vectors*, namely animals that feed on plants, or invading fungi, or by mechanical damage caused by the wind or passing animals, all of which allow viruses to enter directly into cells. In the laboratory, this is mimicked by the gentle application of abrasive carborundum to leaves during local lesion assays of virus-containing material (see Fig. 1.1b). Many plants that become infected naturally do so because virus-carrying animals feed upon them. However, this transmission is not a casual process which occurs whenever any animal chances to feed on an uninfected host plant after feeding on an infected one. Rather, the transmission of most plant viruses is a highly specific process, requiring the participation of particular animals as vectors. Although some viruses, such as tobacco mosaic virus (TMV), require no vector and can be transmitted mechanically in a nonspecific manner, most have a specific association with their animal vectors that feed by piercing plant tissues with their mouthparts (leafhoppers, aphids, thrips, whiteflies, mealy bugs, mites, or nematodes; see Appendix 2).

Once introduced into the plant, virus particle uncoating takes place in the cytoplasm through the capsid binding to cytoplasmic proteins and the influence of divalent cations like calcium. The released genome can move around the plant due to connections (plasmodesmata) between plant cells that make a plant functionally unicellular. This can be shown by infecting plants with infectious genomes from a virus with a mutated, nonfunctional coat protein. The usual signs of disease appear on the leaves, showing that the virus spreads within the plant as normal despite its mutation. To achieve this, plant viruses encode *viral movement proteins* that form channels along the plasmodesmata and facilitate the transmission of viral genomes to adjacent cells.

5.4 INFECTION OF BACTERIA

Bacteria, like plants, have a strong cell wall which a virus must breach in order to infect the cell. Some have evolved the well-known injection equipment which introduces their genome into the cell while the coat protein stays outside (see Section 1.5), but many other bacterial viruses enter cells in other ways.

Attachment of bacteriophages to the bacterial cell

Most bacteriophages attach to the cell wall, but there are other cell receptors on the pili, flagella, or capsule of host cells. In the main, tailed bacteriophages attach to the cell wall, and do so by the tip of their tail. The chemical nature of the cell receptor has been elucidated for some *Salmonella* phages. Figure 5.5 shows the structure of the O antigen of wild-type *Salmonella typhimurium* and certain cell wall mutants derived from it. When the mutants are tested for sensitivity to different phages, it becomes apparent that each has a characteristic sensitivity pattern. For example, the presence of the O-specific side-chains makes a cell resistant to phages 6SR, C21, Br60, and Br2 but sensitive to P22 and Felix O, i.e. the latter bind to the O-specific side-chains. The absence of the O-specific side-chains, as in *rfb* T mutants, makes the cell resistant to P22, Felix O, and C21, and sensitive to 6SR, Br60, and Br2, i.e. the latter bind the terminal N-acetyl galactosamine. This information is also of interest to cell wall chemists. For example, suppose we wished to isolate an *rfc* mutant of *Salmonella*. In the absence of a selective procedure, this would be a tedious task. However, if a bacterial lawn showing plaques of P22 phage is incubated for several days, small colonies arise within the plaques which represent the growth of phage-resistant mutants. It is clear that these resistant mutants could belong to any one of the mutant classes shown in Fig. 5.5. However, by simultaneously selecting for resistance to 6SR, C21, Br60, and Br2, we can eliminate all but the *rfc* class, which should be sensitive only to Felix O (Fig. 5.5, lower panel).

The best-documented example of phage attachment is that of phage T2 and T4. These viruses have a complex structure, including a tail, base plate, pins, and tail fibers (see Chapter 3). The initial attachment of these phages to the receptors on the bacterial surface is made by the distal ends of the long tail fibers (Fig. 5.6). These fibers attach first, bend at their center, and their distal tips contact the cell wall only some distance from the midpoint of the phage particle. After attachment, the phage particle is brought closer to the cell surface. When the base plate of the phage is about 10 nm from the cell wall, contact is made between the short pins extending from the base plate and the cell wall.

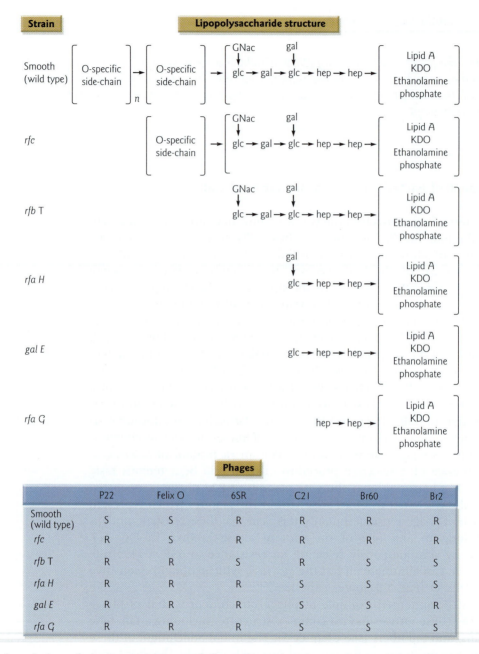

Fig. 5.5 Correlation of phage sensitivity with lipopolysaccharide structure of the cell wall. The upper part of the diagram shows the structure of the lipopolysaccharide in different mutants of *Salmonella typhimurium*. Abbreviations: glc, glucose; gal, galactose; GNac, *N*-acetylglucosamine; hep, heptose; KDO, 2-keto-3-deoxyoctonic acid. The lower panel shows the sensitivity of the different mutants to several bacteriophages. Abbreviations: S, sensitive; R, resistant.

Not all tailed phages attach to the cell wall. Some, such as phage χ(chi) and PBSI, attach to flagella. The tip of the tail of these phages has a fiber that wraps around the flagellum. The phage then slides until it reaches the base of the flagellum (Fig. 5.7). Other tailed phages attach to the capsule of the cell while a further site of important cell receptors is on the sex pili. Bacteria which harbor the sex factor (F) or certain colicins or drug-resistance factors produce pili, and two classes of phage are known to attach to these pili. The filamentous single-stranded deoxyribonucleic acid (DNA) phages attach to the tips of the pili, while many types of spherical ribonucleic acid (RNA) phages attach along the pili (Fig. 5.8). These phages are particularly useful to microbial geneticists because they offer a ready means of establishing whether cells harbor pili of these types.

Entry of the genomes of bacteriophages into bacterial cells

The head-tail bacteriophages

The experiments of Hershey and Chase (see Section 1.5) indicated that mainly nucleic acid entered the cell during infection of the head–tail bacteriophage, T2. The way in which this occurs has now been elucidated and is a complex but fascinating story, which can only briefly be sketched here. The tail of the bacteriophage is contractile and in the extended form consists of 24 rings of subunits surrounding a core. Each ring consists of a number of small and large subunits. Following attachment, the tail contracts, resulting in a merging of the small and large subunits to give 12 rings of 12 subunits. The tail core, which is not contractile, is pushed through the outer layers of the bacterium with a twisting motion, and contraction of the head results in the injection of the DNA into the cell. This process is probably aided by the action of the lysozyme which is built into the phage tail. There are 144 molecules of adenosine 5′-triphosphate (ATP) built into the sheath and the energy for contraction most probably comes from their conversion to adenosine diphosphate (ADP). The phage has been likened to a hypodermic syringe. The various steps in penetration are shown in Fig. 5.9.

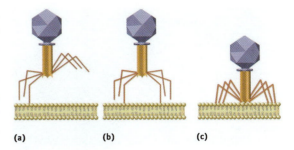

(a)   (b)   (c)

Fig. 5.6 Steps in the attachment of bacteriophage T4 to the cell wall of *Escherichia coli*. (a) Unattached phage showing tail fibers and tail pins (compare Fig. 3.16). (b) Attachment of the long tail fibers. (c) The phage particle has moved closer to the cell wall and the tail pins are in contact with the wall.

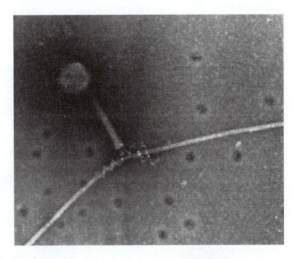

Fig. 5.7 Attachment of a bacteriophage χ to the filament of a bacteria flagellum. (Courtesy of J. Adler.)

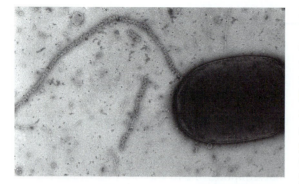

Fig. 5.8 Attachment of many spherical RNA phages to the sex pilus of *Escherichia coli.* (Courtesy of C. C. Brinton.)

The RNA bacteriophages

Most bacteriophages do not possess contractile sheaths (see Appendix 4), and the way in which their nucleic acid enters the cell is not known. Hershey–Chase-type experiments with the filamentous DNA phages, which attach to the sex pili, suggest that both the DNA and the coat protein enter the cell and it has been postulated that, after attachment, the phage–pilus complex is retracted by the host cell. However, when similar experiments are performed on the RNA phages, which also attach to the sex pili, the results obtained are similar to those for T2, i.e. the phage coat protein does not enter the cell. Indeed, following attachment, the RNA phage particles are rapidly eluted. Although many of these particles retain their RNA genomes (i.e. they have failed to infect the cell), they have all undergone a structural change as a result of their interaction with the bacterium so that their genome is now accessible.

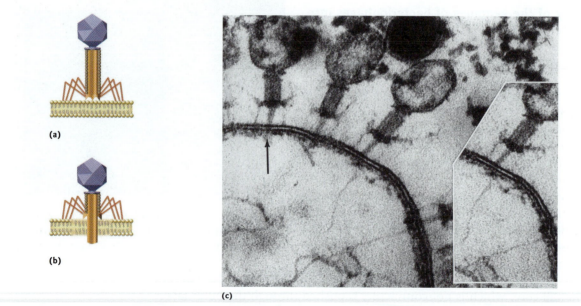

(a)

(b)

(c)

Fig. 5.9 Representation of the mechanism of entry of the phage T4 genome into the bacterial cell. (a) The phage tail pins are in contact with the cell wall and the sheath is extended. (b) The tail sheath has contracted and the phage core has penetrated the cell wall; phage lysozyme has digested away the cell beneath the phage. (c) Negatively stained electron micrograph of T4 attached to an *Escherichia coli* cell wall, as seen in thin section. The needle of one of the phages just penetrates through the cell wall (arrow). Thin fibrils extending on the inner side of the cell wall from the distal tips of the needles are probably DNA.

Box 5.3

Evidence that RNA bacteriophages are altered by their interaction with the target cell

- Phage particles eluted from bacterial cells following attachment and analyzed by centrifugation through sucrose gradients are 70% intact 78S* particles and 30% particles of 42S that lack the RNA genome.
- Treatment of the 78S particles with ribonuclease converts them into 42S particles, whereas phage particles which have not been allowed to interact with pili remain RNase-resistant and 78S.
- Thus interaction with the pilus causes conformational changes to the virus particles that permit the penetration of RNase.
- The fate of the attachment protein during the interaction of phage and bacterium can be followed because it is the only phage particle protein to contain histidine. When ³H-histidine-labeled phages are allowed to attach and elute from pili, the histidine label is completely absent from the 42S particles and only a small amount is present in the 78S particles.
- Thus the A protein is released from the phage particles following attachment.

*"S" stands for Svedberg, the unit of sedimentation rate. A particle that sediments rapidly has a high S value. The factors affecting the rate of sedimentation of a particle are complex but the S value depends mainly on particle size and density.

This change is the loss of the attachment protein, A, from the particles (Box 5.3). It is likely that in a successful infection, the A protein enters the cell taking the RNA with it as viral RNA and the A protein are taken up by the cell in approximately equimolar amounts, and their kinetics of uptake into the cell are similar.

5.5 PREVENTION OF THE EARLY STAGES OF INFECTION

One of the goals of studying virus–cell relationship is to develop methods of aborting viral infections, particularly those of humans and domestic animals, and how better to achieve this than by preventing virus from attaching and releasing its genome. The main approach has been and still is to raise neutralizing antibody through immunization (see Chapters 12 and 21). Alternatively, therapeutic drugs can be sought that will block attachment and/or entry. Cellular receptors are mainly proteins or glycoproteins which are components of the cell and only incidentally serve the needs of the virus. Thus attacking these may endanger the cell itself. From the virus perspective, there is an increasing number of atomic

Table 5.2

Antiviral compounds that inhibit attachment or the entry process.

Antiviral compound	Mode of action	Usage
Pleconaril	Binds to a hydrophobic pocket on the virion that the cell receptor normally recognizes	Several picornaviruses
Neutralizing antibody (passive immunotherapy)	Binds to the surface of the virions and prevents infection (neutralizes infectivity)	Although antibodies specific for all viruses exist, injected antibody is mainly used for life-threatening infections (e.g. Ebola virus)
Enfuvirtide (T-20)	A gp41-derived peptide that binds gp41 and inhibits the fusion of viral and cell membranes; acts after virus binds the CD4 receptor (Fig. 5.3, stage 2) and prevents progression to stage 3	HIV-1
Zanamivir (inhaled); oseltamavir (oral)	These are a special case. They are analogs of the viral neuraminidase substrate (NANA) and prevent progeny virus from detaching from the cell in which it was made, and hence transmission of infection	Influenza A and B viruses; only effective if taken before infection or shortly after infection

resolution, three-dimensional structures of virions/coat proteins being revealed, so the search for antiviral drugs which inhibit the attachment and uncoating processes is now encompassing rational development in addition to random screening (Table 5.2). However, a problem for antiviral drugs in the *prevention* (rather than treatment) of disease is that no one knows that they are infected until clinical signs and symptoms appear, and by then the infection is so well advanced that it responds poorly to antivirals. Antiviral drugs are discussed in Section 21.8.

Virus infection of plants is a serious agricultural problem. For the protection of plants, one of the best preventive measures is to reduce the number of virus vectors with insecticides or by biological control, but this is not necessarily an easy task. The selection of genetic resistance to disease in the plant species is another option. Bacteriophages can also cause problems in industry, since some infect organisms used in industrial fermentations, and result in cell lysis and loss of product. The best measure here is the use of resistant cell strains, or to reduce the concentration of divalent cations in the growth medium, since the latter are frequently important for phage attachment.

KEY POINTS

- There is a paradox: in order to survive in a hostile environment while the virus is being transmitted to a new host, viruses have evolved particles that enclose and protect their genome from degradation. However at the right moment the virus particle has to open up and deliver the viral genome into the target cell. If the virus gets it wrong the genome dies.
- Viruses solved this paradox by evolving to recognize certain molecules on the surface of the target cell; in nature a cell cannot be infected without such recognition (apart from plant viruses).
- These cell surface molecules are called virus receptors, but their normal function is communicating between the cell and its environment.
- Most cell receptor molecules used by viruses are proteins.
- Interaction with cell receptors overcomes natural repulsive forces that would keep a virus particle from making the close contact with a cell that is necessary for infection.
- For the viral genome to gain entry into the cell cytoplasm all virus particles ultimately need to be in intimate contact with the lipid bilayer of the cell plasma membrane.
- Some viruses have a hierarchy of receptors of increasing affinity; binding to one permits recognition of the second, and so on; thus these interactions are like a series of passwords and answers.
- Binding to cell receptors is an active process that causes a virus particle to undergo conformational changes that eventually lead to the viral genome entering the cell.
- The genomes of all enveloped animal viruses enter the cell cytoplasm by fusion of the lipid bilayers of the virus and the cell; this process is activated by interaction of virus protein with cell receptor(s), or by the low pH environment of an endocytic vesicle that has taken up the virus particle.
- All nonenveloped animal viruses are taken up by the cell in an endocytic vesicle as a result of receptor-mediated endocytosis, but the contents of the vesicle are still outside the cytoplasm; the genome of the nonenveloped virus enters the cytoplasm as a result of the low pH environment of the vesicle – this causes rearrangements of the virus particle to form a hydrophobic conduit through which the genome enters the cytoplasm.
- Plant viruses enter the cytoplasm of plant cells through a physical break in the plant cell wall caused by mechanical damage or animal vectors.
- Some bacterial (head–tail) viruses inject their genome into the host cell, but many use other mechanisms.
- Understanding how viruses recognize and enter their target cells is leading to a new generation of antiviral drugs that specifically block one of the steps in these processes.

QUESTIONS

- Discuss the processes used by viruses to enter cells and uncoat their genomes.
- Compare and contrast the mechanisms by which viruses achieve entry into mammalian cells, plant cells, and bacteria.

FURTHER READING

Carson, S. D. 2001. Receptor for the group B cox-sackie viruses and adenoviruses: CAR. *Reviews in Medical Virology* **11**, 219–226.

Clapham, P. R., McKnight, A. 2002. Cell surface receptors, virus entry and tropism of primate lentiviruses. *Journal of General Virology* **83**, 1809–1829.

Dimitrov, D. S. 2004. Virus entry: molecular mechanisms and biomedical applications. *Nature Reviews Microbiology* **2**, 109–122.

Forrest, J. C., Dermody, T. S. 2003. Reovirus receptors and pathogenesis. *Journal of Virology* **77**, 9109–9115.

Greenberg, M., Cammack, N., Salgo, M., Smiley, L. 2004. HIV fusion and its inhibition in antiretroviral therapy. *Reviews in Medical Virology* **14**, 321–337.

Hogle, J. 2002. Poliovirus cell entry: common structural themes in viral cell entry pathways. *Annual Review of Microbiology* **56**, 677–702.

Lindberg, A. A. 1977. Bacterial surface carbohydrate and bacteriophage adsorption. In *Surface Carbohydrates of Prokaryotic Cells* (I. W. Sutherland, ed.), pp. 289–356. Academic Press, London.

Marsh, M., Pelchen-Matthews, A. 2000. Endocytosis in viral replication. *Traffic* **1**, 525–532.

Pierson, T. C., Doms, R. W., Pohlmann, S. 2004. Prospects of HIV-1 entry inhibitors as novel therapeutics. *Reviews in Medical Virology* **14**, 255–270.

Poranen, M., Daugelavicius, R., Bamford, D. H. 2002. Common principles in viral entry. *Annual Review of Microbiology* **56**, 521–538.

Rinaldo, C. R., Piazza, P. 2004. Virus infection of dendritic cells: portal for host invasion and host defense. *Trends in Microbiology* **12**, 337–345.

Schneider-Schaulies, J. 2000. Cellular receptors for viruses: links to tropism and pathogenesis. *Journal of General Virology* **81**, 1413–1429.

Sieczkarski, S. B., Whittaker, G. R. 2002. Dissecting virus entry via endocytosis. *Journal of General Virology* **83**, 1535–1545.

Skehel, J. J., Wiley, D. C. 2000. Receptor binding and membrane fusion in virus entry: the influenza hemagglutinin. *Annual Review of Biochemistry* **69**, 531–569.

Stewart, P. L., Dermody, T. S., Nemerow, G. R. 2003. Structural basis of nonenveloped virus cell entry. *Advances in Protein Chemistry* **64**, 455–491.

Yanagi, Y. 2001. The cellular receptor for measles virus – elusive no more. *Reviews in Medical Virology* **11**, 149–156.

Also check Appendix 7 for references specific to each family of viruses.

6

The process of infection: IIA. The replication of viral DNA

Viruses with deoxyribonucleic acid (DNA) genomes infect animals, plants, and bacteria. Their genomes may be single- or double-stranded, and be either linear molecules or circles. Viral DNA genomes also vary considerably in size. Among the human viruses, genomes range from the hepadnaviruses at around 3000 base-pairs (3 kbp) to the herpes and poxviruses, in the range 130–230 kbp. However, much larger viral genomes are now known that infect other species. Among the poxvirus family, the genome of canarypox, an avian virus, is 360 kbp but, dwarfing that, the genome of Acanthamoeba polyphaga mimivirus, which infects single-cell protozoa, is 1180 kbp. This is larger than the genomes of several bacterial species!

The basic mechanism whereby a new strand of DNA is synthesized in a cell is the same, regardless of whether the DNA is of cellular or viral origin. The fact that all DNA synthesis shares these fundamental features has meant that viral DNA molecules, which can be manipulated with ease, have often been studied by biochemists attempting to unravel the mysteries of DNA replication. However, although these studies have added greatly to our understanding of this process, they have also revealed that viruses employ this basic mechanism in many different ways, in each case to suit the peculiarities of their genome structures. This chapter explores this diversity of viral DNA replication mechanisms.

6.1 THE UNIVERSAL MECHANISM OF DNA SYNTHESIS

DNA strand polarity and the nature of DNA polymerases

The successful replication of a double-stranded DNA molecule requires that two daughter strands be synthesized. Since these must base-pair with the two parental (template) strands, which are antiparallel, the daughter strands must also be antiparallel. Thus, if an enzyme complex were simply to move along a double-strand DNA template synthesizing two daughter strands, one of these strands would need to be made with $5' \rightarrow 3'$ polarity and the other with $3' \rightarrow 5'$ polarity. However, no DNA polymerase yet characterized has the capacity to synthesize DNA in the $3' \rightarrow 5'$ direction, and the biochemistry of DNA synthesis suggests that such activity is not possible.

One solution to this problem is for synthesis to proceed in the $5' \rightarrow 3'$ direction along one parental strand, the *leading* strand, and for discontinuous $5' \rightarrow 3'$ synthesis to occur in the opposite direction along the other, or *lagging*, strand (Fig. 6.1, Box 6.1). In essence, while the enzyme complex moves forwards copying the leading strand template, it reveals an increasing length of uncopied lagging strand template. Once a sufficient length of this template is available, a new daughter strand is initiated and copied from it, synthesis continuing until the polymerase comes up against the 5' end of the previously synthesized lagging strand segment. This is the solution which is adopted in all organisms other than viruses, and by some viruses too, and is discussed later in the context of SV40 genome replication.

The alternative solution to the direction of synthesis problem, which is adopted by some viruses, is to separate in time the production of the two daughter strands, with each therefore being produced separately. Once the requirement to have both strands made by the same enzyme complex is removed, then it is clearly straightforward for both the daughter

Box 6.1

Evidence for discontinuous DNA synthesis

- When *Escherichia coli* cells infected with phage T4 are pulse-labeled with radioactive thymidine, the newly synthesized (i.e. radioactively labeled) DNA found immediately after the pulse is in small fragments of 1000–2000 nucleotides, as measured by velocity sedimentation.
- After similarly labeled cells are incubated for a further minute with unlabeled thymidine (a pulse-chase experiment), the labeled DNA is found in very much larger pieces.
- The small fragments of DNA produced initially are known as Okazaki fragments after Tuneko Okazaki, the person who first published this result.

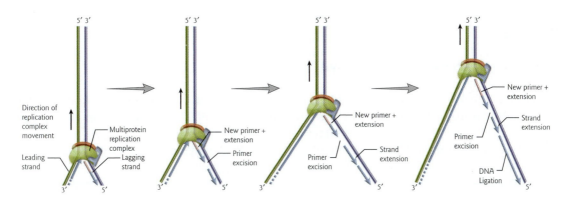

Fig. 6.1 Model for replication of double-stranded DNA through discontinuous synthesis of the lagging strand. Both strands are synthesized in a 5′ → 3′ direction but only the leading strand is synthesized continuously, the other being formed from a series of short DNA molecules, each primed by a piece of RNA. The primers are excised and replaced by DNA before the fragments are joined by DNA ligase. Green, dark blue: parental DNA; blue: new DNA; pink: RNA primer.

strands to be made by continuous synthesis in the 5′ → 3′ direction. An example of a virus employing this replication mechanism is adenovirus, which is discussed in Section 6.4. The problem with this mechanism, which is probably the reason why it has not evolved more generally, is that synthesis of one strand is necessarily delayed relative to the other and therefore, in the interim period after the first replicating enzyme has passed through the template duplex and copied the first strand, there are large amounts of single-stranded DNA (the second template strand) exposed. Since single-stranded DNA promotes recombination by strand invasion, such a mechanism would carry a cost of genetic instability which would be unsustainable for large genomes.

DNA strand initiation and the need for a primer

For any DNA replication mechanism, the process of new synthesis must begin somewhere. The locations on the template molecule where this occurs are specific and are termed origins of replication. For replication to begin, specific proteins must bind to the origin and the two DNA strands must be separated to provide the template for new synthesis. These requirements lead to certain typical features in origin sequences, such as sequence symmetry for specific protein binding and the presence of an AT-rich region for ease of strand separation.

During replication, new DNA strands must be initiated at least to begin the process and, if the discontinuous mechanism applies, repeatedly thereafter. This raises another problem since DNA polymerases are unable to start DNA chains *de novo*; all require a hydroxyl group to act as a primer from which to extend synthesis. The general solution to this problem is

Box 6.2

Evidence for RNA-primed DNA synthesis

Conversion of bacteriophage M13 single-stranded DNA to the double-stranded form in infected *E. coli* cells is inhibited by rifampicin, an inhibitor of *E. coli* RNA polymerase. In rifampicin-resistant *E. coli*, the inhibitor has no effect on phage DNA synthesis.

NB: Eukaryotes, rather than using one of their three RNA polymerases to synthesize primers, have a dedicated enzyme for this purpose known as a primase.

to invoke the action of a ribonucleic acid (RNA) polymerase, since these do not require a primer to start synthesis. The enzyme synthesizes a short RNA primer, copying the DNA template, and this is then extended by DNA polymerase (Fig. 6.1, Box 6.2). Once they have served their purpose, the primers are excised by specific enzymes which recognize and degrade RNA that is duplexed with DNA. The gaps so created are then filled by continuing 5' → 3' DNA synthesis from the adjacent fragment and the fragments finally joined together by DNA ligase.

Why the requirement for an RNA primer?

Why has this complex process of RNA-primed DNA synthesis evolved, when it could have been avoided if DNA polymerases were able to initiate strands *de novo*? The answer probably lies in a key evolutionary benefit of using DNA rather than RNA as genetic material, which is that the fidelity of DNA replication is very much greater (one error in 10^9–10^{10} base-pair replications) than that of RNA replication (one error in 10^3–10^4). The first check on fidelity of synthesis, applicable to both RNA and DNA polymerases, is the strong selection for base-pairing of substrate nucleoside triphosphate to the template. However, DNA polymerases have an additional check, their ability to "proofread" the DNA which they have just synthesized. Proofreading means that DNA polymerases excise any unpaired nucleotides from the 3' end of a growing strand before adding further nucleotides. RNA polymerases do not have this activity since they have to be able to initiate synthesis of new molecules without a properly base-paired primer. Since RNA transcripts are continually turned over and have no long-term role in the organism, errors in their synthesis can be tolerated. This however has consequences for those viruses which have RNA genomes (see Chapter 7).

RNA primers and the "end-replication" problem

The requirement for a primer for DNA synthesis creates a difficulty in achieving complete replication of linear molecules. If synthesis in the

$5' \rightarrow 3'$ direction is initiated with an RNA primer which is later digested away, there is no mechanism for filling the gap left by the primer at the 5' end of the leading strand (Fig. 6.2); to fill this gap would require $3' \rightarrow 5'$ synthesis and we know that this cannot occur. When the replicating fork reaches the other end of the template, a similar problem arises with the lagging strand. Without a solution, the net result would be synthesis of two daughter molecules, each with a 3' single-stranded tail. If such molecules were to undergo further rounds of replication, smaller and smaller 3' tailed duplexes would result.

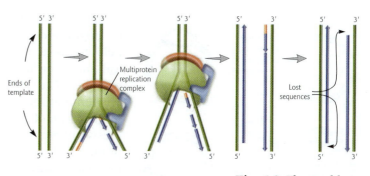

Fig. 6.2 The problem of replicating the ends of linear DNA molecules through the use of RNA primers. Once the first primer has been excised, there is no mechanism for filling the gap. Green: parental DNA; blue: new DNA; pink: RNA primer.

This "end-replication" problem is universal in biology. In eukaryotes, it is solved through the use of telomeres at the ends of each linear chromosome. These are highly repetitive sequences which are replicated reiteratively outside of the normal replication process, so as to maintain the chromosome length from one generation to the next. Prokaryotes, by contrast, have solved the problem by evolving circular genomes. When replicating such a molecule, the first primer can be excised and replaced by extension from the 3' end of the fragment that is eventually synthesized adjacent to it. Some viruses employ this latter strategy to achieve replication of their genomes, while others have unique solutions to the "end-replication" problem. The following sections discuss the genome replication of a series of DNA viruses to illustrate these different mechanisms.

6.2 REPLICATION OF CIRCULAR DOUBLE-STRANDED DNA GENOMES

Important viruses in this category

- *The polyomavirus family, including SV40, an important model for the study of eukaryotic cell processes.*
- *The papillomavirus family – the causative agents of skin and genital warts, and cervical carcinoma (see Chapter 20).*
- *The baculovirus family, insect viruses considered as agents for biocontrol of crop pests and used as laboratory tools for heterologous protein production.*

Simian virus 40 (SV40) has been studied in great detail, initially because of its tumorigenic properties (see Chapter 20) and then as an easily analyzed model system through which to understand eukaryotic cell processes.

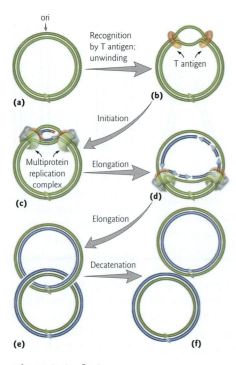

Fig. 6.3 (a–f) A general scheme for SV40 replication (see text for details). Green: parental DNA; blue: new DNA; pink: RNA primer. Arrowheads on newly synthesized DNA fragments represent 3′ ends; the 5′ → 3′ polarity of other DNA strands is indicated by > symbols.

Its genome is a circular double-stranded DNA of 5243 bp in length and the replication of this molecule is widely held to have close parallels, in its detailed biochemical mechanism, to the replication of eukaryotic cell genomes. The overall characteristics of SV40 replication are summarized in Fig. 6.3. The circular genome has a single origin of replication (Fig. 6.3a), at which new synthesis initiates co-ordinately on both strands using RNA primers. These two newly initiated strands are then extended away from the origin in opposite directions, forming the leading strands of two diverging replication forks (Fig. 6.3c). As synthesis proceeds, the lagging strand templates are revealed at each fork (note that different strands of the template constitute the lagging strand template at the two forks) and these are then copied by discontinuous synthesis (Section 6.1). Progress of the two forks around the template gives a structure with the appearance of the Greek letter θ ("theta"), which is therefore known as a theta-form intermediate (Fig. 6.3d). Ultimately, the two forks converge again on the opposite side of the template circle from the origin and when they meet, both strands will have been completely copied (Fig. 6.3e). The two daughter molecules are, like the parent, closed circular duplexes. Initially, they are topologically linked (Fig. 6.3e); this follows necessarily from the closed circular nature of the template, one strand of which ends up in one daughter and one in the other. The final stage in SV40 replication is the separation of these circles (Fig. 6.3f).

Through work in the laboratories of Kelly, Stillman and others, the complete replication of circular DNA molecules has been achieved in the test tube, directed by the SV40 origin and using only purified, characterized proteins. The 10 proteins needed are listed in Table 6.1, with their activities in the replication process. It is noteworthy that only one of these proteins is encoded by the virus; all the others are provided by the cell and perform the same functions in SV40 replication as they are believed to do in host cell replication. The one viral protein, large T antigen (see Chapter 9), provides the key initial recognition of the origin DNA sequence, binding specifically to it (Box 6.3). It then assembles a double hexamer around the DNA that has helicase activity, begins the separation of the DNA strands (Fig. 6.3b), and then recruits cellular single-strand DNA binding protein and DNA polymerase α/primase complex through protein to protein interactions, setting the scene for the initiation reaction. After initiation, T antigen continues to play a role as a helicase, unwinding the template ahead of each replication fork.

Papillomavirus genomes are a little larger than that of SV40, at around 8000 bp, but the mechanism of replication is very similar. The bovine

Box 6.3

Evidence for SV40 T antigen binding to the origin of replication

- Some viruses with point mutations in the origin of replication have a "small-plaque" phenotype, meaning they grow poorly. Selecting for revertants of such mutants gave better-growing viruses with changes to the coding sequence of T antigen that could compensate for the mutation in the origin. This result suggests an interaction between origin DNA and T antigen.
- When SV40 DNA is fragmented with a restriction enzyme, mixed with T antigen, and then the T antigen is purified back out of the mixture using a specific antibody, the DNA fragment containing the origin specifically copurifies with T antigen while other fragments remain behind in the solution.

Table 6.1
Proteins involved in SV40 replication.

Protein	Function(s)	Role in SV40 replication
Large T antigen*	Sequence-specific DNA binding	Initial recognition of the viral replication origin
	DNA helicase	Unwinding the DNA template duplex
RP-A	Single-stranded DNA binding	Stabilization of unwound DNA in the replication bubble at initiation, and within the replication forks
DNA polα/primase	Complex of activities synthesizing primers for initiation of new DNA strands and production of short DNA molecules	Initiation and short distance extension of the leading strands and each lagging strand fragment
DNA polδ	Extension of DNA strands; highly processive[†] in association with accessory factors	Processive[†] extension of leading strands and each lagging strand fragment
PCNA and RF-C	Accessory factors for DNA polδ	Increased processivity[†] of DNA polδ
topoisomerase I	Relieving tortional stress in duplex DNA molecules	Removes the excess supercoiling which builds up in the DNA template ahead of the replication fork due to helicase action
RNase H	Degradation of RNA within RNA : DNA hybrids	Removal of primers
DNA ligase	Joining DNA ends	Links up lagging strand fragments after primer removal and fill-in synthesis
topoisomerase II	Separating topologically linked DNA molecules	Separates the daughter duplexes at the end of the replication process

*Large T antigen is the only virus-specified protein – all others are encoded by the cell.
[†]Polymerase processivity describes how far the polymerase is likely to travel on a given template molecule before dropping off and having to rebind the template (or another template) molecule to recommence synthesis.

papilloma virus (BPV) E1 protein functions analogously to SV40 large T antigen in the replication process, except that its sequence-specific DNA binding activity is very weak, and is only revealed in the presence of the E2 protein, which has much stronger specific DNA binding activity (see Section 9.3). E1 and E2 bind as a complex to the origin of replication of BPV. Thereafter, E1 provides DNA helicase activity whilst all other replication functions are taken from the host as for SV40.

Baculovirus genomes are very large circles, of 100–140 kbp. Their replication is not fully understood, but appears to be a combination of theta-form replication (as for SV40, above) and rolling-circle replication (Section 6.3). The best studied virus of this type is *Autographa californica* nuclear polyhedrosis virus. The genome has multiple origins of replication and encodes most, if not all, of the proteins needed to complete its replication.

6.3 REPLICATION OF LINEAR DOUBLE-STRANDED DNA GENOMES THAT CAN FORM CIRCLES

Important viruses in this category

- *The herpesvirus family, including the human pathogens Epstein–Barr virus and herpes simplex virus types 1 and 2.*
- *Bacteriophage λ, a classic model system for the study of control of gene expression.*

The HSV1 genome and its replication

The HSV1 genome is a linear molecule of about 153 kbp in length. This DNA has a complex structure, with several repeated elements (Fig. 6.4a, Box 6.4). The first of these is a direct repeat of about 500 bp at the ends of the molecule. This is referred to as terminal redundancy, and appears to be a general feature of herpesvirus genomes. These repeats contain no genes, but are crucial in the process of genome packaging during the formation of new virus particles. The second group of repeats in the HSV1 genome are inverted copies of sequences several kilobases in length. These divide the DNA into two parts, each comprising a unique region flanked by a pair of inverted repeats. Each pair of repeats contains genes, for which the virus is therefore diploid. Although some other herpesvirus genomes also have this latter type of repeat organization, it is not a general feature of herpesvirus genomes and does not appear to be essential for HSV1 replication. The final feature of the genome organization that is significant is the presence of single nucleotide 3′ extensions on each end; these are complementary and mediate genome circularization prior to replication, probably through the action of a protein.

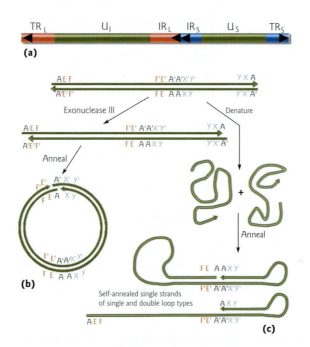

Fig. 6.4 (a) HSV1 genome organization: L, long; S, short; U, unique sequence; TR, terminally repeated sequence; IR, internally repeated sequence. Black triangles represent direct repeats (A) at the genome termini and their internal inverted copies (A′). Red and blue shading represent the remainder of the inverted repeat sequences that flank the long and short unique regions respectively. The genome is not drawn to scale; U_S should represent about 8.5% and U_L about 71% of the total genome. (b,c) The HSV1 genome is depicted by two parallel lines, with terminal direct repeats shown as **A** and its complement **A′**, the repeats flanking the long unique region as **EF** and its complement **E′F′**, and the repeats flanking the short unique region as **XY** and its complement **X′Y′**. 3′ ends are represented by arrowheads. (b) Experimental demonstration of direct repeats at the HSV1 genome termini (Box 6.4). (c) Experimental detection of inverted repeats in HSV1 DNA (Box 6.4). The two forms of self-annealed strand that are seen are shown for the upper strand of the double-stranded genome. The same forms may be adopted by the lower strand.

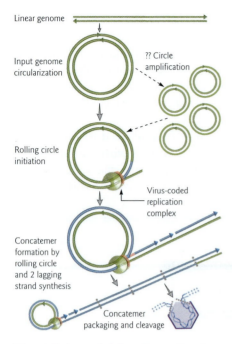

Linear genome

Input genome
circularization

?? Circle
amplification

Rolling circle
initiation

Virus-coded
replication
complex

Concatemer
formation by
rolling circle
and 2 lagging
strand synthesis

Concatemer
packaging and cleavage

Fig. 6.5 A model for the replication of HSV1 DNA (see Section 6.3 for details). Green: parental DNA; blue: new DNA; pink: RNA primer; purple: assembling capsid.

The replication of HSV1 is outlined in Fig. 6.5. Progeny genomes are produced as concatemers (multiple genome-equivalents covalently joined end-to-end) by rolling circle replication from circular templates (Box 6.5). Unless the incoming genome can be amplified in some way, only one such replication complex can exist in the cell, and this is not the case. This leads to the inference of a preceding circle amplification phase in HSV1 replication but there is no direct evidence for this. However, circle amplification during phage λ replication is well understood and another herpesvirus, Epstein–Barr virus (EBV), has a well-documented mechanism for replication as a circle (similar to that used by SV40, Section 6.2). Both these facts support the notion that HSV1 might also use such a strategy. HSV1 rolling circle replication has been well characterized. Three origins of replication have been defined by experiment but mutants possessing only one of these sequences can still grow, indicating redundancy of function. Replication involves a considerable contribution from viral proteins: these provide origin recognition, DNA helicase, single-strand DNA binding, primase, and DNA polymerase activities. The requirement for cellular proteins is correspondingly far more limited but includes host topoisomerases. Concatemeric DNA is only cleaved during incorporation into virions to give encapsidated, unit-length, genomes.

Bacteriophage λ replication

The bacteriophage λ linear genome circularizes upon injection into the cell via its cohesive ends, which are 12 nucleotide complementary exten-

---

Box 6.5

Evidence for rolling circle replication during HSV1 infection

- The presence of concatemeric DNA is diagnostic of rolling circle replication.
- DNA from HSV1-infected cells, analyzed by restriction digestion and Southern blotting, shows that the expected fragments from the genome ends are present in low amounts compared to the remaining fragments, with a correspondingly increased amount of fragments representing covalently joined end fragments. This is evidence for abundant end-to-end joining of viral genomes, i.e. concatemeric DNA.
- Pulse–chase analysis with labeled DNA precursors shows that concatemeric DNA is produced before any labeled unit-length genomes. This discounts the possibility that the concatemers form by ligation of unit-length molecules.

sions on the 5′ ends of the molecule. During the early replication period, λ DNA replicates bidirectionally to generate up to 20 circular progeny copies per cell. Late replication is initiated by the conversion of replication to the rolling circle mechanism, which generates the many hundreds of progeny genomes required. Rolling circle synthesis generates a concatemeric tail, which is cleaved during packaging into molecules of the correct length and possessing the same cohesive ends as the linear genome which initiated infection.

6.4 REPLICATION OF LINEAR DOUBLE-STRANDED DNA GENOMES THAT DO NOT CIRCULARIZE

Important viruses in this category

- *The adenovirus family, which includes a variety of human pathogens affecting the respiratory tract, eye, or gut. These viruses are also being actively developed as vectors for gene therapy.*
- *The poxvirus family, including the eradicated human pathogen, variola virus (smallpox), and vaccinia virus, important as a recombinant vaccine carrier.*

Adenoviruses

Adenovirus genomes are linear double-stranded DNA molecules of around 36 kbp in length. At the two ends there are inverted terminal repeat sequences of 100–150 bp that contain the origins of replication, and on each 5′ terminus there is a covalently attached, virus-encoded, protein known as the terminal protein (TP) (Box 6.6). The replication strategy of adenoviruses is summarized in Fig. 6.7. Understanding the details of this process was made possible through the development of cell-free replication systems for the virus by Kelly and coworkers. To initiate replication, the origins are recognized by a complex of two viral proteins, a DNA polymerase and the TP precursor (pTP). This binding is assisted by two cellular proteins, which bind specifically to sequences adjacent to the origin. These are actually transcription factors which the virus "borrows" to assist its replication. However, it is not their transcription regulatory activity which is needed. Rather, they are used to alter the conformation of the DNA so as to promote binding of the pTP–pol replication complex.

Once the viral DNA polymerase–pTP complex has bound to the origin (Fig. 6.7a), the polymerase initiates DNA synthesis by copying from the 3′ end of the template strand, using an amino acid side chain –OH in the bound pTP as a primer (Fig. 6.7b). This priming mechanism is the key to

---

Box 6.6

Evidence for adenovirus genome structure

• Denaturation of purified linear viral DNA with alkali, followed by neutralization with acid, causes the formation of unit-length single-stranded circles that are visible by electron microscopy and have short double-stranded tails (Fig. 6.6). This is indicative of inverted terminal repeats.

• DNA extracted from particles by a process that includes proteolysis is linear when viewed by electron microscopy whereas when purified without the use of such enzymes, much of the DNA is circular. This result suggests that protein to protein interactions mediate linkage of the genome ends.

• A covalent phosphodiester link between a serine in terminal protein and the 5′ deoxy-cytidine residue in the DNA has been demonstrated biochemically.

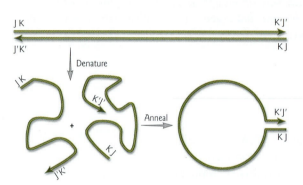

Fig. 6.6 Demonstration of inverted terminal repeat sequences in adenovirus DNA. The genome is depicted by two parallel lines, with inverted terminal repeats shown as JK and its complement J′K′. 3′ ends are represented by arrowheads. Denaturation and annealing of the genome allows formation of single-stranded circles with double-stranded "pan-handle" projections.

adenovirus solving the "end-replication" problem as there is no RNA primer complementary to viral sequence which later has to be excised and somehow replaced. However, the mechanism could be said to break the rule of universal proofreading of DNA synthesis, since initiation does not require a primer which is base-paired to the template; in some way the pTP–DNA interaction must substitute for this requirement. Synthesis then progresses 5′ → 3′ across the length of the template, displacing the non-template strand which is bound and stabilized by a virus-coded DNA binding protein (Fig. 6.7c). This process may occur at similar times from the origins at each end of the molecule (Fig. 6.7d), in which case the template duplex falls apart when the replication complexes meet (Fig. 6.7e); replication is then completed by the polymerases moving on to the ends of their respective templates (Fig. 6.7f). When only one origin is used, the nontemplate strand is completely displaced without being replicated (Fig. 6.7g). Its replication is achieved via formation of a "pan-handle" intermediate (Fig. 6.7h), where the short double-stranded region exactly resembles a genome end and can therefore serve as an origin in the way already described (Fig. 6.7i,j). Since replication of the two strands is separated in time, only one DNA strand being synthesized by a replication complex, all of the DNA can be synthesized continuously – there is no lagging strand.

Poxviruses

Poxvirus replication has been characterized principally through studying vaccinia virus (the smallpox vaccine virus). The vaccinia virus genome is a linear molecule of around 190 kbp that is unusual in having covalently closed ends. This means that the two pairs of nucleotides at the ends of the molecule are each linked by a standard $5' \rightarrow 3'$ phosphodiester bond so that there are no free 5′ or 3′ ends on the molecule. As a result, when the DNA is completely denatured, a circular single-stranded molecule is generated. At the two ends of the genome there are inverted terminal repeats. These are much larger than those in adenovirus, extending for some 10 kbp, and they contain several genes.

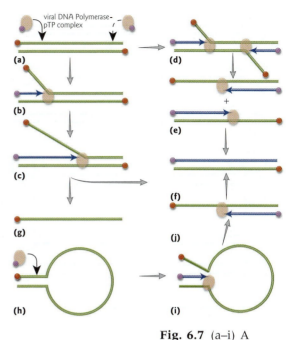

Fig. 6.7 (a–i) A general scheme for adenovirus DNA replication (see Section 6.4 for details). Green: parental DNA; blue: new DNA. Arrowheads on new DNA strands represent 3′ ends. Red circles represent the terminal protein (TP) attached to parental DNA 5′ ends and purple circles the terminal protein precursor molecules (pTP) which prime new DNA synthesis. The viral DNA polymerase is represented in pink.

Vaccinia virus is thought to use a novel strategy to avoid problems in replicating its genome ends. The current model for its replication is summarized in Fig. 6.8; however, the detailed biochemistry of the process is not yet established. The process occurs in the cytoplasm of infected cells using exclusively virus-coded proteins. Replication is thought to initiate through an enzyme recognizing and nicking one template strand at a specific site within the inverted repeats. These therefore constitute origins of replication. Either or both of these sites may be used simultaneously (Fig. 6.8b). DNA polymerase can then extend from the 3′ end that is produced, displacing the nontemplate strand, until the end of the template is reached (Fig. 6.8c). However the polymerase does not have to stop at this point, since the terminal repetition means that the molecule can base-pair in an alternative way which once again presents a template for extension. The newly synthesized strand folds back and base-pairs to itself (Fig. 6.8d), so that the paired 3′ end is now directed back towards the center of the molecule and synthesis can continue. As with adenovirus, exactly the same events can proceed from both ends of the duplex template. When the replication forks meet in the center of the molecule, the two halves fall apart (Fig. 6.8e) and replication is completed when the polymerases run up against the base-paired 5′ end and the ends are ligated (Fig. 6.8f). Notice how nicking the parental DNA to provide a primed template complex avoids the need to use RNA primers that would have to be excised later.

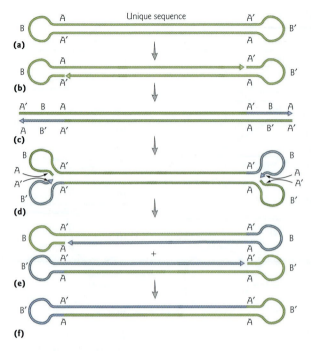

Fig. 6.8 (a–f) General scheme for vaccinia virus DNA replication (see Section 6.4 for details). Green: parental DNA; blue: new DNA. Arrowheads on new DNA strands represent 3′ ends. Complementary sequences are denoted **A**, **A′** etc. Note that the two alternative forms of the terminal loop sequence, **B**, **B′** are exchanged with each of replication; individual molecules may have loops with identical or complementary sequences.

What happens if, on a template molecule, only one of the two origins is activated? The replication fork, as it proceeds back through the template, will not meet one coming the other way. Neither will it come up against a base-paired 5′ end. Instead, it will continue to synthesize a complementary strand from its template all the way up to and then around the covalently closed end of the genome and back along the other side. What results is a concatemer – two unit length molecules covalently joined. In fact, concatemers can be produced even when both origins are used together. When the polymerase reaches the base-paired 5′ end ahead of it, it does not have to stop. Its situation exactly resembles that at the initiation event, and just as it did then, it can continue on, displacing the nontemplate strand and beginning the whole cycle of events again. How then are these concatemers resolved? They can be cleaved by the same site-specific nicking enzyme which created the 3′ end on which synthesis initiated. These nicked molecules can then refold into unit-length molecules which can be closed by DNA ligase to provide progeny genomes.

6.5 REPLICATION OF CIRCULAR SINGLE-STRANDED DNA GENOMES

Important viruses in this category

- *Bacteriophage φX174, an early target for molecular and structural characterization.*
- *Bacteriophage M13, a crucial laboratory tool for DNA sequencing prior to the development of PCR techniques.*

The first step in replicating a single-stranded genome is the creation of a complementary strand. Once this is achieved, replication mechanisms will resemble those for double-stranded genomes. The best-studied example of single-stranded circular DNA replication is bacteriophage φX174. The incoming single strand is converted to a double-stranded replicative form (RF) via RNA-primed DNA synthesis. This primer is synthesized by host RNA polymerase using as template a short hairpin duplex formed in the single strand genome by intramolecular base-pairing. The parental RF is then amplified before switching to rolling circle replication for the production of a single-strand concatemer from which progeny genomes are excised and circularized. A virus with this form of genome has been described that affects humans, TT virus, however little detailed information is available so far.

6.6 REPLICATION OF LINEAR SINGLE-STRANDED DNA GENOMES

Important viruses in this category

* *The autonomous parvovirus B19, a human pathogen*
* *The defective parvovirus, adeno-associated virus, which is being developed as a gene therapy vector.*

The parvovirus family comprises both autonomous and defective viruses. The autonomous parvoviruses, such as minute virus of mice (MVM), package a negative-sense DNA strand while defective parvoviruses, such as the adeno-associated viruses (AAV), package both positive- and negative-sense DNA strands in separate virions, either one being infectious. These defective viruses are almost completely dependent on coinfection with helper virus for their replication (either adenovirus or various herpesviruses can perform this function). Parvovirus genomes contain terminal hairpins (inverted repeats). Whilst these terminal hairpins have distinct sequences in the autonomous viruses, they are complementary in the defective viruses (Fig. 6.9). As discussed below, this difference explains why the two types of virus differ in the polarities of DNA strand which they package.

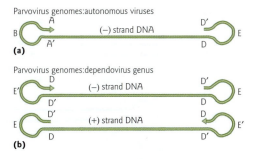

Fig. 6.9 (a,b) Schematic representation of the genome structures of autonomous and defective parvoviruses. Complementary sequences are denoted A, A' etc.

A model for autonomous parvovirus DNA replication is presented in Fig. 6.10, with some evidence for it in Box 6.7. The essential first step in parvovirus replication is conversion of the genome to a double-stranded form by gap-fill synthesis (Fig. 6.10a,b). The terminal hairpin provides a base-paired 3′ OH terminus at which DNA elongation can be initiated without the need for an RNA primer. Once this is achieved, the replication mechanism is quite similar to that of the poxviruses. Displacement of the base-paired 5′ end allows synthesis to continue to the template strand 5′ end (Fig. 6.10c). The structure now undergoes rearrangement to form a "rabbit-eared" structure (Fig. 6.10d). This recreates the hairpin originally present at the 5′ end of the parental genome and also forms a copy of this hairpin at the 3′ end of the complementary strand, which can serve as a primer for continuing synthesis (Fig. 6.10e). Further cycles

Box 6.7

Evidence for the mechanism of parvovirus replication

- Infected cells accumulate double-stranded forms of viral DNA, a large fraction of which cannot be irreversibly denatured, suggesting that the two strands are covalently linked.
- The only virus encoded replication activity so far defined is a site-specific endonuclease.
- Duplex DNA molecules with single-strand tails have been observed in cells infected with MVM whereas no free single-stranded DNA has been detected in such cells.
- The 5' end of the genome, with attached viral endonuclease protein, is exposed outside the assembling virion.

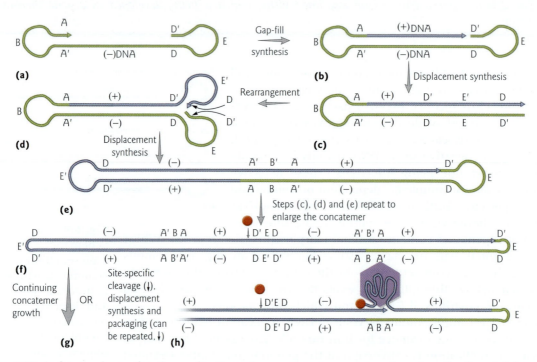

Fig. 6.10 (a–h) A scheme for the replication of an autonomous parvovirus (see Section 6.6 for details). Green: parental DNA; blue: new DNA; purple: assembling capsid. Arrowheads represent 3' ends. The orange circle represents the viral site-specific nicking enzyme. Complementary sequences are denoted A, A' etc.

of rearrangement and strand displacement yield a tetramer-length duplex (Fig. 6.10f), and this process can continue (Fig. 6.10g).

The concatemers created by replication can serve as intermediates from which progeny single-stranded viral DNA can be generated by displacement synthesis. A nick is introduced 5' to a genome sequence within the concatemer by a sequence-specific nicking enzyme, NS1 in MVM and Rep68 in AAV, which remains attached to the 5′ end (Fig. 6.10f,h). The 3′ OH terminus then acts as a DNA primer for displacement synthesis of DNA single strands, which are packaged concomitantly (Fig. 6.10h). After a complete genome has been displaced, excision of the progeny genome is completed by endonuclease cleavage, resulting in the release of a complete virus particle and termination of the displacement synthesis.

The defective parvovirus AAV differs from MVM in having terminal hairpin sequences that are complementary to each other. The result is that the 5′ ends of both negative- and positive-sense strands are identical and so, within the model just described for linked strand-displacement and packaging, site-specific cleavage will occur equally at the 5′ end of both positive and negative strands and therefore both will be packaged. It is still unclear why this subset of the parvoviruses show dependence on helper functions for growth, since these required functions do not necessarily include any that are directly involved in DNA synthesis. The helper functions can be provided by various unrelated viruses and the dependence on them can also be overcome in cell culture by treating cells with genotoxic agents (chemicals which damage DNA). It is also significant that AAV is now known to be able to establish a latent infection, integrating into a specific site in the host chromosomes; this might be the preferred strategy of the virus for ensuring perpetuation of its genetic material. Thus "helper" functions may alternatively be viewed as viral functions which alter the cell environment so as to block establishment or maintenance of latency and hence favor AAV productive replication.

6.7 DEPENDENCY VERSUS AUTONOMY AMONG DNA VIRUSES

The autonomy of viruses from their hosts as regards the replication of their DNA varies between wide limits and is a function of the size of the viral genome. At one end of the scale are large viruses, such as the poxviruses. Such viruses require little more from their host cells than an enclosed environment, protein synthesizing machinery, a supply of amino acids and deoxyribonucleotide triphosphates, and an energy source. Some may not even require this much; herpes simplex and vaccinia viruses both specify a thymidine kinase and several other enzymes. At the other end of the scale are viruses, such as minute virus of mice and SV40, whose genomes can specify only a few proteins. Since some of these are needed to form the virus coat, not many genes are left to code for functions essential to replication. These viruses rely on the host not only for nucleic acid precursors but also for polymerases, ligases, nucleases, etc. The extreme in terms of lack of autonomy is represented by viruses such as adeno-associated virus, which requires assistance for replication not only from the host but also from another virus.

KEY POINTS

- DNA as genetic material has the advantage over RNA of increased replication fidelity, but this means that DNA viruses adapt and evolve more slowly than RNA viruses.
- Viruses must obey the cellular rules for DNA synthesis, which means using a primer and having to solve the "end replication" problem on linear genomes.
- Double-stranded DNA genomes that are intrinsically circular (SV40, papillomavirus) or which can circularize before replication (herpes simplex virus) avoid the "end replication" problem. Linear genomes with covalently closed ends (poxviruses) also avoid this problem.
- Circular double-strand DNA can replicate either via a theta-form intermediate to produce daughter molecules that exactly resemble the parent, or via a rolling circle mechanism to give concatemers that must then be resolved to unit length molecules by specific cleavage events before or during packaging.
- Adenovirus is unique in replicating its linear genome from the ends, using a protein primer.
- Viral single-strand DNA genomes must be converted to double-strand forms before replication can proceed as for double-stranded genome types.
- There is a roughly inverse relationship between genome size and dependence on the host for essential functions in DNA replication.
- SV40 replication is a good model for events during eukaryotic cell DNA replication.

QUESTIONS

- Compare and contrast the strategies adopted by members of the various virus families that infect mammals and belong to Baltimore class 1 for replication of their double-stranded DNA genomes.
- Discuss the basis for the "end-replication" problem for linear double-stranded DNA in biological systems, and the various strategies employed by viruses to overcome this problem.

FURTHER READING

Challberg, M., ed. 1991. Viral DNA replication. *Seminars in Virology* **2**(4).

Cotmore, S. F. & Tattershall, P. 1995. DNA replication in the autonomous parvoviruses. *Seminars in Virology* **5**, 271–281.

Lehman, I. R. & Boehmer, P. E. 1999. Replication of herpes simplex virus DNA. *Journal of Biological Chemistry* **274**, 28059–28062.

Kornberg, A. & Baker, T. 1991. *DNA replication*, 2nd edn. San Francisco: W. H. Freeman.

Liu, H., Naismith, J. H., Hay, R. T. 2003. Adenovirus DNA replication. *Current Topics in Microbiology and Immunology* **272**, 131–164.

Ogawa, T. & Okazaki, T. 1980. Discontinuous DNA replication. *Annual Review of Biochemistry* **49**, 421–457.

Waga, S. & Stillman, B. 1994. Anatomy of a DNA replication fork revealed by reconstitution of SV40 DNA replication in vitro. *Nature* **369**, 207–212.

Waga, S. & Stillman, B. 1998. The DNA replication fork in eukaryotic cells. *Annual Review of Biochemistry* **67**, 721–751.

Also check Appendix 7 for references specific to each family of viruses.

The process of infection: IIB. Genome replication in RNA viruses

The synthesis of RNA by RNA viruses involves replication, which is defined as the production of progeny virus genomes and transcription to produce messenger RNA (mRNA). The process of transcription for RNA viruses is described in Chapter 10 where it is discussed in terms of gene expression. This chapter focuses on the process of replication of RNA viruses.

Chapter 7 Outline

The replication of RNA genomes requires the action of RNA-dependent RNA polymerases which are not encoded by the genome of the infected host cell but instead are synthesized by the virus. During the process of replication of RNA virus genomes, as for all other processes which involve synthesis of nucleic acid, the template strand is "read" by the polymerase travelling in a $3' \rightarrow 5'$ direction with the newly synthesized material starting at the $5'$ nucleotide and progressing to the $3'$ end. Most RNA viruses can replicate in the presence of DNA synthesis inhibitors, indicating that no DNA intermediate is involved. However, this is not true for the retroviruses (Baltimore class 6) and these will be considered separately (see Chapter 8).

The polymerases which carry out RNA replication are encoded by the virus and are either transported into the cell at the time of infection or are synthesized very soon after the infection has begun. Frequently, the polymerases involved in RNA replication are referred to as "replicases" to differentiate them from the polymerases involved in transcription. However, both processes are carried out by the same enzyme exhibiting different synthetic activities at different times in the infectious cycle.

Box 7.1

Evidence for RNA as the genetic material for some viruses

• The presence of uracil instead of thymidine.
• The presence of ribose instead of deoxyribose sugars.
• Buoyant density of the nucleic acid: RNA is more dense than DNA.
• Sensitivity to RNase and not DNase.
• Some purified RNA genomes when introduced into cells yield infectious virus.

7.1 NATURE AND DIVERSITY OF RNA VIRUS GENOMES

As demonstrated in the Baltimore classification scheme (see Section 4.4), RNA genomes can be single-stranded or double-stranded and the former may be of either positive (mRNA-like) or negative sense (Box 7.1). RNA molecules which act as virus genomes exist only as linear molecules, although infectious circular RNA molecules form the genomes of a specialized type of agent, the viroid, described in Section 4.6. An unusual feature of many RNA viruses is that their genomes consist of multiple segments, analogous to chromosomes of host cells, and these viruses must ensure that at least one copy of each segment is present in the mature particle to generate a full complement of genes. During the replication process the RNA molecules remain linear, and covalently closed circular molecules are never observed. The RNA genomes of different viruses vary greatly in size, though they do not display the range seen with DNA virus genomes. The largest single molecule RNA virus genomes known are those of the coronaviruses which are approximately 30,000 nt, with the smallest animal virus RNA genomes being those of the picornaviruses at approximately 7500 nt. Several bacteriophage have RNA genomes smaller than this, one of the smallest being MS2 with a single-stranded positive sense RNA genome of 3569 nt. The nodaviruses which infect insects and some animals have a very small genome which consists of two RNA molecules of approximately 3000 nt and 1400 nt.

All enzymes involved in RNA synthesis, whether of virus or host cell origin, are unable to "proofread" (i.e. to correct incorrectly inserted bases). This contrasts with DNA synthesis in which correction may occur (see Section 6.1). The lack of proofreading means that the RNA genomes mutate more rapidly than DNA genomes. Estimates suggest that the error rate during RNA virus replication is estimated to be approximately 10^{-3} to 10^{-5} per base per genome replication event (one mutation per 1000 to 100,000 bases per genome replication event). This has implications for the evolution of RNA viruses (see Chapter 17) and is likely to place a limit on the maximum size of an RNA-based genome, since as the size of the genome increases the probability of it containing a mutation in an important region will also increase.

7.2 REGULATORY ELEMENTS FOR RNA VIRUS GENOME SYNTHESIS

Certain features are common in the process of replication for all RNA viruses. In order to make a faithful copy of the genome, the RNA-dependent RNA polymerase must begin synthesis at the 3′ terminal nucleotide of the template strand. The 3′ terminus must therefore contain a signal to direct initiation of synthesis. As indicated by the principles which underpin the Baltimore classification scheme (see Section 4.4), all RNA viruses must replicate via a dsRNA intermediate molecule. For example, a virus with a ssRNA genome must produce a dsRNA intermediate by synthesizing a full-length "antigenome" strand. The antigenome strand will then, in turn, be used as a template to synthesize more genomes for packaging into progeny virions. As before, synthesis using an antigenome as template must begin at the 3′ terminal nucleotide if a faithful, full length, copy is to be made, and the 3′ end of the antigenome must also contain a signal to direct the polymerase to begin synthesis. For most viruses the mechanism of initiation of RNA synthesis during replication is only poorly understood, if at all, but a combination of old and new analyses have identified the termini of RNA viruses as containing the regulatory elements which direct RNA synthesis. The most convincing evidence has come from the study of the genomes of defective-interfering viruses and, more recently, from reverse genetics studies on RNA viruses.

For some RNA viruses the sequences at the immediate termini of the genomes are almost completely complementary to each other. These complementary regions, which can range from approximately a dozen to over fifty nucleotides in length in different viruses, are referred to as inverted repeat sequences. For genomes with complementary termini both the genome and antigenome strands will contain nearly the same sequence at the 3′ terminus so that the initiation of synthesis with these molecules as templates can occur using the same mechanism. This is seen for many viruses such as paramyxoviruses and influenza viruses. For viruses whose genomes do not contain inverted repeat sequences, the initiation of synthesis of antigenome- and genome-sense molecules must each be controlled by different processes. In some cases different features may be present at the termini of the RNA genomes. These include covalently attached proteins at the 5′ end in the case of picornaviruses, such as poliovirus, and long homopolymeric polyadenylate (polyA) tracts at the 3′ ends of many positive sense RNA genomes such as those of picornaviruses, alphaviruses, flaviviruses, and coronaviruses.

The generation and amplification of defective–interfering (DI) virus RNA

A common phenomenon for viruses is the production of DI particles as the result of errors in their nucleic acid synthesis. Here we shall consider only DI RNA viruses, about which more is known. DI viruses are mutants in which the genomes have large deletions, leaving RNA which may comprise as little as 10% of the infectious genome from which they were replicated. DI viruses are unable to reproduce themselves without the assistance of the infectious virus from which they were derived (i.e. they are defective). The portions of the genome which are retained in DI viruses contain all of the elements necessary for replication, and packaging, of the genome. Analysis of DI genomes provided the first evidence to show the location of replication sequences, and their subsequent analysis coupled with modern molecular biological techniques has assisted in identifying key elements in virus replication. Propagation of DI virus is optimal at a high multiplicity of infection when all cells contain an infectious virus genome. DI genomes depress (or interfere with) the yield of infectious progeny, by competing for a limited amount of one or more product(s) synthesized only by the infectious parent, referred to as the helper virus. Interference only takes place when the ratio of DI to infectious genomes reaches a critical level. Up to this point both infectious and DI genomes are replicated to the fullest extent. Many DI viruses contain genomes which are deleted to such an extent that they synthesize no proteins and some have no open reading frame. Because they depend upon parental virus to provide those missing proteins, DI and parental viruses are composed of identical constituents, apart from their RNAs. Thus it is usually difficult to separate one from the other. A notable exception is the DI particle of rhabdoviruses such as VSV whose particle length is proportional to that of the genome. When centrifuged these short particles remain at the top of sucrose velocity gradients and are thus called T particles to distinguish them from infectious B particles, which sediment to the bottom. Some biological implications of DI particles are discussed in Section 13.2.

A clue leading to one hypothesis explaining how DI RNAs are generated came from electron microscopic examination of genomic and DI RNAs from single-stranded RNA viruses. Both were found to be circularized by hydrogen bonding between short complementary sequences at

Fig. 7.1 Hypothetical schemes to explain the generation of DI RNAs having sequences identical with both the 5′ and 3′ regions of the genome (a) and with the 5′ region only (b).

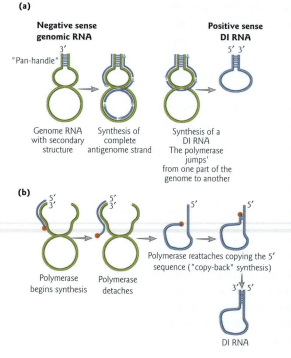

(a)

Negative sense genomic RNA | Positive sense DI RNA

"Pan-handle"

Genome RNA with secondary structure | Synthesis of complete antigenome strand | Synthesis of a DI RNA The polymerase 'jumps' from one part of the genome to another

(b)

Polymerase begins synthesis | Polymerase detaches | Polymerase reattaches copying the 5′ sequence ("copy-back" synthesis)

DI RNA

the termini, forming structures called "panhandles" or "stems" (Fig. 7.1a). The deletion that results in DI RNA may arise when a polymerase molecule detaches from the template RNA strand and reattaches either at a different point of the genome or to the newly synthesized, incomplete, strand (Fig. 7.1b). Thus the polymerase begins to replicate faithfully but fails to copy the entire genome. Most VSV DI viruses are of the latter type and lack the 3' end of the standard virus genome, having instead a faithful copy of the 5' end and a complementary copy of the 5' end at the two termini. There are no DI viruses known which lack the 5' terminus. In other DI viruses parts of the genome can be duplicated, often several times over, during subsequent replication events, making complex structures which bear little resemblance to the standard genome from which they were derived. The three classes of DI genome are summarized in Fig. 7.2. A key point for the DI RNA viruses is that their genomes always contain the same termini as those of the parent virus, or consist of inverted repeats of sense and antisense copies of the 5' end of the normal genome, while the remainder of the genome can be substantially deleted without impairing the ability of the DI genome to replicate. This is taken to indicate that the genome termini are essential and contained the regulatory elements to direct genome synthesis.

Usually, interference only occurs between the DI virus and its parent. This is because the DI virus lacks replicative enzymes and requires those synthesized by infectious virus. Specificity resides in the enzymes, which only replicate molecules carrying certain unique nucleotide sequences. Intuitively, it can be seen that in a given amount of time an enzyme will be able to make more copies of the smaller DI RNA. Thus, as time progresses, the concentration of DI RNAs increases relative to the parental RNA in an amplification step. However, this is not the whole story, as some large DI RNAs interfere more efficiently than smaller ones. Such DI RNAs seem to have evolved a polymerase recognition sequence which has a higher affinity for the enzyme than that of the infectious parent and hence confers a replicative advantage. This may indicate that, while the essential minimal sequences for directing genome synthesis are located at the termini, sequences located elsewhere may also play an enhancing role. The only other sequence that all DI genomes must retain is a packaging or encapsidation sequence, because without this they cannot be recognized by virion proteins and form viral particles.

Reverse genetics of RNA viruses

Over recent years one of the most exciting developments has been the generation of reverse genetic systems for a wide range of RNA viruses, particularly those with negative sense ssRNA genomes. The absence of a DNA intermediate in the replication cycle of RNA viruses limited

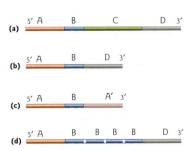

Fig. 7.2 Comparison of the genome organizations of normal and DI virus. (a) A representation of a normal virus genome. The colored segments, labelled A to D, represent different regions of the genome (not to scale). (b) The genome of a 5', 3' DI virus in which a large portion of the genome (region C) has been deleted. (c) The genome of a 5' DI virus. Large portions of the genome (regions C and D) are not present and the stippled red block (region A') represents the complement of the normal 5' end sequence of region A. (d) A complex DI virus genome containing multiple copies of a short portion (region B) of the normal genome with other sequences (region C) deleted.

Box 7.2

Reverse genetics systems for viruses with RNA genomes

Reverse genetics systems have been described for several viruses in Baltimore classes 3, 4, and 5 including:
Class 3 viruses – birnaviruses
Class 4 viruses – picornaviruses, alphaviruses, coronaviruses
Class 5 viruses – rhabdoviruses, filoviruses, paramyxoviruses, bunyaviruses, orthomyxoviruses

research as there are no tools to manipulate RNA molecules in the same way that there are for DNA. However, for many RNA virus systems it is now possible to generate DNA copies of virus genomes and convert these back into RNA, and ultimately into infectious virus particles. The DNA can be manipulated in a variety of ways including the generation of deletions or specific mutations which are then mirrored in the synthetic replicas of the virus genomes. Analysis of these mutated genomes has shown, for representatives of all RNA virus families, that the immediate termini of the genome, or each genome segment, contain the elements which are essential to direct RNA synthesis for replication. This is true whether the virus genome contains inverted repeat sequences or different, unique sequences at the termini of the genome. Mutation of the sequences at the termini has serious deleterious consequences for the ability of the RNA molecules to be replicated by virus proteins.

7.3 SYNTHESIS OF THE RNA GENOME OF BALTIMORE CLASS 3 VIRUSES

Important viruses in this category

The class 3 viruses contain a dsRNA genome.
* *Reovirus – the type member of the reovirus family. Studied as the example typical of the family. May cause mild infections of the upper respiratory and gastrointestinal tract of humans.*
* *The rotavirus family – the most common viral cause of severe diarrhea among children. Animal rotaviruses cause significant economic losses.*
* *Bluetongue virus, an orbivirus – transmitted by the biting fly* Culicoides variipennis. *Infects sheep, cattle, goats, and wild ruminants.*

Class 3 viruses contain multiple segments of dsRNA. In reoviruses the genome consists of 10 segments, each of which replicates independently

Box 7.3

Evidence for conservative replication of the reovirus genome

Schonberg, in 1971, provided evidence that the reovirus genome was replicated conservatively.
Infected cells were labeled with ^{3}H-uridine for 30 minutes at various times during the replication phase of the reovirus infectious cycle. Total dsRNA was then isolated and hybridized to excess unlabeled positive sense RNA which had been prepared *in vitro*. Three results could be obtained, depending on the mode of replication (Fig. 7.3):

1 If replication was semiconservative, then the dsRNA should be equally radioactively labeled in both strands. Hybridization to an excess of unlabeled positive sense strands would occur with only 50% of the label.

2 If the positive sense strand was used as template, then hybridization of the radioactively labeled negative strand would result in 100% of the label forming a hybrid.

3 If the negative sense strand was used as template then radioactively labeled positive sense RNA would be produced and hybridization would generate dsRNA which did not contain any radioactivity.

The result showed that all of the label was associated with the negative strand and thus replication utilizes the positive sense mRNA as template. In cells exposed continuously to ^{3}H-uridine, the label was found to be divided between both strands. Varying the time of the short pulse of radioactivity showed that there is a lag between the synthesis of the positive template strand, and the onset of replication of the negative sense strand. Presumably this allows the mRNA to be translated to produce the polymerase before being used as a template in replication. When cells are labeled for long periods at later times the label is distributed between positive and negative sense RNA, as the newly synthesized positive sense mRNA is used as template to make new negative sense strands.

of the others. By analogy with DNA replication, this dsRNA could replicate by a *semiconservative* mechanism such that the complementary strands of the parental RNA duplex are displaced into separate progeny genomes, or the parental genome could be conserved or degraded. In fact, dsRNA genomes are replicated conservatively (Box 7.3).

Following initiation of infection, several proteins in the reovirus particles are removed by protease digestion during the uncoating process to form a subviral particle which is found in the cytoplasm where replication takes place. The dsRNA genome is retained within the subviral particle and does not leave it during the infectious cycle. The observation that the genome is not completely uncoated to liberate both negative and positive strands of the genome segments in the infected cell indicated that replication of the dsRNA could not occur in the normal semiconservative way as seen for DNA. The only virus nucleic acid found in the cytoplasm

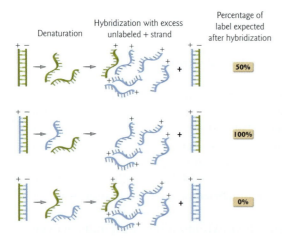

Percentage of label expected after hybridization

50%

100%

0%

Fig. 7.3 Diagrammatic representation of the experiment that demonstrates that the reovirus genome is replicated conservatively. Radioactively labeled RNA strands are represented by red lines and unlabeled strands by black lines. See Box 7.2 for details.

outside the subviral particles is mRNA, which is generated by transcription using the particle-associated RNA-dependent RNA polymerase as described in Chapter 10. Since both strands of the genome RNA are retained in the subviral particle, it was clear that the single-stranded mRNA transcripts must be the sole carriers of genetic information from parent to progeny. This means that only one strand of each of the 10 genome segments is used as template and the newly synthesized RNA is then replicated to form a new dsRNA genome segment.

Newly synthesized negative sense RNA is only found as part of a dsRNA molecule. The dsRNA segments are never found free in infected cells but are always associated with an immature virus particle. Each particle must contain a single copy of each of the 10 reovirus genome segments. The mechanism by which a virion specifically packages one of each of the 10 RNA segments is not yet understood.

7.4 SYNTHESIS OF THE RNA GENOME OF BALTIMORE CLASS 4 VIRUSES

Important viruses in this category

Class 4 viruses contain a positive sense ssRNA genome.
* *The picornaviruses which include poliovirus, hepatitis A virus, foot and mouth disease virus, and rhinoviruses which cause the common cold.*
* *The coronaviruses which includes important veterinary viruses and the SARS virus.*
* *Rubella virus, the causative agent of German measles.*
* *Hepatitis C virus.*
* *Yellow fever virus.*

While there are many differences in the details of the replication cycles of class 4 viruses in terms of gene expression (see Sections 10.4–10.6) and assembly, the process by which their positive sense ssRNA genomes are replicated is very similar. A great deal of information is available about the mechanism of picornavirus genome replication, and this will be described here in detail. In principle, the process used by picornaviruses is applicable to all class 4 viruses with the generation of the same type of intermediate molecules *in vivo* and *in vitro*. A similar process is also applicable to the production of the coronavirus subgenomic mRNAs described in Section 10.5.

Following infection, the picornavirus genome RNA is the only nucleic acid which enters the infected cell so it must act as a template for both translation and transcription, just as for the reovirus mRNAs. The picornavirus genome RNA has a 3' polyA tail and a small, 22 amino acid, protein, called VPg, covalently attached to the 5' end. Only picornavirus genomes have a VPg protein attached to the RNA. Translation is necessary as a first step in the infectious cycle to produce the polymerase enzyme which will synthesize the new genomes. Aspects of translation of picornavirus genome RNA are discussed in Section 10.4. Replication takes place on smooth cytoplasmic membranes in a replication complex and, as indicated earlier, must involve the generation of a dsRNA intermediate. Initiation of replication occurs at the 3' end of the positive sense, genome/mRNA polyA tail, the presence of which is essential for replication to occur. An interesting feature of picornavirus replication is that the genome RNA is held in a circular form due to the action of both virus and host proteins (Fig. 7.4). Cells contain a polyA binding protein which plays a role in translation of mRNA and this protein binds to the 3' end of the picornavirus genome RNA. A second protein which has an affinity for polyC tracts in RNA binds to a region of the virus genome RNA near the 5' end. This region, which is rich in pyrimidine residues, forms a complex three-dimensional structure referred to as a cloverleaf. The third protein in the complex is the picornavirus 3CD protein, the precursor of the polymerase, which is activated by proteolytic cleavage. In this way the polymerase is located near the 3' end of the genome where replication will begin. It is not known whether other class 4 virus genomes are circularized in a similar way as a necessary requirement for replication. The process of initiation of picornavirus RNA replication is not yet understood, nor is how the VPg protein becomes attached to the 5' end of the newly synthesized RNA. One possibility is that the attachment of VPg and initiation of RNA synthesis are coupled. If so, this presumably happens on initiation of both positive and negative strand synthesis, since both strands have VPg at their 5' ends. Once RNA synthesis begins, the polymerase proceeds along the entire length of the template RNA. The replication process requires concurrent protein synthesis, since addition of protein synthesis inhibitors, even when replication has started, inhibits any further rounds of replication.

Analysis of virus-specific RNA isolated from picornavirus-infected cells shows that the replication complex contains an RNA molecule which is partially ssRNA and partially dsRNA. This is called the replicative intermediate (RI). Negative sense RNA is only ever found in the cell in association

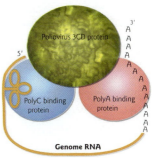

Fig. 7.4
Circularization of picornavirus genome RNA in the replication complex.

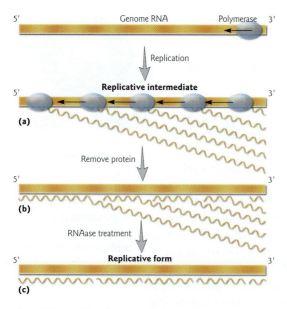

Fig. 7.5 Replication of picornavirus genome RNA. Note that the replication takes place using a template which is held in a circular form by the action of three proteins but is represented here as linear for simplicity. (a) Proposed structure for a molecule of replicative intermediate before deproteinization; (b) after deproteinization to generate a replicative intermediate (RI); (c) the effect of treating deproteinized RI with RNase to produce a replicative form (RF) RNA.

with positive sense in the RI. Deproteination and ribonuclease (RNase) treatment of the RI generates a pseudo dsRNA the same length as the virus genome and with nicks in the newly synthesized strand of RNA. This is called the replicative form (RF). The RF is produced by removing the ssRNA tails found in the RI (Fig. 7.5). A dsRNA RF complex is also found in poliovirus-infected cells treated with inhibitors of host cell RNA polymerase but it is not known whether this is involved in the replication of the virus RNA or is an artefact of the drug treatment.

The suggested mode of replication is that, initially, the positive sense genome ssRNA is used as a template to generate negative sense, antigenome RNA with VPg at the 5′ end. In the RI the nature of the association of the positive and negative strand is not known, but it is likely that they are only loosely linked and the two strands may only associate at the region where synthesis is occurring. Once the polymerase complex has moved along the template the 3′ end will become available for a further round of replication even though the preceding complex has not yet completed copying the template. In fact, multiple initiations may occur on the template before the first replication complex has completed its work and it is estimated that up to five functional replication complexes may be associated with a picornavirus template at any one time (Fig. 7.5). The newly synthesized RNA is in turn used as template for production of new copies of the genome RNA, with VPg attached to the 5′ end, in a similar way. Since much more positive sense genome ssRNA than negative sense ssRNA is produced, the synthesis process must be biased, or asymmetrical. Unlike the negative sense RNA, the completed positive sense RNA is released from the replication complex for packaging into virions, use as template for further replication, or for translation (with concomitant loss of VPg).

For the class 4 viruses other than the picornaviruses, and which lack a VPg on the genome, RNA synthesis is also initiated at the 3′ end of the genome and antigenome RNAs. In all cases RI structures are produced suggesting that the basic principles of the replication process are the same.

7.5 SYNTHESIS OF THE RNA GENOME OF BALTIMORE CLASS 5 VIRUSES

Important viruses in this category

Class 5 viruses contain a negative sense ssRNA genome.
- *Measles virus, a paramyxovirus, responsible for many child deaths worldwide.*
- *Influenza virus, an orthomyxovirus, responsible for severe respiratory disease. Worldwide pandemics have killed many millions of people.*
- *Rabies virus, a rhabdovirus.*
- *Ebola and Marburg viruses, filoviruses.*

Class 5 viruses are of two types, those with genomes consisting of a single molecule and those with segmented genomes. The former are grouped together in a taxonomic order, the Mononegavirales, indicating that they have a single, negative sense ssRNA molecule as genome. The Mononegavirales include the paramyxo-, rhabdo-, filo-, and bornavirus families. The viruses with segmented genomes include the orthomyxo-, arena-, and bunyaviruses. The details of many aspects of RNA replication of class 5 viruses are not yet known, but the general principles appear to be common to all, irrespective of the number of molecules which make up the genome. An important feature for the replication of all class 5 virus genome RNAs is the presence of complementary sequences at the termini as described above (Section 7.1). Since the negative sense genome RNA cannot be translated into protein to produce the required polymerases, as seen with positive stranded RNA genomes, class 5 virus RNA synthesis can only occur using a pre-existing RNA-dependent RNA polymerase present in the virus particle. Polymerase activity can be detected for some class 5 viruses *in vitro* after partial disruption of the virus with detergent in the presence of the four ribonucleoside triphosphates and appropriate ions. Viruses "activated" in this way in the absence of whole cells synthesize RNA at a linear rate for at least 2 hours. In general, the *in vitro* systems yield only mRNA, not positive sense antigenome RNA. However, the virus-associated polymerase is responsible for both replication and transcription in infected cells.

Replication of the RNA genome of nonsegmented class 5 viruses: rhabdoviruses

Most information is available about the process of replication of rhabdoviruses, especially vesicular stomatitis virus (VSV), and this has been used to generate a model of replication applicable to many other class 5 viruses with either a single or multiple genome segments, including all

Mononegavirales, and the bunya- and arenaviruses. Recent studies have confirmed that the general principles of the model derived for VSV are true for all members of the order Mononegavirales.

The ssRNA genome (and antigenome produced during replication) of VSV is always found closely associated with three virus proteins in a helical complex. The most abundant protein is the nucleoprotein (NP), with smaller amounts of a phosphoprotein (P), and only a few molecules of a large (L) protein. The L protein is the catalytic component of the replication complex responsible for carrying out the RNA synthesis but the NP and P proteins are essential for its activity. The complex of NP, P, and L proteins together with genomic RNA, referred to as the nucleocapsid, also carries out transcription to produce mRNA (see Section 10.9). It is not known what causes the complex to transcribe mRNA at some times and to replicate the genome at others. The relative amounts of the three proteins appear to be critical for the function of the nucleocapsid.

The replication of class 5 viruses requires continuous protein synthesis, and addition of inhibitors of protein synthesis results in an immediate cessation of replication. Class 4 viruses are the same in this respect. Consequently, virus mRNA and protein synthesis occur prior to the onset of replication. As for all nucleic acid synthesis, replication begins at the 3' end of the negative sense RNA template where the polymerase binds to a specific sequence, continues to the 5' end, generating a positive sense, antigenome, RNA. Due to the complementarity of the terminal sequences, a very similar polymerase-binding sequence is present at the 3' end of the antigenome where the polymerase begins synthesis to generate new genome RNA molecules. Circular nucleocapsids are often observed within infected cells, suggesting that the complementary termini interact to form a panhandle structure, similar to that shown in Fig. 7.1. The interaction of the termini may occur either by direct base-pairing or in association with the nucleocapsid proteins. This circularization may be a necessary step in the replication process, though this is not yet clear. During replication the polymerase ignores the signals for termination of mRNA synthesis which are recognized during transcription.

The molecular structure of the RI generated during replication of rhabdoviruses is similar to that of picornaviruses. The RI is closely associated with the three replication proteins and antigenome RNA is only found in the RI. Purification of the RI and treatment with RNase generates a double-stranded RF RNA analogous to that generated from the picornavirus RI (Fig. 7.5).

The antigenome RNA is used as template to produce negative sense ssRNA for progeny virus genomes. The negative sense RNA is found as a nucleocapsid structure which can be used in further rounds of replication prior to incorporation into virus particles. Since considerably more negative sense RNA is produced than positive sense RNA, the replication process must be asymmetrical to favor production of one strand over the other.

Replication of the RNA genome of segmented class 5 viruses: orthomyxoviruses

As for the Mononegavirales, the genome RNA of class 5 viruses that are comprised of multiple segments also form helical nucleocapsid structures. These often appear as circles probably due to the formation of pan-handle structures held together by the nucleocapsid proteins. The major protein of the influenza virus nucleocapsid is the nucleoprotein (NP). The NP protein interacts directly with the genome RNA, binding to the sugar phosphate backbone and leaving the nucleotide bases exposed on the surface of the structure. The location of the other nucleocapsid proteins PA, PB1, and PB2 (which are also involved in transcription, Chapter 10) is less clear, but they are thought to be associated with the nucleotide bases on the outside of the helix. Each segment replicates independently.

The positive sense RNA which is generated from the genome template is probably initiated *de novo*. This contrasts with the use of a primer derived from the host cell mRNA during transcription of mRNA (see Section 10.8). The reason for the different activities of the nucleocapsid complex during replication and transcription is not known. The replication complex is dependent on the continued synthesis of at least one viral protein. This is similar to the situation for the other RNA viruses.

During replication the synthesis of RNA does not stop in response to the transcription termination and polyadenylation signal in the genome segments that is used in mRNA synthesis, but continues to the end of the template molecule. The production of the positive sense RNA is thought to occur by way of an anti-termination event, i.e. in the absence of a cap structure on the RNA being synthesized, NP protein acts in an unknown way to prevent termination at the polyadenylation signal, before the end of the template is reached. This may be achieved by direct interaction of the NP with the RNA and the PB and PA proteins in the nucleocapsid structure. This complex may be different depending on whether or not a cap has been used to initiate RNA synthesis. However, the details of this process are not yet fully understood.

The newly synthesized positive sense RNAs, which are present in nucleocapsid complexes, are used as templates for the production of negative sense ssRNAs, which in turn can be used for further rounds of replication prior to formation of new progeny virions.

The mechanism used to control the acquisition of genome segments by assembling influenza viruses is unknown. It is possible that each particle can package more than one copy of each segment into new virus particles. Recombinant viruses generated *in vitro* can be forced to accept additional segments, though this does not appear to be a stable situation and the additional segments are quickly lost. Such potential flexibility would mean that there is less need for specificity in packaging to ensure a complete complement of genome segments is present but the limit to the number of segments which can be packaged, and the relevance to the situation *in vivo*, is not clear.

7.6 SYNTHESIS OF THE RNA GENOME OF VIROIDS AND HEPATITIS DELTA VIRUS

Replication of viroids

The covalently closed ssRNA genomes of plant viroids do not encode any proteins and so there is no positive sense mRNA made. By convention, the strand of RNA which is most abundant in the infected cell is termed positive sense, though this has no practical meaning. When the naked RNA enters the cell it must be replicated by the proteins already present in the host plant. The enzyme most likely to be responsible for replicating the genome RNA is the host cell DNA-dependent RNA polymerase. It is not known how this enzyme functions on an RNA template, but this may be due, at least in part, to the extensive base-pairing structure of the genome RNA (see Section 4.6).

The replication of viroid genome RNA begins by adopting a rolling circle mechanism as described for circular DNA (see Section 6.5). The RNA polymerase II begins replication at a precise point on the genome but the nature of the initiation event is not known. The replication process generates a linear concatemeric RNA, of opposite sense to the genome, from which genome length RNA molecules are excised. The excision relies on the action of an unusual RNA sequence within the newly synthesized molecule, called a ribozyme. These ribozyme sequences adopt a complex three-dimensional structure and autocatalytically cleave the RNA at a specific site, generating genome length linear ssRNA molecules. These molecules are then converted into circles.

It is not known how circular plant viroid antigenome RNAs are formed. In the covalently closed circular conformation the RNA adopts an extensively base-paired rod-like structure, analogous to that of the infecting genome, and the ribozymes cannot adopt their active conformation. Consequently, the circular RNA is not cleaved. The circular antigenome RNA is used as template to produce more genomes by the rolling circle model. The genome sense RNA contains a ribozyme sequence which cleaves the concatemer to produce linear genome-length molecules which are circularized as before.

Hepatitis delta virus

Hepatitis delta virus (HDV) has a circular RNA genome of 1.7 kb. The genome is extensively base-paired and has a rod-like appearance in the electron microscope. Unlike the plant viroids, HDV encodes two proteins, the large and small delta antigens. These are produced from the same mRNA but the larger protein results from translation of a proportion of the mRNA molecules which is modified by a host cell enzyme to allow translation beyond the normal stop codon that terminates synthesis of the small delta antigen. This means that the two proteins share amino acid sequence but the large delta antigen is extended by 19 amino acids at the carboxy terminus. HDV can only replicate in cells also infected with hepatitis B virus. The process of HDV genome replication follows the same process as described for the viroids. The host cell RNA polymerase replicates the HDV RNA by a rolling circle mechanism and the linear genomes which are produced by cleavage of the concatemer by the HDV ribozyme sequence are circularized by an unknown process. The small delta antigen has been implicated in the initiation of replication, but this remains an area of debate and research. The genome is packaged into particles using hepatitis B virus structural proteins together with both delta antigens. The resultant particle is indistinguishable from those of hepatitis B virus.

KEY POINTS

- Viruses with RNA genomes use novel, virus-encoded enzymes to synthesize their RNA.
- Class 3 viruses have segmented genomes and each segment replicates independently using a polymerase contained within the virion. Replication occurs within virus-like structures in the cytoplasm of the infected cell.
- The genome of class 4 viruses is translated immediately after infection to synthesize the polymerase and other proteins.
- The picornavirus genome is held in a circular form by a combination of host and virus proteins.
- Class 5 virus RNA genomes have complementary sequences at the termini and these sequences are critical for replication.
- As with class 3 viruses, class 5 viruses carry a polymerase in the virion which enters the infected cell.
- The replication of RNA molecules, either whole genomes or individual genome segments, follows the same process of formation of a replication intermediate (RI) which consists of partially ssRNA and partially dsRNA in a complex with the replication proteins.
- Replication of viroids uses the host cell DNA-dependent RNA polymerase in a rolling circle process, followed by autocatalytic cleavage by ribozyme sequences and subsequent circularization of the new genome RNA.

QUESTIONS

- Compare and contrast the mechanisms of replication of the RNA genomes of viruses belonging to Baltimore classes 4 and 5.
- Discuss the different roles of reovirus mRNA in translation and as the template for genome replication.

FURTHER READING

Conzelmann, K-K. 1998. Nonsegmented negative-stranded RNA viruses: genetics and manipulation of viral genomes. *Annual Review of Genetics* **32**, 123–162.

Curran, J., Kolakofsky, D. 1999. Replication of paramyxoviruses. *Advances in Virus Research* **54**, 403–422.

Dimmock, N. J. 1991. The biological significance of defective interfering viruses. *Reviews in Medical Virology* **1**, 165–176.

Karayannis, P. 1998. Hepatitis D virus. *Reviews in Medical Virology* **8**, 13–24.

Lai, M. M. C., Cavanagh, D. 1997. The molecular biology of coronaviruses. *Advances in Virus Research* **48**, 1–100.

Marriott, A. C., Easton, A. J. 2000. Paramyxoviruses. In, reverse genetics of RNA viruses. *Advances in Virus Research* **53**, 312–340.

Portela, A., Digard, P. 2002. The influenza virus nucleoprotein: a multifunctional RNA-binding protein pivotal to virus replication. *Journal of General Virology* **83**, 723–734.

Roux, L., Simon, A. E., Holland, J. J. 1991. Effects of defective interfering viruses on viral replication and pathogenesis *in vitro* and *in vivo*. *Advances in Virus Research* **40**, 181–211.

Spence, N., Barbara, D. 2000. Viroids and other sub-viral pathogens of plants: the smallest living fossils? *Microbiology Today* **27**, 168–170.

Taylor, J. M. 1992. The structure and replication of hepatitis delta virus. *Annual Review of Microbiology* **42**, 253–276.

Taylor, J. M. 2003. Replication of human hepatitis delta virus: recent developments. *Trends in Microbiology* **11**, 185–190.

Also check Appendix 7 for references specific to each family of viruses.

8

The process of infection: IIC. The replication of RNA viruses with a DNA intermediate and vice versa

Some viruses switch their genetic material between RNA and DNA forms during their infectious cycles. The idea of DNA synthesis from an RNA template was once regarded as heresy to the doctrine of information flow from DNA to RNA to protein. However, it now has an established place in molecular biology. Indeed many mammalian genome DNA sequences (some pseudogenes, many highly repetitive sequences and certain types of transposable elements) are known to have been created in this way.

This chapter discusses the replication of the retroviruses and the hepadnaviruses, two important virus families that have their genetic information in both RNA and DNA forms at different stages of their life cycles. The form of nucleic acid packaged into particles differs between the viruses, being RNA in most retroviruses and DNA in hepadnaviruses. The process by which DNA is copied from an RNA template is known as reverse transcription. This step is essential in the replication of both virus families but, on its own, does not achieve genome amplification. Instead, an increase in genome number, suitable for progeny particle formation, only comes when RNA copies are transcribed from the DNA.

8.1 THE RETROVIRUS REPLICATION CYCLE

Important viruses in the retrovirus family

- *Human immunodeficiency virus types 1 and 2, members of the lentivirus genus in this family, are the cause of the global AIDS pandemic (see Chapter 19).*
- *Human T-cell lymphotropic virus type 1 is associated with a neuromuscular condition, tropical spastic paraparesis, and adult T-cell leukaemia (see Section 20.7).*
- *Animal retroviruses have been important model systems in studies to understand the events that occur during cancer development (see Section 20.6).*
- *Other retroviruses, and even HIV itself, are also important as gene therapy vectors – carriers to get therapeutic DNA into cells (see Section 23.2).*

The various stages in the cycle of retrovirus replication are considered in detail in the following sections of this chapter. However, it is useful to see the bigger picture of how these stages fit together before trying to understand them in detail. All retroviral particles contain two identical single-stranded genome RNA molecules, typically 8000–10,000 nucleotides in length, that are associated with one another (Box 8.1). These RNAs have the same sense as mRNA and also have the characteristic features of eukaryotic mRNA, with a 5′ cap structure and a 3′ poly-adenylate (poly-A) tail. Despite these features the genome RNA is never translated after the particle enters a cell. Instead, it is used as a template for the synthesis of a double-stranded DNA molecule. This event, known as reverse transcription, occurs in the cytoplasm within the incoming virus particle. The DNA then moves to the nucleus where it is integrated into the host genome. Only then can progeny genomes be produced by transcription of the DNA to give mRNA molecules. These can either be translated to give protein or packaged into progeny particles that then leave the cell, so completing the cycle.

Box 8.1

Evidence for dimerization of retroviral genomes

- Native genomes sediment with a size of 70S by ultracentrifugation whereas the size after denaturation is 35S.
- Electron microscopy analysis of several retroviral genomes shows similar pairs of RNA molecules linked together close to one end.
- Mutation analysis maps a dimerization function to one or more stem-loop structures near to the 5′ end of the genome RNA.

8.2 DISCOVERY OF REVERSE TRANSCRIPTION

Reverse transcriptases allow DNA copies to be created from RNA molecules. This process is crucial to the replication of important human pathogens, has shaped the structure of large parts of our own genome, and is key to the study of gene function in the laboratory.

The hypothesis of a DNA intermediate in retroviral replication was developed by Howard Temin in the 1960s (Box 8.2). His "provirus" theory postulated the transfer of the information of the infecting retroviral RNA to a DNA copy (the provirus) which then served as a template for the synthesis of progeny viral RNA. It is now clear that this theory is correct.

Temin's theory required the presence in infected cells of an RNA-dependent DNA polymerase or "reverse transcriptase" but, at the time, no enzyme had been found that could do this. If such an enzyme existed, then the retrovirus must either induce a cell to make it or else carry the enzyme into the cell within its virion. A search was begun for a reverse transcriptase in retrovirus particles, and David Baltimore and Howard Temin each independently reported the presence of such an enzyme in 1970, work for which they were subsequently awarded the Nobel prize.

Reverse transcriptase has since become a cornerstone of all molecular biology investigations because it provides the means to produce complementary DNA (cDNA) from mRNA in the laboratory and so allows cDNA cloning. It has also become clear that reverse transcription has played a big part in the shaping of the genomes of complex organisms such as ourselves. Within our DNA there are many intronless pseudogenes (non-functional genes) and also lots of repetitive elements, each of which has the hallmarks of a reverse-transcribed and integrated sequence.

Box 8.2

Evidence for a DNA intermediate in retroviral replication

- Infection can be prevented by inhibitors of DNA synthesis added during the first 8–12 hours after exposure of the cells to the virus, but not later.
- Formation of virions is sensitive to actinomycin D, an inhibitor of host RNA polymerase II which is a DNA-dependent enzyme.
- Rous sarcoma virus (a virus of birds) infection of cells* confers stably inheritable changes to their appearance and growth properties (see Sections 20.1 and 20.6). The details of these changes are virus-strain specific, indicating heritability of viral genetic information in the cells, a DNA property.

*Rous sarcoma virus infection is not cytolytic so the cells do not die.

Box 8.3

The reverse transcriptase (RT) protein

- The active form of the RT enzyme is a dimer formed of two related polypeptides (i.e. a heterodimer), but the details of its composition vary between retroviruses.
- RT from avian retroviruses is composed of α (60,000 M_r) and β (90,000 M_r) subunits. The β subunit comprises the α polypeptide with another enzyme, integrase, still linked to its C-terminus (for the production of retroviral proteins, see Section 9.9).
- RT from the human retrovirus, HIV, is fully cleaved from integrase and instead the subunits of the heterodimer differ by the presence (p66) or absence (p51) of the RNaseH domain of RT. Although the protein domains needed to form the reverse transcriptase/polymerase active site are present in both subunits, there is no catalytic activity in p51 because it adopts a very different structure from p66 in the heterodimeric enzyme (Fig. 8.1).

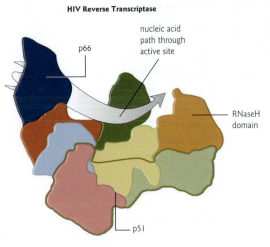

HIV Reverse Transcriptase

Fig. 8.1 Retroviral heterodimeric reverse transcriptase. The enzyme from human immunodeficiency virus is formed of p66 and p51 subunits. P66 has five structural domains, colored blue, red, green, yellow, and orange in order from the N-terminus. The last of these is the RNaseH domain, not present in p51. The remaining four domains found in p51 are colored in pale versions of the equivalent p66 domains. The structural arrangement of these domains in p51 is distinct from their organization in p66, where they form the active site for DNA synthesis. (Drawn from data presented in Ding C. H. *et al.* (1998). *Journal of Molecular Biology* **284**, 1095–1114.)

8.3 RETROVIRAL REVERSE TRANSCRIPTASE

The reverse transcriptase protein (RT) provides three enzymatic activities: (i) reverse transcriptase – synthesis of DNA from an RNA template; (ii) DNA polymerase – synthesis of DNA from a DNA template; and (iii) RNaseH activity – digestion of the RNA strand from an RNA : DNA hybrid to leave single-stranded DNA. An enzyme with RNaseH activity is also required to excise the primers during DNA replication (see Sections 6.1 and 6.2). For more detail about the RT protein, see Fig. 8.1 and Box 8.3.

The RT enzyme, like all other DNA polymerases, requires a primer from which to initiate DNA synthesis (see Section 6.1). The primer used to reverse transcribe the viral RNA is a host tRNA that is carried in the virus particle in association with the viral genome. The 3′ end of the tRNA is base-paired to the genomic RNA near to its 5′ terminus (Fig. 8.2). Each retrovirus contains a specific type of tRNA, e.g. Rous sarcoma virus has tryptophan tRNA and Moloney murine leukemia virus has proline tRNA. It is not known how these tRNAs are selected, although it is probable that the specific sequence of the genomic RNA at the tRNA binding site is the important determinant.

To perform its function in the virus life cycle, RT must enter a cell together with the genomic RNA. Mutant retroviral particles that do not contain active RT cannot establish proviral DNA in a cell even if the cell has been engineered to contain RT already. This indicates that reverse transcription occurs without full uncoating of the genome and explains why the genome, although having all the features of mRNA, is never translated; host ribosomes do not have access to it.

8.4 MECHANISM OF RETROVIRAL REVERSE TRANSCRIPTION

Comparing the structures of the genome RNA and proviral DNA

If you compare the sequences of a viral genomic RNA and a proviral DNA that is

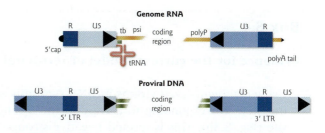

Fig. 8.2 Comparison of the structures of retrovirus genome RNA (top) and the proviral DNA created from it by reverse transcription. U5 and U3 are unique sequences at the 5′ and 3′ ends of virion RNA; R is a directly repeated sequence at the RNA termini. Short inverted repeat sequences are represented as ◀, ▶. tb is the binding site for a transfer RNA and polyP is a polypurine region, both significant in reverse transcription (Section 8.4 and Fig. 8.3). Long terminal repeats (LTRs) comprise duplications of the sequences U3, R, U5. psi is the specific packaging signal for RNA genomes. Not to scale.

copied from it, the two molecules are not precisely co-linear (Fig. 8.2). The directly repeated R sequences, found at the 5′ and 3′ ends of the genome RNA adjacent to the cap and polyA tail, are internal in the proviral DNA molecule. In other words, additional sequence has been added outside each R sequence in the double-strand DNA provirus as compared with the genomic RNA. Where do these additional sequences come from? A search of the genome RNA sequence shows that they are present, but are internal to the R repeats. The sequence U_5, which is copied to the 3′ end of the provirus, lies just inside the 5′ end of the genome, adjacent to R. Conversely U_3, which is copied to the 5′ end of the provirus, lies originally just in from the 3′ end of the genome, again adjacent to R. Therefore the process of reverse transcription has to duplicate sequences from one end of the genome and place the copy at the other end. The result is long directly repeated sequences at each end of the provirus, comprising the elements U_3, R, and U_5; these are known as the long terminal repeats or LTRs. The proviral DNA has one further significant feature. At the outer ends of the LTRs are short *inverted* repeats that derive from sequences originally present at the internal ends of the U_5 and U_3 sequences in the genome. These are important in retroviral integration (Section 8.6).

The model for reverse transcription to form proviral DNA

Figure 8.3 presents a model for proviral DNA synthesis, some evidence for which is highlighted in Box 8.4. The tRNA that is base-paired to the

Box 8.4

Evidence for the current model of retroviral reverse transcription

- Using detergent-lyzed virions in DNA synthesis reactions *in vitro*, a major product is a short fragment corresponding to a copy of the 5′ end of the genome from the tRNA binding site (Fig. 8.3b). This is termed negative strong-stop DNA.
- In infected cells, a discrete DNA intermediate is detected that corresponds to the (+)DNA fragment in Fig. 8.3g. This is termed positive strong-stop DNA.

genome provides the primer for synthesizing the DNA negative sense strand ((−)DNA). It is positioned exactly adjacent to the U_5 sequence (Fig. 8.3a). This positioning effectively defines U_5 as it is the sequence between this point and R that becomes duplicated during DNA synthesis. Synthesis initiated at the tRNA extends to the 5′ end of the template (Fig. 8.3b), forming an RNA : DNA hybrid duplex. This duplex is a substrate for the RNaseH activity of RT (Section 8.3), which degrades the template RNA that has been copied (Fig. 8.3c). As a result of this RNaseH action, the newly synthesized DNA sequence (R′) is now free to base-pair with another R sequence (Fig. 8.3d). This could be at the 3′ end either of the same RNA molecule (as shown) or of the second genome copy in the particle. This template switch is often referred to as a jump; once it has occurred, synthesis of the (−)DNA can then proceed along the body of the genomic RNA (Fig. 8.3e) with the RNaseH activity of RT continuing to degrade the template RNA as it goes (Fig. 8.3f).

A key question is then how synthesis of the positive sense strand of the provirus ((+)DNA) is begun, as this too needs a primer. A comparison of genome and provirus sequences predicts that (+)DNA initiation must occur immediately 5′ to the U_3 sequence, as this is the sequence that will form the 5′ end of the provirus. Adjacent to U_3 in the genomic RNA, all retroviruses have a conserved purine-rich region (polyP in Fig. 8.2). During degradation of the genome by RNaseH following (−)DNA synthesis, polyP is relatively resistant to degradation. It remains base-paired to the (−)DNA for long enough to provide the required primer for (+)DNA synthesis (Fig. 8.3f) but is eventually removed by RNaseH. Synthesis proceeds rightwards from this primer using the newly synthesized (−)DNA as a template (Fig. 8.3g). Some retroviruses, such as HIV, additionally prime (+)DNA synthesis from other RNaseH-resistant genome oligonucleotides and so produce a (+)DNA strand that is fragmented. These pieces are presumably joined together later by host DNA repair enzymes.

The 5′ end of the (−)DNA strand is still attached to the primer tRNA. RT uses the 3′ segment of this, which had been paired with the genome originally and therefore has the exact complementary sequence, as template for further (+)DNA synthesis (Fig. 8.3g). In the meantime, synthesis

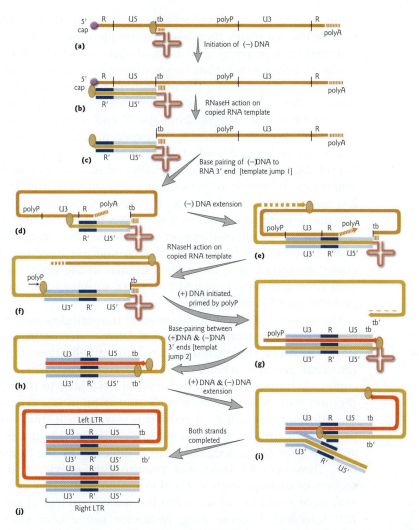

Fig. 8.3 (a–j) A scheme for the synthesis of retroviral linear proviral DNA by reverse transcriptase (Section 8.4 for details). Orange lines: RNA; yellow lines: (–)DNA; red lines: (+)DNA; green ellipses: reverse transcriptase enzyme. Arrowheads represent 3′ ends. The color coding of the U3, R, and U5 elements of the LTRs is carried forward from Fig. 8.2 to illustrate how the LTRs are created. Abbreviations and other symbols as for Fig. 8.2.

of the (–)DNA continues to the end of the available template RNA (Fig. 8.3g), which will be the sequence which originally bound the tRNA primer. Both of these RNA sequences have been spared from RNaseH degradation up to this point as they have been paired as an RNA : RNA hybrid, which is not a substrate for this enzyme, but once they have been reverse-transcribed they too are degraded. This exposes single-stranded DNA sequences at the 3′ ends of the (+)DNA and (–)DNA that are complementary (tb and tb′). These then base-pair (Fig. 8.3h), allowing RT to make the second template jump that occurs during reverse transcription. This jump gives the molecules of RT that are synthesizing (+)DNA and (−)DNA strands the templates they need to complete the synthesis of a double-stranded DNA provirus with LTRs at each end (Fig. 8.3i,j).

The existence of template jumps during reverse transcription suggested a possible explanation for the presence of two genome RNA copies in the particle. Perhaps the RT could not jump between ends of the same molecule but had to jump from the 5′ end of one genome copy to the 3′ end of the other. However, the experimental evidence is that proviral DNAs can be formed solely by intramolecular jumps, as shown here. It is probable though that the possibility of making intermolecular jumps minimizes the effects of genome damage on virus viability and so confers an evolutionary advantage. It would, of course, have been easier to describe reverse transcription if the tRNA primer binding site was at the 3′ end of the genome! However, this would have made the provirus an exact copy of the genome and so it would not have had LTRs. As discussed in Section 8.6, these are essential to the generation of progeny viral RNA genomes. It would also leave unsolved the "end-replication" problem (see Section 6.1).

The action of RNaseH during reverse transcription means that the process can only generate one proviral DNA molecule from each genome RNA, i.e. it is a conversion rather than amplification step. The cycle of retroviral genome replication is only completed when multiple RNA copies are transcribed from the proviral DNA later in the infectious cycle (Section 8.6). Nonetheless, reverse transcription is a crucial first step in the infection of a cell by a retrovirus – without it, the infection cannot progress any further. Thus it is not surprising that this process is an important target for drugs designed to treat infection by the human retrovirus, HIV. However, RT lacks any proofreading activity and so retroviral sequences undergo high levels of mutation with every cycle of infection. This rapid evolution of sequence is a major problem in the treatment of HIV infection, because drug resistance can evolve very rapidly (see Section 19.8). The resulting antigenic variation is also one of the reasons why an effective vaccine against this virus has not so far been developed.

8.5 INTEGRATION OF RETROVIRAL DNA INTO CELL DNA

Proviral DNA is produced in the cytoplasm within the partly uncoated virus particle. This complex of DNA with residual virion proteins, including the virus-coded integrase enzyme that entered the cell in the particle, is termed the pre-integration complex or PIC. It migrates to the nucleus where the DNA is integrated into cellular DNA. Successful integration requires the short inverted repeat sequences at the proviral termini (Fig. 8.2) and the integrase enzyme. As with RT, this enzyme must enter the cell in the infecting particle to be able to act; it cannot be added later.

There are three steps in the integration process (Fig. 8.4, Box 8.5). First, two bases are removed from both 3′ ends of the linear proviral DNA molecule by the action of the viral integrase (Fig. 8.4b). Second, the 3′ ends are annealed to sites a few (four to six) bases apart in the host genome (Fig. 8.4c). These sites are then cleaved by the integrase with the ligation of the proviral 3′ ends to the genomic DNA 5′ ends (Fig. 8.4d). This part of the integration reaction requires no input of energy from ATP, etc., because the energy of the cleaved bonds is used to create the new ones (i.e. it is an exchange reaction), and thus is reversible. Finally, gaps and any mismatched bases at the newly created junctions are repaired by host DNA repair functions (Fig. 8.4e); this renders the integration irreversible.

Infected cells typically contain 1–20 copies of integrated proviral DNA. There are no specific sites for integration, although there is some preference for relatively open regions of chromatin, i.e. regions with active genes. For most retroviruses, integration only occurs in cells which are moving through the cell cycle. This is thought to reflect an inability of the PIC to cross the intact nuclear membrane; breakdown of the membrane at mitosis therefore allows the PIC to access the host chromosomes. However, HIV does not suffer this restriction; proteins within its PIC can mediate nuclear uptake and hence allow infection of quiescent cells.

While most retroviral integration events during an infection will occur in somatic cells, at various points in evolutionary history there have clearly been retroviral integrations into the germ line of humans and other species. These events have led to the establishment of endogenous retroviruses that are inherited just like standard genetic loci. Most of these loci have suffered mutations that prevent the production of virus from them, but some are still capable of producing virus particles under appropriate conditions (see Section 23.3).

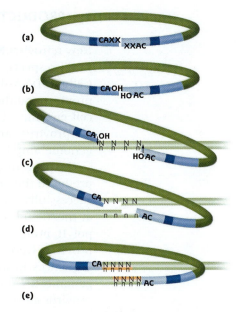

Fig. 8.4 (a–e) Integration of retroviral DNA into the host genome. For details see Section 8.5. Mid, dark, and light blue: the U3, R, and U5 elements of the retroviral LTRs; dark green: retroviral DNA; yellow: host DNA random integration target; dark pink: DNA repair synthesis.

Box 8.5

Evidence for the current model of retroviral integration

- Integrated proviruses always lack terminal nucleotides as compared with unintegrated DNA and the integrated DNA is always flanked by a short duplication of cellular sequence.
- Appropriate integration intermediates have been characterized from *in vitro* integration reactions.

8.6 PRODUCTION OF RETROVIRUS PROGENY GENOMES

New retroviral RNA genomes are transcribed from integrated proviral DNA. In all respects, this process resembles the production of cellular mRNA. Cellular DNA-dependent RNA polymerase II (RNA pol II) transcribes the provirus and the primary transcript is capped and polyadenylated by host cell enzymes. Clearly, many RNA copies can be transcribed from a single provirus, and it is this, rather than reverse transcription, that gives genome amplification during the replication cycle.

The progeny genomes must resemble exactly the parental genome that produced the provirus, otherwise the virus will not have reproduced itself successfully. To achieve this, the transcription start site must be exactly at the 5′ end of the R element within the left-hand LTR. However, RNA pol II promoters lie upstream of the transcription start point. If the provirus were simply a copy of the genome, there would be no viral sequences upstream of this start point to provide the promoter and so genome RNA synthesis would depend on fortuitous integration of the provirus adjacent to a host promoter. This is why the creation of the LTRs during reverse transcription is so important. With the creation of the left LTR, a virus-coded sequence, U_3, is placed upstream of the required start site at R and this provides the necessary promoter elements for RNA pol II. Equally, at the other end of the genome, the polyA addition site must be fixed exactly at the 3′ end of the R element within the right-hand LTR. Since sequences both upstream and downstream of a polyadenylation site are important in determining its position, it is again essential that proviral sequences extend beyond the intended genome 3′ end; these sequences are provided by the U_5 element within the right-hand LTR. Thus, the construction of the LTRs is crucial to retroviral replication. By virtue of the duplications of sequence that occur during reverse transcription, a retroviral genome manages both to encode, and to be encoded by, an RNA pol II transcription unit.

8.7 SPUMAVIRUSES: RETROVIRUS WITH UNUSUAL FEATURES

The spumaviruses of the retrovirus family (also known as the foamy viruses) have only been studied in detail relatively recently. Most work has been done on human foamy virus, although this virus is actually a chimpanzee virus which crossed into man as a dead-end zoonotic infection; there is no evidence for the existence of a genuine human foamy virus. Unlike the standard retrovirus replication cycle, reverse transcription in spumaviruses at least begins (and may even be completed) within assembling progeny particles before they are released from a cell. In other words, DNA synthesis occurs at the end of the replication cycle rather than at the beginning. Thus the genetic material within at least a proportion of

spumavirus particles is DNA rather than RNA. In this sense, the spumaviruses somewhat resemble the hepadnaviruses, which are also reverse-
transcribing viruses with DNA in their particles (Section 8.8). Other details
relating to gene expression (see Section 9.9) also suggest that the spumaviruses have similarity to both standard retroviruses and hepadnaviruses.
These differences from standard retroviruses mean that the spumaviruses
are now regarded as a subfamily of the retroviruses, the *Spumavirinae*,
rather than a genus in that family.

8.8 THE HEPADNAVIRUS REPLICATION CYCLE

Important viruses in this family

*Human hepatitis B virus creates a large global burden of chronic liver disease and is the cause of
many cases of hepatocellular carcinoma (see Section 20.8).*

Human hepatitis B virus (HBV) particles contain a partially double-
stranded circular DNA genome. This comprises two linear DNA strands
that form a circle through base-pairing (Fig. 8.5a). It is possible to des-
ignate positive and negative strands because all the genes are arranged
in the same direction (see Section 9.9). The (−)DNA strand is covalently
linked to the virus-coded P protein at its 5′ end and extends the full circumference of the circle and beyond. The (+)DNA strand overlaps the
5′–3′ junction of the (−)DNA and acts as a cohesive end to circularize the
genome; it is always incomplete.

When an HBV particle enters the cell, the genome is transported to
the nucleus where it is completed to give an intact double-stranded circle. This is then transcribed to give a variety of mRNA (see Section 9.9),
including one type that can be packaged into progeny capsids as an alternative to being translated. Once in the capsid, this pregenome RNA
serves as a template for reverse transcription, giving rise to the partially
double-stranded DNA that is found in particles after they have left the
cell. Not all the DNA-containing particles leave the cell. They can also
reinfect the nucleus of the cell that produced them. Since the infection
itself does not kill the cell, this process leads to an amplification of the
number of copies of viral DNA in the nucleus that are available for viral
gene expression and progeny production.

8.9 MECHANISM OF HEPADNAVIRUS REVERSE TRANSCRIPTION

Molecular analysis of HBV replication has been slow in coming as the
virus has been very difficult to grow in culture. However, a scheme for

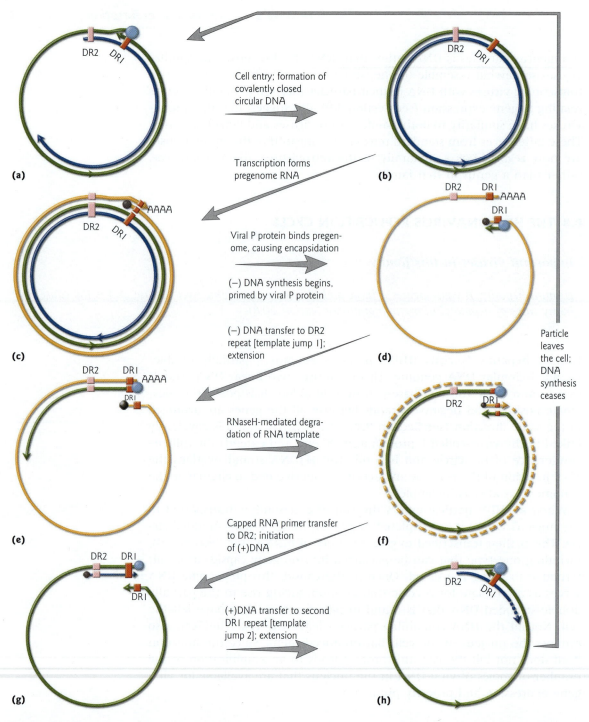

Fig. 8.5 (a–h) Replication of the hepadnavirus genome by reverse transcription (see Section 8.9 for details). Orange: RNA; blue: (+)DNA; green: (−)DNA; purple sphere: RNA cap structure; AAA: RNA polyA tail. DR1, DR2 represent two copies of a short directly repeated sequence in the genome. The viral P protein (terminal protein and reverse transcriptase) attached to the genome 5′ end is shown as a blue sphere. Filled arrowheads represent 3′ ends. Other arrows indicate the polarity (5′ → 3′) of nucleic acid strands.

Box 8.6

Evidence for the current model of hepadnavirus replication

- Formation of covalently closed viral DNA circles precedes the appearance of viral RNA.
- There is a substantial excess of (−)DNA over (+)DNA in infected cells.
- Synthesis of (−)DNA is insensitive to actinomycin D and some of this DNA can be found in RNA : DNA hybrid molecules.
- Radioactive precursor incorporation into viral DNA is associated with immature particles in the cytoplasm.

replication has been derived from studies involving HBV and its relatives, woodchuck and duck hepatitis viruses (Fig. 8.5), some evidence for which is summarized in Box 8.6.

After the HBV particle enters the cell, the DNA genome is transported to the nucleus, where the attached P protein is removed and both strands are completed and ligated by host DNA repair systems to give a covalently closed circle, CCC (Fig. 8.5b). The (−)DNA strand of the CCC then provides a template for transcription by host RNA pol II. Note that, in contrast to the retroviral provirus, there is no requirement for the hepadnavirus CCC to integrate into the host genome for it to be transcribed. Transcription produces various mRNAs (see Section 9.9) which are exported to the cytoplasm. The longest class of mRNA (Fig. 8.5c), which has a terminal repetition because it extends over more than the full circumference of the circular template, encodes the P protein and also serves as the pregenome, i.e. the template for genome DNA synthesis. P protein is multifunctional, with terminal protein, reverse transcriptase/DNA polymerase and RNaseH domains. As soon as it is synthesized, the P protein interacts with a sequence, known as ε, close to the 5′ end of the RNA which encoded it, and directs the packaging of this RNA into particles by core protein. Once this has occurred, creation of the DNA genome can begin.

The terminal protein domain of P protein serves as a primer for synthesis of (−)DNA by the reverse transcriptase domain of P. This explains why P is found attached to the 5′ end of this genome strand in the particle. DNA synthesis begins at the ε sequence in the pregenome RNA 5′ end (Fig. 8.5d). The short (−)DNA fragment produced then moves, with the associated P protein, to base-pair with the second copy of its template sequence at the RNA 3′ end. By doing this, the polymerase makes the first template jump of hepadnavirus reverse transcription. DNA synthesis then continues (Fig. 8.5e), with RNaseH activity degrading the template as it does so (Fig. 8.5f). A specific positive sense RNA oligonu-cleotide from the very 5′ end of the RNA, containing the repeat sequence DR1, is spared this degradation (Fig. 8.5f). It transfers to base-pair with the second repeat, DR2, near the 5′ end of the new negative strand, where it primes synthesis that extends to the very 5′ end of its template (Fig. 8.5g). To continue synthesis, a second template jump is then needed. The newly synthesized (+)DNA copy of DR2 switches its base-pairing to the second copy of DR1 at the other end of the (−)DNA template. Synthesis of (+)DNA can then continue onwards into the body of the genome (Fig. 8.5h). It is very unusual for this positive strand to be completed before the particle exits the cell, depriving the particle of substrates for DNA synthesis and so terminating further strand extension. Hence, the double-stranded genomes that are seen in virus preparations are normally incomplete.

8.10 COMPARING REVERSE TRANSCRIBING VIRUSES

Use of a reverse transcriptase is not restricted to the animal retro- and hepadnaviruses. The caulimoviruses are the only truly double-stranded DNA virus family in the plant kingdom. Investigation of the represent-ative virus, cauliflower mosaic virus (CaMV), has shown that it is a reverse-transcribing virus, with properties intermediate between retroviruses and hepadnaviruses. Like HBV, the CaMV genome is a double-stranded DNA circle, with a complete but gapped negative strand and an incom-plete positive strand. However, like retroviruses, negative strand DNA syn-thesis is primed by a host cell tRNA which base-pairs to the template RNA close to its 5′ end.

The retroviruses, hepadnaviruses, and caulimoviruses are in most senses completely unrelated; their protein coding strategies are different and only the retroviruses carry a specific integration function. However, some molecular aspects of their replication show considerable similarity. All three viruses have reverse transcription mechanisms which involve shifts or jumps of the extending polymerase from the 5′ end to the 3′ end of a template molecule, mediated through sequences repeated at the two ends of the template. Also, in each type of virus, host RNA polymerase II is used to produce RNA which serves as either genome or pre-genome. It is the timing of this event in the viral life cycle that varies, leading to the difference observed in the nature of the nucleic acid in mature virions. This variation can be seen even within a virus family, as the spumavirus subfamily of the retroviruses illustrates. Thus, in essence, the replication cycles of all these viruses are temporally permuted ver-sions of each other.

KEY POINTS

- Reverse transcriptase enzymes can use RNA as template to generate new DNA strands, and thus are able to reverse the classical flow of genetic information.
- Retroviruses and hepadnaviruses use reverse transcription as an obligatory step in their replication cycles and encode reverse transcriptases.
- Reverse transcription provides genome conversion for these viruses, not replicative amplification. This latter event is provided by RNA synthesis using the DNA created by reverse transcription as a template.
- Retroviruses use reverse transcription at the beginning of their replication cycle, immediately after entry into a cell, whereas hepadnaviruses use this process at the end of the cycle, within maturing virions.
- Integration of retroviral DNA (proviral DNA) is essential to the virus life-cycle and is catalyzed by a virus-coded integrase carried in the particle. By contrast, hepadnavirus DNA integration is not required and there is no specific mechanism provided for this to occur.

QUESTIONS

- Explain the molecular events that allow retroviruses to utilize the machinery of a eukaryotic host cell for mRNA production, despite the viruses having positive sense single-stranded RNA genomes.
- Compare and contrast the mechanisms of genome replication employed by hepadnaviruses and retroviruses.

FURTHER READING

Acheampong, E., Rosario-Otero, M., Dornburg, R., Pomerantz, R. J. 2003. Replication of lentiviruses. *Frontiers in Bioscience* **8**, S156–S174.

Delelis, O., Lehmann-Che, J., Saib, A. 2004. Foamy viruses – a world apart. *Current Opinion in Microbiology* **7**, 400–406.

Goff, S. P. 1992. Genetics of retroviral integration. *Annual Review of Genetics* **26**, 527–544.

Linial, M. L. 1999. Foamy viruses are unconventional retroviruses. *Journal of Virology* **73**, 1747–1755.

Nassal, M., Schaller, H. 1993. Hepatitis B virus replication. *Trends in Microbiology* **1**, 221–228.

Paillart, J. C., Shehu-Xhilaga, M., Marquet, R., Mak, J. 2004. Dimerization of retroviral RNA genomes: an inseparable pair. *Nature Reviews Microbiology* **2**, 461–472.

Seeger, C., Mason, W. S. 1996. Reverse transcription and amplification of the hepatitis B virus genome. In *DNA replication in Eukaryotic Cells* (M. DePamphilis, ed.), pp. 815–831. Cold Spring Harbor Laboratory Press, Cold Spring Harbor, NY.

Wilhelm, M., Wilhelm, F. X. 2001. Reverse transcription of retroviruses and LTR retrotransposons. *Cellular and Molecular Life Sciences* **58**, 1246–1262.

Also check Appendix 7 for references specific to each family of viruses.

9

The process of infection: IIIA. Gene expression in DNA viruses and reverse-transcribing viruses

The process of gene expression by the various DNA viruses closely parallels that of their host organisms, with many using host enzymes for transcription and translation. Most of the viruses impose a temporal phasing on their gene expression. Initially, their focus is on producing proteins either that are required for genome replication or that modify the host environment to make it more favorable for the virus to grow. Later, the emphasis switches to the production of large amounts of the proteins needed to form new virions.

To be expressed, the genomes of DNA viruses must be transcribed to form positive sense mRNA, and this must then be translated into polypeptide. For viruses of eukaryotes, further steps intervene: RNAs are usually capped and polyadenylated, and may also need to be spliced, to give functional mRNA. If it has been produced in the nucleus, the mRNA must be moved to the cytoplasm to allow translation to occur. Each stage in the pathway of gene expression, from transcription of RNA through to post-translational modification of protein, represents a potential point of control and DNA viruses exploit these possibilities in various ways so that each one achieves an organized program of gene expression. Retroviruses, which have RNA genomes, also carry out gene expression from a DNA intermediate (see Chapter 8). Thus their gene expression processes have considerable similarity with those of some DNA viruses.

For most DNA viruses, gene expression is phased, with early genes being expressed before DNA synthesis begins and late genes only being activated after this event. In some cases, further temporal divisions are also apparent. These patterns of expression have been characterized in considerable detail and as well as furthering our understanding of virus growth cycles, such studies have increased greatly our understanding of the molecular biology of eukaryotic cells. This chapter, and the accompanying one on RNA virus gene expression (see Chapter 10), demonstrates these points by examining gene expression and its control in a variety of virus systems.

9.1 THE DNA VIRUSES AND RETROVIRUSES: BALTIMORE CLASSES 1, 2, 6, AND 7

All DNA viruses synthesize their mRNA by transcription from a double-stranded DNA molecule; those from Baltimore class 1 can express their genes directly while classes 2 and 7 have to convert their genomes to double-stranded form or complete the second strand before transcription begins. The retroviruses of class 6 can also be considered here since they must create a dsDNA version of their genome RNA before transcription can take place. For each of the viruses featured in this chapter, there is detailed information available about the specific mRNAs produced, although this information is only included here for the simpler viruses. Evidence for these viral transcription patterns has come from applying a common set of techniques (Box 9.1).

Among those viruses that infect eukaryotes, all except the pox-, irido- and asfarviruses carry out transcription in the cell nucleus. Thus, for the majority of DNA viruses, the cell's own transcription and RNA

Box 9.1

Evidence for detailed patterns of viral transcription

- The viral mRNA species that come from each part of the genome are revealed by probing Northern blots of infected cell RNA with labeled probes from each genome region.
- The colinearity of mRNA and genome can be measured by R-loop mapping in the electron microscope.
- Precise mRNA 5′ and 3′ end positions on the genome, also the positions of splice donor and acceptor sites, are determined by S1 nuclease mapping or RNase protection analysis. Both assays use the encoding DNA (or its derivative) as a probe and map the point of discontinuity between the mRNA and the probe.
- Primer-extension analysis can also be used to map RNA 5′ ends.
- Promoter elements in the genome capable of directing transcription are detected and mapped using reporter gene assays (see Box 9.5).

modification machinery is available and the viruses make use of this to produce their mRNA. These viruses are therefore good model systems for studying the synthesis of cellular mRNA, and several significant milestones in understanding eukaryotic gene expression have come from this work. For example, transcription enhancers and transcription factors which regulate RNA polymerase II (RNA pol II) activity were first identified during studies of SV40 gene expression, and splicing was discovered by analyzing adenovirus mRNA.

The amount of genetic information varies by more than 100-fold between different DNA virus families, as considered in Chapter 6. Some have severe restrictions on their coding capacity and depend greatly on their host for essential replicative functions, despite having evolved to maximize their use of the available genetic information, whereas others have had the opportunity to evolve or acquire many functions that are not essential for virus multiplication. The larger viruses are less dependent upon the cell and can, for example, multiply when cells are not in S phase (synthesizing cellular DNA), whereas the smallest viruses are unable to do so. These larger viruses can also use inhibition of host macromolecular synthesis as a strategy to target all of the host's resources towards their own replication. Several of the viruses have the capacity to induce the host cell to enter S phase to promote their replication. Such functions may not be crucial in rapidly growing experimental cell cultures, but are probably essential during natural infections, when the virus infects cells that are not dividing. As discussed in Chapter 20, several of the DNA animal viruses are capable of transforming cells. Transformation can result when the virus functions that regulate cell division become permanently expressed in cells that have not died as a result of virus infection.

9.2 POLYOMAVIRUSES

The polyomaviruses, e.g. SV40, use host mechanisms to achieve temporally phased expression from their two transcription units. The SV40 enhancer was the first DNA element of this type to be recognized and production of the two SV40 early proteins was the first characterized example of differential splicing. It is now clear that both are widespread in host gene expression. The SV40 early protein, large T antigen, is a classic example of a multifunctional protein.

Polyomavirus (of mice) and SV40 (of monkeys) exemplify the polyomavirus family; there are also related human viruses BK and JC. These viruses each have double-stranded circular genomes, of around 5.3 kbp in length, which encode early and late genes on opposite strands of the template, each occupying about half the genome (Fig. 9.1). The genes are transcribed by RNA pol II to produce a single primary transcript in each case; these are then differentially spliced to form multiple mRNA species.

Differential splicing means that there is more than one accept-able splicing pattern for the primary transcript. Which pattern is used on a specific RNA molecule will depend on the relative affinity of the splicing machinery for the alternative splice sites available on the molecule. Altering the balance of use of altern-ative splice sites is a potential method for regulation of gene expression during infection (see adenoviruses and influenza viruses, Sections 9.4 and 10.7 respectively), although it does not appear to operate in the polyomaviruses.

The two mRNA from the SV40 early region encode a major early protein, the large T (for tumor-specific) antigen, and a second protein called small t (Fig. 9.2a); BK and JC viruses are similar. Each protein is made from a discrete spliced mRNA (Box 9.2). The polyoma early region codes for three proteins, called large T, middle T, and small t antigen, using a similar strategy to SV40 (Fig. 9.2b). The late mRNAs produced by each virus also encode three virion proteins, VP1, VP2 and VP3, by differential RNA splicing (Fig. 9.3). VP1 is the major capsid protein, with VP2 and VP3 being minor components. There is no viral protein produced to complex with the DNA in the virion; histones H2A, H2B, H3, and H4 from the host cell are used for this purpose. The Agno protein (whose gene gained its name because, when it was defined, it had no known protein

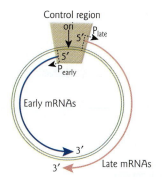

Fig. 9.1 Map of the genome of SV40 virus showing the positions of the early and late transcription units. P denotes promoters of transcription and ori denotes the origin of replication. The mRNAs are shown in detail in Figs 9.2 and 9.3, and the control region in Fig. 9.4.

Box 9.2

Expression of the SV40 early proteins

- The longer early mRNA is translated to give the 174 amino acid small t protein; a short intron, which is removed during processing of the primary transcript for this mRNA, lies downstream of the small t antigen open reading frame (ORF).
- The shorter early mRNA is formed by removal of a larger intron, which includes the C-terminal half of the small t antigen ORF. This alternative splicing event fuses together the first 82 codons of the small t antigen ORF in frame with 626 codons from a much longer ORF located downstream. Translation from this mRNA generates the 708 amino acid large T antigen.
- Although the 626 amino acid ORF is also present in the small t mRNA, it has no initia-tion codon and so is not accessed by ribosomes.
- SV40 large T antigen is an excellent example of a multifunctional protein, with roles in DNA replication, viral gene expression, and modulation of host functions. These are medi-ated through a combination of intrinsic enzyme activity and the ability to interact with viral DNA and an array of host cell proteins.

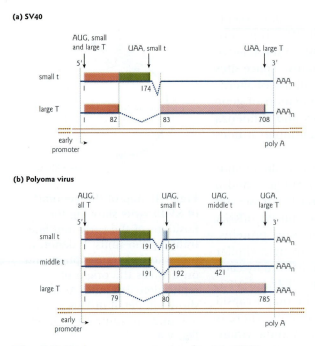

Fig. 9.2 Early gene expression by (a) SV40 and (b) polyoma virus. Differential splicing of the primary transcript in each case produces mRNAs with distinct coding capacities (see text and Box 9.2 for details). Red lines denote DNA and blue lines, RNA; dotted lines denote introns and thin lines untranslated RNA regions. Proteins are represented in linear form, with color coding to indicate the sequences that are common between two or more proteins and numbers indicating amino acid residues.

product) functions very late in the virus life cycle; its role is not yet well understood.

The SV40 early and late promoters lie close together, flanking the origin of replication, in a region of the genome known as the control region (Fig. 9.4, Box 9.3). The early promoter is activated by host cell transcription factors and by its product, large T antigen, when this is present at low concentrations. However, in larger amounts, large T antigen inhibits early gene expression by binding to sites 1, 2, and 3 in its own promoter, contributing to the transition from early to late gene expression; binding to site 2 is also required to initiate DNA replication (see Section 6.2). The early promoter is controlled by an enhancer, the first of this type of control element to be described. The late promoter is activated by DNA replication and by T antigen, which acts here through its effects on host cell transcription factors. As well as its functions in DNA replication (see Section 6.2) and in controlling viral gene expression, large T antigen (together with small t) is able to alter host gene expression and hence activate cell cycle progression, driving cells into S phase. These effects on cell cycle regulation are considered in Section 20.3.

9.3 PAPILLOMAVIRUSES

The papillomaviruses all infect epithelial surfaces in the body. They show phasing of gene expression not only through the time course of virus reproduction but also spatial phasing of expression through the different layers of the epithelium. The mechanisms of gene expression are, like those of SV40, similar to those of the host organism.

Papillomaviruses are exemplified by bovine papillomavirus type 1 (BPV) and the many types of human papillomavirus (HPV). Details of the gene expression of these viruses have been much harder to work out than those of the polyomaviruses because of the difficulty of growing them in cell culture systems (Box 9.4). A genome map of the widely studied bovine papillomavirus type 1 is shown in Fig. 9.5. As with the polyomaviruses, all transcription occurs in the nucleus mediated by host RNA pol II, however a striking difference from SV40 is that the genes are all organized in the same orientation, i.e. only one genome strand is transcribed. The slightly longer genome (8 kbp) encodes probably 11 proteins and mRNA for these is produced from seven different promoters. Transcription for the early and late genes is largely overlapping, covering most of the genome, but they are expressed from different promoters (six early and one late). Alternative splicing and polyadenylation are used to give multiple mRNAs from a primary transcript, especially in the case of the late transcripts where three dif-

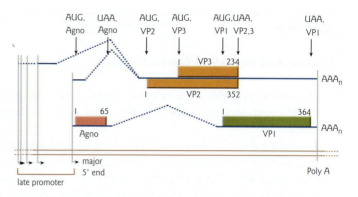

Fig. 9.3 Late gene expression by SV40 virus. The late mRNAs have heterogeneous 5′ ends because the late promoter lacks a TATA box, which normally serves to direct initiation to a specific location. Legend details as for Fig. 9.2. Polyoma virus is very similar but the virion protein (VP1) coding region does not overlap with that of VP2/VP3.

ferent proteins are encoded (E4, L1, and L2). The HPVs show an essentially similar picture, but with some detailed differences.

Regulation of BPV transcription occurs through the E2 gene, whose full length product, E2TA dimerizes to form a DNA-binding transcription factor; this same protein also facilitates E1 protein binding to DNA during replication (see Section 6.2). There are two E2-dependent enhancer elements in the long control region (Box 9.5), and also binding sites associated with most/all of the viral promoters. As well as E2TA, two truncated E2 forms are produced, one initiated within the E2 reading frame (E2TR) and the other an alternatively spliced form (E8E2); each contains only the C-terminal half of the E2TA protein. E2TR and E8E2 can form homo- and heterodimers, with each other or E2TA, which retain specific DNA binding activity but lack the transcription activation function of the E2TA dimer. Thus the truncated forms of E2 modulate the level of E2-mediated transactivation in two ways: by inhibiting the formation of active E2TA through competition for dimerization; and by competing with active dimers for DNA binding.

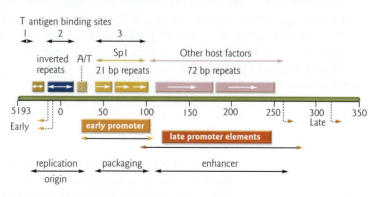

Fig. 9.4 The control region from laboratory strains of SV40. A number of sequence features are evident in the DNA and these correspond to functional units as indicated. Sites 1, 2, and 3 are three DNA regions, each containing multiple binding sites for large T antigen.

Box 9.3

Features of the SV40 control region

- The early promoter includes a TATA box which recruits general transcription factor TFIID, and binding sites for the host cell transcription factor Sp1 within the 21-bp repeats.
- The late promoter lacks a TATA box and hence there are start points for transcription spread over some distance on the genome.
- The enhancer comprises the 72-bp repeats plus some adjacent sequences, but only one of the two repeat copies is needed for activity. Wild strains of SV40, in contrast to laboratory-adapted isolates, only contain one copy of this repeat element.
- The inverted repeat at the center of the origin of replication, with its four binding sites for T antigen, is essential for DNA replication.
- Progeny genome packaging is initiated by an interaction between capsid proteins and Sp1 bound to the 21-bp repeats.

Box 9.4

Systems for studying papillomavirus gene expression

- Transfection of cloned DNA into permanent cell lines has revealed the mRNAs expressed in the early phase of gene expression.
- Analysis of RNA from wart tissue identified novel mRNAs made only during productive infection.
- *In situ* hybridization to sections through a wart showed that so-called early mRNAs are produced in different regions of the lesion from late mRNAs.
- The full productive life cycle of papillomavirus can now be achieved, on a small scale, in epithelial raft cultures (cells grown on a solid support at an air–medium interface that form a pseudo-epithelium).

The biology of papillomavirus gene expression is particularly fascinating. These viruses cause warts, i.e. benign growths in epithelia, and expression of their early and late genes is separated in time and space. In infected epithelia, early genes are expressed in the dividing cells of the basal cell layer and, as the cells divide, the viral DNA is maintained at an approximately constant copy number per cell by limited replication. The E6 and E7 proteins act to alter cell cycle control within these cells using mechanisms similar to those shown by SV40 large T antigen (see Section 20.3). The late events (accelerated replication and late protein

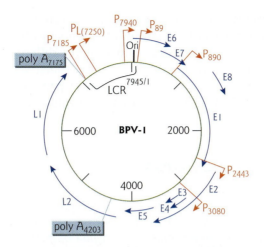

Fig. 9.5 The genome organization of bovine papillomavirus type 1. E1–E8, L1, and L2 represent open reading frames. Reading frames were originally designated E or L based on the observation that a 69% fragment of the genome was sufficient for cell transformation *in vitro* and was hence defined as the early region by analogy with SV40. However, within the early region, the E4 gene is now known to be expressed primarily within the late temporal class of mRNA. Primary transcripts can undergo differential splicing to produce multiple mRNAs (not shown), that in some cases fuse parts of reading frames together. The region lacking open reading frames is known as the long control region (LCR). P_{89} etc., RNA polymerase II promoters; poly A, sites of mRNA polyadenylation; Ori, origin of DNA replication.

Box 9.5

Evidence that the BPV E2 protein is a sequence-specific transcription factor

- Reporter gene assays test for promoter or enhancer activity in pieces of DNA, by linking them in a plasmid with the coding sequence of a protein that is easy to assay, and measuring the amount of protein produced when the plasmid is introduced into cells.
- Using the long control region (LCR) DNA of BPV as an enhancer to increase expression of the reporter gene chloramphenicol acetyl transferase (CAT) from a very weak promoter, significant CAT expression was only found when the cells also contained BPV DNA, i.e. the LCR could act as an enhancer provided one or more proteins from BPV was present.
- Using a series of plasmids in place of the complete BPV DNA, each designed to express just one of the BPV proteins, the E2 gene alone was found to cause upregulation of CAT expression.
- The response of the CAT reporter construct to E2 protein was shown to depend on a specific sequence motif in the LCR by creating a series of LCR fragments containing different sequence deletions and testing each one for enhancer activity in the assay.
- Purified E2 protein produced in bacteria was found to bind specifically to the sequences from the LCR already shown to be crucial for enhancer activity.

synthesis) only begin once an infected cell has left this layer and is committed to terminal differentiation. Thus the pattern of expression of the virus genes is determined by the differentiation state of the host cell. The factors which turn on this so-called vegetative phase in the growth cycle are not known. The human papillomaviruses are also of interest for their involvement in the development of certain human cancers (Section 20.4).

9.4 ADENOVIRUSES

Adenoviruses have been widely studied, not just because some of them are significant human pathogens, but also because of what they can tell us about the workings of mammalian cells. The first description of RNA splicing was the recognition that the 5′ region of mRNA from the adenovirus major late transcription unit was not collinear with the genome. Detailed study of the E1A proteins has been crucial to shaping our understanding of how transcription can be controlled in eukaryotes. Mechanisms of adenovirus gene expression resemble those of the host, but differential RNA processing is taken to an extreme, with large numbers of distinct mRNAs and hence proteins arising from most of the transcription units.

Adenovirus gene expression has been best studied in the closely related human serotypes 2 and 5. In these viruses, transcription occurs from both strands of the 36-kbp linear genome. All mRNA is produced by host RNA polymerase II; there are also two short RNAs (the VA RNAs) produced by RNA pol III. Gene expression shows temporal regulation (Fig. 9.6a) and the terms early (pre-DNA synthesis) and late (post-DNA synthesis) were used originally to classify the genes as E or L. However, it is now appreciated that expression of the E1A gene commences before the other early genes, and so should be classified as immediate-early, and that the small IVa2 and IX protein genes form an intermediate class rather than being true late genes. The observation which defines E1A as "immediate early" is that, unlike the other early genes, it is transcribed in an infected cell when protein synthesis is blocked by a chemical inhibitor. This result also shows that E1A proteins are needed for expression of the remaining viral early genes. Controlling promoter activity is just one of the ways in which the pattern of gene expression is modulated during adenovirus infection. Some of this complexity is summarized in Box 9.6, although a detailed discussion is beyond the scope of this book. Overall, the mechanisms achieve production of each of the viral proteins at the time, and in the amount, it is required.

The full gene expression map of adenovirus 5 is complex and can appear daunting (Fig. 9.6b). However, the key principle is that a relatively modest number of transcriptional promoters drive expression of a much larger number of mRNAs (which then encode different proteins) through differential splicing and polyadenylation. In other words, several distinct

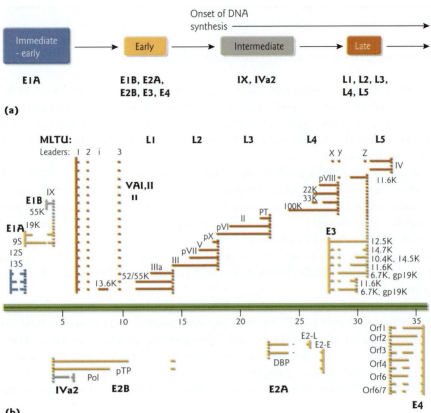

Fig. 9.6 Gene expression by adenovirus type 5. (a) The phases of gene expression. The numbers E1A, L1, etc. refer to regions of the viral genome from which transcription takes place. (b) A transcription map of the adenovirus 5 genome. The genome is represented at the center of the diagram as a line scale, numbered in kilobase pairs from the conventional left end, with rightwards transcription shown above and leftwards transcription below. Genes or gene regions are named in boldface. Promoters of RNA polymerase II transcription are shown as solid vertical lines and polyadenylation sites as broken vertical lines. VAI and VAII are short RNA polymerase III transcripts. Individual mRNA species are shown as solid lines, color-coded according to the temporal phase of their expression (panel a), with introns indicated as gaps. The protein(s) translated from each mRNA is indicated above or adjacent to the RNA sequence encoding it. Structural proteins are shown by roman numerals (major proteins: II, hexon; III, penton; IV, fiber; pVII, core; see Chapter 11). PT, 23K virion protease; DBP, 72K DNA binding protein; pTP, terminal protein precursor; Pol, DNA polymerase. (Reproduced with permission from Leppard, K. N. (1998) *Seminars in Virology* **8**, 301–307. Copyright © Academic Press).

open reading frames are arranged downstream of a single promoter and, after transcription occurs, splicing is used to bring a downstream reading frame up to a 5′ proximal position in the RNA, so that it can be translated (according to the normal rules of translation in eukaryotes, only the 5′ proximal open reading frame in a mRNA is expressed). Picking out some of these features in Fig. 9.6b in detail, each of the five early (E) regions

Box 9.6

Control of adenovirus gene expression at multiple levels

- Transcription is regulated so it takes place in an ordered sequence. The E1A proteins, especially the 13S mRNA product (see Fig. 20.3), activate expression of the remaining genes, but significant transcription from the intermediate and late genes awaits the onset of DNA synthesis; the molecular mechanism of this switch is not fully understood. Host transcription factors are also important.
- RNA processing is regulated so that the relative proportions of the possible mRNA products of a gene change as the infection proceeds. For example, when the major late gene is first transcribed, RNA processing produces L1 52/55K mRNA almost exclusively while later on, mRNA is produced from all of the regions L1–L5 and the predominant L1 mRNA is that which encodes the IIIa protein.
- Movement of mRNA within the cell is regulated so that viral mRNA reaches the cytoplasm in preference to cellular mRNA. Early proteins from E1B and E4 are required to achieve this.
- Translation is switched from a cap-dependent to a cap-independent mode during the late phase of infection through inactivation of cap recognition. This shuts off host protein synthesis whilst permitting continued viral late protein production since the late viral mRNAs, although capped, can recruit ribosomes in a cap-independent manner due to features of their 5′ untranslated regions.

has its own promoter. Two of them (E2 and E3) have two alternative polyadenylation sites A and B, leading to two families of mRNA in each case, while transcription from the major late transcription unit (MLTU) produces five families of mRNA (L1–L5), as a result of polyadenylation at any one of five possible sites. At the same time, alternative splicing within each mRNA family produces multiple mRNAs, each ending at the same polyadenylation site. Except for E1A, where the principal products are closely related in sequence (see Section 20.3), these mRNAs generally encode unrelated proteins.

The multiple proteins encoded by an adenovirus gene region tend to have related functions. The E2 region, for example, encodes the three viral proteins which are directly involved in DNA replication (see Section 6.4), while splice and polyA site choice within the five families of MLTU mRNA allows synthesis of at least 14 different proteins, mostly involved in forming progeny particles. The E3 region is the most variable when different adenoviruses are compared. It encodes proteins involved in avoiding aspects of the host immune response.

RNA processing from the MLTU is particularly wasteful of the cell's resources; each late mRNA, each one 1–5 kb long, is made from a precursor up to 28 kb in length. There is no possibility of making multiple mRNAs from the same precursor molecule as all the mRNAs include the same three RNA segments from the precursor's 5′ end; these are spliced together to form the so-called tripartite leader sequence. Thus, a lot of newly synthesized viral RNA, both from the major late gene and other genes, is removed as intron sequence and degraded. However, the advantage of this system is that the virus can achieve coordinated expression of functionally related proteins using a minimum of genome space to direct transcription.

9.5 HERPESVIRUSES

Herpesviruses, including the eight known human viruses, all establish life-long latent infections in specific tissues. To achieve this, they have to have two alternative programs of gene expression, for lytic infection and latency, both of which employ host mechanisms for mRNA production but with very little splicing involved in the lytic program. Whilst the productive or lytic program is probably quite similar for all herpesviruses, the latent programs are completely different. The fact that reactivation of herpesvirus infections, such as the periodic appearance of cold sores due to herpes simplex virus type 1, is a significant clinical problem means it is important to try to understand how these programs of gene expression are controlled. The process of latency is considered in Chapter 15.

Herpes simplex virus types 1 and 2

Herpesvirus genomes are very much larger than those of adenoviruses and polyomaviruses, and show considerable diversity in length (130–230 kbp) and hence gene content. However, there are conserved blocks of genes that are common among the herpesviruses and that are probably key genes in virus reproduction. Herpesvirus gene expression has been studied principally using herpes simplex virus type 1 (HSV1), which productively infects epithelial cells *in vivo* and shows a productive, cytolytic infection of cell lines in culture. Gene expression during HSV1 lytic infection is considered here while gene expression in relation to latency, which is very different, is considered in Chapter 15.

HSV1 has around 70 genes interspersed on both strands of its 153-kbp genome. An immediate contrast with the smaller DNA viruses is that transcription of most HSV1 genes is controlled separately by specific promoter elements and there is very little use of splicing during mRNA production. A further contrast is that about half of the genes can apparently be mutated without affecting the growth of the virus in cell culture. These dispensable genes are presumed to be important *in vivo*.

As with the other animal DNA viruses, HSV1 gene expression is arranged into temporal phases. These are generally termed α, β, and γ, corresponding to immediate early, early, and late phases. α, β, and γ genes are not grouped together but are found scattered throughout the genome. The α genes of

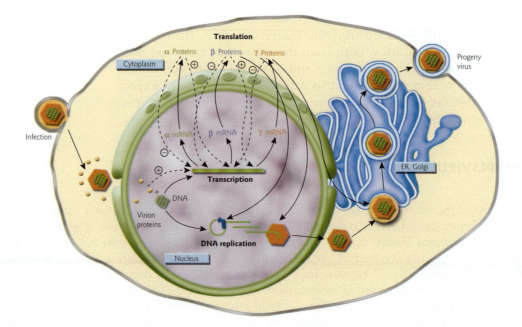

Fig. 9.7 Gene expression in herpes simplex virus type 1 infection. The three successive phases of gene expression are denoted α, β, γ corresponding to immediate-early, early, and late (see text). Solid arrows indicate the flow of material; dashed arrows indicate regulatory effects on gene expression; + indicates activation; – indicates repression.

HSV1 produce various regulators of gene expression, the β genes produce proteins required for DNA replication, and the γ genes produce mainly structural proteins for progeny particle formation. Although expression of the γ genes accelerates with the onset of replication, it is in fact more or less independent of this process. Thus, when DNA synthesis is inhibited, both early and late transcripts are detected in the nucleus.

The different phases of HSV1 gene expression are linked through the production of proteins in one phase which activate the next (Fig. 9.7). Thus, in the same way that adenovirus E1A proteins activate other viral genes, so an HSV1 α gene product, known variously as ICP4 or IE175, turns on transcription of the β genes. One or more β gene product(s) then inhibit expression of α genes and induce γ gene expression, proteins from which inhibit expression of the β genes. A particularly interesting feature of the cascade of regulated HSV1 gene expression is that it is circular. A protein produced during the late phase of infection (VP16, a γ gene product) is packaged into the particle, in the layer between the capsid and the envelope that is known as the tegument, and serves as an activator of α gene expression in the next round of infection (Box 9.7). Overall, this pattern of gene expression, with a series of genes controlled by activators and inhibitors, ensures that the relevant proteins are present when required and enhances the efficiency of the virus replication cycle.

> **Box 9.7**
>
> **Action of the tegument protein VP16 in activating herpes simplex virus α genes**
>
> - VP16 has a potent transcriptional activation function when recruited to a promoter.
> - Two host proteins, Oct1 and HCF, are required for VP16 to upregulate α gene expression.
> - HCF associates with VP16 and mediates its transit into the nucleus.
> - Oct1 is a DNA binding transcription factor that binds to α gene promoters adjacent to a VP16 binding site. VP16/HCF complex can only bind DNA in the presence of Oct1.
> - Why the virus has evolved this indirect mechanism for activation by VP16 is uncertain; it may relate to the switch into and out of latency.

The tegument of HSV1 also carries at least one other active protein, UL41, also known as vhs, standing for virion host shutoff. This protein is a ribonuclease that degrades cytoplasmic mRNA that is being translated. It is not specific for host mRNA, but serves to decrease the half-life of all mRNA in the cell, so favoring the expression of protein from genes that are being actively transcribed, i.e. viral genes. Later in the infection, another more specific host shut-off function is expressed by the virus while the large amounts of UL41 that the ongoing infection produces are kept inactive via binding to VP16, so allowing late viral mRNA to accumulate.

Epstein–Barr virus

Like HSV1, Epstein–Barr virus (EBV) productively infects epithelial cells. However, it establishes latency in B lymphocytes rather than sensory neurons (see Sections 13.3 and 15.6). The molecular events of the productive phase of the EBV life cycle are poorly characterized, because productive infection is hard to achieve in cell culture, however, they are believed to resemble those defined for HSV1. By contrast, the pattern of EBV gene expression during latency is well defined; this is described in Chapter 15.

9.6 POXVIRUSES

Poxviruses replicate in the cytoplasm and are independent of the host for mRNA production. Like all viruses, they use host ribosomes for protein synthesis.

Unlike the other families of DNA viruses of eukaryotes discussed in this chapter, poxviruses multiply in the cytoplasm. The molecular details of their gene expression have been studied using vaccinia virus. Its linear double-stranded genome (190 kbp) encodes a large number of proteins and these allow the virus to replicate with a considerable degree of autonomy from its host. It appears that viral gene expression is totally independent of the host nucleus since infection can proceed in experimentally enucleated cells. This cytoplasmic life cycle requires that the infecting virion carries the enzymes needed for mRNA synthesis, including a DNA-dependent RNA polymerase and enzymes that cap, methylate, and polyadenylate the resulting mRNAs. Splicing activities are not known to be used by these viruses.

Using the enzymes from the infecting particle, viral transcription begins in cytoplasmic replication "factories." Only a subset of viral genes (the early genes) is transcribed initially. Early gene products include proteins needed for replication and to activate the intermediate genes, which then activate late genes. Thus, infection leads to a phased sequence of gene expression punctuated by viral DNA synthesis, much as is seen for the other DNA viruses. Independence from the cell's own transcription machinery allows the virus the opportunity to shut the cell nucleus down, so that all the metabolic resources of the cell are devoted to the virus. Thus, unlike the other families of DNA viruses considered here, all of which can establish chronic, persistent or latent infections *in vivo*, poxviruses are exclusively cytolytic. Poxviruses make a large array of proteins involved in modulating the host immune response to infection.

9.7 PARVOVIRUSES

Adeno-associated virus (AAV) is the best characterized parvovirus. Its linear single-stranded genome, once converted to double-stranded form (see Section 6.6) contains three promoters, each of which produces at least two related mRNAs by differential splicing (Fig. 9.8). The proteins produced by expression from P_5 and P_{19} (the Rep proteins) are all sequence-related and can be regarded as early proteins, because they activate the third promoter, P_{40}. The unique functions of the Rep proteins have not yet been fully defined but include the site-specific DNA cleavage activity which is essential to genome replication (see Section 6.6). The P_{40} mRNAs code for the structural proteins, VP1, 2, and 3, from which new virions are assembled. These three proteins are all sequence-related, the major capsid protein being VP3. This protein is produced by internal initiation at an AUG codon within the shortest mRNA. Read-through to this initiation point occurs in the majority of translation events because the

Fig. 9.8 Adeno-associated virus gene expression. The 4.6-kb genome is represented as a line scale, with mRNAs beneath it with $5' \to 3'$ polarity from left to right. ITR, inverted terminal repeat (see Section 6.6). Encoded proteins are shown as colored boxes: blue – early; green – late. See Section 9.7 for further details.

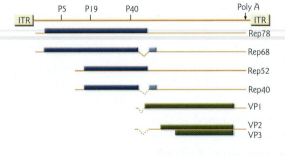

alternative initiation event, specifying VP2, uses a noncanonical ACG codon. As noted in Sections 6.6 and 13.3, AAV requires a helper virus for growth under most circumstances and can establish latency when such help is not available. However, although transcriptional activators from the helper (such as adenovirus E1A, see Section 9.4) can upregulate AAV gene expression and other proteins may facilitate gene expression at a post-transcriptional level, there does not appear to be an obligatory requirement for any helper function to achieve AAV gene expression.

9.8 RETROVIRUSES

Retroviruses are RNA viruses but, having converted their genomes into a DNA form, the mechanisms of their mRNA production exactly resemble those of the host and hence many DNA viruses. The principal proteins, common to all retroviruses, are made as polyproteins and cleaved later into their functional components. Translation of the polyprotein that includes the crucial enzymes for reverse transcription and integration of the viral DNA requires the ribosome to make a programmed "error" that causes it to avoid a stop codon.

Retroviral gene expression occurs exclusively from the provirus, a DNA copy of the viral RNA genome (typically 8–10 kbp), which is normally integrated into the host chromosome (see Chapter 8). The promoter for transcription lies in the upstream LTR of the provirus, and this is regulated by a variety of cell transcription factors. In more complex retroviruses, such as human immunodeficiency virus, there are additional virus-coded regulators of gene expression (see Chapter 19). The mechanism of retrovirus mRNA production exactly mirrors the process for any host cell gene, i.e. it involves host RNA pol II, capping, splicing, and polyadenylation functions. Subsequently, mRNAs are transported to the cytoplasm where they are translated. Unlike the true DNA viruses, simple retroviruses have no temporal phasing to their gene expression. However, the additional functions of complex retroviruses do allow them to control their gene expression in this way.

The mRNAs and protein-coding strategy for a typical simple retrovirus, avian leukosis virus (ALV), are shown in Fig. 9.9a. The two ALV mRNAs, produced from a start site in the 5′ LTR, have identical 5′ and 3′ ends. One is equivalent to the full-length genome and the other has a section removed by splicing. The spliced mRNA has *env* as its 5′ proximal open reading frame and is translated to produce the Env polyprotein. This is subsequently cleaved by a host protease into the associated surface (SU) and transmembrane (TM) components of the mature trimeric envelope protein during its passage through the endoplasmic reticulum/Golgi body. Spliced mRNAs are also used by complex retroviruses for expression of their extra genes which are located between *pol* and *env* and/or between *env* and the 3' LTR (see Section 19.2).

The full length ALV mRNA encodes both *gag* and *pol* gene products. This requires a special mechanism since translation in eukaryotes is normally initiated at the first AUG in the message, scanning from the 5' end, and such a mechanism would produce exclusively Gag protein. In fact no free Pol protein is produced by ALV and the protein is only synthesized fused to the C-terminus of Gag. The *pol* gene is in the −1 reading frame with respect to *gag*. Ribosomes translating *gag* avoid its termination codon in about 5% of translation events by slipping backwards one nucleotide, an event known as a ribosome "frameshift" (Box 9.8). Once this has occurred, the translation termination codon for the *gag* gene becomes invisible to the ribosome because it is no longer in the reading frame being used, so translation continues into the *pol* gene, producing a Gag-Pol fusion protein. The extremely high frequency of ribosome "error" at this position at the 3' end of *gag* is caused by a specialized sequence context. This comprises a slippery sequence, on which the

---

Box 9.8

Evidence for expression of a Gag-Pol fusion from avian retroviruses by ribosomal frameshifting

- The *pol* gene was known to be expressed as a C terminal fusion with the Gag protein, through characterization at the protein level, before the mechanism for achieving this was established.
- The idea that a spliced mRNA was produced that encoded the fusion protein was disproved by showing that RNA transcribed from Gag-Pol DNA *in vitro* (where no splicing machinery is present) and then translated *in vitro* produced both Gag and Gag-Pol proteins in a ratio similar to that seen *in vivo*. This result shows that the fusion arises as an option at the translation stage – i.e. ribosome frameshifting.
- A sequence common to the region of overlap between Gag and Pol reading frames in a number of retroviruses was shown to be necessary for Gag-Pol fusion protein expression by making point mutations in the sequence. This is the "slippery sequence."
- The amino acid sequence of the fusion protein exactly matched that expected for a −1 frameshift at the mapped sequence.
- Deletion/mutation analysis of the region downstream of the slippery sequence defined an RNA secondary structure element (the pseudoknot) as being required for frameshifting at the slippery sequence.

frameshift event actually occurs, and a pseudo-knot downstream of the slippery sequence, which causes the ribosome to pause on the slippery sequence and so promotes frameshifting. As a result, lower levels of *pol* gene expression are achieved than of *gag*, reflecting the relative amounts in which these proteins are needed to form progeny particles. This mechanism closely resembles that operating in coronaviruses (see Section 10.5). Other retroviruses face the same problem in achieving expression of their *pol* genes, but they find a variety of solutions. In the murine equivalent of ALV, known as MLV, the *gag* and *pol* genes are in the same frame separated only by a stop codon. This is misread by a specific loaded tRNA, producing a Gag-Pol fusion, again with about 5% efficiency.

For all retroviruses, the *gag* and *pol* genes code for polyproteins that together contain the proteins needed to build the internal parts of the virion (Fig. 9.9b). The protein encoded at the *gag-pol* boundary is a specific protease (PR), which is required to process them in a post-assembly maturation event (see Section 11.7). In ALV, PR is coded at the C-terminus of *gag*, whereas in MLV it is at the N-terminus of *pol*. A further variant mechanism for *pol* expression then arises in other retroviruses, such as human T-cell lymphotropic virus (HTLV), where the protease is encoded in a reading frame distinct from both *gag* upstream and *pol* downstream. One frameshift takes the ribosome from the Gag reading frame into the Pro reading frame and a second shift takes it on into the Pol reading frame; three different polyproteins, Gag, Gag-PR, and Gag-PR-Pol, are therefore produced. Finally, in the least well characterized retrovirus genus, the spumaviruses, there is evidence for a separate spliced mRNA being used to encode Pol, so that no Gag-Pol fusion protein is made.

The spumaviruses actually represent an interesting intermediate state between standard retroviruses and the hepadnaviruses (Section 9.9), based on features of both their genome replication (see Section 8.7) and their gene expression. As well as expressing Pol protein independently of Gag (in the same way as hepadnavirus P and C proteins are separate), they also make use of a second promoter to express two proteins from reading frames 3' to *env*. Although other complex retroviruses have additional genes in this same position, they use splicing to express them from the same LTR promoter as is used for all gene expression in other retroviruses.

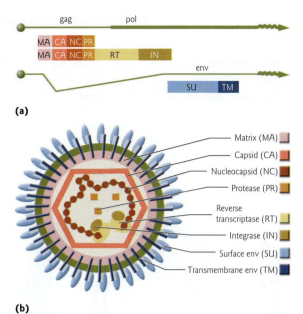

(a)

(b)

Matrix (MA)
Capsid (CA)
Nucleocapsid (NC)
Protease (PR)
Reverse transcriptase (RT)
Integrase (IN)
Surface env (SU)
Transmembrane env (TM)

Fig. 9.9 (a) The mRNAs and proteins synthesized in avian leukosis virus-infected cells. In nonavian retroviruses, the protease is encoded as part of Pol and so is only present in the Gag-Pol fusion protein (see text for details). (b) The location of retroviral proteins within the mature virion. The role of the protease in the maturation of retrovirus particles is considered in Section 11.7. Note that the complexes of SU and TM in the envelope are actually trimers of each protein.

9.9 HEPADNAVIRUSES

Hepadnaviruses use host mechanisms for gene expression. They manage to make a lot of proteins from a very small genome by a combination of multiple promoters, overlapping coding regions and the use of alternative translation start codons on an mRNA.

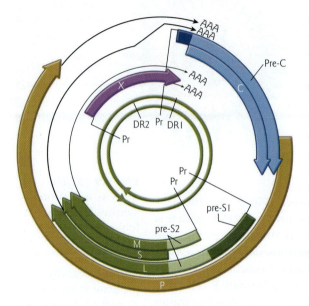

Fig. 9.10 Hepatitis B virus genome map. The closed circular DNA is transcribed from four promoters (Pr); arrowheads indicate 3′ ends. For details of protein expression, see text. Pre-S1, pre-S2 are the designations given to the upstream portions of the S reading frame which, when added to the N-terminus of S, give rise to the longer L and M surface proteins. DR1, DR2 are repeated sequences shown to orient the map (see Fig. 8.5).

The hepadnaviruses have very small circular genomes, e.g. hepatitis B virus, 3.2 kbp. The particular feature of note regarding the gene expression of this virus is the density at which information is compressed onto the genome, with protein encoded by every part of the DNA and two unrelated proteins being encoded in alternative reading frames over a significant part of its length (Fig. 9.10). There are four RNA pol II promoters which produce pregenomic and three classes of subgenomic mRNA respectively. None of these mRNAs is spliced and all end at the same polyadenylation site. Transcription is regulated by cell-type-specific factors and this is part of the basis for the tropism of this virus for the liver.

mRNAs from the four promoters encode seven proteins (Fig. 9.10). All except the X and pre-C proteins are structural components of the virion. Pre-C is secreted from the cell (this is known as e antigen) and probably serves to modulate the immune response to the virus, whereas X is a transcriptional activator that affects a wide range of viral and cellular promoters. The mRNAs transcribed from the genomic and S promoters show heterogeneous 5′ ends. This means that some molecules include the translation start sites for the pre-C and M proteins respectively, while the remainder do not. Translation from these latter mRNAs then begins at downstream AUGs, encoding the C (core) and S (surface) proteins. Synthesis of the P protein presents a problem since there is no mRNA in which its reading frame is proximal to the 5′ end. Although the position of the P reading frame, overlapping the C-terminus of the core protein sequence, is reminiscent of the retroviral arrangement of *gag* and *pol* genes, there is no C-P fusion protein produced. Instead, a modified ribosome scanning mechanism is believed to allow a subset of ribosomes, which load onto the mRNA at its 5′ end, to initiate translation at the start codon for the P protein. Again, this achieves a lower level of P expression relative to C, reflecting the differing requirements for these proteins.

9.10 DNA BACTERIOPHAGES

As with the DNA viruses of eukaryotes, bacteriophages with DNA genomes show increasing sophistication of control of gene expression as the genome size increases, with temporal phasing of production of the proteins required early and late in the infection. Bacteriophage gene expression is controlled mainly at the level of transcription, since RNA processing is not a feature of prokaryotic gene expression and the short half-life of prokaryotic mRNA makes transcriptional control particularly effective. Regulated transcription of phage genes can occur either through modification to the specificity of the host RNA polymerase, as occurs in infections by the *Escherichia coli* bacteriophage T4, or through the provision of a new phage-specified RNA polymerase, as in *E. coli* bacteriophage T7. The unique promoter specificities of the T7 RNA polymerase and similar enzymes from phages SP6 and T3 make them valuable tools for RNA transcription from cloned genes in the test-tube. One of the best studied examples of control of gene expression in a prokaryotic virus is bacteriophage λ, some information on which is provided in Chapter 15, focusing on lysogeny of this phage.

KEY POINTS

- Most DNA viruses of animals replicate in the nucleus and use host systems for mRNA production and protein synthesis.
- Those that replicate in the cytoplasm must provide their own enzymes for mRNA production.
- Retroviruses can be considered as DNA viruses when it comes to gene expression.
- Gene expression for all but some of the smaller DNA viruses is temporally phased, with different genes expressed at different stages of the infection.
- Where phasing occurs, proteins made early in an infection are required for DNA synthesis and to modify the host environment, while proteins made later are directly involved in progeny particle formation or facilitate particle assembly.
- The polyoma-, papilloma-, adeno-, parvo-, and retroviruses all make extensive use of RNA splicing in the expression of viral genes.
- Herpesviruses make little use of RNA splicing for lytic gene expression but extensive use of splicing during latency.
- Poxviruses and other cytoplasmic DNA viruses do not use RNA splicing.

QUESTIONS

- Discuss the strategies employed by animal DNA viruses to maximize the coding capacity of their genomes for the production of viral proteins.
- Using appropriate examples, explain how many viruses achieve a temporally regulated pattern of gene expression over the course of an infection.

FURTHER READING

Flint, J., Shenk, T. 1997. Viral transactivating proteins. *Annual Reviews of Genetics* **31**, 177–212.

Fuchs, D. G., Pfister, H. 1994. Transcription of papillomavirus genomes. *Intervirology* **37**, 159–167.

Wysocka, J., Herr, W. 2003. The herpes simplex virus VP16-induced complex: the makings of a regulatory switch. *Trends in Biochemical Sciences* **28**, 294–304.

Gene expression of retroviruses and hepadnaviruses is considered in the further reading suggested in Chapter 8.

Some aspects of gene expression for various viruses covered in this chapter are included in the further reading suggested in Chapter 20.

Also check Appendix 7 for references specific to each family of viruses.

10

The process of infection: IIIB. Gene expression and its regulation in RNA viruses

The presence of RNA as genome means that the process of gene expression by RNA viruses differs from that of the host and of DNA viruses. However, an RNA genome must also be transcribed to form positive sense mRNA which is translated into proteins using host cell ribosomes. RNA viruses control their gene expression using a number of processes.

All RNA viruses synthesize their mRNA from a negative sense RNA molecule; those from Baltimore class 3 use the double-stranded genome as template while class 4 viruses must replicate their genomes to generate a replicative intermediate (RI; see Section 7.3) which contains the negative sense RNA template, before transcription can occur. Class 5 viruses can use the genome RNA directly as a template for transcription. Host cell enzymes cannot transcribe an RNA template, and so all RNA viruses must encode their own RNA-dependent RNA polymerases. Viruses of classes 3 and 5 carry the polymerase as an essential, internal, component of the virus particle and introduce it into the infected cell so that transcription can begin immediately (Box 10.1).

Class 4 virus particles do not carry a polymerase and the enzyme is generated instead by translation of the positive sense RNA genome. The enzymes which carry out transcription of RNA viruses are frequently referred to as "transcriptases" to differentiate the process from replication. However, as indicated in Chapter 7, both transcription and replication are carried out by the

Box 10.1

Evidence that viruses of Baltimore classes 3 and 5 carry a polymerase in the virus particle

- Treatment of cells with specific inhibitors of cellular RNA polymerases, such as actinomycin D, prior to and throughout infection with class 3 viruses, does not inhibit virus RNA synthesis. Class 5 viruses which replicate in the cytoplasm are not affected by actinomycin D, though those replicating in the nucleus may be affected.
- Treatment of cells with inhibitors of translation, such as the drug cycloheximide, prior to and throughout virus infection prevents synthesis of both host and virus proteins. However, protein synthesis inhibitors do not prevent class 3 and class 5 virus mRNA synthesis.
- For some viruses, purified virions can transcribe RNA *in vitro*.

Taken together these data indicate that the polymerase responsible for class 3 and class 5 virus transcription is not a cellular enzyme. Since no protein synthesis is required for virus mRNA synthesis to occur the enzyme must be introduced in an active form by the virus at the time of infection.

same enzyme for each virus, though the enzymes may be modified in different ways to function in the two separate processes. Messenger RNAs are synthesized and translated in a $5' \rightarrow 3'$ direction. Consequently, their $5'$ termini are of some interest, particularly with regard to their ability to regulate gene expression.

10.1 THE RNA VIRUSES: BALTIMORE CLASSES 3, 4, AND 5

- *Class 3 viruses have a dsRNA genome.*
- *Class 4 viruses have a positive sense ssRNA genome.*
- *Class 5 viruses have a negative sense ssRNA genome.*

In general, temporal control of gene expression in RNA viruses, if present, is not usually as pronounced as it is for many DNA viruses. Commonly, the difference in gene expression is a quantitative one, with all genes expressed simultaneously but some more abundant than others at different times, and where there is true temporal regulation, such as with the alphaviruses, the early genes are also expressed at late times during infection. Thus, unlike the DNA viruses, and despite the necessity of using the host cell ribosomes for translation, many RNA viruses produce mRNA which is either not capped, or not polyadenylated, or neither capped nor polyadenylated. The lack of these structures, particularly the 5' mRNA cap, requires the virus to use novel mechanisms to ensure translation of its mRNA.

While most RNA viruses replicate in the cytoplasm of the host cell, some do so in the nucleus. Viruses cannot deviate from this "choice." Since, unlike the DNA viruses, RNA viruses do not use the host cell DNA-dependent RNA polymerase, there are other reasons why some RNA viruses replicate in the nucleus which are explained below. Despite the fundamental difference in the nature of the template nucleic acid, control of gene expression by RNA viruses frequently shows similarities with that seen in DNA viruses.

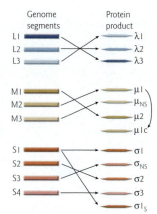

Fig. 10.1 Genome organization of reovirus. Each reovirus genome segment encodes a single major protein, but post-translational modification leads to the functional protein. Segment S1 also encodes an additional protein σ1s. The proposed functions of the proteins are shown in Table 3.2.

10.2 REOVIRUSES

Transcriptional regulation of gene expression

The genome of mammalian reoviruses, the type member of the Family of reoviruses, is double-stranded RNA. Electrophoretic analysis of RNA extracted from reovirus particles showed that the dsRNA genome consists of 10 different segments of RNA that fall into three size classes: L (large; L1–L3 of approximately 4500 bp), M (medium; M1–M3 of approximately 2300 bp), and S (small; S1–S4 of approximately 1200 bp). These 10 segments showed no base sequence homology in hybridization tests and so could not have arisen by random fragmentation of the genome. Nucleotide sequence analysis of all 10 segments has confirmed that they are unique. The sum total of the lengths of the 10 segments corresponds to the size of the viral genome estimated by chemical means, suggesting that the intact virus contains one copy of each segment. Each segment is perfectly base-paired, with no overlapping single-stranded regions at either end of the molecule.

The reovirus proteins fall into three size classes, called λ, μ, and σ by analogy with the L, M, and S genome segments, and this implied that there might be a 1 : 1 relationship between each segment and a gene product. This was confirmed and additional proteins were shown to be generated by post-translational modification of the precursor generated from each segment. The genome organization, with the proteins encoded by each segment and proposed functions of the proteins, are summarized in Fig. 10.1.

The mechanism by which reoviruses initiate the expression of their genome presented a problem to early investigators since it was known that double-stranded RNA could not function as mRNA. The first virus mRNAs in the cell therefore must result either from transcription of the parental genome or by separation of the two strands. However, attempts to demonstrate strand separation were unsuccessful, and all cellular

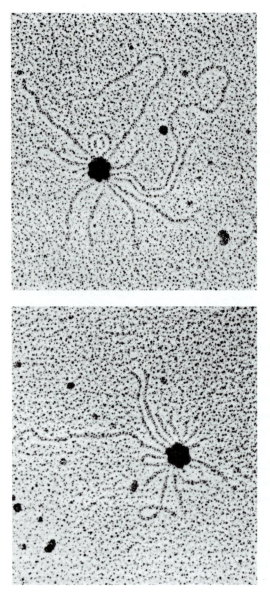

Fig. 10.2 Electron micrographs showing nascent mRNA leaving reovirus cores via the spike proteins. (From Bartlett N. M., *et al.* 1974. *Journal of Virology* **14**, 315–326.)

polymerases tested were unable to transcribe the dsRNA genome. This was not unexpected as reoviruses replicate exclusively in the cytoplasm, while host cell RNA polymerases are confined to the nucleus.

A dsRNA-dependent RNA polymerase has now been identified as an integral part of the virion particle. The polymerase is inactive in intact virions, but becomes activated at the second stage of uncoating. The activation step appears to be regulated by an outer capsid protein, called µ1c. This protein, which is modified by phosphorylation and glycosylation, is cleaved specifically during the second stage of uncoating, to leave a residual component, the ∂ polypeptide, attached to the cores. The polymerase is probably a multi-subunit complex intimately associated with the structure of the core, since it has proved impossible to separate from the virus particles. See the reovirus structure in Chapter 3.

Reovirus transcription is copied using only one strand of RNA as template, and the mRNA produced is *exactly* the same length as the genome RNA. Added dsRNA is not recognized by the polymerase, suggesting that the genome is transcribed while still within the virus core. Newly synthesized mRNA leaves the cores of the virions which initiate the infection (the "parental" cores) by way of channels through the external protein spikes. The newly synthesized mRNA from transcribing core particles has been visualized by electron microscopy (Fig. 10.2).

The reovirus mRNAs made at the beginning of the infection are modified at their 5′ end by the addition of a cap structure, but do not contain the 3′ poly(A) tail found on normal cellular mRNAs. In prokaryotes, and viruses of prokaryotes, the 5′ end of many mRNAs is a triphosphorylated purine corresponding to the residue that initiated transcription. By contrast, most eukaryotic cellular and viral mRNAs, as well as native nucleic acid from some RNA viruses of eukaryotes, are modified at the 5′ end. The modification consists of a "cap" that protects the RNA at its 5′ terminus from attack by phosphatases and nucleases and promotes mRNA function at the level of initiation of translation. In eukaryotic cells translation of mRNA is "cap-dependent," i.e. only RNA molecules with a 5′ cap are accepted by ribosomes for translation. The functional ribosome contains a number of initiation factors

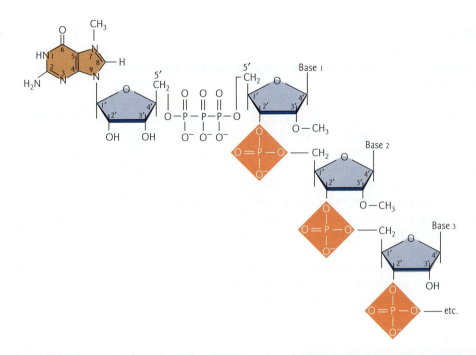

Fig. 10.3 Structure of a capped RNA molecule. Note the 5'–5' phosphodiester linkage. Bases 1, 2, and 3 can be any of the four nucleic acid bases. If the 2' positions of the ribose of bases 1 and 2 are not methylated the structure is referred to as a cap 0 structure, if only the 2'-O-methyl group on the ribose of base 1 is present it is called a cap 1 structure, and if a 2'-O-methyl group is present on both bases 1 and 2 it is a cap 2 structure.

and one of these, eIF4E, binds to the cap structure and brings mRNA into the translation complex in association with other proteins, including the factors eIF4GI and eIF4GII. If the cap structure is not present the RNA is not recognized as a potential mRNA.

The general structural features of the 5' cap are shown in Fig. 10.3. The terminal 7-methylguanine and the penultimate nucleotide (the first nucleotide copied from the genome template) are joined by their 5' hydroxyl groups through a triphosphate bridge. This 5'–5' linkage is inverted relative to the normal 3'–5' phosphodiester bonds in the remainder of the polynucleotide chain and is formed post-transcriptionally. The normal cellular capping enzymes are found only in the nucleus but the reovirus particle contains several viral methylase activities which carry out this reaction. All of the reovirus mRNAs have the sequence GCUA at the 5' end.

Following replication of the genome (see Chapter 7) new virus cores containing one of each of the 10 genome segments are generated in the cytoplasm and these also synthesize mRNA from all segments. However, the mRNA produced from these progeny particles, in what is termed late transcription, as it occurs after genome replication, does not contain a cap structure at the 5' end because the methylases in these immature virus

---

Box 10.2

Evidence that temporal control of reovirus gene expression is regulated by a host function

- When cores expressing the pre-early genes preferentially are isolated from infected cells and allowed to continue transcription and translation *in vitro* in the absence of additional virus proteins, the pattern of transcription alters and all segments are transcribed efficiently. This indicates that the presence of a pre-early protein is not required to activate transcription of the remaining segments.
- If cores are generated *in vitro* by chymotrypsin treatment they transcribe all segments with equal efficiency, but when introduced into cells they exhibit the pre-early pattern of transcription. This indicates that a factor in the cells reduces transcription from some segments.

The change from the preferential transcription of segments 1, 6, 9, and 10 to equal transcription from all segments occurs by a de-repression event.

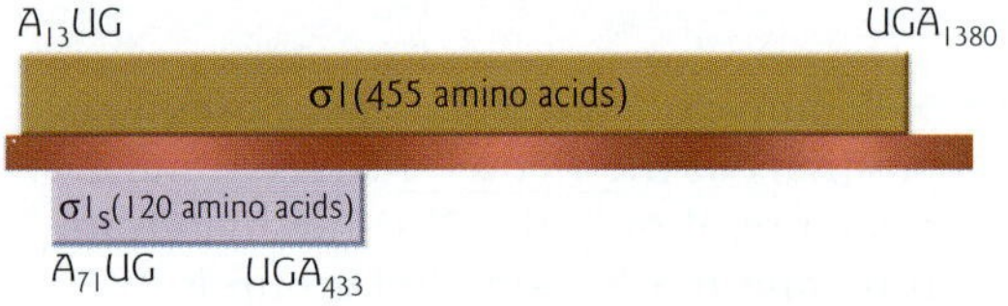

$A_{13}UG$ UGA_{1380}

σ1(455 amino acids)

σ1$_S$(120 amino acids)

$A_{71}UG$ UGA_{433}

Fig. 10.4
Arrangement of the overlapping open reading frames of the reovirus S1 gene mRNA. The relative locations and protein products of the two ORFs are indicated. The numbers refer to the position of the first nucleotide of the relevant codon.

particles are inactive. This has important consequences for translation of these mRNAs as described below.

At 2–4 hours after infection, all of the genome segments are transcribed, but some studies suggest that the mRNA transcribed from segments 1, 6, 9, and 10 (L1, M3, S3, and S4) are predominant. While these results remain the point of some debate, these are called *pre-early* mRNAs. However, by 6 hours after infection all of the genome segments are transcribed with equal efficiency. Thus, there appears to be some degree of temporal control of transcription, and transcription of each segment is independent of any other. The mechanism by which this control is achieved is not known, but is likely to be regulated through a host cell repressor molecule which allows preferential transcription (Box 10.2). One, or more, of the virus proteins would then cause de-repression.

Translational regulation of gene expression

From the earliest sequence analyses of virus genomes it has been known that open reading frames (ORFs) in different reading frames may overlap each other, as seen in Chapter 9 for SV40 and polyomavirus. Similar features are seen with some RNA genome viruses. An exception to the one segment–one protein rule for reovirus is seen with the S1 segment. Segment S1 contains a large ORF with an AUG translation initiation codon located near the 5′ end of the mRNA. This ORF directs the synthesis of a structural protein of 455 amino acids called σ1. Further from the 5′ end is a second ORF which begins with another AUG codon and which is in a different reading frame (Fig. 10.4). Occasionally, ribosomes which have failed to initiate translation at the first AUG do so at the second AUG, producing a protein of 120 amino acids called σ1s. Because σ1 and σ1s are encoded from different reading frames they are unique and do not share any amino acid sequences. The relative level of the two proteins is determined by the efficiency of initiation of translation at the AUG initiation codons of the two ORFs, with the 5' proximal AUG being utilized most to generate the σ1 protein. This arrangement ensures that less of the protein encoded by the second ORF, σ1s, is produced.

While the 10 reovirus mRNAs are present in approximately equimolar amounts late in the infectious cycle, the proteins are nevertheless made in greatly differing amounts, from very low levels of polymerase to high levels of other, major, structural proteins. This means that the various mRNAs must be translated with different efficiencies. How is this achieved? The sequences surrounding the AUG initiation codon for cellular mRNAs is important in determining the efficiency of translation. The consensus sequence, identified by Marilyn Kozak, which is most favored for initiation of translation is G/AnnAUG(G), where n is any nucleotide, and mRNAs bearing sequences closest to this are most readily recognized by ribosomes. In the case of the S1 segment, the Kozak rules appear to apply in as much as the sequence surrounding the second AUG which initiates the σ1s protein agrees very closely with the proposed consensus while the sequence surrounding the AUG for σ1 (the first AUG) is not so good. This counteracts the preference for ribosomes to intiate translation at the 5' proximal AUG codon in mRNA, encouraging translation of the second ORF and enhancing the production of σ1s, although the levels of production still do not come up to those of σ1. However, the rate of translation of the other reovirus mRNAs does not correlate strongly with the Kozak sequence each contains, and this alone is not sufficient to explain the different rates of translation.

In infected cell lysates, the level of translation of each of the virus mRNAs *in vitro* is equal. However, if the constituents of the lysate are altered such that one factor, for example an ion, becomes limiting, the situation becomes very much like that seen *in vivo*. It seems possible, therefore, that the level of translation of reovirus mRNAs is determined, not only by the sequence surrounding the AUG as pointed out by Kozak, but also by the ability of each mRNA to compete with each other for limiting factors. This ability is not related to the presence of a cap structure since competition occurs at both early and late times in infection. The efficiency of competition may be determined by the sequence of the RNA, or its secondary structure which is a consequence of the sequence. This is an example of control of gene expression at the level of translation.

In reovirus-infected cells, there is no rapid shutoff of host protein synthesis and host synthesis is only gradually reduced so that not until 10 hours after infection do virus proteins comprise the majority of the translation products. Thus, for much of the infection virus mRNAs are competing

Box 10.3

Evidence that the translation process in reovirus-infected cells is altered from cap-dependent to cap-independent

During translation the initiation factor eIF4E binds to the 5′ cap structure found on cellular mRNAs. The molecule 7-methyl guanosine triphosphate (m7G(5′)ppp) is an analog of the cap structure and binds to eIF4E, preventing cap-dependent translation by competition. Using cellular lysates from uninfected and reovirus-infected cells to translate capped or uncapped mRNA in the presence or absence of m7G(5′)ppp, the requirement for a cap during translation can be assessed. The data below summarize this experiment:

Source of extract	mRNA	m7G(5′)ppp present	Translation seen
Uninfected cells	None	No	No
	None	Yes	No
	Capped	No	Yes
	Capped	Yes	No
	Uncapped	No	No
	Uncapped	Yes	No
Infected cells	None	No	No
	None	Yes	No
	Capped	No	No
	Capped	Yes	No
	Uninfected	No	Yes
	Uninfected	Yes	Yes

These data show that in reovirus-infected cells translation is cap-independent. When the lysates are prepared at various times after infection it can be seen that there is a gradual change from cap-dependence to cap-independence as the infection proceeds.

with existing host mRNAs – and winning. A striking observation is that the late, uncapped reovirus mRNA is translated extremely efficiently. This is surprising, as normal cellular ribosomes do not translate uncapped mRNA. However, during the infection reovirus proteins specifically inactivate the cap-binding protein complex which forms part of the functional ribosome (Box 10.3). This has the result that the affected ribosomes can now translate only uncapped mRNA. Since the virus late mRNA is the only uncapped RNA present this means that the altered ribosomes translate only virus mRNA. This is the reason for the decline in host cell mRNA translation and preferential translation of virus mRNA and is referred to as a change from cap-dependent to cap-independent translation. Eventually, as all of the initiation factors become altered, only virus mRNA is translated in the cell.

10.3 PICORNAVIRUSES

The picornavirus genome is a positive sense ssRNA molecule of approximately 7500 nt. It has a polyA tract at the 3′ end, resembling that seen in host cell mRNA, but the 5′ end

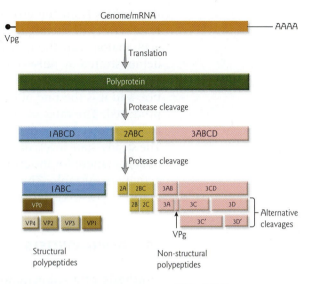

Fig. 10.5 Representation of the cleavages of the poliovirus polyprotein and the smaller products to yield mature viral proteins. Note that most poliovirus mRNA lacks the 5′ VPg protein present on virion RNA. 3C′ and 3D′ are produced by cleavage of 3CD at an alternative site to that producing 3C and 3D.

has a small covalently attached protein, VPg, rather than a cap structure. The first event following uncoating of the picornavirus genome is translation using host cell ribosomes. The presence of VPg and the absence of a cap means that the picornavirus genome RNA cannot be translated directly on host ribosomes. To overcome this barrier to gene expression, picornavirus genomes contain a region of RNA, 5′ of the AUG initiation codon, that adopts a specific three-dimensional conformation which directs ribosomes to initiate translation that is independent of a cap structure and internal to the mRNA. This region is referred to as the internal ribosome entry site (IRES). During the replication cycle the host translation system is altered due to proteolytic cleavage and consequent inactivation of both eIF4GI and eIF4GII. The cleavage prevents the interaction with the cap binding protein eIF4E. The result of this inactivation is that an initiation complex can form but that it does not do so on capped mRNA, translation becomes cap-independent, and the cell is unable to translate its own mRNA. As a result of the virus-induced changes to the translation machinery (together with virus-encoded cleavage inactivation of the host cell TATA-binding protein in the nucleus which arrests host cell polymerase II transcription), host cell protein synthesis is inhibited and only virus-specific synthesis takes place. Production of more mRNA is by replication of the genome as described in Section 7.4.

Control of gene expression by post-translational cleavage

The entire genome of picornaviruses is translated as a huge polyprotein approximately 2200 amino acids in length, which is cleaved in a series of ordered steps to form smaller functional proteins (Fig. 10.5), both structural and nonstructural. Cleavage starts while the polyprotein is still being synthesized and is carried out by virus-encoded proteases which also cleave

themselves from the growing polypeptide chain. Only by inhibiting protease activity or altering the cleavage sites through the incorporation of amino
acid analogs can the intact polyprotein be isolated. Cleavage can also be
demonstrated by pulse–chase analysis. In theory, all picornavirus proteins
should be present in equimolar proportions, but this is not found in practice, and it is the rate of cleavage that controls the amount of each protein
produced. The rates of cleavage of the precursors vary markedly and this
allows a great degree of control over the amounts of each protein produced.
The steady-state levels of the virus proteins are also affected by different rates
of degradation; for instance, it is known that the virus-specified polymerase
activity is unstable. The best known is protease 3C which is produced by
autocatalytic excision when P3 achieves the required conformation.

10.4 ALPHAVIRUSES

Synthesis of a subgenomic mRNA

The genome of alphaviruses consists of one positive sense ssRNA molecule of approximately 11,400 nt with a sedimentation coefficient of 42S.
In contrast to the situation with picornaviruses, *in vitro* translation of the
positive sense genomic RNA of alphaviruses, such as Semliki Forest virus
generates nonstructural but not virion proteins. In virus-infected cells two
types of positive sense RNA are produced: one of 42S identical to the
genome RNA and one with a sedimentation coefficient of 26S. The 26S
RNA is approximately one-third (3800 nt) the length of the genomic RNA.
Both of these positive sense RNA molecules have a cap at the 5′ end, a
polyA tail at the 3′ end, and function as mRNAs. The 26S mRNA represents the 3′ terminal portion of the genomic RNA and is translated into
the structural proteins (Fig. 10.6a). The primary product of translation of
each mRNA is processed by proteolytic cleavage to produce functional
proteins in a manner similar to that described for picornaviruses. Pulse-
chase experiments and tryptic peptide maps show clearly that the small
functional proteins are derived from larger precursor molecules. (Fig. 10.6b).

Since the sequence of the 26S mRNA is contained within the genomic
RNA, and the latter does not direct the synthesis of the structural proteins,
this infers that there is an internal initiation site for the synthesis of virion
proteins within the genomic RNA which cannot be accessed by ribosomes.

The only negative sense virus RNA found in infected cells sediments
at 42S and this must be the template from which both the 42S and 26S
positive sense RNA are transcribed. To achieve this the virus polymerase
must be able to initiate transcription at both the 3′ terminus of the negative sense template and also at a point within the template approximately
one-third of the distance from the 5′ end (Fig. 10.6). These two mRNAs
are co-terminal at the 3′ end.

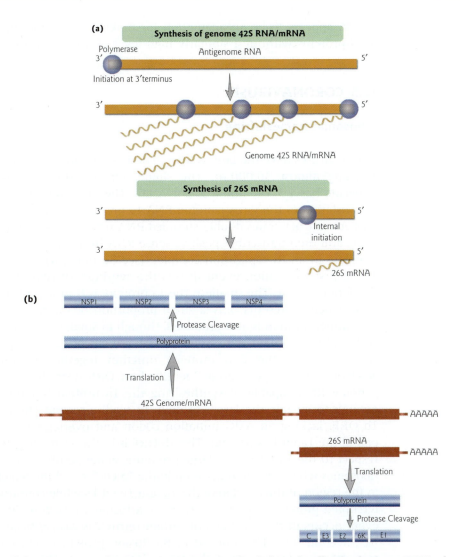

Fig. 10.6 Alphaviruses synthesize two mRNAs in the infected cell. The larger mRNA also acts as the genome; the smaller, 26S mRNA has the same sequence as the 3′ end of the genome. (a) The mechanism for production of the two mRNAs is shown. (b) The proteins encoded by each mRNA and the pattern of proteolytic cleavage that results in the generation of functional proteins.

It is not known how transcription is controlled but the levels of 26S mRNA are very high, indicating that the internal initiation of transcription is efficient and this ensures that the structural proteins are abundant. Since the 26S mRNA can only be produced following replication of the genome this is a late mRNA and control of gene expression in this system is determined at the level of transcription. Synthesis of the 26S mRNA appears to be an adaptation for making large amounts of structural proteins without a similarly high level of nonstructural proteins.

10.5 CORONAVIRUSES

Ribosomal frameshifting

Coronaviruses have the largest RNA genomes described to date with some reaching almost 30,000 nt. The single, linear, positive sense, ssRNA molecule has a 5′ cap and a 3′ polyA tail. The first event after uncoating is translation of the virus genome RNA to produce the virus polymerase. This enzyme generates double-stranded RNA using the plus-strand as template, and then transcribes positive sense RNA from a negative sense RNA template in an RI replication complex (see Section 7.4).

The first translation event directs the synthesis of two proteins from the genome RNA. The smallest, the 1a protein, is translated from the first ORF which represents only a small proportion of the molecule. A second, larger, protein is also produced, though in smaller amounts, and this is thought to contain the enzymatic component of the polymerase. The way in which the two proteins function together is not known. Nucleotide sequence analysis identified an ORF near the 5′ end of the genome RNA capable of synthesizing the 1a protein but no ORF long enough to encode the larger protein. Instead, a second ORF, called the 1b ORF, lacking an AUG initiation codon and overlapping with the 3′ end of the 1a ORF is present. The 1b ORF is in the −1 reading frame with respect to the 1a ORF. The large protein is generated by a proportion of ribosomes which initiate translation in the 1a ORF switching reading frames in the region of the overlap of the 1a and 1b ORFs while continuing uninterrupted protein synthesis to generate a fusion 1a–1b protein. A similar event occurs in the synthesis of certain retrovirus *gag-pol* fusion proteins (see Section 9.8). The amount of the fusion protein produced is determined by the frequency of the frameshifting event. This is an example of translational regulation of gene expression.

The frameshifting event in coronaviruses is determined by the presence of two structural features in the genomic RNA. The first is a "slippery" sequence in the region of overlap between the ORFs at which the frameshift occurs and the second is a three-dimensional structure called a pseudoknot in which the RNA is folded into a tight conformation. These two features combine to effect the frameshift.

Functionally monocistronic subgenomic mRNAs

Coronaviruses carry the alphavirus strategy of producing a subgenomic mRNA to extremes, producing not two but several mRNAs. For mouse hepatitis virus (MHV), six mRNAs in addition to the genome are found in infected cells. The sizes of these mRNAs added together exceed that of the total genome. This observation was explained by sequence analysis which showed that the sequences present in a small mRNA were also present in the larger mRNAs, and that all of the mRNAs shared a common 3′ end with each other and the genome RNA. This is described as a nested set of mRNAs. In addition, each mRNA has a short sequence at the 5′ end which is identical for each (Fig. 10.7). Each mRNA has a unique first AUG initiation codon allowing translation of a unique protein. Although the largest (genome) mRNA contains all of the coding sequences for all of the proteins, the first ORF is used preferentially, in accordance with the normal process of eukaryote translation. For some MHV mRNAs more than one protein is produced by each individual mRNA though the second protein is present in low levels, as with certain reovirus and paramyxovirus mRNAs. The mRNAs are present in different quantities with respect to each other, some abundant, some less so, but the ratios of each do not alter during the infectious cycle. There is therefore no temporal control of gene expression.

The subgenomic mRNAs cannot be produced using the cell's splicing enzymes since these are located in the nucleus and coronaviruses replicate in the cytoplasm. The mechanism by which they are generated is novel and there is

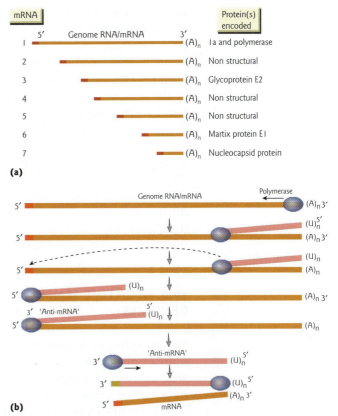

Fig. 10.7 Control of gene expression in the coronaviruses mouse hepatitis virus. (a) The products of transcription of mouse hepatitis virus and the proteins encoded by each mRNA. The common sequence found at the 5′ end of the genome RNA and all mRNAs is shown in red. (b) Negative sense RNAs are produced by a novel mechanism in which the virus-encoded RNA polymerase copies the 3′ end of the genome RNA template and completes synthesis by "jumping" to finish copying the immediate 5′ end. The negative sense RNAs are then used to synthesize individual mRNAs. Each mRNA contains the same 5′ end sequence.

some controversy over the details. While it is possible that the negative sense antigenome RNA is used as the template for synthesis of mRNA, most evidence points to the genome RNA as the primary template. The polymerase molecule begins transcription at the 3′ end of the genome RNA, producing a negative sense RNA, and at specific points the polymerase terminates and the complex of protein and newly synthesized negative sense RNA dissociates, at least partially, from the template. The nascent

RNA is then taken to a point near the 5′ end of the genome RNA and synthesis continues to the end. This produces a series of negative sense RNA molecules which contain sequences complementary to the 3′ end of the genome and to a small region from the 5′ end. The transfer mechanism is not understood and may be related to the secondary structure of the template RNA. The frequency with which the polymerase terminates within the template determines the abundance of each RNA. The negative sense RNA molecules are then used as templates to make faithful positive sense copies which act as mRNAs (Fig. 10.7). In effect, each mRNA is the product of the replication of a subgenomic negative sense RNA, produced using a replication intermediate as described for the ssRNA virus genome replication process (see Sections 7.4 and 7.5). Each mRNA is then transcribed from the subgenomic negative sense RNA and translated into the appropriate protein. Coronavirus gene expression is therefore controlled primarily at the level of transcription.

10.6 NEGATIVE SENSE RNA VIRUSES WITH SEGMENTED GENOMES

The genomes of viruses belonging to the Baltimore class 5 vary in the number of ssRNA segments of which they are comprised (Box 10.4), but gene expression is predominantly, but not exclusively, from monocistronic mRNAs. For viruses with segmented genomes each segment must be transcribed into mRNA offering the opportunity for transcriptional control of gene expression. Normal cells do not contain enzymes capable of generating

---

Box 10.4

Baltimore class 5 viruses have a variety of numbers of genome segments which are packaged into the virus particle

Family	No. of segments/genome
Orthomyxoviridae	8
Bunyaviridae	3
Arenaviridae	2
Rhabdoviridae	1
Paramyxoviridae	1
Bornaviridae	1
Filoviridae	1

mRNA from an RNA template and all viruses with nonsegmented negative sense RNA genomes synthesize mRNA using virus-encoded transcription machinery which is carried into the cell with the infecting genome. Almost all of the negative sense RNA viruses replicate in the cytoplasm. The exceptions are the orthomyxoviruses and Borna disease virus.

10.7 ORTHOMYXOVIRUSES

Transcriptional regulation of gene expression

Of the orthomyxoviruses, most is known about the type A influenza viruses of humans and other animals. An unusual feature of influenza virus replication is that it occurs in the nucleus of the infected cell. Immediately after infection, the uncoated virus particle is transported to the nucleus, where viral mRNA is synthesized. The mRNA is exported from the nucleus to the cytoplasm for translation and later in infection certain of the newly synthesized proteins migrate from their site of synthesis in the cytoplasm into the nucleus. Since passage of molecules across the nuclear membrane is a highly selective process, this constitutes a level of control which is unique to eukaryotic cells.

Influenza viruses can only replicate in cells with a functional cellular DNA-dependent RNA pol II, though this enzyme cannot transcribe the virus genome. Two types of positive sense RNA are generated in infected cells – mRNA and template for replication (antigenome RNA). These two types of RNA can be differentiated from each other by a number of criteria (Box 10.5).

Box 10.5

Differences between the influenza virus mRNA and antigenome positive sense RNA

mRNA	Antigenome RNA
Shorter than the template genome segment	Exact copy of the template genome segment
Contains 3′ poly(A) tail	No 3′ poly(A) tail
Contains 5′ cap	No 5′ cap
Synthesis does not require protein synthesis (insensitive to protein synthesis inhibitors)	Synthesis requires continuous virus protein synthesis (sensitive to protein synthesis inhibitors)

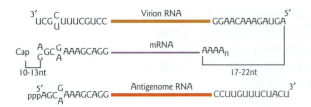

Fig. 10.8 The three primary types of virus RNA found in influenza virus type A infected cells are shown, with the key differentiating features indicated. All eight genome segments have the three forms. The genome RNA has conserved terminal sequences which are present in complementary form in the antigenome which is an exact replica of the genome segment from which it was copied. The mRNA contains additional sequences at the 5′ end not present in the genome template and the mRNA is terminated before the end of the template is reached at which time a polyA tail is added by reiterative transcription (Box 10.5).

The production of mRNA from the genome initiating the infection, termed primary transcription, uses the virion-associated pol merase and it is this step which requires active pol II. Each of the eight genomic RNA segments are used as templates for transcription of a monocistronic mRNA using a virus-encoded polymerase complex. Each genomic RNA segment contains its own signals to initiate transcription. The first 13 bases at the 5′ end of every segment are identical, and similarly the last 12 bases at the 3′ end of every segment are identical, but different from bases at the 5′ end. The signal for initiation of transcription of the genome segments is contained in the 3′ terminal sequences. The features of the three types of virus RNA, genome, antigenome, and mRNA found in influenza virus infected cells, are shown in Fig. 10.8.

The method of influenza virus transcription is unusual and explains the reliance on host pol II. The ribonucleoprotein (RNP) complex which carries out the transcription process has no capping or methylation activities associated with it, but the virus mRNAs have a cap 1 structure at the 5′ end (Fig. 10.3). Protein PB2, a component of the RNP complex, binds to the capped 5′ end region of newly synthesized host cell mRNAs. The host mRNA is then cleaved 10–13 bases from the cap, preferably after an A residue though occasionally after a G. This small segment of RNA is then used as a primer for influenza virus mRNA synthesis. The proteins PB1 and PA of the RNP complex initiate transcription and extend the primer, respectively. Transcription is terminated at a specific point on the template at a homopolymer run of uracil residues 17–22 residues from the template 5′ end. These are copied into A in the mRNA and the virus polymerase continues to add A residues, in a reiterative fashion, to generate a poly(A) tail before the mRNA dissociates from the template. By contrast, synthesis of the antigenome positive sense RNA does not use a primer and generates an exact copy of the template RNA. Influenza virus transcription is shown in Fig. 10.9.

Initially, similar amounts of each influenza virus mRNA are produced (as are the proteins), but within an hour of infection beginning, the levels of mRNAs encoding the nucleoprotein (NP) and NS1 (nonstructural protein 1) increase greatly with respect to the others. Later, after genome replication has started, the levels of mRNAs encoding the hemagglutinin (HA), neuraminidase (NA), and matrix (M1) proteins predominate. Thus, although all of the genes are expressed throughout infection, there is some temporal control of transcription determined by the relative rates of transcription of some segments.

Control of gene expression by mRNA splicing

Superimposed on the basic pattern of transcription of a single mRNA from each influenza A virus genome segment is a system which permits the synthesis of two additional proteins from influenza virus segments 7 and 8 by the process of splicing. This mechanism is available because the virus RNA is transcribed in the nucleus. Transcription of each of these segments produces an mRNA which encodes a specific protein: M1 and NS1, respectively. An internal section is removed from a proportion of the primary transcripts with the two remaining molecules ligated (Fig. 10.10). In both cases the splicing events delete a substantial portion of the first ORF, leaving the AUG initiation codon in place. The result is that an alternative ORF, in a different reading frame and not previously accessed by ribosomes, is fused to the first few codons of the M1 or NS1 ORFs. The newly generated ORF directs the synthesis of novel proteins, called M2 and NS2. This is similar to the situation for the polyomaviruses (see Section 9.2). An additional, rare, mRNA is generated from the segment 7 primary transcript by an alternative splicing process. The splice removes the first AUG codon in the primary transcript which is used to synthesize the M1 and M2 proteins. The first ORF in the alternatively spliced segment 7 mRNA contains only nine codons. The putative protein produced by translation of this mRNA is called M3 but it has not been detected in infected cells and its role in the infectious cycle is not known. The levels of expression of NS2 and M2 are determined by the frequency of the post-transcriptional splicing event.

Translational regulation of gene expression

Some human and animal influenza A viruses encode an additional protein from the mRNA transcribed from genome segment 2 that also encodes PB1. The protein, called PB1-F2, is encoded in a short second ORF which is contained within the PB1 ORF but is in a different reading frame. The PB1-F2 proteins from various influenza viruses differ in size between strains, ranging from 57 to 90 amino acids in length. The function of the PB1-F2 protein is not known but after translation it is transported to the host mitochondria where it is rapidly degraded. Due to the mitochondrial location and other

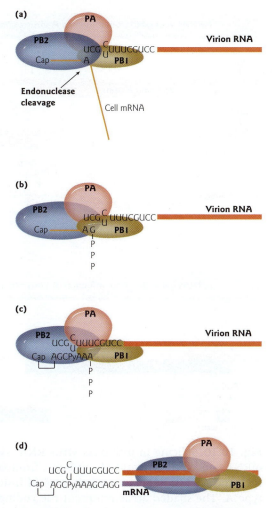

Fig. 10.9 The process of transcription from influenza virus genome segments. The three virus proteins, PB2, PB1, and PA, act together (a) to remove cap structures together with 10–13 nucleotides from the 5′ end of host cell mRNA (PB2) and (b) use these as primers to initiate transcription (PB1) followed by (c) extension of the newly synthesized RNA (PA) to produce (d) an mRNA molecule.

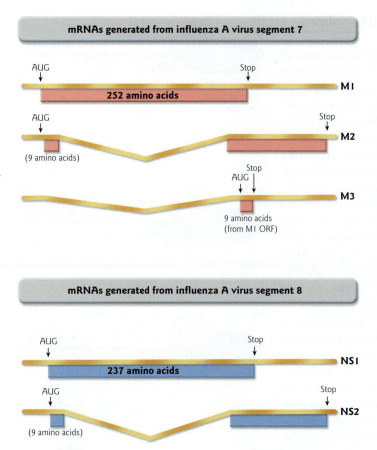

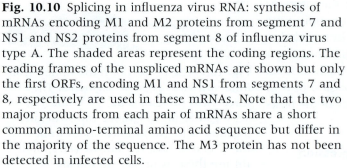

Fig. 10.10 Splicing in influenza virus RNA: synthesis of mRNAs encoding M1 and M2 proteins from segment 7 and NS1 and NS2 proteins from segment 8 of influenza virus type A. The shaded areas represent the coding regions. The reading frames of the unspliced mRNAs are shown but only the first ORFs, encoding M1 and NS1 from segments 7 and 8, respectively are used in these mRNAs. Note that the two major products from each pair of mRNAs share a short common amino-terminal amino acid sequence but differ in the majority of the sequence. The M3 protein has not been detected in infected cells.

experimental evidence it is believed that the PB1-F2 protein may induce death of infected cells by the process of apoptosis.

Post-translational protein cleavage

While some viruses, such as the picornaviruses, encode proteases to produce functional proteins, most viruses do not encode enzymes to specifically cleave their proteins and it might be expected that post-translational cleavage would not be involved. However, post-translational cleavage is frequently the final maturation event for viruses; for example, after its synthesis, the influenza virus HA glycoprotein undergoes a single protease cleavage. The HA protein is synthesized as a precursor HA0 which is cleaved to give two products, the largest is HA1 which is derived from the carboxy terminus of the HA0 and the smaller HA2, which comes from the amino terminus of the HA0 precursor. Uncleaved molecules occur in cells that are deficient in protease activity but are still incorporated into virus particles which are morphologically normal and can agglutinate red blood cells. However, such virus particles are noninfective until the HA protein is cleaved. The cleaved HA changes conformation and a hydrophobic domain, referred to as the fusion peptide, is now free at the N terminus of HA2 (see Section 5.1). The protease which cleaves the influenza virus HA protein is a host cell enzyme located in the Golgi network through which the HA is translocated. The restriction of such enzymes to specific tissues such as the respiratory tract determines the site of productive virus replication. If HA is cleaved by a more ubiquitous protease such as is seen in avian influenza virus infection in birds the virus can spread systemically. Some viruses, such as Sendai virus, require cleavage activation of proteins which can occur after the virus has left the host cell rather than inside the cell as for influenza virus.

10.8 ARENAVIRUSES

Ambisense coding strategy

Although arenaviruses appear similar to conventional segmented negative-strand viruses with just two (L and S) genomic RNAs, they have an unusual gene organization. A virus-encoded enzyme transcribes only the 3′ part of the genome RNAs into a capped, polyadenylated subgenomic mRNA. In the case of the S RNA this mRNA encodes the nucleocapsid (N) protein, while for the L RNA it encodes the large polymerase protein. Following replication, the 3′ end of the antigenome copy of each RNA segment is also used as a template for transcription. This generates a subgenomic mRNA from the S antigenome RNA which encodes a protein called GPC, the precursor to the virus structural glycoproteins. A protein called Z is encoded by the mRNA copied from the L antigenome RNA. Since the GPC and Z mRNAs can only be synthesized after replication they are, by definition, late mRNAs. This mechanism means that both the genome and antigenome act as negative sense RNA templates for transcription and therefore are termed "ambisense" RNA (Fig. 10.11). Some bunyaviruses such as Punta Toro virus also use an ambisense strategy for expression from their S genome segment.

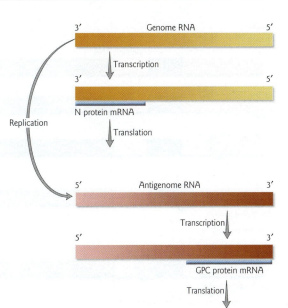

Fig. 10.11 The ambisense strategy of the S RNA segment of the arenavirus, lymphocytic choriomeningitis virus.

10.9 NEGATIVE SENSE RNA VIRUSES WITH NONSEGMENTED, SINGLE-STRANDED GENOMES: RHABDOVIRUSES AND PARAMYXOVIRUSES

Transcriptional regulation of gene expression

Of the rhabdoviruses, most is known about vesicular stomatitis virus (VSV). The proposed model for VSV transcription is used as a paradigm for all other viruses with nonsegmented negative sense RNA genomes, though the precise molecular details of each virus in terms of regulatory sequences etc. differ. Transcription of the VSV genome yields five separate monocistronic mRNAs and gene expression is controlled at the level of transcription, with the abundance of the mRNAs determining the abundance of the proteins. A similar pattern of transcription is seen with the paramyxoviruses, though these may encode six, seven, or ten mRNAs, depending on the virus.

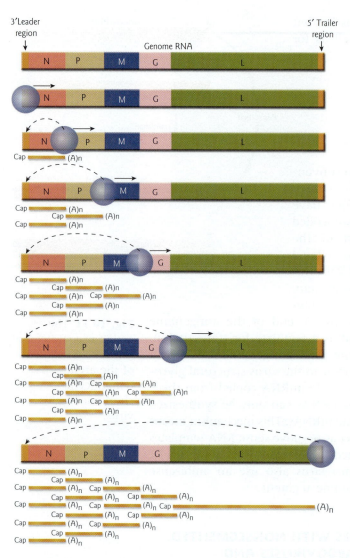

Fig. 10.12 Diagrammatic representation of the sequential transcription process of the rhabdovirus vesicular stomatitis virus (VSV). Transcription occurs in a sequential, start-stop process described in the text. mRNAs are separated by nontranscribed regions, the lengths of which are virus-specific. As the polymerase complex moves towards the 5' end of the template fewer transcripts are produced.

The ribonucleoprotein (RNP) complex which carries out the transcription (and replication: see Section 7.5) process of VSV consists of the genome RNA in association with large amounts of the virus nucleoprotein (N), lesser amounts of the virus phosphoprotein (P), and a few molecules of a large (L) protein. The L protein is thought to contain the catalytic sites for RNA synthesis, and possibly also capping of the mRNA, and is often referred to as the polymerase. The RNP complex can only initiate transcription at the 3' end of the genomic template RNA. For VSV, transcription begins with the production of a small, 49-nucleotide uncapped, nonpolyadenylated RNA which is an exact copy of the 3' terminus. This is the leader RNA for which no function is known. Each of the remaining five transcription units on the genome RNA are flanked by consensus sequences which direct the polymerase to firstly initiate and subsequently terminate transcription. The initiation sequences can only be recognized by polymerases travelling from the 3' end, the polymerase cannot bind to the internal transcription initiation sites. As the polymerase moves along the genome it meets a transcription initiation signal and begins mRNA synthesis. At some point during this process a cap is added, presumably by a component of the RNP complex since no capping enzymes are found in the cytoplasm where VSV transcription and replication occur (though there are some rhabdoviruses, notably some infecting plants, which replicate in the nucleus and so in these cases capping may use host processes). At the consensus transcription termination signal, which includes a homopolymer uracil tract, the polymerase adds a poly(A) tail as described for influenza virus. At this point the mRNA dissociates from the template and is removed for translation.

The polymerase then does one of two things. A proportion, possibly 50%, of the polymerase molecules also dissociate from the template and, having done so, can only rebind at the 3′ terminus where the transcription process begins again. The remaining polymerases move along the genome without transcribing until they encounter the next consensus transcription initiation signal where they begin to transcribe a region of the genome. At the end of this transcription unit the polymerase exercises the same two options of either dissociation, or translocation followed by reinitiation. The result of this process is that the mRNAs are synthesized sequentially in decreasing proportions as the polymerase moves towards the 5′ terminus of the genome (Fig. 10.12). The relative abundances of the proteins reflect those of the mRNAs. A similar strategy is used by paramyxoviruses such as Sendai virus.

Translational regulation of gene expression

The mRNA encoding the phosphoprotein (P protein) of several, but not all, paramyxoviruses contains two ORFs. The 5′ proximal AUG codon directs translation of a large ORF encoding the P protein (Fig. 10.13). The second ORF, initiated by the next AUG, directs the synthesis of a protein called C. The amino acid sequences of the P and C proteins are completely different, similar to the situation for the reovirus σ1 and σ$_s$ proteins (see Section 10.2). Analysis of the proteins produced by translation of the Sendai virus P mRNA indicates that several other proteins are also generated. Initiation of translation at the second and third AUG codons in the C protein ORF generates proteins called Y1 and Y2, respectively (Fig. 10.13a). Since these proteins are translated from the same reading frame as that used for the C protein, the Y1 and Y2 proteins are identical in sequence to the carboxy-terminus of the C protein. However, the

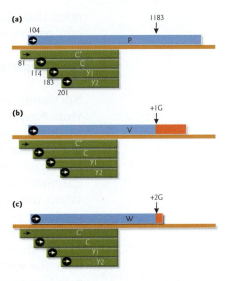

Fig. 10.13 Arrangement of the overlapping open reading frames of the Sendai virus P gene mRNA and the various potential protein products. (a) The relative locations and protein products of the ORFs are indicated. The arrow shows the insertion site for nontemplated G residues at nucleotide 1183 in the mRNA. The altered reading frames leading to the novel carboxy termini of the V (b) and W (c) proteins are indicated in red. Initiation of translation at AUG (white arrow) codons and the ACG (black arrow) for the C′ protein are indicated. The numbers refer to the nucleotide positions of the first nucleotide of the relevant codon.

differences at the amino termini of the three proteins mean that they are likely to have different functions in the virus replication cycle. This is similar to the papillomavirus E2 proteins described in Section 9.3.

The Sendai virus P mRNA also encodes a protein which is initiated at a codon ACG upstream of the first AUG codon which initiates synthesis of the P protein. Initiation of translation at this point uses the same reading frame as that for the C protein and the protein produced, called the C′ protein, contains additional amino acids at the amino terminus when compared to the C protein. As with the Y1 and Y2 proteins this difference may be sufficient to give the C′ protein a unique function. Use of a codon other than AUG for initiation of translation is rare but is also seen in certain retroviruses and parvoviruses.

Nontemplated insertion of nucleotides during transcription

Nucleotide sequence analysis of mRNAs transcribed from the Sendai virus P gene identified a novel strategy for gene expression which is also used by many other paramyxoviruses. During transcription of any gene it is important that the polymerase makes a faithful copy of the template to ensure fidelity of the protein sequence. However, RNA synthetic enzymes have no proofreading capacity so any errors cannot be corrected. When the Sendai virus P gene is being transcribed most of the mRNA produced is a faithful copy of the template. However, in approximately 30% of molecules an additional G nucleotide is inserted at a precise point, nucleotide 1183 in the mRNA, which lies in the ORF encoding the P protein. The mRNA with the additional G residue directs the synthesis of a novel protein which has the same amino terminal sequence as the P protein but a unique carboxyl terminus (Fig. 10.13b). The novel protein is called V. Strangely, some paramyxoviruses encode the V protein from the mRNA faithfully copied from the genome template and the P protein from the mRNA with the additional base(s). The function of the V protein is not yet known but it is thought to be involved in the disease process. In approximately 10% of transcription events from the Sendai virus P gene, two G residues are inserted at the same location (Fig. 10.13c). Translation of the resulting mRNA generates yet another protein, the W protein, which has the same amino terminal sequence as the P and V proteins but a unique carboxy terminus. The frequency of the nontemplated insertion event determines the relative abundance of the novel proteins.

Post-translational protein cleavage

In paramyxoviruses two glycoproteins, the attachment and fusion (F) proteins, are inserted into the lipid bilayer of the cell. The F protein is inserted with the carboxyl terminus on the external surface of the cell and, eventually, the virus (a type I membrane protein), while the amino terminus of the attachment protein is external (a type II membrane protein). The F protein is synthesized as a precursor, F_0, which has an amino-terminal hydrophobic signal sequence cleaved off by a host cell protease during insertion into the cell membrane. A further cleavage event, also carried out by a host cell protease, this time in the Golgi network, is required for the F protein to become functional. The two components of the F protein, called F_1 and F_2, generated by this cleavage activation event are covalently attached to each other by disulfide bonds. This is analogous to the cleavage of the influenza virus HA protein (see Section 10.8). The attachment protein of paramyxoviruses is not cleaved.

KEY POINTS

• Some RNA viruses may regulate their gene expression at the level of transcription (mRNA abundance, nontemplated nucleotide insertion).
• Some RNA viruses may regulate their gene expression at the level of post-transcription (RNA splicing)
• Some RNA viruses may regulate their gene expression at the level of translation (alternative open reading frames, alternative translation initiation codons, ribosomal frameshifting).
• Some RNA viruses may regulate their gene expression at the post-translational level by protein cleavage.
• Viruses may alter normal cellular process to gain an advantage in expression of their genes. The alteration of translation on ribosomes from cap-dependent to cap-independent is seen with reoviruses and picornaviruses.
• While all of these options are available in principle, each virus utilizes only a selection to suit its own circumstances and requirements.

QUESTIONS

• Compare and contrast the strategies adopted by alphaviruses and coronaviruses for generation of mRNA from a single-stranded positive sense RNA genome.
• Compare and contrast the mechanisms used by coronaviruses and rhabdoviruses to control the relative levels of their various mRNAs.

FURTHER READING

Bartlett N. M., Gilles, S. C., Bullivant, S., Bellamy, A. R. 1974. Electron microscopy study of reovirus reaction cores. *Journal of Virology* **14**, 315–326.

Bishop, D. H. L. 1986. Ambisense RNA genomes of arenaviruses and phleboviruses. *Advances in Virus Research* **31**, 1–51.

Conzelmann, K-K. 1998. Nonsegmented negative-stranded RNA viruses: genetics and manipulation of viral genomes. *Annual Review of Genetics* **32**, 123–162.

Curran, J., Kolakofsky, D. 1999. Replication of paramyxoviruses. *Advances in Virus Research* **54**, 403–422.

Curran, J., Latorre, P., Kolakofsky, D. 1998. Translational gymnastics on the Sendai virus P/C mRNA. *Seminars in Virology* **8**, 351–357.

Kormelink, R., de Haan, P., Meurs, C., Peters, D., Goldbach, R. 1992. The nucleotide-sequence of the M RNA segment of tomato spotted wilt virus, a bunyavirus with 2 ambisense RNA segments. *Journal of General Virology* **73**, 2795–2804.

Lai, M. M. C., Cavanagh, D. 1997. The molecular biology of coronaviruses. *Advances in Virus Research* **48**, 1–100.

Lamb, R. A., Horvath, C. M. 1991. Diversity of coding strategies in influenza-viruses. *Trends in Genetics* **7**, 261–266.

Portela, A., Zurcher, T., Nieto, A., Ortin, J. 1999. Replication of orthomyxoviruses. *Advances in Virus Research* **54**, 319–348.

Strauss, J. H., Strauss, E. G. 1994. The alphaviruses – gene-expression, replication, and evolution. *Microbiological Reviews* **58**, 491–562.

11

The process of infection: IV. The assembly of viruses

Assembly of virus particles is a highly ordered process which requires the bringing together of various particle components, including the genome nucleic acid, in a controlled sequence of events.

In infected cells virus-encoded proteins and nucleic acid are synthesized separately and must be brought together to produce infectious progeny virus particles (virions) in a process referred to as assembly or morphogenesis. For some viruses there is an additional aspect when assembly takes place in the nucleus and proteins synthesized in the cytoplasm must be moved to that location. Little information about assembly is available for most viruses. Despite the diversity of virus structure (see Chapter 3), there appears to be only three ways in which virus assembly occurs. Firstly, the various components may spontaneously combine to form particles in a process of self-assembly. In principle this is similar to crystallization, in which the final product represents a minimum achievable energy state. Secondly, the assembly process may require specific, virus-encoded, proteins which do not ultimately form a structural part of the virus. Finally, a particle may be assembled from precursor proteins, which are then modified, usually by proteolytic cleavage, to form the infectious virion. In the last two cases it is not possible to dissociate and then reassemble a mature particle from its constituent parts. The presence of a lipid envelope in some viruses, surrounding a nucleocapsid core, introduces an additional aspect to assembly of an infectious particle since this must be acquired separately from the main assembly process.

Introduction of the virus genome into the progeny virus particle is a critical step in the assembly of viruses. There are two processes by which this can be achieved. Reconstitution and other studies have shown that for some viruses the genome which forms part of the infectious particle plays an integral role in assembly, acting as an initiating factor with the particle components forming around the nucleic acid. For other viruses the particle, or a precursor of the particle, is formed first and the nucleic acid is then introduced at a late stage in the assembly process. The first process obligatorily involves the interaction of virus-encoded structural proteins with specific nucleotide sequences in the genome, referred to as packaging signals. While the second process may also require packaging signals to be present in the genome this is not always the case.

The structure of virus particles, features of which are considered in Chapter 3, determines, to some extent, the process of assembly. Recent advances in the determination of the three-dimensional structures of virus particles by cryoelectron microscopy and X-ray crystallography has shed more light on the precise nature of interactions between virion components. This has led to suggestions of how some of these components may come together during the assembly process.

11.1 SELF-ASSEMBLY FROM MATURE VIRION COMPONENTS

Absolute proof of self-assembly requires that purified viral nucleic acid and purified structural proteins, but no other proteins, are able to combine *in vitro* to generate particles which resemble the original virus in shape, size, and stability, and are infectious. A critical step in demonstrating assembly *in vitro* is the process used to effect disassembly. Disassembly should release the subunits in such a way that they retain the ability to reassociate in a specific manner to assemble the virus particle. Ideally, the disassembled constituent monomers of the virion should not be denatured by the disassembly process. This has been demonstrated for very few viruses to date.

For many viruses it is not possible to demonstrate spontaneous self-assembly even though it is suspected that it is an integral aspect of the process. For example, the assembly of a virus may be spontaneous but acquisition of infectivity then requires a maturation event which modifies one of the structural proteins after assembly has occurred. Following dissociation of purified infectious particles, the modified protein may not be able to interact spontaneously with the other components to reform the virion. Similarly, where a virus is enclosed in a lipid envelope, the disassembly process is likely to destroy this structure so an infectious particle cannot be regenerated.

11.2 ASSEMBLY OF VIRUSES WITH A HELICAL STRUCTURE

The best studied example of self-assembly is the in vitro reconstitution of tobacco mosaic virus (TMV) which has been used as a paradigm for the assembly of many other viruses with helical symmetry.

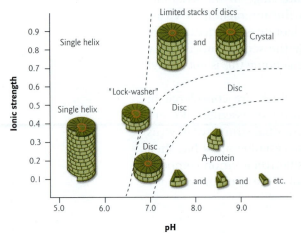

Fig. 11.1 Effect of pH and ionic strength on the formation of aggregates of TMV A protein.

Assembly of TMV

The TMV particle consists of a single molecule of positive sense ssRNA, embedded in a framework of small, identical protein molecules (A protein), arranged in a right handed helix (see Chapter 3), with each protein binding to three nucleotides of RNA. As already outlined in Chapter 1, TMV can be disassembled to yield protein and RNA components, which can be reassembled *in vitro* to yield active virus. However, the isolated protein, free from any RNA, can also be polymerized into a helical structure, indicating that bonding between the subunits is a specific property of the protein. While the most likely model for the assembly of the virus would be for the protein molecules to arrange themselves like steps in a spiral staircase, enclosing the RNA as a cork-screw-like thread, research has indicated that the assembly of TMV is a much more complicated process, in which the genome RNA plays an essential role.

In solution, TMV A protein forms several distinct kinds of complex, depending on the environmental conditions, particularly ionic strength and pH (Fig. 11.1). The complexes differ in the number of individual proteins which make them up. Of these complexes, the disc structure is considered the most important, since it is the dominant one found under physiological conditions. Each disc consists of two rings with 17 subunits per ring. This is close to the 16.34 A protein subunits per turn of the helix seen in the virus particle, so the bonding between the subunits is probably very similar in discs to that seen in the virus. While it is tempting to suggest that the discs could simply align to form a helix which would have a slightly different packing arrangement for the A protein subunits, this does not occur, indicating that the process is not that simple. The key components in the spontaneous assembly of TMV are "lock-washer" structures. These are very short helical structures, slightly more than two turns of a helix in dimension (Fig. 11.1). The lock-washers are produced from the discs by subtle changes in the conformation of the A protein subunit (Box 11.1). However, while lock-washers interact to form

Box 11.1

Evidence that the lock-washer structure is the intermediate in TMV assembly from subunit to virus particle

Analysis of electron micrographs of the discs formed by A protein *in vitro* found that the structure at the edges of the subunits is slightly different to that seen in the virus particle. The two ring disc structure presents a slightly different region to the underside of any further disc adding to it than would be the case in the assembled virus. This very small difference in structure appears to prevent the addition of further discs, and so no virus-like helices are formed. When the pH of the solution containing the discs is reduced, aggregation progressively occurs, with the appearance of short rods made up of imperfectly meshed sections of two helical turns. After many hours, these short rods combine to give the regular virus-like structure (Fig. 11.1). This is thought to be due to the lower pH reducing the repulsion between the carboxylic acid side chains of two adjacent amino acids in the A protein subunits leading to a conversion of the disc into a "lock-washer". It is the lock-washers which then associate to form the helix. However, this takes many hours indicating that the process is not as efficient as assembly seen *in vivo* and suggesting that another component may be involved.

helices like those in the virus particle, this is a very slow process indicating that there is another component in the system which catalyzes the assembly process. This essential catalyst is the TMV genome RNA.

When A protein subunits and small aggregates of dimers and trimers, etc. are mixed with TMV RNA, assembly is slow and formation of virus particles requires about 6 hours. However, when discs are mixed with RNA under the same conditions, assembly is rapid and mature virus forms within 5 minutes. Addition of small aggregates as well as discs to the RNA does not increase the rate of assembly. A possible model for the assembly of TMV is shown in Fig. 11.2 (but note that an alternative model is presented later). In this model the interaction of the genome RNA with a disc neutralizes the charges on the adjacent carboxyl group in the A protein subunits. This causes conversion of the disc into a lock-washer entrapping the RNA in the groove between successive turns of the helix (Box 11.1). Following conversion to the lock-washer form, a second disc can join, undergoing the same structural conversion to a lock-washer as the helix extends. The genome RNA is therefore a catalyst for the rapid assembly of a helix and becomes contained within the helical structure. Subsequent analysis has shown

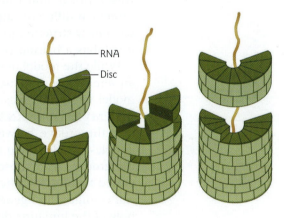

RNA
Disc

Fig. 11.2 (a–c) Simple model for assembly of TMV. See text for details.

Box 11.2

Evidence that TMV genome RNA contains a specific packaging site

When purified TMV RNA is mixed with limited amounts of viral coat protein in the form of discs, the RNA is incompletely encapsidated. Treatment of the incomplete structure with nucleases to destroy RNA not protected by the A protein identified a unique region of the RNA which is resistant to digestion. The protected RNA consists of a mixture of fragments up to 500 nucleotides long. The shortest fragments define a core about 100 residues long common to all the fragments. Analysis of the larger fragments show that they are not equally extended in both directions from the packaging site. Rather the helix is extended more rapidly in one direction than the other. These data are interpreted as showing that assembly is initiated at a unique internal packaging site on the RNA, and that the helix is extended bi-directionally but is more rapid in the 3′ → 5′ direction with extension 5′ → 3′ being much slower. This is acceptable as the packaging site is close to the 3′ end of TMV RNA.

that TMV genome RNA contains a specific region near the 3′ end, called the packaging site, at which the first disc binds and is converted into lock-washer form, and from where the helix is extended in both directions, though at different rates in each direction (Box 11.2). Computer-assisted secondary structure prediction of the packaging site suggests strongly that it exhibits a hairpin configuration (Fig. 11.3).

While the model described above is compatible with the available data, an alternative explanation is also possible. This is the "travelling loop" model which suggests that the hairpin structure at the packaging site in the TMV genome RNA inserts itself through the central hole of the disc into the groove between the rings of subunits. The nucleotides in the double-stranded stem then unpair and more of the RNA is bound within the groove. As a consequence of this interaction, the disc becomes converted to a lock-washer structure trapping the RNA (Fig. 11.4). The special configuration generated by the insertion of the RNA into the central hole of the initiating disc could subsequently be repeated during the addition of further discs on top of the growing helix; the loop could be extended by drawing more of the longer tail of the RNA up through the central hole of the growing virus particle. Hence, the particle could elongate by a mechanism similar to initiation of packaging, only now instead of the specific packaging loop there would be a "travelling loop" of RNA at the main growing end of the virus particle (Fig. 11.4). This loop would insert itself into the central hole of the next incoming disc, causing conversion to the lock-washer form and continuing the growth of the virus particle.

A feature of the first model shown in Fig. 11.2 is that discs have to be threaded on to the RNA chain and this would obviously be the rate-limiting step. However, the "travelling loop" model shown in Fig. 11.4

overcomes this problem as far as growth in the 5′ direction is concerned, for incoming discs would add directly on to the growing protein rod. Discs would still have to be threaded on to the 3′ end of the RNA and thus elongation in this direction would be much slower, as has been observed experimentally. One prediction of the "travelling loop" model is that both the 5′ and 3′ tails of the RNA should protrude from one end of partially assembled TMV particles. Electron micrographs of such structures have been observed. Currently it is not known which model is correct.

11.3 ASSEMBLY OF VIRUSES WITH AN ISOMETRIC STRUCTURE

As indicated in Chapter 3, the particles of all isometric viruses are icosahedral with 20 identical faces. The process of triangulation describes the mechanism used by isometric viruses to increase the size of their capsid, with a concomitant increase in the number of capsomere subunits. As indicated above, reconstitution experiments have had only limited success and the most detailed study so far reported on the self-assembly and reconstitution of an isometric virus is that of cowpea chlorotic mottle virus, a plant virus (CCMV; a bromovirus of Baltimore class 4). The formation of infectious CCMV particles from a stoichiometric mixture of initially separated CCMV RNA and protein affords proof of the ability of this virus to self-assemble.

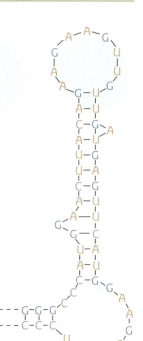

Fig. 11.3 The packaging site of TMV RNA. The loop probably binds to the first protein disc to begin assembly. The fact that guanine is present in every third position in the loop and adjacent stem may be important in this respect.

Assembly of picornaviruses

Over recent years the three-dimensional structures of a number of viruses have been determined by X-ray crystallography. Many of these have been picornaviruses, including poliovirus, which have icosahedral particles, of approximately 30 nm in diameter. Poliovirus particles consist of 60 copies of each of the four structural proteins VP1–4. The proteins associate together in a complex and these complexes are arranged in groups of three on each of the 20 faces of the icosahedron (see Fig. 3.7). Much information is available about the sequence of events in the assembly of poliovirus which illustrates how an icosahedral virus particle can be generated.

The entire genome of poliovirus is translated as a single giant polypeptide, which is cleaved as translation proceeds into smaller polypeptides (see Section 10.3). The first cleavage generates a polypeptide called P1 (1ABCD in Fig. 10.5), which is the precursor to all four virion proteins. Synthesis of P1 is directed by the 5′ end of the genome and it is synthesized completely before being cleaved, suggesting that folding is necessary for cleavage. Subsequent cleavages of P1 give rise to proteins called VP0, VP1, and VP3. Poliovirus assembly begins from these three proteins, which associate with each other in infected cells to produce a complex

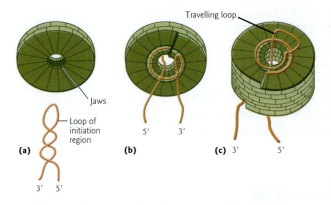

Fig. 11.4 The "travelling loop" model for TMV assembly. Nucleation begins with the insertion of the hairpin loop of the packaging region of TMV RNA into the central hole of the first protein disc (a). The loop intercalates between the two layers of subunits and binds around the first turn of the disc. On conversion to the lock-washer (b) the RNA is trapped. As a result of the mode of initiation the longer RNA tail is doubled back through the central hole of the rod (c), forming a traveling loop to which additional discs are added rapidly.

with a sedimentation coefficient of 5S. Five of the 5S complexes come together to form a 14S pentamer complex. Twelve of the 14S complexes in turn aggregate together to form an empty 73S capsid. At this point the genome positive sense ssRNA is added. The mechanism by which the genome is recognized and then inserted into the capsid is not known. The VPg protein covalently attached to the 5′ end of the genome RNA may be involved in the genome acquisition but this cannot be the only factor since the antigenome RNA produced during replication (see Section 7.4) also has VPg at the 5′ end and it is not encapsidated. Consequently, it is believed that a packaging signal sequence must be present in the genome RNA. After the RNA is encapsidated, VP0 is cleaved to yield the virion proteins VP2 and VP4 (Section 11.6), causing an alteration in the sedimentation coefficient of the particle. VP4 is located inside the particle with the other three proteins on the particle surface. Because of this late cleavage event, dissociation and reassembly to form infectious particles *in vitro* is not possible. These steps, summarized in Fig. 11.5, clearly show that the poliovirus particle can spontaneously assemble in the cell and that the genome RNA is added to an immature particle rather than being an integral component in the assembly process such as is seen for TMV.

Assembly of adenoviruses

Fig. 11.5 Summary of the steps involved in the assembly of poliovirus.

Adenovirus particles range in size from 70 to 90 nm in diameter, depending on the strain being studied, and appear as icosahedra in the electron microscope (see Fig. 3.13). The particle is more complex than that of the picornaviruses, containing at least 10 proteins. The 252 capsomeres which form the external surface consist of 240 proteins arranged such that they have sixfold symmetry (called the hexon capsomeres) and the remaining 12 are arranged at the vertices of the icosahedron with fivefold symmetry (penton proteins). Fiber structures project from the

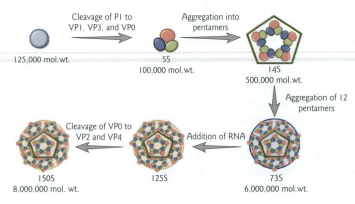

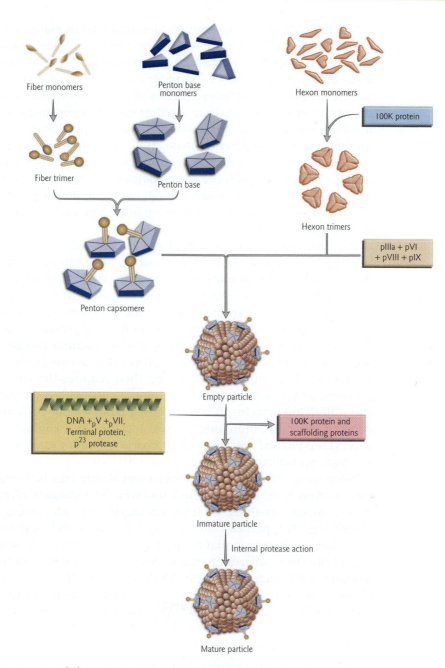

Fig. 11.6 Summary of the steps involved in the assembly of adenovirus.

pentons. While many details of the assembly of adenoviruses remain to be established, those available indicate that, unlike the picornavirus assembly process, the individual components of the adenovirus particle are assembled independently of each other and are brought together in a directed fashion (Fig. 11.6).

The later stages of assembly of adenovirus takes place in the nucleus of the infected cell where the virus genome is replicated and so the proteins which form the virion must be translocated there. The proteins which form the fiber, the base of the penton capsomere, and the hexon capsomere are synthesized independently of each other in the cytoplasm. The fiber and hexon proteins come together to form independent trimer intermediates while the penton monomers form a pentameric penton base. The fiber trimers and penton base then associate to give a penton capsomere. The formation of the hexon trimer unit and its import into the nucleus require the presence of an additional adenovirus protein with an M_r of 100,000 which interacts directly with the hexon proteins. The M_r 100,000 protein is not found in the mature virus particle and is termed a "scaffolding" protein.

The remaining steps in the adenovirus assembly process have been inferred from detection and analysis of putative intermediate structures. A key aspect is the formation of immature virus particles which do not contain the genome dsDNA or core proteins but which do contain at least three proteins which are not found in the infectious particle. These three scaffolding proteins may be removed, in part, by proteolytic degradation. The hexon capsomeres come together in a nonamer complex and subsequently 20 of the nonamer complexes interact to produce an icosahedral cage-like lattice. This structure then acquires the virus DNA, core proteins, and remaining structural components, with concomitant loss of the scaffolding proteins. It is not known whether the DNA and the core proteins enter the immature particle together in a complex or in rapid succession. Finally, a protease in the particle cleaves various components to create an infectious virion.

Adenovirus DNA contains a protein covalently attached to the termini (see Section 6.4), but this is not involved in packaging of the genome into particles. Analysis of DNA packaged into adenovirus defective–interfering (DI) particles and the generation of deletion mutants has identified a region essential for packaging. Approximately 400 bp at one end of the DNA, adjacent to the E1A gene, must be present for the DNA to enter the immature particle. The conclusion that this is a packaging signal responsible for the specific acquisition of DNA by the particle is supported by evidence that the genome enters the particle in a polar fashion, one end first, and that this polarity is lost when the sequence is duplicated at the other end of the DNA.

11.4 ASSEMBLY OF COMPLEX VIRUSES

The assembly process for animal viruses with particles which are neither helical nor isometric is not well understood. From our understanding of asssembly it is likely that the processes for these viruses are also highly organized. In the case of phage, such as phage λ, which have a complex structure consisting of an isometric head attached to a helical tail the process has been well characterized using mutants which are unable to proceed beyond certain points in assembly. The isometric head structure of phage λ is similar to that of adenovirus and also acquires the DNA genome after capsid assembly. A head-like structure, called procapsid I, is produced with the assistance of scaffolding proteins. The "entrance" to the procapsid is formed by the presence of a portal protein complex (Fig. 11.7). The scaffolding proteins are removed causing a change in shape to produce a procapsid II structure. A protein within the phage procapsid II head recognizes the *cos* sequence in the concatemeric DNA and cleaves at that point to produce the single-stranded overhang described in Section 6.3. The DNA is then brought into the empty head until it is full, followed by a second cleavage to produce a mature capsid (Fig. 11.7).

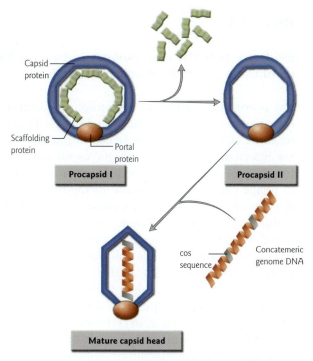

Fig. 11.7 Diagram of the morphological changes in the head structure of bacteriophage λ. The procapsid I structure loses the scaffolding proteins to produce a procapsid II structure which in turn acquires the phage genome DNA to form a mature head.

The phage λ tail and tail fiber components assemble independently of each other (Box 11.3). The tail takes the form of a helical structure which includes the head–tail connector, the core to which it is attached, and the surrounding sheath. For both the tail and tail fibers, scaffolding proteins are involved in assembly. The tail and tail fibers then spontaneously come together, with the possible involvement of one or more scaffolding protein. Following the independent assembly of the heads and tails, the two components come together in an unknown way. It is possible that one end of the phage DNA protrudes through a vertex, disturbing the fivefold symmetry, and that this promotes tail addition. Indeed, protrusion of the DNA a short way into the tail may be a necessary structural feature for successful injection following contact of the phage with a susceptible bacterium.

11.5 SEQUENCE-DEPENDENT AND -INDEPENDENT PACKAGING OF VIRUS DNA IN VIRUS PARTICLES

The adenovirus assembly process indicates clearly that packaging of genome DNA is sequence-dependent. Many other viruses also rely on the presence of specific sequences to select the virus DNA or RNA though the processes differ in detail. In the case of herpes simplex virus the assembly process is coupled to the excision of the genome from the concatemeric replication intermediate (see Section 6.3).

Box 11.3

Evidence that the components of tailed phage are assembled separately before combining to produce progeny particles

The process of assembly of bacteriophage lambda and T4 are very similar, with the head and tail components being assembled separately before coming together. Several conditional lethal mutants of T4 are available which are unable to produce infectious particles when bacteria are infected under nonpermissive conditions. Electron microscopic examination of extracts from these bacteria revealed the presence of structures readily recognizable as components of phage particles. While some of the mutants were defective in one or other of the genes which encoded the structural proteins, many of these mutants were not. This indicated that additional, nonstructural gene products were involved in the assembly process. The ability of the mutants to synthesize some recognizable structural components of the phage T4 particle, such as the head, tail, and tail fibers, also showed that these are assembled independently of each other, similar to the situation for adenovirus capsid components (Section 11.3).

A significant advance came with the discovery that the morphogenesis of phage T4 from partially assembled components could be made to occur *in vitro*. In one experiment, purified fiberless phage isolated from cells infected with a tail fiber mutant were mixed with an extract from cells infected with a mutant which could not synthesize heads. Infectivity rapidly increased by several orders of magnitude, indicating that the "headless" mutant extract was acting as a tail fiber donor, whilst the other extract supplied the heads.

The herpes simplex virus particle is covered by a lipid envelope which is acquired as the final stage in assembly in a process described in Section 11.6, but the internal component of the particle, which is an isometric nucleocapsid, is formed first in a process probably similar to that described for adenovirus. The herpes simplex virus assembly process generates an immature form of the nucleocapsid lacking the virus DNA. As indicated in Section 6.3 the HSV genome contains direct repeats (the "a" sequence) at the ends of the DNA. Each "a" sequence contains within it two short regions, called *pac*1 and *pac*2, which are required for packaging. During assembly the immature nucleocapsid recognizes the *pac* sequences within a DNA concatemer and cleaves the DNA at the precise termini while at the same time inserting it into the interior of the particle. The presence of the "a" sequence alone cannot be sufficient to specify cleavage and ensure that the entire genome is taken into the particle since an additional copy of the "a" sequence is located within the genome at the boundary of the internal long and short repeat elements (IR_L and IR_S; Fig. 6.4). The most likely explanation is that the length of the DNA also plays a role in assembly, with constraints on the minimum and maximum permissible lengths of the DNA accepted into a particle.

A similar process occurs with bacteriophage λ which also generates con-catemeric DNA from a circular intermediate. The correct length of DNA within the phage head ensures that the next *cos* sequence in the con-catemer is readily accessible to the cleavage enzyme and the DNA is cut to leave an overhang. If two *cos* sequences are brought close together by a deletion of the genome no *cos* sequence is available to the enzyme when the phage head is full and particle assembly aborts. Similarly, if an inser-tion occurs to increase the distance between *cos* sequences the head becomes full before the next cleavage site is reached and again the assembly pro-cess aborts. In this way phage λ regulates the size of the genomic DNA which can be packaged and ensures that all of the genes are present. However, the phage λ packaging system tolerates some flexibility in genome length which permits the virus to carry additional DNA derived from the host during the process of specialized transduction (see Section 15.4).

For a long time it was considered essential that dsDNA animal viruses had controls on the assembly of particles to ensure that only virus DNA could be packaged. For those viruses which have a packaging signal this is the only control present and the rest of the sequence is irrelevant. An example is SV40 which shows specific packaging via the six GC boxes in its con-trol region. These sequences bind to the host transcription factor SP1 which then interacts with viral capsid proteins to initiate packaging around the genome. However, it has been shown that polyomavirus can, under certain conditions, package foreign DNA. This indicates that the control mechanisms for some viruses may not be as precise as was originally thought and it may be possible to exploit this in the development of new ways to transfer DNA into cells for gene therapy. True sequence-independent packaging of DNA is seen in phages which exhibit generalized transduction.

11.6 THE ASSEMBLY OF ENVELOPED VIRUSES

A large number of viruses, particularly viruses infecting animals, have a lipid envelope as an integral part of their structure. These include her-pes-, filo-, retro-, orthomyxo-, paramyxo-, corona-, arena-, pox-, and iri-doviruses, the tomato spotted wilt virus group and rhabdoviruses of plants and animals (see Appendix). For each virus, the component held within the envelope, the nucleocapsid, is of a predetermined morphology which may be helical, isometric or of a more complex nature. For most enveloped viruses the nucleocapsid is formed in its entirety prior to acqui-sition of the lipid envelope.

Assembly of helical nucleocapsids

The assembly process of TMV is usually considered to be a model for the generation of all helical virus structures. However, the structures of nucleocapsids of enveloped viruses differ in many ways. This is particularly true of the nucleocapsids of the negative sense ssRNA viruses such as filo-, paramyxo-, rhabdo-, and orthomyxoviruses. For members of some of these virus groups the RNA genome in the nucleocapsid is protected from degradation by nucleases *in vitro* while for the others it is not. This indicates that the structures must be arranged differently. For all Baltimore class 5 viruses except the orthomyxoviruses, the basic structure of the nucleocapsid consists of the genome RNA encapsidated by a nucleoprotein, called N or NP, in association with smaller numbers of a phosphoprotein and a few molecules of a large protein. This complex carries out the processes of replication (see Chapter 7) and transcription (see Chapter 10). For orthomyxoviruses, the nucleoprotein associates with three proteins, PA, PB1, and PB2 (see Chapter 10), which carry out RNA synthesis. While some of the steps in the assembly process for nucleocapsids of specific viruses are understood, no precise details are available to describe all aspects.

Analysis of genomes of negative sense ssRNA DI viruses (see Chapter 10) and the generation of synthetic genomes for reverse genetics studies have shown that the termini of the genomes are essential for the encapsidation process. It is thought that specific sequences are located in the terminal regions of the genome (and antigenome) which initiate the generation of nucleocapsid structures, and ensure that virus RNA is packaged. Expression of the measles virus nucleoprotein in bacteria, in the absence of virus genome RNA, results in the production of short nucleocapsid-like RNA–protein complexes, indicating that the nucleoprotein has the inherent ability to form nucleocapsids around any RNA. Presumably the signal in the genome termini gives a significant advantage to the genome in competition with any other RNA for binding nucleoprotein. For the paramyxovirus Sendai virus, the phosphoprotein which is part of the nucleocapsid associates strongly with the nucleoprotein prior to the interaction with RNA. This association prevents the nucleoprotein from interacting nonspecifically with RNA and gives specificity to the encapsidation process. This may be a general feature of many class 5 viruses.

The molecular details of the interaction between the nucleoprotein and genome RNA have not yet been established for any helical nucleocapsid complex. For several paramyxoviruses, such as Sendai virus and measles virus, the genome RNA must always have a number of nucleotides that is divisible by 6. This "Rule of Six" is interpreted to be the result of the nucleoprotein binding to groups of six nucleotides in the RNA. If the genome does not conform to the Rule of Six, unencapsidated nucleotides will be present at one, or both, termini and this will prevent replication and hence propagation of the genome. However, for most viruses there is no such strict length requirement for the genome and the nature of the interaction between the nucleoprotein and the genome is not known.

A significant difference between the structure of the TMV capsid and the nucleocapsid of some negative sense ssRNA viruses lies in the location of the genome RNA. In the TMV particle the RNA is located entirely within the capsid. For influenza virus and the rhabdovirus vesicular stomatitis virus (VSV) biochemical analysis suggests that the RNA is wound around the outside of the nucleocapsid complex with the nucleotide bases exposed. This leaves the nucleotides available to be used as templates during replication and transcription. It is not yet known whether this is a common feature of the class 5 viruses, but the difference between this structure and that of the TMV capsid reflects their different roles: in TMV the capsid serves to protect the genome RNA as the particle does not contain the extra, lipid, layer found in the class 5 viruses.

Assembly of isometric nucleocapsids

In the absence of detailed information it is assumed that the assembly of isometric nucleocapsids is similar to the processes described for non-enveloped viruses. Thus, the various components are assembled either around the virus genome in response to a packaging signal, as is the case for the togaviruses, or an immature particle is formed and the genome is added subsequently, as in herpesviruses (Section 11.5).

Acquisition of the lipid envelope

Cells contain a large quantity of lipid bilayer membranes. These membranes define boundaries between cellular compartments, such as the nucleus and cytoplasm, as well as making up the external surface of the cell itself. The majority of enveloped viruses acquire their envelope by budding from the plasma membrane of the infected cell (Fig. 11.8). Four events leading to budding have been identified. Firstly, the nucleocapsids form in the cytoplasm. Secondly, viral glycoproteins, which are transmembrane proteins, accumulate in patches of cellular membrane. Thirdly, the cytoplasmic tails of the glycoproteins which protrude from the membrane interact with the nucleocapsid. This interaction may be direct, or indirect via an intermediary matrix protein which becomes aligned along the inner surface of the modified patch of membrane. Once these interactions have taken place the virion is formed by budding. A simple way of envisioning the process is that the glycoproteins in the membrane progressively interact with the nucleocapsid, or the matrix protein and nucleocapsid where appropriate. As the number of interactions increases, the membrane with the glycoproteins inserted in it is "pulled" around the nucleocapsid. When the nucleocapsid is completely enclosed the lipid bilayer pinches off to release the virus particle. As a result of this, viruses which bud from the plasma membrane are automatically released when the budding process is complete. An unusual feature of

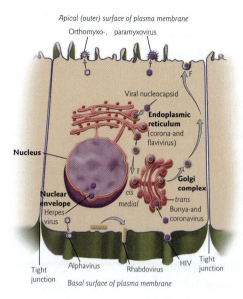

Apical (outer) surface of plasma membrane

Orthomyxo-, paramyxovirus

Viral nucleocapsid

Endoplasmic reticulum (corona-and flavivirus)

Nucleus

Golgi complex

cis
medial

Nuclear envelope
Herpes virus

trans
Bunya-and coronavirus

Tight junction

Alphavirus

Rhabdovirus

HIV

Tight junction

Basal surface of plasma membrane

Fig. 11.8 Sites of maturation of various enveloped viruses. F, fusion of a vesicle with a membrane.

the budding process is that host membrane proteins are excluded entirely from viral envelopes except for the retroviruses. The mechanism by which this exclusion is achieved is not fully understood but it is likely that the virus glycoproteins are inserted into membranes in specific areas where host proteins are not readily found. Retrovirus envelopes contain specifically selected host proteins.

Some viruses, such as corona- and bunyaviruses, bud into the endoplasmic reticulum (ER), acquiring cytoplasmic membranes as their envelopes, and are then released to the exterior via the Golgi complex. First, the virus particle buds into the ER, acquiring an envelope in a process similar to that described for budding with the plasma membrane. The enveloped virus then exits the ER within a vesicle, following the normal mechanism and route for trafficking of material from the ER to the Golgi complex. The vesicle containing the enveloped virus moves to and fuses with the *cis* Golgi complex. The virus then moves through the Golgi and exits in another vesicle which buds off from the concave *trans* Golgi network surface. This vesicle is transported to the plasma membrane where it fuses to release the enveloped particle to the exterior of the cell. The virus envelope does not fuse with any of the vesicle membranes, so the original lipid bilayer is retained throughout this process.

Herpesviruses assemble in the cell nucleus, with many of the viral proteins synthesized in the cytoplasm being transported into the nucleus. After entry of the virus genomic DNA into the nucleocapsid, the virus must acquire an envelope. The source from, and process by, which the envelope is obtained is still not completely clear. The most likely process is that the virus buds through the inner nuclear membrane and thus becomes enveloped. This envelope then fuses with the outer nuclear membrane, releasing the nucleocapsid into the cytoplasm. The nucleocapsid then undergoes a second budding event into cytoplasmic vesicles so acquiring the envelope that will stay with them. The vesicles progress through the normal Golgi system and fuse with the cell membrane, releasing their contents including the enveloped HSV virions outside the cell (see Fig. 9.7). This is similar to the release of the bunya- and coronaviruses.

For other viruses the process by which the lipid envelope is acquired by the nucleocapsid is unclear. Very little is known about the assembly of lipid-containing phages and the iridoviruses, but it appears that the envelope is not incorporated by a budding process. In this respect, they resemble poxviruses, whose morphogenesis has been studied extensively by electron microscopy of infected cells. In thin sections, poxvirus particles initially appear as crescent-shaped objects within specific areas of cytoplasm, called "factories," and even at this stage they appear to contain the trilaminar membrane which forms the envelope. The crescents are then completed into spherical structures. DNA is added, and then the external surface undergoes a number of modifications to yield mature virions. The origin of the lipid in the virus is not known since it does not appear to be physically connected to any pre-existing cell membrane.

One complication with envelope acquisition is that many differentiated cells *in vivo* are polarized, meaning that they carry out different functions with their outer (apical) surface and their inner (basolateral) surface. For instance, cells lining kidney tubules are responsible for regulating Na^+ ion concentration. They transport Na^+ ions via their basolateral surface from the endothelial cells lining blood vessels to the cytoplasm and then expel them from the apical surface into urine. Some cell lines, such as Madin–Darby canine kidney (MDCK) cells, retain this property in culture, but this can only be demonstrated when they form confluent monolayers and tight junctions between cells. The latter serve to separate and define properties of the apical and basolateral surfaces. These properties reside in cellular proteins which have migrated directionally to one or other surface. The lipid composition of a polarized cell is also distributed asymmetrically between apical and basal surfaces. Directional migration and insertion into only one of the cell surfaces also occurs with some viral proteins. For instance, if a cell is dually infected with a rhabdovirus and an orthomyxovirus, electron microscopy shows the rhabdovirus budding from the basal surface, which is in contact with the substratum, and the orthomyxovirus from the apical surface (Fig. 11.8). This property is determined by a molecular signal on the viral envelope glycoproteins which directs them to one surface or the other; tight junctions are needed or proteins diffuse laterally and viruses can then bud from either surface. This aspect of virus assembly has implications for viral pathogenesis (see Section 18.8).

11.7 MATURATION OF VIRUS PARTICLES

For most animal and bacterial viruses, formation of infectious virions requires the cleavage of precursor protein molecules into functional proteins. These cleavages may occur before or after the precursors have been assimilated into the virus particle. Examples of cleavage of a precursor prior to assembly into the virion is seen with the glycoproteins of orthomyxo- and paramyxoviruses, as well as many others. For the orthomyxoviruses, like influenza virus, the hemagglutinin (HA) protein is synthesized as an inactive precursor (HA1). Depending on the host and the specific type of influenza virus, cleavage of the HA0 protein into HA1 and HA2 is carried out either by a host protease immediately prior to insertion into the plasma membrane of the infected cell, or by

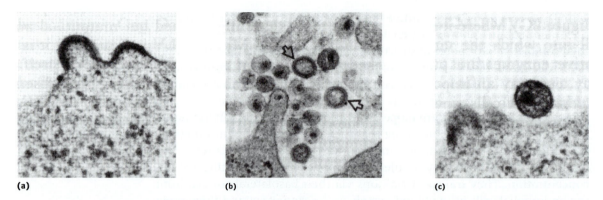

Fig. 11.9 Morphogenetic changes seen in HIV particles following budding from the surface of the host cell. (a) Budding of an HIV capsid. (b) Mixture of both immature (arrows) and mature HIV particles. (c) Mature HIV virion with characteristic cone-shaped core. (Taken from Hunter, E. 1994. *Seminars in Virology* **5**, 71–83.)

an extracellular protease after the virus has been released from the cell. The HA protein is acquired by the virus during budding when the envelope is added. The cleavage of HA protein is essential, as uncleaved protein cannot fuse with the host cell membrane to initiate the next round of infection. A similar situation is seen with the fusion proteins of the paramyxoviruses. For Sendai virus and the economically important avian paramyxovirus, Newcastle disease virus, the protease responsible for the cleavage of the fusion protein is restricted only to cells of the respiratory tract and this distribution prevents the viruses replicating elsewhere in the body. It is for this reason that they cause only respiratory disease in their host.

An example of a maturation cleavage occurring after particle formation is seen with poliovirus where the polypeptide VP0, which is assembled into a particle (Section 11.3), is cleaved to form VP2 and VP4. Without this cleavage the particle is not infectious. A more dramatic example of morphogenetic alterations in a virus particle following assembly is seen with HIV. The HIV virion consists of a nucleocapsid surrounded by a lipid envelope. The HIV nucleocapsid in newly budded particles does not show the distinctive structure that is seen with infectious, mature, particles. Electron microscopic examination of particles at various times after budding has shown that the core of the particle undergoes considerable alteration in structure before achieving the final, infectious, form. This process is shown in Fig. 11.9. Its basis is the action of a virion protease which cleaves the assembled gag precursor proteins (see Section 9.8) into their functional products, the matrix, capsid, and nucleocapsid proteins, after which the morphology of these particle components can mature. The HIV envelope contains two glycoproteins, gp41 and gp120, which are cleaved from a precursor by a cellular protease similar to the influenza virus HA protein and the paramyxovirus fusion proteins. This cleavage is also essential for infectivity of the progeny particle.

KEY POINTS

- Virus assembly occurs spontaneously within the infected cell.
- Assembly is a highly ordered process involving intermediates.
- Assembly may involve scaffolding proteins that are essential components of the process but which do not form part of the final particle.
- The genome may form an integral part of the particle and be a component in the assembly process, or it may be added to preformed immature virus particles.
- Genome acquisition may be sequence specific, involving packaging signals, or, rarely, it may be sequence independent.
- Viruses acquire the lipid envelope from one of the several membrane structures in the cell.

QUESTIONS

- Compare the processes of virion assembly used by bacteriophage λ and adenovirus.
- Discuss the role of packaging signals in the process of acquisition of virus genomes during particle assembly.

FURTHER READING

Butler, P. J. G. 1984. The current picture of the structure and assembly of tobacco mosaic virus. *Journal of General Virology* **65**, 253–279.

Compans, R. W. 1991. Protein traffic in eukaryotic cells. *Current Topics in Microbiology and Immunology* **170**, 1–186.

Earnshaw, W. C., Casjens, S. R. 1980. DNA packaging by the double-stranded DNA bacteriophages. *Cell* **21**, 319–331.

Fujisawa, H., Hearing, P. 1994. Structure, function packaging signals and specificity of the DNA in double-stranded DNA viruses *Seminars in Virology* **5**, 5–13.

Henrik Garoff, H, Sjöberg, M., Cheng, R. H. 2004. Budding of alphaviruses. *Virus Research* **106**, 103–116.

Homa, F. L., Brown, J. C. 1997. Capsid assembly and DNA packaging in herpes simplex virus. *Reviews in Medical Virology* **7**, 107–122.

Hunter, E. 1994. Macromolecular interactions in the assembly of HIV and other retroviruses *Seminars in Virology* **5**, 71–83.

Jones, I. M., Morikawa, Y. 1998. The molecular basis of HIV capsid assembly. *Reviews in Medical Virology* **8**, 87–95.

Mettenleiter, T. C. 2002. Herpesvirus assembly and egress. *Journal of Virology* **76**, 1537–1547.

Nayak, D. P., Hui, E. K., Barman, S. 2004. Assembly and budding of influenza virus. *Virus Research* **106**, 147–165.

Part III

Virus interactions with the whole organism

12

The immune system and virus neutralization

- *The immune system is a complex interacting mixture of cells and soluble components that protect us from infections and neoplasms.*
- *The immune system is clearly imperfect as we all experience infections, some short term and some long term.*
- *No microorganism passively suffers the onslaught of the immune system; all have evolved mechanisms to combat or evade it. How these work depends on the microorganism's survival strategy.*
- *The immune system has evolved to kill cells, and this lethal capability has to be tightly controlled to ensure that it is only deployed against foreign cells or host cells invaded by pathogens.*
- *The immune system is a finely balanced compromise – if set too narrowly it attacks harmless foreign matter that we eat or inhale, and causes bystander damage through "friendly fire" (e.g. allergy, hypersensitivity), or even autoimmunity; if set too slack then some microorganisms go unrecognized.*

Infections of multicellular animals are complicated by the variety of cell types present in an individual and the possession of an elaborate system – the *immune system* – that defends the organism against infection by any foreign invaders. There are two types of host response to virus infection: *innate immunity* and *adaptive immunity*. A key point to be borne in mind is that infection is a battle between the virus and the immune system, with the virus doing everything it can to counter, subvert, and evade immune responses. The two-way nature of a host–virus encounter cannot be emphasized too strongly, and the outcome of an infection – subclinical infection, acute disease, long term infection, or death – depends on which comes out on top. The following chapter is provided as

Box 12.1

An alphabetical glossary of essential immunological terms

Adaptive immunity: a defence system that is not ready to act until it is activated by infection; epitope-specific B and T lymphocytes are stimulated to divide furiously and differentiate into effector cells and memory cells upon encountering specific antigen.

Antibody: a soluble, epitope-specific protein synthesized and secreted by a B effector cell.

Antigen: a molecule that reacts, though its epitopes, with B or T cell receptors.

Antigen presentation: display on the cell surface of a peptide complexed with an MHC I or MHC II protein that is capable of reacting with a T cell receptor.

Antigen processing: digestion of self and non-self proteins to peptides in normal cells that then complex with MHC I or II proteins.

B cell receptor (BCR): integral plasma membrane protein of B cells that recognizes an epitope (any type of molecule); essentially antibody with an added membrane anchor; antibody made by that B cell has the same epitope specificity as the BCR.

B effector cell: short-lived cell that makes antibody; also known as a plasma cell, an activated B cell, and an antibody-forming cell.

B and T lymphocytes: the main components of adaptive immunity; inactive cells, each of which has an epitope receptor of one specificity.

CD4: an integral membrane protein on the surface of T cells that recognizes MHC class II proteins; diagnostic of helper T cells; T cells have *either* CD4 or CD8.

CD8: an integral membrane protein on the surface of T cells that recognizes MHC class I proteins; T cells have *either* CD8 or CD4.

Cell-mediated immunity (CMI): immunity mediated by various T cell responses.

Chemokines: a superfamily of small (*c.* 10,000 M_r) soluble protein mediators and communicators involved in a variety of immune, inflammatory and other processes; involved in the development of dendritic, B and T cells, and lymphoid cell trafficking.

Cytokines: a superfamily of soluble protein mediators and communicators released from cells by specific stimuli; e.g. when activated lymphocytes bind antigen (these are also known as lymphokines); includes the interleukins (IL-1, IL-2, etc.).

Cytotoxic T lymphocyte: activated T cells that can kill target cells by lysing them.

Epitope: part of a molecule that binds the paratope of a cognate T or B cell receptor.

Humoral immunity: antibody-mediated immunity.

Immunogen: a molecule that stimulates adaptive immunity to its epitopes.

Immunological memory: the ability to remember a foreign antigen experienced by the body any time earlier leading to a more easily activated, more rapid, and greater immune response on subsequent encounters with that antigen; physically it is an expanded population of antigen-specific B and T lymphocytes, with evolved high affinity BCRs.

Innate immunity: a defence system that is always ready to act and is unchanged by infection; its action is not epitope-specific.

> **Box 12.1 (*Cont'd*)**
>
> **MHC (major histocompatibility) protein** (in man usually referred to as human leuko-cyte antigen (HLA)): MHC class I – integral plasma membrane protein found on nearly all cells of the body; MHC class II – integral plasma membrane protein that is normally restricted to cells of the immune system.
> **Monoclonal antibody (MAb):** antibody of a single sequence and hence specific for a single epitope made by a clonal population of immortalized B effector cells (a hybridoma).
> **Non-self:** any antigen that is not normally present in that individual.
> **Paratope:** region of a B or T cell receptor that recognizes and binds a cognate epitope.
> **Self (antigen):** any antigen normally present in that individual.
> **T cell receptor (TCR):** integral plasma membrane protein of T cells that recognizes a T cell epitope; the epitope is a peptide complexed with an MHC I or MHC II protein on the surface of another cell of the same person; also known as antigen or epitope presentation.
> **T effector cell:** short-lived cells that directly act on other cells of the body.

a refresher course in immunology as it relates to viruses and virus infections. A number of immunological terms are defined in Box 12.1. Readers are referred to specialized immunology texts for a more comprehensive treatment.

12.1 VIRUSES AND THE IMMUNE SYSTEM – AN OVERVIEW

The immune response has three main components: innate immunity, adaptive T cell-mediated immunity, and adaptive B cell-mediated immunity as summarized in Fig. 12.1 in the context of virus infections. The cells of the immune system that patrol the body are single, and mobile. Cells of the innate immune system are ready for action, but T and B lympho-cytes are sentinels that cannot carry out offensive action until they have clonally expanded and differentiated. There are two main targets during virus infection – *virus particles* and the factories that make them – *virus-infected cells*.

   Innate immunity is the first line of defence and is possessed in some form by all animals. It is composed of soluble components (e.g. interfer-ons, cytokines, chemokines, complement), and cells (e.g. macrophages, dendritic cells, polymorphonuclear leukocytes (PMNL), natural killer (NK) cells). These do not have virus-specific recognition elements and thus act nonspecifically, but they have the advantage of being present constitutively and thus there is no delay in mobilizing them following infection. Their activity is not permanently increased as a result of experiencing an infection, although it may rise transiently.

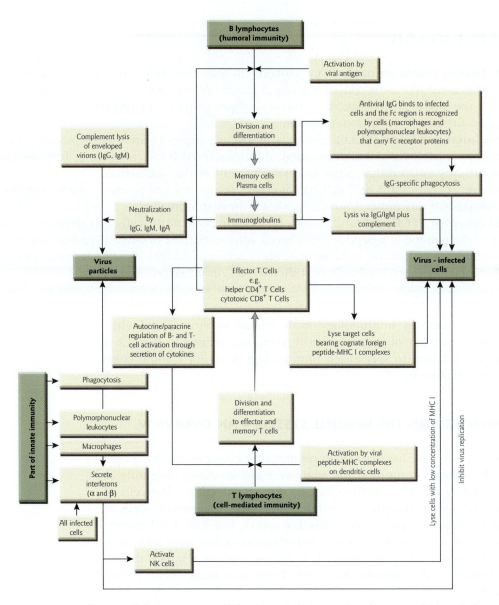

Fig. 12.1 Summary of some of the responses of the immune system to viruses and virus-infected cells.

Adaptive immunity is specific for foreign molecules, in our case virus products (see below). It is mediated by the activation of antigen-specific *B and T lymphocytes*. These are inactive, nondividing (resting) cells which cannot exert any immune response, and are present in very low numbers. On the surface of B cells are epitope-specific receptor proteins called BCRs, and on T cells are TCRs. The epitope is defined as the part of the antigen to which the BCR or TCR binds. Ten or so overlapping epitopes constitute an *antigenic site*, and several such antigenic sites are typically present on a protein on the surface of a virus particle (Fig. 12.2). Each B and T cell synthesizes a receptor that is specific for just one epitope – the cognate epitope. When a receptor binds to its cognate epitope, the cell is stimulated to undergo a large number of divisions to form an expanded clonal population of identical cells. These daughter cells then differentiate into *effector* cells which mount immune responses, or *memory* cells (see below). Thus virus-specific lymphocytes *adapt* to the ongoing infection, and can deal with it. Their receptors also evolve to better fit their cognate epitope. However all this takes a few days, during which time the virus can multiply and may cause clinical disease. T and B effector cells are nondividing, and short-lived. A large population of T and B effector cells is maintained in the body only for as long as their cognate antigen is present to stimulate the production of new effector cells. After the virus is vanquished, the population of T and B effector cells falls but not to the original level, as an expanded population of memory cells remains (Section 12.3). The long-term presence of T effector or B effector cells making antibody probably indicates that long-lived antigen is present. This is not necessarily in the form of infectious virus and it is not known where exactly it is located.

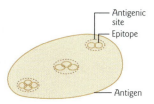

Fig. 12.2 Schematic relationship between the epitope (or antigenic determinant), antigenic site (a collection of overlapping epitopes), and antigen. BCRs can recognize almost any chemical type of molecule, but TCRs react only with peptide epitopes. Each epitope shown is unique.

12.2 INNATE IMMUNITY (FIG. 12.1, LOWER LEFT)

The innate immune system has two types of cells that carry out phagocytosis as their main function (these are sometimes called *professional phagocytes* to distinguish them from other cells that all have endocytic activity). These are the very first line of defence against infection. There is a strict division of territory, with the *polymorphonuclear leukocytes* (PMNL) in blood and the *macrophages* patrolling tissues outside the blood stream. Both types of cell are constitutively capable of phagocytosing virus particles, although their activity can be transiently increased by chemical messengers. Phagocytes destroy virus particles in their lysosomal vesicles by means of a battery of powerful enzymes and so prevent them from reaching their target cells. However, some viruses can avoid being killed within the phagocyte and turn the tables by infecting that cell, sometimes with dire consequences. This happens with HIV-1 in macrophages (see Chapter 19) and with Dengue fever complexed with non-neutralizing antibody (see Section 5.1). In the latter case the Fc region of antibody attached to virus binds Fc receptor proteins in the plasma membrane of the macrophage. The Fc receptors then act as surrogate virus receptors to permit infection of otherwise uninfectable macrophages.

If this arm of innate immunity is effective, an infection is aborted with no trace of it ever having happened in terms of B or T cell responses. However if infection is initiated, other parts of the innate system are called in. Amongst these are the cellular proteins interferon-α and interferon-β. These are so similar in structure and function that they are often called interferon-α/β. Interferons are cytokines (cell-encoded proteins) that are released by cells in response to virus infection. Interferons block virus multiplication at many levels without killing the cell, and this involves cross-talk with other pathways that regulate apoptosis, inflammation, and cell stress. They can also upregulate expression of MHC I proteins, enhancing the antiviral activity of T cells by increasing the concentration of foreign peptides presented on the cell surface (see below and Section 12.3).

The *natural killer* (NK) cell is a further element of innate immunity. It detects the presence of abnormally low concentrations of MHC proteins on the surface of cells, often a reflection of the activity of an infecting virus within the cell. The NK cell then lyses and destroys such cells. *Dendritic cells* are important sentinels that initiate immune responses. They are potent inducers of innate immunity and secrete type 1 interferons when infected. In addition they are the major interface between the innate and adaptive immune systems and are key to the processing and presenting of foreign antigens to T cells. However we are learning how viruses, particularly HIV-1, have evolved mechanisms to subvert dendritic cells to their advantage (see Chapter 19). Finally, although cells of the innate immune system do not have the epitope-specific cell-surface receptors of the adaptive immune system, they do have receptors that are capable of recognizing certain molecular motifs. These are the *Toll-like receptors* (TLRs) that were initially discovered by their signaling role during development. However TLRs also recognize motifs on viral single-stranded and double-stranded RNAs and viral envelope proteins, and transduce signals to the cytoplasm that stimulate various pathways of the innate immune system, including the interferon protein kinase pathway.

Viruses and interferons

Interferon was discovered in 1957 by Alick Isaacs and Jean Lindenmann. They incubated chorio-allantoic membranes from embryonated chicken eggs with heat-inactivated noninfectious influenza virus in buffered saline, washed the membrane to remove any nonadsorbed virus, and continued the incubation for 24 hours. The membranes were then discarded and the buffer solution tested for antiviral activity. This was done by placing a fresh chorio-allantoic membrane in the buffer and inoculating with infectious influenza virus. Membranes so treated did not support the growth of active virus, in contrast to untreated membranes. It was concluded that an extracellular, virus-inhibitory product had been liberated in response to the heat-killed virus, and this substance was named *interferon*. The antiviral activity of interferons can be quantitated by inhibition of the incorporation of radioactive uridine into viral ribonucleic acid (RNA) in cells

Table 12.1

Basic properties of interferons.

Type 1 interferon	Type 2 interferon
Interferons-α and -β	Interferon-γ
14 interferon-α genes and 1 interferon-β gene	1 interferon-γ gene
Glycoproteins made and secreted by all cells in response to virus infection	Glycoprotein secreted by activated T cells when TCRs bind cognate antigen
Acid stable	Acid labile
Interferons-α/β and interferon-γ differ antigenically and can be assayed by ELISA	
All act by binding a common interferon-α/β receptor	Acts by binding an interferon-γ receptor

Receptor binding by all interferons stimulates cellular transcription which results in antiviral activity and upregulation of MHC proteins

infected by an interferon-sensitive virus, or by ELISA. Its antiviral mechanism of action is described in Section 21.8.

We now know that there are two types of interferon: type 1 comprises interferons-α and interferon-β that are synthesized by most cells, are similar in structure, sequence and function, and are made in response to double-stranded RNA made by replicating or abortively replicating virus. The various α and β interferons are stimulated under different conditions, the control of which is not understood. Interferon-γ, has a completely different sequence, and is made and secreted when activated T cells are stimulation by their cognate antigen (Fig. 12.1). Properties of interferons-α, -β and -γ are compared in Table 12.1.

Now that cloned interferon is available, it is clear that the antiviral activity is only one of several physiological activities. Interferons enhance the activity of NK cells and macrophages, and serve as both positive and negative regulatory controls in the expression of innate and adaptive immune responses. Interferons-α and -β upregulate expression of MHC class I proteins, while interferon-γ upregulates expression of both MHC class I and class II proteins. Interferon-γ also induces the *de novo* transient expression of MHC class II proteins on epithelial cells, fibroblastic cells, endothelial cells (which line blood-vessels), and astrocytes (which provide nutrition for neurons in the central nervous system). Upregulation of MHC I proteins on cells increases the efficacy of their interaction with CD8$^+$ T cells, and the *de novo* expression of MHC class II proteins allows these cells to interact with CD4$^+$ T cells. Interferon-α is now part of standard antiviral therapies for hepatitis B virus and hepatitis C virus infections (see Section 21.8).

The production of interferon-γ by activated T cells can be demonstrated by infecting an animal with a virus and then some days later, reacting its T cells *in vitro* with antigen from the same virus that contains no nucleic acid. Interferon-γ is found in the culture fluid. Production of interferon-γ is thus a property of the adaptive immune response. When infected, cells of the immune system also synthesize interferon-α/β, like any other cell.

Further understanding of interferon function has come from the use of stem cell technology to specifically disrupt (knock-out) genes in mice. Initially the animals are chimeras composed of knock-out and normal cells; these are mated with a normal mouse to give rise to progeny that are uniformly heterologous for the knock-out, and then heterozygous animals are mated to produce the homozygous knock-outs that lack a specific gene function. Mice have been produced with no functional interferon-γ gene or no functional interferon-γ receptor gene. Such mice develop normally, but have impaired macrophage and NK cell function, and reduced amounts of macrophage MHC class II protein. CD8$^+$ cytotoxic T cell activity, CD4$^+$ T helper cell, and antibody responses were normal, but the mice showed increased susceptibility to vaccinia virus, and to the bacteria *Mycobacterium bovis* and *Listeria monocytogenes*. Thus interferon-γ is important to the functioning of innate immunity and in combatting these micro-organisms.

12.3 ADAPTIVE IMMUNITY

T cells were the first type of lymphocyte to evolve, and their activity complemented and enhanced the defence capability of the innate immune system. B lymphocytes evolved later and, as will be seen, provided the immune system with another type of weapon to counter infection. They made their first appearance in primitive fish. A B lymphocyte differentiates into a B effector cell that synthesizes antibody (immunoglobulin) of one sequence, and hence one specificity. B cells are responsible for *humoral* or *antibody-based immunity*. Antibody made by a B cell is identical in sequence to that of BCR, except that the BCR has an extra domain that anchors it in the plasma membrane. In other words an antibody is a soluble BCR. Each T or B cell has in its plasma membrane many copies of the TCR or BCR respectively. A receptor recognizes just one small region (an *epitope*) of a foreign molecule. Embryologically we develop with a random set of TCRs and BCRs but, early in life, cells bearing receptors that can react with our own epitopes are eliminated, so that by definition, all remaining BCRs and TCRs recognize only foreign epitopes. This is the immunological doctrine of *self* and *non-self*.

Antibody can recognize an epitope that is protein, lipid, carbohydrate, or nucleic acid. In respect to proteins, an antibody or BCR binds to a small region that constitutes its epitope. This is a planar surface of around 16 amino acid residues which interacts with a complementary surface that forms the binding site (or *paratope*) of the cognate antibody.

The TCR is much more restricted in its scope and recognizes *only* peptides. These are 8–22 amino acids in length, and are derived by intracellular proteolysis of proteins (see below). The peptides are always found complexed with cellular plasma membrane proteins, called major histocompatibility complex (MHC) proteins. A TCR only recognizes its cognate peptide when it is complexed with an MHC protein.

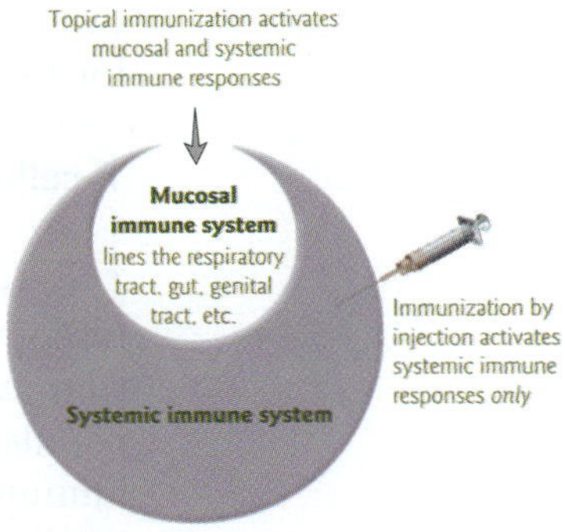

Fig. 12.3 Schematic relationship of the compartmentation of the systemic and mucosal immune systems.

Although immune cells circulate as single cells, after meeting their cognate antigen they migrate to one of the lymphoid centers. There are many of these loosely organized aggregates of cells, and the best known are the *lymph nodes*, located at strategic sites around the body (e.g. at the junction of limbs and torso, torso and head), and the *Peyer's patches* in the gut wall. Here they mature into effector cells. Activated B cells stay there and secrete antibody, but activated T cells patrol the body.

The adaptive immune system can be usefully thought of as consisting of two parts: the systemic immune system (which looks after the body as a whole) and the mucosal immune system (which looks after mucosal surfaces) (Fig. 12.3). The latter comprise the surfaces of the respiratory tract, the intestinal tract, the urinogenital tract, and the conjunctiva of the eye. These are important as most viruses (and indeed most other microorganisms) gain entry through these sites. Mucosal surfaces are particularly vulnerable because they are extensive in area, and their physiological activities require them to be composed of naked epithelium, unlike the skin which is impermeable to viruses unless broken. The problem is that the cells and antibodies that constitute systemic immunity do not circulate to the mucosae. However the mucosae have their own reservoir of lymphocytes, and these are stimulated when antigen is in contact with a mucosal surface. This is particularly relevant to nonreplicating vaccines, which have been traditionally administered by subcutaneous or intramuscular injection and do not reach the mucosae and stimulate local immunity. However, when the mucosa is stimulated with antigen, this activates immunity in both the mucosal and the systemic immune systems. Further, the mucosal areas are connected, so that immunity raised at one mucosal site will be found also at other mucosal sites, through the migration of activated T and B cells.

All innate and adaptive immune responses act in an interlocking and orchestrated fashion. It is a daunting thought that only one specific immune response can overcome a particular infection, even though many other responses may be stimulated. Only the key response is effective, the other responses being less effective or completely ineffective. The problem is that it is difficult to determine which element of the immune system is the key response needed to overcome a particular virus infection. The concept of key immune responses has clear and important implications for vaccines, especially to HIV (see Sections 21.1–21.7).

T-cell-mediated or cell-mediated immunity (Fig. 12.1, lower center)

T cells are crucial to determining the extent and character of the response of this second arm of the immune system. As discussed above, stimulation of a T lymphocyte by its cognate epitope results in a cloned population of T effector cells that is responsible for a particular type of immunity called *cell-mediated immunity* that can kill infected cells. Any one T cell has many copies of a TCR, a single sequence, epitope-specific receptor (see above), which recognizes a foreign peptide complexed with an MHC protein on the surface of the cell. All cells except red blood cells express MHC class I proteins, but a few cell types, notably dendritic cells and B lymphocytes, also express MHC class II proteins. There are three major loci encoding MHC I proteins (HLA-A, HLA-B, HLA-C in humans); each encodes a single polypeptide that forms a dimer with the same β_2-microglobulin protein. Another three loci encode MHC II proteins (HLA-DP, HLA-DQ, HLA-DR in humans); these proteins are dimers of α and β polypeptides encoded by the three loci. The genes that code for MHC proteins are highly polymorphic, and are expressed codominantly. Unless complexed with a peptide, MHC proteins cannot mature.

An MHC protein combines specifically with its cognate peptide by recognizing two terminally situated amino acid residues that anchor the peptide in a groove on the MHC protein. The epitope presented to T cells by this complex is the peptide sequence situated between the anchor residues. Thus an MHC protein can present a variety of epitopes providing they all have the appropriate anchor residues. In the cell, peptides with the appropriate specificity associate with an MHC molecule, so in the noninfected cell these will all be self-peptides and in the virus-infected cell, a proportion will be viral peptides. These peptides are formed by intracellular proteolysis (*antigen processing*) by proteasomes, and complex with MHC proteins rather like an egg in an egg cup. Class I MHC proteins complex with a peptide of 8–10 amino acids and class MHC II proteins with peptides of 17–22 amino acids. There is a further distinction as peptides associating with MHC I proteins are derived by digesting proteins made inside the cell while peptides complexing with MHC II proteins are derived from proteins brought into the cell from the surroundings. All MHC–peptide complexes are then transported to the plasma membrane, where the peptide is displayed on the outside of the cell, bound to the integral membrane MHC protein (*antigen presentation*).

The TCR of a T lymphocyte recognizes an infected cell by interacting with a foreign peptide–MHC complex on the cell surface, and this results in its activation (*clonal expansion and differentiation*) to become a functional *T effector cell*. There are two types of T lymphocytes that express either

Table 12.2

Some functions of activated T cells.

	Cytotoxicity	DTH	Help	Suppression
CD4+ T cell	+	++	+++	+
CD8+ T cell	+++	++	−	++

+ to +++ indicates activity from low to high. DTH, delayed-type hypersensitivity.

CD8 proteins or CD4 proteins on their plasma membranes, and are hence referred to as *CD8+ T cells* or *CD4+ T cells*. CD8 is a key ligand for MHC class I proteins and CD4 for MHC class II proteins. This recognition is additional to that of the TCR for the foreign peptide, but equally important. All cells of the body have MHC class I proteins and hence are under constant immune surveillance by CD8+ T cells. There is a quantitative element to the recognition of MHC–peptide complexes by a T cell, and the sensitivity of this process is proportional to the MHC concentration. Hence the importance of the interferons that upregulate MHC I protein expression in facilitating an antiviral immune response. Conversely, many viruses have evolved functions that downregulate expression of MHC proteins, but when this reaches a level that T cells cannot recognize, the cell is identified and killed by NK cells (Section 12.2). One of the major functions of activated CD8+ T effector cells is to destroy MHC I positive cells that present foreign, i.e. viral, peptides (Table 12.2). However in some infections (measles) CD4+ T effector cells predominate, but these kill only cells that express MHC II proteins. T cells carrying out this killing activity are known as *cytotoxic T cells* or *activated cytotoxic T lymphocytes* (CTLs).

The major function of activated CD4+ T cells is *help*, and only these T cells can provide it. Help is a positive regulatory function which is essential for the activation of all T and B lymphocytes by their cognate foreign epitope. Thus the entire immune response depends on such *helper T cells*. This is also the main type of cell infected by human immunodeficiency virus (HIV) and explains why their destruction is so devastating (see Chapter 19). One specific helper function of CD4+ T cells is to provide chemical messengers (cytokines) that control immunoglobulin heavy chain gene switching, and drive an IgM-synthesizing cell to switch to synthesizing another immunoglobulin (IgA, IgG, IgD, or IgE; Section 12.4) while retaining its epitope specificity. Help and many other functions of T cells are mediated by *cytokines*. These are soluble proteins through which cells of the immune system communicate with each other. Cytokines act like hormones except that they operate over a very short distance, and their effects are said to be *paracrine*, affecting neighboring cells, or *autocrine*, affecting the cytokine-secreting cell itself. The effects

Box 12.2

The roles of CD4$^+$ Th-1 and Th-2 cells

Th-1 cells: cytotoxic responses
- Have evolved to deal with intracellular parasites (viruses and others).
- Bias the immune response to cell-mediated immunity (CD8$^+$ cytotoxic T cells, and other immunologically active cells) and isotype switching to IgG2a (which can strongly activate the cytotoxic complement system).
- Are classified by their cytokine profile: secrete primarily interferon-γ and IL-2 (but also IL-12 and tumor necrosis factor (TNF)-β).
- Evoke DTH (see text) responses with macrophage activation.

Th-2 cells: inflammatory responses
- Have evolved to deal with extracellular parasites.
- Bias the immune response to antibody immunity, notably isotype switching to IgE and IgG1 that cannot activate the complement system.
- Can cause an excess of eosinophils to invade the infected area (eosinophilia) or allergic responses through the interaction of IgE with mast cells (atopy).
- Secrete primarily IL-4 and IL-5 (but also IL-9, IL-10, and IL-13).

are quantitative and can be either positive (i.e. upregulate an activity) or negative (downregulate an activity). In this way fine regulation of the immune system is achieved. There are a large number of cytokines (>100), and their effects may be overlapping, additive, synergistic, or redundant. Understanding them is key to being able to manipulate the immune system, e.g. in terms of making more effective vaccines, and treating the excesses of the immune response as it manifests itself in pathological conditions such as autoimmunity, allergy, and hypersensitivity reactions. One cytokine is the multifunctional protein, interferon-γ. Like interferon-α/β it can inhibit virus replication, but this is their only common attribute (Section 12.2).

There are two sorts of activated CD4$^+$ cell, T helper 1-type (Th-1) cells and T helper 2-type (Th-2) cells, and these have different types of action which decide the character of the resulting immune response. This differentiation step is controlled by cognate antigen plus the balance of cytokines: a preponderance of interferon-γ and transforming growth factor (TGF)-β favors the development of Th-1 cells, while a preponderance of interleukin (IL)-4 and IL-10 favors Th-2 cells. The cytokines are produced by cells of the innate immune system (interferon-γ, TGF-β, and IL-4) and by activated CD8$^+$ T cells (interferon-γ and IL-10). What controls the cytokine balance is not clear. Box 12.2 gives more information about the roles of Th-1 and Th-2 cells. This is a complex area, as Th-1 and Th-2 cells cross-regulate each other in regard to their establishment and effector function. Particularly relevant to virology is that an excess of Th-2 cells can give rise to immune-mediated pathology. More usually there will be a balance of Th-1 and Th-2 cells, without any pathological effects.

The inflammation referred to in Box 12.2 is most apparent when it occurs immediately below the skin. It is defined as an area that is swollen, red, hot, and painful. Part of the inflammatory reaction results from a transient increase in permeability of blood vessels in that area, and a resulting influx into the tissue of plasma, antibody, and cells of the immune system. The nature of the inflammation depends on the immune components (Box 12.2). Inflammation also occurs in deep lying tissues of, for example, the lung. Inflammation is a positive and helpful response designed to enhance local immune reactivity. In moderation it accelerates clearance of the infection, but in excess can interfere with normal tissue function, and cause pathological tissue damage – like any immune response.

Delayed-type hypersensitivity or *DTH* reactions are mediated when T cells locate cognate antigen and secrete cytokines that are detected by macrophages and PMNLs and cause them to follow the cytokine concentration gradient and migrate to the site of antigenic stimulation. Th-1 cells are strongly involved in DTH responses (Box 12.2).

Humoral, or antibody-mediated, or B cell-mediated immunity (Fig. 12.1, upper center)

Antibodies (immunoglobulins) are made by a cell variously called B-effector cell, a plasma cell, an activated B cell, or an antibody-forming cell. Its immediate progenitor is the B lymphocyte, a resting cell. When the BCR recognizes and binds its cognate epitope, the B lymphocyte is stimulated to divide and differentiate. One of the key events that is triggered is mutation of DNA encoding the variable regions of the antibody that together comprise the paratope (Fig. 12.4). It proceeds at a stunningly high frequency (hypermutation). This creates a huge pool of cells each with a variant paratope, and Darwinian evolution and natural selection follows. B cells with the best fitting paratope are preferentially stimulated by binding the cognate epitope; in turn their BCR mutates and affinity and antibody activity is further increased.

The plasma cells (now located in a lymph node) synthesize and secrete huge numbers of antibody molecules, which react specifically with the stimulating epitope. The basic unit of an antibody comprises two heavy and two light chains (Fig. 12.4). These can be considered as a Y-shaped molecule, with paratopes at the two tips and a constant Fc domain forming the stalk. There are five types of immunoglobulin distinguished

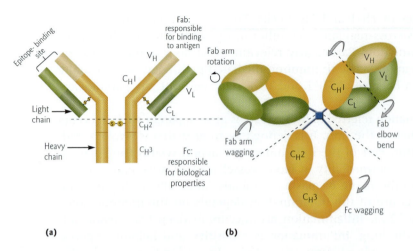

(a)   (b)

Fig. 12.4 Generalized immunoglobulin molecule, showing (a) the outline structure, consisting of two identical dimers formed of H–L polypeptides. Note the sequence-variable (V) region of the H and the L chains. Each contains three hypervariable sequences of 10–20 residues (not shown) that are folded together to form the unique epitope-binding site (paratope). The remaining H and L sequences are relatively constant. The constant region is subdivided into domains with sequence homology (C_H1 to C_H3). IgM and IgE both have an additional domain (C_H4). The molecule is also divide into an N-terminal half that binds epitope (Fab), and a C-terminal half (Fc) that is reactive with various cell mediators. IgM and IgA form 5-mers and 2-mers respectively linked by their C ends with a joining (J) polypeptide. (b) The globular domains, with arrows indicating the flexibility which allows the molecule to bind to two (identical) epitopes, which can be different distances apart in three dimensions. SS, disulfide bond.

by having different heavy chains that are called alpha (α), delta (δ), epsilon (ε), gamma (γ), and mu (μ), and these define IgA, IgD, IgE, IgG, and IgM, respectively. Their light chains are either kappa (κ) or lambda (λ). IgM and IgA are further distinguished by being oligomers of the basic immunoglobulin unit. There has evolved a division of labor among the antibodies; they have specialized immune functions and are found in certain locations in the body that make the immunological defence more efficient. The major types of immunoglobulin are shown in Box 12.3.

Some of these high affinity mutated cells do not mature into plasma cells but become memory cells instead. These have an affinity-enhanced BCR, and on reinfection by the *same* virus there is now a sufficiently large population of cognate memory cells to rapidly provide effector cells that can repulse the infection before it can cause disease. Such cells constitute *immunological memory* (Box 12.4). Memory cells closely resemble lymphocytes in morphology and lack of effector functions. They require stimulation by the cognate epitope just like lymphocytes, but are activated more readily than B lymphocytes as they respond to antigen at a

Box 12.3

Properties of the major immunoglobulins

- **IgM** is composed of five IgM monomers covalently linked at their C-termini. It binds its epitope with low affinity but, being pentameric, the overall IgM avidity (binding strength) is high. It is found only in blood; IgM and IgG are the only antibodies capable of activating the complement system (see below). IgM is the first class of antibody synthesized when lymphocytes are activated, so detection of virus-specific IgM in serum is diagnostic of a recent infection. Contributes to neutralization of virus infectivity.

- **IgG** is the collective name for IgG1, IgG2, IgG3, and IgG4, which each have distinct γ chains synthesized from separate genes, and differ in specific ways that allow them to combat a variety of infections. IgG is a monomer; its small size gives it mobility and allows it to leave the blood stream and enter tissues. It can also enter the body of the fetus by crossing the placenta and protects it from infection. The immune system of the newborn infant develops slowly and maternal IgG remains in circulation for several months until immunity is functional. IgG activates complement and binds to Fc receptor proteins on the surface of phagocytic cells. Antigen stimulation of mucosal surfaces gives rise to the synthesis of "local IgG." IgG is important for neutralization.

- **IgA** is a key defender of the mucosal surfaces, and exceeds IgG in both total amount produced and in local concentration. It has the unique property of being secreted across the mucosal epithelium, so that it occurs outside the body in the lumen of the gut, respiratory tract, and urinogenital tract. Maternal IgA is secreted into milk and protects the gut of the newborn from infection. IgA is particularly resistant to degradation by digestive enzymes. The most abundant type of IgA is a dimer, but monomers also exist. Both neutralize virus infectivity.

Box 12.4

The basis of immunological memory in B cells

Epitope stimulation results in a clonal population of cognate B memory cells which:
- Comprises resting cells that do not secrete antibody.
- Is at a higher concentration than B lymphocytes.
- Has higher affinity B cell receptors. (Note that T cell receptors do not undergo hypermutation and affinity maturation.)
- Applies to all immunoglobulin classes except IgM.
- Has a lower threshold of activation by the cognate epitope.
- Protects against reinfection by killing virus before it can reach a disease-causing concentration.
- Is an evolutionary trade off – mutation gives an infinite variety of BCRs in exchange for suffering the consequences of infection the first time we meet a virus.

lower receptor to virus ratio. Finally it is interesting that only in the immune system is somatic (i.e. non-germ line) mutation permitted – all other somatic mutations are rapidly repaired.

Immunoglobulins may combat virus infectivity in more than one way. Some immunoglobulins are directly neutralizing, meaning that they can bind to viruses and cause them to lose infectivity (Section 12.5). Other immunoglobulins are *non-neutralizing* and they attach to *non-neutralizing epitopes*. These act indirectly, either by inducing phagocytosis of bound virus via their Fc domains or by recruiting complement. The activation of *complement* by IgG or IgM bound to viruses can enhance the activity of neutralizing antibody, or enable non-neutralizing antibody to become neutralizing. Complement is a nine-component system of soluble proteins found in blood. Activation of the first complement component activates the next component by proteolytic cleavage and so on, with progressive amplification. It is a classic biochemical cascade system like blood clotting. Usually, each component is cleaved into two parts, one of which adheres to the antigen–antibody complex or close to where the antibody has bound, and a second diffusible part, which forms a chemical gradient and attracts cells of the innate immune system into the vicinity of antigen. The final stage is the insertion of pore structures (the membrane attack complex) into the cell membrane, which, in sufficient number, create an ionic imbalance and kill that cell. Thus complement-enhanced neutralization of nonenveloped viruses is brought about by a build-up of complement proteins on the surface of the virus that sterically prevent attachment of the virus to its cell receptors, while enveloped viruses in addition can be permeabilized by having complement pores inserted into their lipid bilayers that permit the entry of nucleases, etc. However while complement is important in com-bating certain bacterial infections, it is not certain that it is essential to antiviral defence *in vivo*. For example, people who have a congenital complement deficiency do not have an increased number or severity of virus infections. However it may be that other aspects of their immune response are enhanced and compensate for the lack of complement activity.

An antibody molecule is approximately the same size as a single membrane protein, and not surprisingly does not harm an infected cell even when many molecules are bound to virus antigen on the cell surface. Also antibodies cannot cross membranes and enter cells. However, bound IgG or IgM can activate complement and in sufficient amounts this can lyse the cell as described above. Cells can repair minor damage and it needs about 10^5 complement pore structures inserted into the plasma membrane to kill a cell. Alternatively, bound IgG can act as a ligand for phagocytes that have Fc receptor proteins on their surface, and this leads to the phagocytosis and destruction of the infected cell. In this way the adaptive immune system endows the innate immune system with antigen specificity.

12.4 UNDERSTANDING VIRUS NEUTRALIZATION BY ANTIBODY

Neutralization is the loss of infectivity which ensues when antibody binds to a cognate epitope on the virus particle. Viruses are unusual as neutralization is usually mediated by antibody alone, whereas larger organisms, like bacterial cells, also require the action of secondary effectors like complement as described above. The antibody–antigen reaction is so specific that it is unaffected by the presence of other proteins. Hence antibodies need not be extracted from crude serum, and impure virus preparations can be used to observe neutralization. However not all antibodies which bind to a virus particle are capable of neutralizing its infectivity. Neutralization is an epitope-specific phenomenon. Chapters 14 and 21 discuss the role of antibody in the recovery from virus diseases and prevention of reinfection, but it is appropriate here to discuss how antibody neutralizes or renders virus noninfectious.

Antibodies are usually obtained from venous blood as an antiserum several weeks after an infection has resolved (convalescent serum) or after at least two immunizations of a person or an experimental animal – usually a mouse rat, rabbit, or guinea pig. Antibodies may be formed to the several different viral epitopes present on several different virus proteins, and they join the pool of antibodies to other foreign antigens that the animal has experienced earlier. A further complication is that the isotype distribution of antibody in serum differs from that of antibodies present in mucosal secretions – in serum IgG > IgA, and in secretions IgA > IgG, so care must be taken in extrapolating from one situation to the other (Box 12.5).

Understandably, analysis of such a complex mixture of antibodies is difficult. For unambiguous analysis of viral epitopes monoclonal antibodies (MAbs) are used (see below). An antibody synthesizing B cell makes an antibody of just one sequence (i.e. it is monoclonal), but such cells do not divide, and single cells to do not make enough antibody to be of practical use. However, in 1975 Köhler and Milstein devised a means to immortalize an antibody-synthesizing cell by fusing it with a cell from a B cell tumor which no longer makes its own antibody. The resulting hybrid cell (*hybridoma*) can then be grown to large numbers in the laboratory. Each cloned cell line synthesizes antibody of a single sequence, a *monoclonal antibody* or MAb. In practice a crude mixture of B cells is used in the fusion reaction, and many clones are generated. The desired hybridoma is identified by the reaction of its antibody with the desired antigen.

Early work assumed that neutralizing antibody acted solely by preventing virus from attaching to receptors on the cell surface. However, while this was the mechanism operating with most rhinovirus-specific MAbs, the majority of influenza A virus- and poliovirus-specific, neutralizing MAbs did not block cell attachment. Thus there is more than one way of killing

Box 12.5

Sources of antibody for virus neutralization

- Serum is the fluid portion of clotted blood, i.e. blood less cells and clotting components.
- Plasma is the fluid portion of blood in which clotting has been inhibited and cells removed.
- Serum containing a desired antibody reactivity is called an antiserum.
- Serum and plasma contain a population of polyclonal antibodies, i.e. from many different B cells.
- Mucosal antibody can be relatively easily obtained from nasal secretion (collected by chemically irritating the nasal cavity), milk, and from extracts of feces.
- A monoclonal antibody (MAb) is the product of a single immortalized B cell that has formed a clonal population.
- All MAbs from a given cell clone have exactly the same properties – including the same sequence, epitope, affinity, and isotype.
- In effect, an antiserum is a population of several different monoclonal antibodies, and may well contain several MAbs that react with different epitopes on the same antigen. In this case, antibody reactivity will be an average of all the reacting antibodies.
- By contrast, reaction of an antigen with a single MAb is unambiguous, so MAbs make great reagents for research and bioassays.

a virus with antibody. The mechanism of neutralization is antibody-specific, so a virus can be neutralized in a variety of different ways, and neutralization is determined largely by the epitope that the antibody binds to. Surprisingly, no antibody is made to the attachment sites of influenza A virus, poliovirus, or rhinovirus. These sites are contained in depressions on the surface of the virus, where they are hidden from the immune system, presumably as a result of evolution trying to evade the host's immune response. Rhinovirus-specific neutralizing antibody attaches to and bridges amino acids on either side of the rhinovirus attachment site and blocks it indirectly. Antibody attached to virions can be visualized by EM as a fuzzy outer layer. However, antibody can be diluted so that it is no longer detected by EM but can still thoroughly neutralize infectivity. So few molecules of antibody bound to a virus particle are unlikely to interfere with attachment. Interference with attachment of influenza virus would be particularly inefficient, as there is a high density (500–1000) of trimeric attachment proteins (the hemagglutinin) per virion, and an IgG molecule is slightly smaller than one hemagglutinin protein. In addition, the existence of non-neutralizing antibodies demonstrates just how specific the neutralization reaction is, and means that antibodies bound to the surface of a virus do not necessarily interfere with the attachment process. There is a further dimension to this, since a non-neutralizing antibody bound close to a neutralizing epitope can sterically prevent the binding of neutralizing antibody, and allow the virus to evade the immune response.

So how do viruses lose infectivity? Antibodies (MAbs) to certain epitopes neutralize by blocking attachment of a virion to its cell receptors, but equally it has been demonstrated that neutralizing MAbs to other specific epitopes do not inhibit virus attachment. In the case of poliovirus, the neutralized virion attaches to cells, is taken up in an endocytic vesicle, but is unable to uncoat. In a second example, an influenza virus attaches to cells, is endocytosed, but fusion of the viral and cell membranes does not take place. Current thinking is that there are as many mechanisms of neutralization as there are processes that a virus has to undergo before its genome can enter a cell and be expressed. Any definition of the mechanism of neutralization must include the virus and the MAb, and this may change depending on other factors including the antibody isotype, number of antibody molecules per virion, cell receptor, and the cell type.

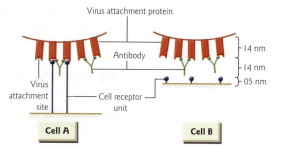

Fig. 12.5
Hypothetical scheme based on the length of the cell receptor to explain inhibition by antibody of attachment of virus to cell B but not cell A.

At first sight it is puzzling that a MAb will neutralize influenza virus by a mechanism that does not prevent the attachment of neutralized virus to cells in culture, but will prevent virus from attaching to red blood cells (as used in the hemagglutination–inhibition test; Fig. 2.8). The explanation probably lies in the nature of the cell receptor unit, *N*-acetyl neuraminic acid (NANA), or rather the molecule that carries the carbohydrate moiety of which NANA is a component. Most of the NANA on the surface of the red blood cell is carried by a small protein that protrudes only a short distance from the cell surface. Thus antibody can sterically prevent virus from attaching to the red blood cell. Presumably the (unknown) molecule that carries NANA on the surface of other cells that are targets for influenza virus infection is longer, and interdigitates with the bound antibody, enabling the NANA to interact with the virus attachment site (Fig. 12.5).

Of some practical importance are *neutralizing antibody escape mutants*. These are viruses that have acquired a point mutation that results in the substitution of a crucial amino acid within the epitope and this abrogates binding of antibody. Such mutations occur naturally at a frequency of about 1 in 100,000. None of the progeny of an antibody-escape mutant is neutralized either. Thus wild-type virus is neutralized and the escape mutant replaces it in the population. The location of the mutation can be determined by comparing sequences of the *wt* and mutant genes encoding the neutralization protein. In this way, using sufficient MAbs, all the epitopes and hence the antigenic sites of a virus particle can be mapped.

There are other conditions under which virions can bind neutralizing antibody and yet not be neutralized. This occurs when the concentration of antibody is too low, or the affinity of antibody is so low that it dissociates rapidly, or because virions are aggregated and protected from contact with antibody. In addition, there are a type of neutralizing antibody escape mutant that has a mutation that does not alter the epitope but abrogates neutralization. This probably acts by preventing a downstream event in the neutralization pathway.

12.5 AGE AND IMMUNITY

The very young and the very old suffer a greater number of episodes of infectious disease than people of other ages, and these infections are sometimes of greater severity. This arises because the young and the old have a less developed or less active immune response respectively. The development of the immune system lags behind other body systems, although the infant gets some protection from maternal transplacental IgG and from immunoglobulins (mainly secretory IgA) from maternal milk. However, by the age of 9 months maternal IgG has disappeared. The developing immune response increases quantitatively, and by 12 months an infant produces 60% of the adult level IgG but only 20% of its adult level IgA; by around 2 years of age it is effectively mature. In the elderly, the immune system, like other body systems, declines with age. With immunological memory the elderly have less of a problem in combatting repeat infections, but their immune system is less able to deal with new infections, such as pandemic influenza A virus. The young and the old both pose a special problem for vaccination, as most clinical trials employ people from the age group in between. For example, in one clinical trial, 46% of those over 65 years of age failed to respond to influenza vaccination.

KEY POINTS

- The immune system is a complex interacting mixture of cells and soluble components that has evolved to protect us from infection.
- The immune system has evolved layers of protection that operate at different times and different locations, and by different means. Thus it can be envisioned as several separate but communicating systems.
- Cytokines are soluble proteins that orchestrate and coordinate the immune system.
- The innate immune system is immediately and constantly available to protect us from infection.
- Viruses that evade the innate immune system are dealt with by the adaptive immune system, comprising B cells (antibodies) and T cells. Each cell of the adaptive immune system is specific for one epitope only. Cells of adaptive immunity are inactive and in low number until stimulated by specific epitopes, and take some days to reach their peak. Thus a virus causing a primary infection can establish infection and clinical disease. This results in immunological memory – a heightened cellular presence with a shorter lag period, which is effective in preventing a second infection by the same agent.
- Adaptive immunity functions through epitope-specific receptors: BCRs and TCRs.
- Antibodies are soluble BCRs that can recognize any type of molecule; they can directly neutralize virus infectivity, but can also signal to other mediators to kill cells.

- TCRs are cell-surface proteins that recognize only peptides complexed with MHC proteins.
- Adaptive immunity has subsystems that operate semi-independently in the body (systemic immunity) and at mucosal surfaces (mucosal immunity) – the main route of infection for all microorganisms. It is necessary to stimulate the mucosae directly to get mucosal immunity.
- Viruses have evolved a variety of mechanisms for evading immune responses; without them there would probably be no viral disease and, depending on their efficacy, a virus remains in the body for a few days or a lifetime.
- Much still remains to be discovered about how the immune system functions; we need to achieve this so that we can manipulate the immune system to overcome viral evasion mechanisms and to make more effective vaccines.

QUESTIONS

- Discuss the ways in which the humoral (antibody) arm of an adaptive immune response can counteract virus infection.
- What roles do T cells play in the adaptive immune response?
- What is the innate immune response and how can it act against virus infections?

FURTHER READING

Ahmed, R., ed. 1996. Immunity to viruses. *Seminars in Virology* **7**, 93–155 (several articles).

Alcamí, A., Koszinowski, U. H. 2000. Viral evasion of immune responses. *Immunology Today* **21**, 447–455.

Biron, C., Nguyen, K. B., Pien, G. C., Cousens, L. P., Salazar-Mather, T. P. 1999. Natural killer cells in antiviral defense: function and regulation by innate cytokines. *Annual Review of Immunology* **17**, 189–200.

Boehme, K. W., Compton, T. 2004. Innate sensing of viruses by Toll-like receptors. *Journal of Virology* **78**, 7867–7873.

Burton, D. R., Williamson, R. A., Parren, P. W. H. I. 2000. Antibody and virus: binding and neutralization. *Virology* **270**, 1–3.

Globerson, A., Effros, R. B. 2000. Ageing of lymphocytes and lymphocytes in the aged. *Immunology Today* **21**, 515–521.

Goldsby, R. A., Kindt, T. J., Osbourne, B. 2000. *Kuby Immunology*, 4th edn. W. H. Freeman, New York.

Jankovic, D., Liu, Z., Gause, W. C. 2001. Th1- and Th2-cell commitment during infectious disease: asymmetry in divergent pathways. *Trends in Immunology* **22**, 450–457.

Katze, M. G., He, Y., Gale, M. 2002. Viruses and interferon: a fight for supremacy. *Nature Reviews Immunology* **2**, 675–687.

McIntosh, E. D., Paradiso, P. R. 2003. Recent progress in the development of vaccines for infants and children. *Vaccine* **21**, 601–604.

Rinaldo, C. R., Piazza, P. 2004. Virus infection of dendritic cells: portal for host invasion and host defense. *Trends in Microbiology* **12**, 337–345.

Slifka, M. K. 2004. Immmunological memory to viral infection. *Current Opinion in Immunology* **16**, 443–450.

Takeda, A., Kawaoka, Y. 2003. Antibody-dependent enhancement of viral infection: molecular mechanisms and *in vivo* implications. *Reviews in Medical Virology* **13**, 387–398.

Tortorella, D., Gewurz, B. E., Furman, M. H., Schust, D. J., Ploegh, H. L. 2000. Viral subversion of the immune system. *Annual Review of Immunology* **18**, 861–926.

Also check Appendix 7 for references specific to each family of viruses.

13

Interactions between animal viruses and cells

Properties of a virus and the cell it infects together determine if the cell lives, dies, or remains infected for ever.

The common perception of virus infections is that the only outcome is death of the infected cell through lysis. However, there are several outcomes of the interaction between a virus and a host cell, ranging from no infection to a long-lived infection. Virus–cell interactions can be classified into acutely cytopathogenic, persistent, latent, transforming, abortive, and null infections. Initial studies rely on analysis of infections of cells in the laboratory and the data obtained are used to pave the way for the eventual understanding of infection of the whole organism (see Chapter 15). Two points should be borne in mind: firstly, that a prerequisite of any of these types of infection is the initial interaction between a virus and its receptor on the surface of the host cell and hence any cell lacking the receptor is automatically resistant to infection; and secondly, that both virus and cell play a vital role in determining the outcome of the interaction; a virus may exhibit, for example, an acutely cytopathogenic infection in one cell type and latency in another.

13.1 ACUTELY CYTOPATHOGENIC INFECTIONS

Acutely cytopathogenic infections are those that result in cell death. These have also been called "lytic" infections, but this term is not entirely accurate, as in some infections cells die without being lysed, i.e. by apoptosis or programmed cell death (see next section). The viruses which cause acutely cytopathogenic infections are the ones most commonly studied in the laboratory, since cell killing is an easy effect to observe. In these infections production of infectious progeny can usually be monitored without difficulty, and the time scale is usually measured in hours. The one-step growth curve (see Section 1.3) describes the essential features of any eukaryotic or prokaryotic virus–host interaction which results in cell death, and Fig. 13.1 shows the successive appearance of cell-associated virus infectivity, infectivity that has been released from the cell into the tissue culture fluid, and the cytopathic effect (CPE). The pathological effects are always last to appear, just as in human infections (see Chapter 15). CPE is observed under the microscope and is often manifested as a change in cell shape from a spread-out, flattened cell morphology to a rounded-up structure. Quite early on in infection viruses often inhibit the synthesis of cellular proteins, DNA, or RNA, but frequently cell death occurs sooner than can be explained by these inhibitory events. Exactly how an animal cell is killed in most cases is still not certain (Section 13.7), but it is nothing to do with lysis by lysozyme which is restricted to a minority of bacteriophages and bacteria. Lysis is one of the ways that a nonenveloped virus may be liberated from the infected cell (e.g. poliovirus, Fig. 13.1). In other infections the cell is not lysed, and viruses enter a cytoplasmic vesicle which then releases its contents into the tissue culture fluid by fusion with the plasma membrane. All membrane-bound virus bud from cellular membranes (see Section 11.6).

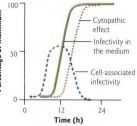

Fig. 13.1 An acutely cytopathogenic infection of a HeLa cell line by poliovirus, a picornavirus. Cells were inoculated with an MOI (multiplicity of infection) of 10 infectious units per cell, so that nearly all cells were infected and a one-step growth curve results. Note that intracellular infectivity declines at later times as cells die.

Apoptosis or programmed cell death of the infected cell

Apoptosis is the normal process which regulates cell numbers during development. Familiar manifestations are the separation of webbed digits in the human embryo to give fingers and toes, the loss of the tail of the amphibian tadpole, and the removal of self-reactive T cells during the development of the immune system. A key unique feature of apoptosis is that the dying cell remains intact and its contents stay within the plasma membrane. This contrasts with cell lysis (or *necrosis*) where the cell disintegrates and its contents are released. In a whole animal these necrotic products are inflammatory and have to be tidied up by scavenger cells, particularly those of the immune system, whereas a cell undergoing apoptosis is not inflammatory. During apoptosis a cell undergoes profound internal changes which include fragmentation of its chromosomal DNA. These processes follow a clear, well-regulated pattern. Ultimately, the cell rounds up and is disposed of by being engulfed by a phagocyte, within which it is hydrolyzed. It is likely that when an infected cell undergoes apoptosis it is the cell's response to infection rather than being induced by the virus as, by committing suicide, the cell prevents the production

Box 13.1

Some virus families which have members that can modulate apoptosis

Viruses which express proteins that promote cell death by apoptosis
- Retroviruses: human immunodeficiency virus type 1.
- Papovaviruses: SV40, human papillomaviruses.
- Hepadnaviruses: hepatitis B virus.
- Adenoviruses.
- Togaviridae: Semliki Forest virus, Sindbis virus, rubella virus.
- Flaviviruses: hepatitis C virus, classical swine fever virus.

Viruses which express proteins that inhibit cell death by apoptosis
- Retroviruses: human immunodeficiency virus
- Papovaviruses: SV40, human papillomaviruses
- Adenoviruses
- Herpesviruses: herpes simplex virus, Kaposi's sarcoma virus, Epstein–Barr virus, cytomagalovirus
- Poxviruses: vaccinia virus, ectromelia virus, myxomavirus

of virus progeny. Some viruses have proteins that trigger the apoptosis pathway as an inevitable consequence of their interactions with the cell, but then make other proteins that inhibit apoptosis while virus replication proceeds (Box 13.1). This is relevant to the mechanisms whereby viruses transform cells or establish persistent infections (Sections 13.2 and 13.4).

13.2 PERSISTENT INFECTIONS

Persistent infections result in the continuous production of infectious virus and this is achieved either by the survival of the infected cell or by a situation in which a minority of cells are initially infected and the spread of virus is limited, so that cell death is counterbalanced by new cells produced by cell division, i.e. no net loss. Persistent infections result from a balance struck between the virus and its host either: (i) through the interaction of virus and cells alone; (ii) the interaction of virus and cells with antibody or interferon to limit virus production; (iii) the interaction of virus and cells with the production of defective-interfering (DI) virus (see Section 7.2); or (iv) a combination of these events.

The suggested explanation for the ability of certain virus–cell combinations to establish a persistent infection is that viruses evolve to a state of peaceful coexistence with their host. In other words the virus gains no advantage in killing its host, rather the reverse as the virus depends absolutely on the host for its very survival.

Persistent infections resulting from the virus–cell interaction

Simian virus (SV) 5 causes an acutely cytopathogenic infection with cell death in the BHK (baby hamster kidney) cell line (Fig. 13.2a), but when it infects a monolayer of monkey kidney (MK) cells, it establishes a persistent infection. The virus multiplies at the same rate in both cell types, and multiplies with a classical one-step growth curve in MK cells (Fig. 13.2b), but the MK cells show no CPE, remain healthy, and produce progeny virus continuously (Fig. 13.2c). Infection by simian virus 5 does not damage the MK cell, in the sense that it does not perturb the synthesis of cellular protein, RNA, or DNA, and cell division continues as normal. Calculations show that this virus infection makes little demand on the host's resources, e.g. total viral RNA synthesis is less than 1% of cellular RNA synthesis (even though each cell is producing about 150,000 particles/day). Thus, in monkey cells, SV5 causes a harmless persistent infection, while in BHK cells it causes an acutely cytopathogenic infection from which all cells die. The origin of the cell is thus all important in determining the outcome of this relationship.

Persistent infections can also arise when viruses are able to inhibit the apoptotic response that normally gives rise to an acutely cytopathogenic event. Many viruses express gene products which can inhibit apoptosis (Box 13.2). For example, human cytomagalovirus (a herpesvirus) encodes a protein called UL37x1 which inhibits apoptosis of infected cells, permitting the virus to establish a long-lasting infection. Since the persistent infection allows the virus to multiply for a long period of time, it is to the advantage of the virus to have evolved the means to suppress apoptosis.

Persistent infections resulting from interactions between viruses, cells, and interferon, or viruses, cells, and antibody

A persistent infection may be established as a result of an equilibrium between the generation of new infectious particles and the antiviral effect of interferon (see

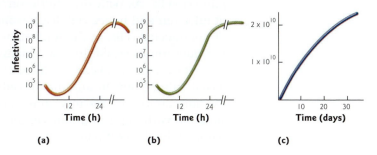

Fig. 13.2 Different types of infection caused by simian virus 5 in BHK (baby hamster kidney) cells and MK (monkey kidney) cells. (a) The acutely cytopathogenic infection in BHK cells. Note that virus yield drops after 24 hours. (b) The initial one-step growth curve in MK cells which kills no cells and becomes persistent. (c) The cumulative yield from MK cells infected in (b) over 30 days. Cells grow normally during this infection and have to be subcultured at intervals of approximately 4 days.

Box 13.2

Evidence that viruses can inhibit apoptosis to establish a persistent infection

Infection of normal cell lines with Sindbis virus (a togavirus) results in lysis. However, infection of primary cultures of neurons results in the establishment of a persistent infection. Analysis of gene expression from the neuronal cell genome showed that the Sindbis virus infection activated the expression of the host cell gene, *bcl*-2. This gene is responsible for preventing apoptosis in normal cells. By stimulating *bcl*-2 expression Sindbis virus prevents the normal apoptotic response of the cell to infection and the virus is able to persist.

Section 14.2). Interferon supresses virus growth, and when only a few cells in a culture are infected naturally produced interferon will induce an antiviral state in the uninfected cells. As a result, virus production is significantly reduced, with an associated reduction in interferon induction. The reduction in interferon levels allows the virus to multiply, infecting a proportion of the cells in culture, until the level of interferon response rises to a point where it inhibits virus production again. In this way a cycle of virus production and repression is set up and a steady state can be achieved.

A persistently infected culture may also be established in the laboratory when only a few cells are infected initially, and a small amount of specific neutralizing antibody (low-avidity antibody is most effective) is added. The antibody decreases the amount of progeny virus available to reinfect cells. As with the interferon situation, the result is that the rate of infection and hence cell death is matched or exceeded by the division of noninfected cells, so that on balance the cell population survives. This is a question of establishing a dynamic equilibrium tilted in favor of the cell. Of course, the overall net production of cells is less than in an uninfected culture but, in an animal, cell division would be upregulated by the normal homeostatic mechanisms that control cell number. Indeed this situation is thought to mimic certain sorts of persistent infections in the whole animal (see Chapter 14).

Persistent infections resulting from interactions between viruses, cells, and defective-interfering virus

DI genomes are produced by all viruses as a result of errors in replication which delete a large part of the viral genome (see Section 7.1) making them *defective* for replication. The DI genome retains the sequences that are needed for recognition by viral polymerases and for the packaging of the genome into a virus particle, but little or nothing else. Thus, the DI genome is replicated only in a cell that is infected with infectious virus of the type from which the DI genome was generated, as this is needed to supply replicative enzymes and structural proteins. In this sense the DI genome is parasitic on infectious virus. The DI virus particles that result from this collaboration are usually indistinguishable in appearance from infectious particles. *Interference* between the DI and infectious virus comes about because more copies of the shorter DI genome can be made in the same time it takes to synthesize the full-length genome. For example, if the DI genome represents one-tenth of the infectious genome, then for every full length genome synthesized, there will be 10 copies made of the DI genome. Since viral polymerases are made in modest amounts, the large number of DI genomes produced will eventually sequester all the polymerase and synthesis of infectious genomes and virus particles will cease. At this point, production of DI genomes declines in consequence of the reduced availability of essential proteins. In practice the situation is more complex, as the generation of DI genomes is very much dependent on the type of cell infected, so that both cell and virus contribute to a balanced situation.

Persistent infections result when there is an equilibrium between the multiplication of infectious virus, the multiplication of DI virus, and cell division. In the beginning, a DI genome is generated *de novo*, and the increase in amount of that DI genome follows that of infectious virus, upon which it is dependent (Fig. 13.3). Initially there is sufficient polymerase to allow replication of both infectious and DI genomes, so there is no interference. Interference is apparent only when the number of DI genomes is so great that they sequester the majority of the polymerase proteins. At this point, there is interference with the multiplication of infectious virus, which in turn results in a concomitant decrease in the dependent DI virus. When the amount of infectious virus reaches a low point, cell numbers recover. As they do so, infectivity increases in the relative absence of DI virus, but then the cycle of events is repeated. In this very dynamic way infectious virus persists under conditions where, without DI virus, it would produce a short-lived acutely cytopathogenic infection. In some systems the cycles get progressively smaller, until there is only a low level production of infectious virus and DI virus and a steady-state persistent infection results (Fig. 13.3).

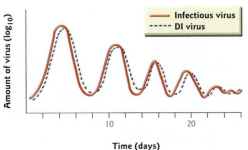

Fig. 13.3 A persistent infection established between a virus that normally causes an acutely cytopathogenic infection, and defective-interfering virus. The dynamic cycles of production of infectious and DI virus eventually give way to a low-level steady state persistent infection.

13.3 LATENT INFECTIONS

The term "latent" is defined as existing, but not exhibited. Thus in the context of a virus-infected cell, this means that the viral genome is present but no infectious progeny are produced. However, latency is an

Table 13.1

Examples of latent infections.

Virus	Synthesis of:			
	State of virus genome	At least one transcribed RNA*	Viral protein(s)	Infectious progeny
Herpes simplex virus	Episomal	+	–	–
Epstein–Barr virus	Episomal	+	+	–
Adeno-associated virus	Integrated	+	+	–

*The amount and nature of gene expression is limited but varies between virus systems, and is strictly controlled.

active infection and some virus-coded products are always expressed (Table 13.1). Lysogeny by temperate phages (see Chapter 15) is clearly a latent infection, and in animal cells, for example, herpesviruses and adeno-associated virus (AAV) exhibit latency. However, understanding of the molecular control processes is still at an early stage compared with phage λ. Bacterial viruses which achieve latency and AAV (Box 13.3) do so by integrating a DNA copy of their genome into the host's genome. This ensures that the viral genome will be replicated, together with host chromosomal DNA, and transmitted to daughter cells, and will be protected from degradation by nucleases. On the other hand, the DNA of herpesviruses is not integrated, but remains episomal, although the normally linear molecule is circularized.

It is in fact inaccurate to call a virus latent, as latent infections always commence and finish as acutely cytopathogenic infections. The initial acutely cytopathogenic infection is converted into a latent infection, and then delicate molecular controls operate to maintain the latent state. Eventually latency is broken by certain external stimuli, and there is full expression of the virus genome, production of infectious virus, and re-establishment of the acutely cytopathogenic infection, giving the virus the opportunity of infecting new hosts. Thus latency can be seen as an evolutionary strategy to remain in a host for a long period of time. In humans, latent infections with herpesviruses last for a lifetime, interspersed with periodic reversals into an acutely cytopathogenic infection. It is apparent that bacteriophage latency is fundamentally different from that of animal viruses, as the former is maintained primarily by virally encoded repressors of lytic replication (see Chapter 15), while the latter is controlled by host factors which are necessary for the expression of early virus gene products (see Section 15.6). The presence of a latent virus can sometimes be shown by using labeled antibody specific for the relevant viral proteins or, in all cases, by polymerase chain reaction (PCR) amplification of virus genome sequences.

Box 13.3

Evidence that adeno-associated virus establishes a latent infection by integration into the host genome

- The virus genome can be detected as a tandem, head-to-tail, pair integrated in the host chromosome.
- Analysis of the integrated DNA shows that it is present at the 19q13.3-qtr region of human chromosome 19.
- Site-specific integration is dependent on the presence the virus Rep protein which binds to the virus genome terminal repeat region (see Chapter 9). In the absence of the Rep protein integration is random.

Epstein–Barr virus (EBV), a member of the herpes virus family, becomes latent when it infects B lymphocytes (nondividing, non-antibody-producing cells that become antibody-synthesizing cells when activated by antigen) *in vitro*. Its very large 172 kbp of linear double-stranded DNA is circularized and not integrated. As a result of infection the B cell is stimulated into continuous cell division. The viral DNA replicates semi-conservatively, once each cell cycle, and each molecule segregates to a daughter cell like a host chromosome. During latency up to nine of the 100 viral genes are expressed (see Section 9.6). However about 1 in 10^3–10^6 cells convert spontaneously to the acute phase of infection, where the full set of genes is expressed, and infectious virus is synthesized and released. EBV is also associated with various malignancies (see Section 17.5).

Adeno-associated virus (AAV) has a small single-stranded linear DNA genome of 4680 nt. This is an unusual virus in that it is usually dependent on coinfection of the same cells with an adenovirus or a herpes virus for its replication (see Section 6.6). In the absence of a helper virus, AAV becomes latent with the help of the virus-encoded Rep protein, integrating into chromosome 19 of the host genome, and is replicated as part of the host genome. There is minimal gene expression during latency. On superinfection with helper virus, and in the presence of Rep, latency is broken. Viral DNA is excised from the host genome and the production of infectious AAV commences.

13.4 TRANSFORMING INFECTIONS

As a result of infection with a variety of DNA viruses or some retroviruses, a cell may undergo more rapid multiplication than its fellows, concomit-ant with a change in a wide variety of its properties, i.e. it is transformed.

Table 13.2

Types of virus that (very rarely) cause transformation of the infected cell.

Family	Genome	Proportion of genome integrated	Progeny
Retrovirus	RNA	Whole	+
Polyomavirus	DNA	Part	−
Papillomavirus	DNA	Part	−
Adenovirus	DNA	Part	−
Herpesvirus	DNA	None; episomal	−
Hepadnavirus	DNA	Part	−

This is often preceded by integration of at least part of the viral genome with that of the host. Chapter 20 deals in detail with transformation and other aspects of tumor viruses. However the term "tumor virus" is a misnomer, as transformation is a rare event caused by a virus that usually causes acutely cytopathogenic or persistent infections. Nevertheless, transformation is a significant event, as one immortalized cell can take over the whole population. For DNA viruses, transformation has no evident evolutionary significance, as the transformed cell contains only a fragment of the genome and cannot give rise to any progeny (Table 13.2).

13.5 ABORTIVE INFECTIONS

Any cell that possesses the appropriate receptors will be infected by a virus, but different cells do not replicate that virus with equal efficiency. There may be a reduction in the total yield of virus particles (sometimes to zero), or the quality of the progeny may be deficient as shown by their particle to infectivity ratio. Both of these reflect a defect in the production or processing of some components necessary for multiplication, be it DNA, RNA, or protein. One example is avian influenza virus growing in the mouse L cell line, in which both the amount of progeny and its specific infectivity are reduced, probably due to the synthesis of insufficient virion RNA. Another is infection of other nonpermissive cells with human influenza viruses which gives rise to normal yields of virions, which are noninfectious. This is because these cells lack the type of protease required to cleave the hemagglutinin precursor protein into HA1 and HA2 (see Section 11.7). This can be reversed by adding small amounts of trypsin to the culture or to released virions. Abortive infections present difficulties when trying to propagate viruses, but have been used to advantage when research into the nature of the defect has furthered understanding of productive infections. In natural infections, abortiveness contributes to the character of the infection, by effectively restricting virus to only those cells in which a productive infection takes place.

13.6 NULL INFECTIONS

This category represents cells which do not have the appropriate receptors for a particular virus, and thus cannot interact with a virus particle. Often this is the sole block to infection, as shown when infectious viral nucleic acids are artificially introduced into such cells in the laboratory and produce progeny virions.

13.7 HOW DO ANIMAL VIRUSES KILL CELLS?

The recognition that animal viruses often kill the cells in which they multiply gives the simplest criterion of infectivity, that of viral CPEs (see Chapter 9). However, such killing is not a property of the virus alone, but of a specific virus–cell interaction, and as exemplified by simian virus 5 above, a virus does not necessarily kill the cell in which it multiplies. Surprisingly, it is still by no means clear exactly how a cell is killed, except that every virus needs to undergo at least part of its multiplication cycle for killing to occur. Thus, it seems that a toxic product is produced by the virus and that viruses with different replication strategies are likely to invoke different mechanisms of toxicity. One problem in these studies is in distinguishing between an effect on a cell function which operates early enough to be responsible for toxicity and those which appear late on and are a consequence of those toxic effects. Viruses can inhibit synthesis of host proteins, RNA, and DNA, but here we shall deal only with the proteins, about which more is known (Table 13.3).

Studies of the inhibition of host cell protein synthesis by poliovirus suggest that this results from inactivation of the translation initiation factors which are responsible for recognizing capped messenger RNAs (mRNAs). Poliovirus mRNA itself is not capped and relies on a special mechanism of translation initiation. Its translation is therefore unaffected (see Section 10.4). In Semliki Forest virus-infected cells, there is evidence that the virus inhibits host protein synthesis by affecting the plasma membrane Na^+/K^+ pump, which controls ion balance. As a result, intracellular Na^+ concentration increases to a level where viral but not cellular mRNA is translated. The rotavirus nonstructural protein, nsP4, has been shown in a number of studies to affect the ion balance in infected cells by altering the permeability of the plasma membrane and other cell membranes, and is toxic even when expressed in a cell by itself. It has thus been claimed as the first known viral endotoxin. Cells infected by some viruses die through apoptosis (Section 13.1).

Table 13.3

Some suggested mechanisms of viral cytopathology.

Mechanism	Virus
Loss of ability to initiate translation of cellular mRNA	Polio, reo, influenza, adeno
Imbalance in intracellular ion concentrations	Semliki Forest, rota
Competing out of cellular mRNA by excess viral mRNA	Semliki Forest
Degradation of cellular mRNA	Influenza, herpes
Failure to transport mRNA out of the nucleus	Adeno
Apoptosis	Sindbis, Semliki Forest, influenza A, B, C, HIV-1, adeno, measles

HIV-1, human immunodeficiency virus type 1.

Evidently, viruses do not kill cells by any one simple process, and we do not understand the complex mechanisms involved. However, cells would not die immediately when their macromolecular synthesis is switched off, unless there was rapid turnover of some vital molecule which could not then be replaced. Thus, the mechanisms discussed above, with the exception of the upset in Na^+/K^+ balance and apoptosis, seem more similar to death by slow starvation than to acute poisoning. Lastly, it is by no means clear what advantage, if any, accrues to the virus in killing its host cell, as most progeny viruses leave the cell by exocytosis or by budding from the cell membrane. However, there is a distinction to be made between cell death that leads to the death of the animal host, and death of cells (e.g. in the gut) that can readily be replaced by the host – often without the host realizing that they are infected. The latter does no harm, but the former suggests that the infection represents a very new virus–host interaction which is poorly evolved in terms of survival of virus and host (e.g. HIV-1 – see Chapter 19; myxoma virus – see Section 17.4) or the invasion of the "wrong" type of cell (as in dengue hemorrhagic fever).

KEY POINTS

- All animal virus–cell interactions can be classified under just six headings.
- Classification depends on the type of cell infected, and as a result a virus can be classified in more than one category.
- Any definition of the virus infection should specify the circumstances of the infection.

QUESTIONS

- Describe three long-lasting interactions between viruses and cells in culture and consider the mechanisms by which these occur.
- Discuss the potential outcomes of an acute cytopathogenic infection of a cell and describe the processes which are responsible for death of the host cell, when this occurs.

FURTHER READING

Carrasco, L. 1995. Modification of membrane permeability by animal viruses. *Advances in Virus Research* **45**, 61–112.

Levine, B., Huang, Q., Isaacs, J. T., Reed, J. C., Griffin, D. E., Hardwick, J. M. 1993. Conversion of lytic to persistent alphavirus infection by the *bcl*-2 cellular oncogene. *Nature (London)* **361**, 739–742.

O'Brien, V. 1998. Viruses and apoptosis. *Journal of General Virology* **79**, 1833–1845.

Oldstone, M. B. A. 1991. Molecular anatomy of viral persistence. *Journal of Virology* **65**, 6381–6386.

Roulston, A., Marcellus, R. C., Branton, P. E. 1999. Viruses and apopotosis. *Annual Review of Microbiology* **53**, 577–628.

Also check Appendix 7 for references specific to each family of viruses.

14

Animal virus–host interactions

When an animal is infected with a virus a range of outcomes are possible, ranging from no disease to long-lasting disease states. The various types of interaction can be placed in one of seven possible categories.

Chapter 14 Outline

14.1 Cause and effect: Koch's postulates
14.2 A classification of virus–host interactions
14.3 Acute infections
14.4 Subclinical infections
14.5 Persistent and chronic infections
14.6 Latent infections
14.7 Slowly progressive diseases
14.8 Virus-induced tumors

The study of animal virus–host interactions is essential for our understanding of the capacity of viruses to cause disease. It must always be borne in mind that viruses are parasites and that the biological success of a virus depends absolutely upon the success of the host species. Hence, the evolutionary strategy of a virus in nature must take into account that it is counterproductive for the virus to eliminate the host species (although killing a small proportion will not affect the host's survival) or to impair its reproductive ability. Viruses demonstrate a number of different, and often complex, interactions with whole animals. A number of different factors are involved in determining the ultimate outcome of the virus–animal interactions. Many of these interactions involve components of the immune system which are described in Chapter 12.

14.1 CAUSE AND EFFECT: KOCH'S POSTULATES

Over a century ago, the bacteriologist Robert Koch enunciated criteria for deciding if an infectious agent was responsible for causing a particular disease, and was not merely a passenger. With the continuing evolution of new diseases, these are equally relevant (with some modifications) today. In essence, Koch postulated that: (i) the suspected agent must be present in particular tissues in every case of the disease; (ii) the agent must be isolated and grown in pure culture; (iii) pure preparations of the agent must cause the same disease when they are introduced into healthy subjects. As understanding of pathogenesis grew the following modifications were made:

Postulate 1. Koch originally thought that the agent should not be present in the body in the absence of the disease; this was abandoned when he realized that there were asymptomatic carriers of cholera and typhoid bacteria. As will be seen below, many viruses also cause such subclinical infections.

Postulate 2. Viruses were not known in Koch's time and some still cannot be grown in culture today, so this postulate was modified to say that it is sufficient to demonstrate that bacteria-free filtrates induce disease and/or stimulate the synthesis of agent-specific antibodies.

Postulate 3. Clearly, it is impossible to fulfil the third postulate when dealing with serious disease in humans, although accidental infection can sometimes provide the necessary evidence. For example, unfortunate laboratory accidents have directly demonstrated that human immunodeficiency virus (HIV) is responsible for the acquired immune deficiency syndrome (AIDS).

14.2 A CLASSIFICATION OF VIRUS–HOST INTERACTIONS

A classification of the various types of virus–host interactions with some named examples is given in Table 14.1. However, the table is only intended as a guide, as there is a continuous spectrum of virus–host interactions, and divisions are imposed solely for the convenience of classification. The categories are distinguished on four criteria: the production of infectious progeny, whether or not the virus kills its host cell, if there are observable clinical signs or symptoms, and the duration of infection. It can be seen that cell death does not necessarily correlate with disease, as many types of cell can be replaced without harming the individual. It is also apparent that a single virus can appear in several of the categories, depending on the nature of its interaction with its host, and that the duration of infection appears to correlate inversely with the need for efficient transmission. HIV-1 exemplifies the difficulty of attempting to classify infections (Box 14.1). It is also important to appreciate the three-way interaction between virus, host cell and immune system that determines the outcome of all infections (Fig. 14.1). Thus, in acute and subclinical infections, the balance favors the host (i.e. the virus is cleared from the body), whereas in persistent and chronic infections it is tilted towards the virus (which is not cleared and does not have to face the hazards of finding a new susceptible host for a long time). Persistent, chronic, and latent infections have in common the feature that the

Table 14.1

A classification of virus infections at the level of the whole organism.

Type of infection	Production of infectious progeny	Cell death	Clinical signs of disease	Duration of infection*	Transmission	Examples†
Acute	+	+	+	Short	Efficient	Measles virus, polio-virus (1% of infections), HIV-1
Subclinical	+	+	–	Short	Efficient	Poliovirus (99% of infections)
Persistent	+	– or +	–	Long (+ immune defect?)	Many opportunities	Rubella virus, HIV-1
Chronic	+	+	+	Long (+ immune defect?)	Many opportunities	Hepatitis B virus
Latent	–	–	–	Long	Many opportunities	Herpes viruses, HIV-1
Slowly progressive disease						
(a)	+	+	Eventually	Long	Many opportunities	HIV-1, TSE agents
(b)	–	+	Eventually	Long	None	Measles virus (SSPE)
Tumorigenic	+	–	+	Long	Many opportunities	Retroviruses
	–	–	+	Long	None	Hepatitis B virus, human papilloma-viruses, Epstein–Barr virus

*Short, approximately 3 weeks or less; long, up to a lifetime.
†Examples are given of viruses that have the given type of infection at some point in their life history, and are not intended to convey that they cannot also be classified elsewhere. e.g. herpesviruses switch between latency and acute infection (see text).
HIV-1, human immunodeficiency virus type 1; SSPE, subacute sclerosing panencephalitis; TSE, transmissible spongiform encephalopathy (includes bovine spongiform encephalopathy and Creutzfeld–Jacob disease) – these are not virus infections but prion diseases (see Chapter 22).

Box 14.1

The many faces of HIV-1 infection

Acute infection
There is initially a short-lived, self-limiting, flu-like disease.

Persistent/chronic infection
An asymptomatic infection continues for approximately 10 years, during which time the patient is highly infectious; this may become symptomatic (chronic) from time-to-time, and eventually causes the immunodeficiency that is responsible for clinically recognizable AIDS.

Latent infection
Drug therapy suppresses virus replication, but there is always a reservoir of latently infected cells that can reactivate if therapy is stopped.

Slowly progressive
Because of its long duration HIV-1 infection can also be considered under this heading.

immune system cannot clear the viruses responsible from the body. However all infections interact with the immune system and modulate one or more of its functions, so it never operates at its full potential.

Immune-mediated disease

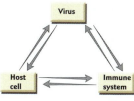

Fig. 14.1 Diagram to show the three-way interactions that decide the outcome of infection.

While attempting to classify virus *infections* of animals, it must be borne in mind that this is inextricably entwined with virus *disease*, e.g. the only difference between an acute infection and a subclinical infection, and the main difference between chronic and persistent infections is the disease that may or may not ensue. Disease itself may be caused directly by the virus destroying target cells (poliovirus and motor neurons), but an important component of disease results from the immune response to that infection. This seems strange when the immune system is meant to be fighting the infection, but it is one of the imperfections of that response, and is brought about by the many and conflicting demands put upon it; other downsides are allergic reactions and autoimmunity. The immune system is immensely powerful and contains enough firepower to destroy the body that houses it, if this is not properly directed or controlled. Indeed, there are extreme responses that occur in certain virus infections of animals that result in death – easily proved as immunosuppression prevents the disease and saves the life of that animal. In between are a variety of clinical signs and symptoms, such as the measles rash (due to deposition of antigen–antibody complexes in the skin) and the vaccinia pock (a T cell-mediated response) that are the result of the immune response rather than the virus. It is not just the adaptive immune response that plays a role in this way, as administration of purified interferon gives all the symptoms of influenza (elevated body temperature and muscular aches and pains), and interferon stimulated by infection with influenza virus may even be responsible for flu itself.

14.3 ACUTE INFECTIONS

Acute infections are analogous to acutely cytopathogenic infections *in vitro*, except that the infecting dose of virus is always small, and that the virus goes through many rounds of replication, spreading from the first cells infected to new susceptible cells. Thus, the minimum time-scale is measured in days rather than hours. During many infections viruses circulate around the body and come in contact with many different organs. However, most viruses do not cause generalized infections but attack particular organs or tissues, know as *target organs* or *target tissues*. Thus, hepatitis viruses infect the liver, and influenza viruses infect the respiratory tract, and the reverse never occurs. This specificity is achieved largely through the presence of specific cell receptors on only certain cells in the body, as discussed for *in vitro* infections in Chapter 13, but there may be intracellular restrictions on infection as well.

Infection of organisms can be described in terms of clinical signs and symptoms (Box 14.2) and by a variety of laboratory tests. Without the latter, no identification of the causative agent is complete. Laboratory tests (see Chapter 2) include isolation and titration of infectious virus, detection of viral antigens in blood and other tissues obtained by biopsy by a variety of immunological assays, detection of viral nucleic acids by the polymerase chain reaction for DNA or RT-PCR for RNA, and the direct identification of virus using the electron microscope – possibly in conjunction with antibody which will agglutinate cognate virus particles. Electron microscopy can be used for fecal specimens, nasal wash material, or biopsy materials.

Box 14.2

Definitions of signs and symptoms

- **Clinical signs:** attributes of infection that are objectively assessed, such as elevated body temperature.
- **Clinical symptoms:** attributes of infection that are subjectively assessed, such as pain.

Box 14.3

Key points about acute infections

Infection → incubation period → signs or symptoms → recovery or death

|_____|

infectious virus produced
(= transmission)

The course of acute infection and recovery from infection

An acute infection begins with infection of one or a few cells, infectious progeny are produced, and the infected cells die. Further cycles of multiplication ensue with increasing numbers of infected cells and eventually the first signs and symptoms of disease appear. Consequently, the infection has been progressing for several days before we are even aware of it. Often, virus is being shed too before we are aware that an infection is present, which can only enhance the spread of the virus to the susceptible people we meet. Fortunately, most people recover from acute virus infections within a few days or weeks. There are many different acute infections and the essentials are summarized in Box 14.3. An excellent example is measles virus because almost all the infections that it causes are acute. By comparison, only 1% of poliovirus infections are acute, and 99% are subclinical (Section 14.4). The example shown in Fig. 14.2 is of that familiar infection, the common cold, caused by one of the 100 or so different serotypes of rhinovirus (a picornavirus; see Box 14.4). The damage that a rhinovirus does to its target tissue, the ciliated epithelium of the upper respiratory tract, is shown in Fig. 2.3.

At this point it is appropriate to consider the mechanisms that are responsible for survival from the first encounter with a particular virus, i.e. a primary infection (Box 14.5). It is difficult to obtain unequivocal data from such a complex situation, far removed from the study of cells in culture. However, the innate immune system is the first line of defence. If it is successful, the infection progresses no further, but there is no immunological memory from the encounter. However, if the infection gets past innate immunity, then adaptive immunity is activated (see Chapter 12). T cells, in particular CD8[+] cytotoxic T cells, and antibodies are the main defence against primary infections. Information on local mucosal immune responses in humans is readily obtainable, but studies on people with natural immune deficiencies have been revealing. These have led to the

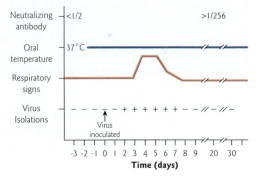

Fig. 14.2 A representation of an experimental infection in humans caused by a rhinovirus following deliberate intranasal inoculation.

Box 14.4

Evidence of an acute common cold infection caused by a rhinovirus

Virus infections are measured by clinical signs and laboratory tests for the presence of virus. Following infection with a rhinovirus there is no fever or rise in body temperature (Fig. 14.2). Respiratory signs are detected only 3 days after the infection and are assessed by the simple expedient of counting the number of paper tissues used per day – each used only once. The respiratory signs persist for 2 days before declining. Laboratory signs include virus isolation, and the generation of specific neutralizing antibody. Virus can be detected at least 1 day before the onset of symptoms and antibody can be detected several days after the clinical signs.

Box 14.5

Key points about combating acute infections – a quasi-military operation

- Constant readiness to detect and repel attack: the sentries and nonspecialized troops (the innate immune system). If this first defensive line is breached . . .
- Call up specialized reinforcements to carry out counter attacks, including search-and-destroy missions to hit the factories where munitions are being manufactured (adaptive T cell-mediated immunity and especially CD8+ cytotoxic T cells). To help in this effort . . .
- Call up more reinforcements with a different specialization to recognize and take out individual virus particles (adaptive antibody-mediated immunity).
- With the enemy defeated, stand down the troops but maintain an increased defensive line to abort any second attack from the same enemy (adaptive antibody-mediated immunity, memory T and B cells).

important conclusion that a virus infection, once established, can only be controlled by one or other of the arms of the adaptive immune response, and that this antiviral effect is not interchangeable. For example, people who have a congenital deficiency in antibody-mediated immunity, but have normal T cell-mediated immunity, are unable to combat primary infections with picornaviruses, orthomyxoviruses, and paramyxoviruses (Table 14.2). In these cases the virus is not cleared from the body and may persist for years, and clinical disease may be exacerbated. Conversely, a congenital inability to mount T cell-mediated immunity means that herpesvirus and poxvirus infections are not cleared, and can become generalized throughout the body and even life-threatening. However, the conclusion that antibody has no role in recovery from these infections is

Table 14.2

Recovery from primary virus infections (when innate immunity has failed).

Immunity responsible for recovery	Virus family
Antibody from activated B cells	Picornavirus
	Orthomyxovirus
	Paramyxovirus
Activated T cells	Herpesvirus
	Poxvirus

only tentative, as the lack of CD4⁺ T-helper cells means that immuno-globulin M (IgM) is the only antibody present.

One important implication of these findings is for vaccines. Vaccines are made empirically in the sense that it is not generally known which immune responses they stimulate, only that they protect from infection. Hence, it is not known which immune response is responsible for the observed protection, or how to stimulate a particular part of the immune response, which may be useful. Thus, if a candidate vaccine does not provide protection, there is no logical way of solving the problem (see Chapter 16). This is the very problem facing vaccinologists who have tried every empirical approach available in an attempt to solve the HIV problem, and signally failed. Thus they now have to identify which immune response(s) provide protection from HIV infection, and then determine how to stimulate them – in effect it is necessary to understand how the entire immune system operates, and this is a significant question.

B cells and T cells continue to be stimulated to divide and differentiate as long as cognate antigen is present, and adaptive immunity is said to be "antigen driven." Once virus antigen is eliminated, immune cell division stops, leaving a population of memory cells. However antibody often persists long after the infection, and this is thought to reflect the presence of long-lived, noninfectious antigen. Memory cells are resting cells with no effector functions, and provide, for the remaining lifetime, a much larger population of cognate B and T cells than were present before the infection. In addition, memory cells have a lower threshold for activation by antigen than lymphocytes. When reinfection with the same virus occurs, all this enables a secondary immune response to be mounted with less delay, more rapidly, and to a higher level than the primary response.

Antibodies are the main defence against reinfection by the same agent. If they are present in sufficient quantity there will be no reinfection, but it is doubted in some quarters that such "sterilizing immunity" actually exists, and that when a virus is encountered for a second time, infection actually occurs, although this time the infection is low-level and subclinical. In fact this may be beneficial, as it will boost immune responses.

Box 14.6

Key points about subclinical infections

Infection → incubation period → *no* signs or symptoms → clearance of virus
|_____|
infectious virus produced
(= transmission)

14.4 SUBCLINICAL INFECTIONS

Subclinical infections are also known as inapparent or silent infections. These are the commonest infections and, as their name implies, there are no signs or symptoms of disease. In all other regards, these are the same as acute infections. Evidence of infection comes only from laboratory isolation of the virus or by a postinfection rise in virus-specific antibody. Any virus which causes a subclinical infection has evolved a favorable equilibrium with its host. The enteroviruses (picornavirus family), which multiply in the gut, are one such group and the classic example is poliovirus, which causes no symptoms in over 99% of infections. A subclinical infection is arguably the expression of a highly evolved relationship between a virus and its *natural* host, since there are many examples of such a virus causing lethal disease when it infects a different host, e.g. yellow fever virus (a flavivirus) causes a subclinical infection in Old World monkeys but results in a severe infection in humans and is fatal in some New World monkeys. Subclinical infections have the same duration and are cleared by the same means as acute infections (Box 14.6).

14.5 PERSISTENT AND CHRONIC INFECTIONS

Both persistent and chronic infections are acute or subclinical infections that are not terminated by the immune response (Table 14.1). However, quite frequently persistent infections infect few cells and produce low levels of progeny virus, whereas chronic infections are more active (Box 14.7). The pertinent question is why the immune response is ineffective. In general, it appears that viruses that cause persistent and chronic infections have major inhibitory effects on some aspect of the immune response itself, or on the expression of major histocompatibility complex (MHC) proteins. Alternatively persistent or chronic infections result when an individual with an abnormality of the immune system contracts a virus that would otherwise cause an acute or subclinical infection. Whatever the cause, the end result is that virus particles and infected cells are not cleared.

Box 14.7

Key points about persistent and chronic infections

- Most result from the virus modulating the immune system in some way (e.g. immuno-suppression, downregulation of interferons, downregulation of MHC proteins, infection early in life); a few result from congenital immunodeficiency.
- Because chronic infections yield large numbers of infected cells and amounts of viral antigen for years, immune responses can have pathological instead of beneficial consequences

Box 14.8

Evidence that an incomplete immune response is associated with the establishment of persistent and chronic infections

Persistent and chronic infections are frequently associated with incomplete immune responses including:

- Immunosuppression by infection and inactivation of functions of macrophages, B cells, T cells, etc.
- Immune deficiency of the host, e.g. agammaglobulinemia leading to persistent infection with live poliovirus vaccine viruses.
- Infection of fetus or neonate before it has a fully competent immune system, resulting in immunological tolerance, e.g. infection of the fetus *in utero* by rubella virus.
- Viruses that are poorly immunogenic, and/or fail to synthesize or display enough antigen on the cell surface.
- Viruses that synthesize excess soluble antigen that binds all available neutralizing antibody.
- Viruses that stimulate non-neutralizing antibody that binds to virions and blocks neutralizing antibody.
- Viruses that stimulate insufficient interferon (and cells that synthesize low amounts of interferon).
- Viruses that generate mutants through inaccurate transcription (includes synthesis of antigenic variants and defective interfering (DI) genomes).

There are many scenarios which have this effect and some are listed in Box 14.8. It is important to appreciate the three-way interaction between virus, host cell, and immune system that determines the outcome of all infections (Fig. 14.1). In persistent and chronic infections this balance is tipped more to the virus, whereas in acute and subclinical infections it favors the host. Many viruses have specific functions that downregulate the expression of specific immune responses, including α/β-interferon. Downregulation is not absolute but is a quantitative phenomenon, and

Table 14.3

Infection of *adults* by hepatitis B virus: chances of progression of liver infection and disease.

Approximate percentage of infected adults who develop the type of infection listed	Type of infection
100 ↓	Acute ↓
10 ↓	Persistent ↓
1.0 ↓	Chronic ↓
0.1	Liver cancer (primary hepatocellular carcinoma)

this makes it harder to analyze. Interferon, for example, must be present in a sufficiently high concentration to be effective. In addition, viruses vary in their inherent sensitivity to the antiviral activity of interferon, so a level that inhibits one virus may not be effective against another.

A classic example of a chronic infection occurs when lymphocytic chorio-meningitis virus (an arenavirus) is inoculated into newborn mice. The animals become immunologically tolerant to the viral antigens, and virus can be found in large quantities in the circulation and every tissue including the brain. One unexplained feature is that tolerance is rarely complete, but presumably the reduced immune response is unable to cope with the infection. Similarly, little interferon is made, although the viruses are sensitive to its action. Neonatally infected animals remain healthy, whereas animals infected as adults suffer an acute infection. They mount an extremely strong T cell response to the virus which is lethal – effectively immunological suicide.

Another example of a chronic infection occurs when humans are infected with hepatitis B virus (HBV, an hepadnavirus), one of the causative agents of viral hepatitis. Infection is transmitted sexually, and from infected blood and saliva. After an initial acute infection of an adult there is disease with liver damage, and most infections are completely cleared. However, in a small proportion of people (10%), virus persists in the liver for a lifetime, although most of these infections are subclinical (Table 14.3). This contrasts with the situation in the Far East where HBV is endemic (i.e. most people are infected), and is transmitted from a carrier mother to her children early in life, possibly via infected saliva. The younger the age at which infection takes place, the higher is the chance of a persistent infection resulting, and in very young infants, it rises to over 90% (Fig. 14.3).

HBV is able to persist by downregulating MHC class I proteins on the surface of infected liver cells (hepatocytes), with the result that CD8$^+$ cytotoxic T cells are unable to act. Increased expression of MHC class I proteins stimulated by the action of interferon α/β leads to a reduction of the number of infected cells through the action of CD8$^+$ cytotoxic T cells (see Section 12.3), but unless all infected cells are destroyed the relief is only transient. A small proportion of the persistent infections evolve to become chronic, with further liver pathology which can develop into total liver failure. This probably results from the balance tilting in favor of the CD8$^+$ cytotoxic T cells which then destroy a large number of infected liver cells. In addition, during chronic infection, co-circulation of large amounts of virus antigens and virus-specific antibody causes the formation of antigen–antibody complexes. When deposited in the kidney, these circulating complexes can activate complement and result in the destruction of kidney tissue leading to immune complex disease. Hence, the appearance of disease in this instance depends not on the cytopathic effects of virus but upon the relative proportions of viral antigens and virus-specific antibody. Finally a very small proportion of the chronic cases (0.1%) develop liver cancer (see Section 18.7) with part of the viral genome integrated into the host DNA. The example of HBV shows how one virus can cause four different types of infection depending on host cell–virus–immune response balance (see Fig. 14.1).

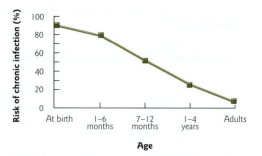

Fig. 14.3 The risk of becoming a carrier who is chronically infected with hepatitis B virus (HBV) depends on the age at which a person is infected.

14.6 LATENT INFECTIONS

By definition, all latent infections start and finish as acute infections. No infectious virus is present during a latent infection, and such infections are always subclinical (Box 14.9). The herpesviruses are a diverse and ubiquitous family which cause latent infections in many different animal species. Several cause common childhood infections. All become latent and such infections endure for life, making these viruses probably the best adapted of all to coexistence with their host. The action of the immune system is important for some viruses in maintaining their latent state. All of us are latently infected with several different human herpesviruses – and some of us are host to eight herpesviruses. One of the commonest and most successful is the human herpes simplex virus type 1 (HSV-1), because by 2 years of age nearly everyone has been infected. This is caused by contact with infectious saliva (e.g. when the baby is kissed by an infected adult). Latency and herpesviruses are dealt with in detail in Chapter 15. Finally, latency is also one of the several states that HIV-1 is able to adopt in some cell types.

Box 14.9

Key points about latent infections

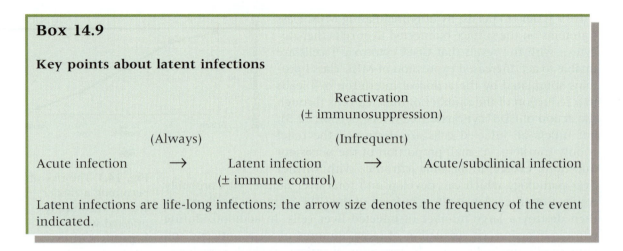

Latent infections are life-long infections; the arrow size denotes the frequency of the event indicated.

14.7 SLOWLY PROGRESSIVE DISEASES

As their name implies these are diseases that take many years to manifest, while virus multiplication proceeds at the normal rate. There are two categories: slowly progressive diseases caused by viruses and those caused by the spongiform encephalopathies. The viruses are subdivided into those that make infectious progeny throughout (e.g. HIV-1) and those whose genomes become defective during the long incubation period and hence are noninfectious (e.g. measles virus causing subacute sclerosing panencephalitis or SSPE). The transmissible spongiform encephalopathies are believed to be caused by a novel type of infectious agent, and are dealt with in Chapter 22.

Slowly progressive virus diseases that are infectious

The classic example of such a virus disease is HIV-1 which predominantly infects CD4$^+$ T cells. From the detailed statistics that are only available for infections in the developed world, a typical adult infection starts with an acute influenza-like infection. The disease is then quiescent for an average of 10 years, although the virus continues to multiply to high titer (10^{10} virions produced per day), with high turnover. During this time there is a progressive loss of CD4$^+$ T cells, and these eventually reach such a low level that there is a virtual collapse of the immune response. The patient is then overwhelmed by various endogenous viral or bacterial infections that are normally kept in check by the immune system. In the developed world these are commonly *Pneumocystis carinii* (a yeast-like organism) or cytomegalovirus. HIV-1 is considered in detail in Chapter 19.

Slowly progressive virus diseases that are noninfectious

Infection with measles virus (a paramyxovirus) demonstrates a different slowly progressive disease. Before a vaccine was available, most children contracted an acute measles virus infection, which was then cleared from the body by the immune system in around 3 weeks. However, in a very small proportion, the virus established itself in the brain and, after a long incubation period, caused a pattern of degenerative changes in brain function, including loss of higher brain activity, and inevitable progression to death (Box 14.10). Two different, but related diseases result from the neural infection. These are measles inclusion body encephalitis (MIBE) and subacute sclerosing panencephalitis (SSPE). MIBE occurs in approximately 1 in 2000 cases of measles infection. It is a chronic progressive encephalitis occurring in children and young adults and is associated with a persistent measles virus infection. MIBE can be fatal with the average survival time after diagnosis being 3 years. It is often seen in immuno-compromised patients and is associated with early (<2 years of age) primary infection. SSPE occurs in approximately 1 in 100,000 cases. A slowly progressing neurological degenerative disease associated with a long term persistent measles virus infection leads to death many years after the primary infection.

In SSPE at post-mortem, areas of "hardening" or "sclerosing" of brain tissue can be seen which give the disease its name. Why only certain individuals contract the disease is not understood. Infectious viruses cannot be isolated from SSPE patient tissue, although virus can be transferred

---

Box 14.10

Comparison of acute and slowly progressive measles virus infections

Acute measles virus infection
- A common acute childhood infection worldwide before mass immunization was established.
- Respiratory transmission.
- Systemic infection.
- 100% disease – a diagnostic smooth skin rash.

Subacute sclerosing panencephalitis (SSPE)
- Very rare, affecting about 6–22 per 10^6 cases of acute measles.
- A sequel to acute measles.
- Brain infection and disease.
- Incubation period of 2–6 years.
- Associated with measles infection early in life (<2 years old).
- Invariably fatal.

by co-cultivation with brain extracts. The virus generated by co-culture is not able to establish a productive infection. Genomes of MIBE and SSPE-measles viruses obtained from the brains of infected individuals have now been sequenced, and these have accumulated many mutations and produce a defective internal virion protein (matrix). The nature of the presumed defect in the immune response which fails to clear the initial acute infection is not known.

14.8 VIRUS-INDUCED TUMORS

All viruses that cause tumors have DNA as their primary genetic material in the cell but some, members of the retrovirus family, have RNA in their virions, i.e. they belong to Baltimore class 6. DNA "tumor" viruses normally cause an acute infection and it is rare that a tumor results, e.g. nearly everyone is infected as a child or young adult by Epstein–Barr virus (EBV; a herpesvirus), which is subclinical or causes infectious mononucleosis (glandular fever) respectively, and yet occurrence of the tumors (Burkitt lymphoma, nasopharyngeal carcinoma) with which the virus is associated is very rare. Demonstration of tumorigenicity in the laboratory is usually made under "unnatural" circumstances which are known to be favorable to the development of the tumors. Important factors are genetic attributes and age at which infection took place: certain inbred lines of animals develop tumors more readily than others, and young animals are more susceptible because their immune system is not fully mature. If no tumors result, it may be necessary to transfer cells transformed by the virus *in vitro* into the animal to demonstrate viral tumorigenicity.

The diagnostic criterion for a tumor virus in the laboratory is the conversion of a normal cell into a transformed cell. This involves many complex changes in cellular properties, which are discussed more fully in Chapter 20. However, not all transformed cells are tumorigenic when transplanted into appropriate animals, and other steps in addition to transformation are needed to produce a cancer cell. Thus, transforming viruses that cause cells to divide more rapidly than normal are at one end of the spectrum and those causing cellular destruction are at the other. Tumors fall into two groups: those that produce infectious virus and those that do not. The latter are the more common and are caused by viruses from several different families. Such tumors contain viruses which are unable to multiply, whose multiplication is in some way repressed, or those in which only a fragment of genome is present. The methods described above for detecting viral components in latently infected cells can be used successfully on tumor cells. However, some tumors are probably caused by a "hit-and-run" infection which leaves no trace of the virus behind.

As for other virus–host interactions that have been discussed above, the induction of tumors depends upon the balance of a complex situation.

This can be viewed as shown in Fig. 14.4, where, initially, infection may be acute, persistent, or latent. The transformation/tumorigenesis process initiated by infection requires additional factors, such as chemical carcinogens. How often transformation of cells takes place in the animal is not known, but it is likely that this is frequent and that on most occasions the immune system recognizes and destroys the transformed cell. Tumor formation probably requires that the immune system is in some way deficient, a state perhaps induced by infection or by aging. The apoptosis mechanisms present in all cells also act as a powerful protection against tumor development.

A number of viruses whose principal interactions with their hosts can be classified into one or more of the categories described in Sections 14.3–14.7, can also, as a rare consequence of infection, trigger a sequence of events leading to the formation of a tumor. The *in vitro* parallel of this is the morphological transformation of cells in culture following virus infection. Examples of some common viruses that can cause cancer are shown in Table 14.4. These are chosen to emphasize that they all cause an acute infection initially that is not cleared, and remain in the body in a variety of ways. Then depending on the circumstances of the infection and the co-factors experienced in life, a cancer may eventually be manifested. All cancers are multifactorial.

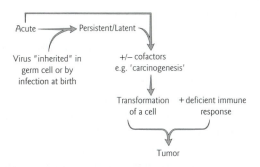

Fig. 14.4 The relationship between virus infection and tumorigenesis.

Table 14.4
Some common human viruses that (rarely) cause cancer.

Virus	Initial infection	Subsequent infection	Cofactors	Cancer
HBV (hepatitis B virus: an hepadnavirus)	Acute (hepatitis)	Persistent or chronic (hepatitis)	Infection early in life, cigarette smoking, alcohol, fungal toxins	Liver (primary hepatocellular carcinoma*)
HPV (human papillomavirus: a papovavirus)	Subclinical	Persistent (anogenital warts)	Early and frequent sexual activity, smoking	Cervix of the uterus (carcinoma)
EBV (Epstein–Barr virus: a herpesvirus)	Child: subclinical Adult: acute (infectious mononucleosis)	Latent	Malaria, MHC genetics	Lymphoid tissue of the jaw (Burkitt's lymphoma), or nasopharyngeal carcinoma

*Carcinoma: a cancer of epithelial cells.
MHC, major histocompatibility complex.

KEY POINTS

- There are seven categories of *infection*: acute, subclinical, persistent, chronic, latent, slowly progressive, and tumorigenic.
- Any one *virus* may be found in more than one category, thus the categorization of a virus and its infection must include the circumstances of the infection.
- The majority of infections result in a satisfactory end for both the host species (if not the individual) and the virus. In other words, infections are dealt with by the host's immune system so that the host recovers and can reproduce and provide new nonimmune hosts for the virus. Either before adaptive immunity has matured or by evading immune responses, the virus multiplies and perpetuates itself by being transmitted to other susceptible individuals.
- Infections that kill (or reproductively incapacitate) large sections of the host species are rare, and these result from an evolutionarily new virus–host interaction (such as HIV-1 in man). Such infections evolve towards genetic compatibility and peaceful coexistence (e.g. myxoma virus in the European rabbit, see Section 17.4).

QUESTIONS

- Discuss the proposal that the outcome of virus infection is regulated both by the virus and by the host.
- Discuss the mechanisms by which antibodies regulate virus infection via both negative and positive effects on virus infectivity.

FURTHER READING

There is an enormous area to cover, but the following provide a way into the subject and more references.

Alcamí, A., Koszinowski, U. H. 2000. Viral evasion of immune responses. *Immunology Today* **21**, 447–455.

Casadevall, A., Pirofski, L.-A. 1999. Host–pathogen interactions: redefining the basic concepts of virulence and pathogenicity. *Infection and Immunity* **67**, 3703–3713.

Craighead, J. E. 2000. *Pathology and Pathogenesis of Human Viral Disease*. Academic Press, New York.

Fujinami, R. S., ed. 1990. Mechanisms of viral pathogenicity. *Seminars in Virology* **1**, (4) (several articles).

Kovarik, J., Siegrist, C.-A. 1998. Immunity in early life. *Immunology Today* **19**, 150–152.

Lederberg, J. 1999. Paradoxes of the host–parasite relationship. *American Society for Microbiology News* **65**, 811–816.

Mims, C. A., Nash, A., Stephen, J. 2000. *Mims' Pathogenesis of Infectious Disease*, 5th edn. Academic Press, London.

Pawelec, G., Solana, R. 1997. Immunosenescence. *Immunology Today* **18**, 514–516.

Preston, C. M. 2000. Repression of viral transcription during herpes simplex virus latency. *Journal of General Virology* **81**, 1–19.

Smith, G. L., ed. 1998. Immunomodulation by viruses. *Seminars in Virology* **8**, 359–442 (several articles).

Stevens, D. L. 1998. Immunity and the host response. *Current Opinion in Infectious Disease* **11**, 269–270.

Timbury, M. C. 1997. *Notes on Medical Virology*. Churchill Livingstone, Edinburgh.

Wright, P. F., ed. 1996. Viral pathogenesis. *Seminars in Virology* **7**, (4) (several articles).

Also check Appendix 7 for references specific to each family of viruses.

15

Mechanisms in virus latency

Latent infections are an important facet of the interaction of certain viruses with their hosts. The ability to establish a latent infection means that the virus can maintain its genome in the host for the entire life of the host, being reactivated periodically to produce new viruses which can infect new hosts.

A common perception of virus infection is that the only possible outcome is an immediate and rapid production of progeny virus particles. While this is the most common result of infection it is not the only one, as described in Chapter 14. Of the possibilities described there, one of the most interesting is where certain, but not all, viruses enter a different type of replicative cycle in which the host cell survives, either for a fixed period or indefinitely. During the infection the virus is quiescent with no progeny being produced and no signs or symptoms of disease in the host. This is termed a latent infection. The appreciation that some viruses have the capacity to establish a latent infection has significant implications for our understanding of their associated diseases and potential therapies. It also has relevance to the potential use of animal viruses, such as the defective parvovirus adeno-associated virus, as vectors for gene therapy. The establishment of latency represents the ultimate parasitic interaction in virology, with the virus able to coexist with a host which, in most cases, gains no obvious advantage.

15.1 THE LATENT INTERACTION OF VIRUS AND HOST

Latent infections begin with an acute infection, which may or may not be associated with a disease, and end with production of infectious virus particles, which again may or may not be associated with disease (see Section 14.6). The new particles produced when latency "breaks down" are then available to infect other hosts. The diseases at the beginning and end of the process may differ; a classic example is varicella zoster virus, a herpesvirus, which causes chicken pox as a result of the initial infection but shingles in the same individual when it reappears later from the latent state. Between the initial infectious event and the ultimate production of progeny virions the virus establishes an intimate association with the host cell and this stage may last for very extended periods of time, even for the lifetime of the host. This offers the virus an opportunity to lie dormant until a new susceptible host presents itself. The new host may be separated by a generation or more from the original infected individual and so latency has important implications for virus survival and transmission.

The group of viruses for which latent infections are best known are the herpesviruses. Herpesviruses infect a wide range of hosts, with eight viruses able to infect humans; herpes simplex virus (HSV) types 1 and 2, Epstein–Barr virus (EBV), varicella zoster virus (VZV), cytomegalovirus (CMV), and human herpesviruses 6, 7, and 8. The herpesviruses which infect mammals all establish latent infections as part of their normal processes, and once the latent infection is established the virus remains associated with the host for the remainder of its life. It is likely that the herpesviruses infecting other animals can also establish latent infections but we have only little information about these. While in the latent state, the herpesviruses do not produce any infectious progeny but at various times the virus may be reactivated, go through a complete replication cycle, and produce progeny viruses. The stimuli which activate reactivation are very varied and can be very individual for each infected person.

The first indication that two alternative outcomes of infection by the same virus are possible was with bacteriophage λ (Box 15.1). This was followed by the discovery of other bacteriophage which can also adopt two different types of infectious process. Bacteriophage that are able to adopt either a lytic or a lysogenic replication cycle are called temperate phage. Several studies have shown that the phage DNA is inserted in the bacterial host genome during lysogeny. In this state the virus DNA is replicated with the host chromosome and is called a prophage. The bacteria carrying the phage DNA are called lysogens. Phage λ remains a key example of latency and our understanding of the molecular events which occur during both lytic and lysogenic replication have formed the basis for our understanding of gene expression and latency. Phage λ is a paradigm for the processes involved in latency and will be considered in detail first here.

Box 15.1

Evidence that infection with bacteriophage λ can have two outcomes

Following infection of a culture of *E. coli* with phage λ, most of the bacteria in the culture lyse, releasing more infectious phage. This is the result of a lytic infection. However, a small number of bacteria survive the infection and can be propagated. The surviving bacteria are resistant to infection by phage λ and also by related phage. When grown in culture, the resistant individual bacteria can spontaneously produce infectious phage. Treatment of the resistant bacteria with ultraviolet irradiation at levels sufficient to stop the culture growing induces this event in all cells, causing lysis, with release of large amounts of phage λ. These observations indicate that the phage λ introduced initially is present in the resistant bacteria and can be propagated from one generation to the next. However, the phage must be present in a noninfectious form which can spontaneously revert to produce virus.

15.2 GENE EXPRESSION IN THE LYTIC CYCLE OF BACTERIOPHAGE λ

When phage λ enters a bacterium it must initiate either a lytic or a lysogenic cycle. For the latter to occur the phage DNA must insert itself into the host chromosome and be maintained. In order to understand fully the mechanism by which lysogeny is established and maintained, and how the virus can regain the lytic replication cycle from the lysogenic state, it is necessary first to consider the genes involved in a lytic infection.

Following attachment, the phage λ linear genomic DNA is introduced by an injection mechanism similar to that used by phages T2 and T4 (see Section 5.3) into the *E. coli* and is immediately converted into a covalently closed dsDNA circle by host enzymes. Circularization is possible because the phage λ dsDNA genome contains 12 bases of single-stranded DNA at either end of the linear molecule which are complementary. The single-stranded regions are called *cos* sequences, and on infection they come together by base pairing to form a circular DNA molecule which is covalently closed by the host DNA ligase enzyme (Fig. 15.1). Having circularized, the phage λ genes are then switched on and off in a tightly regulated, coordinated, manner which determines the outcome of the infection. During lytic replication the circular phage DNA replicates independently of the host DNA and does not integrate into the genome of the host.

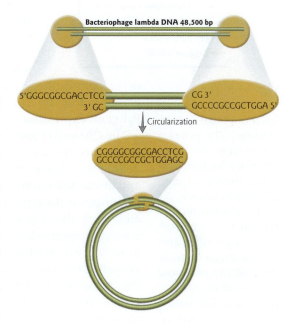

Fig. 15.1 Formation of circles of phage λ DNA after infection. The base composition of the complementary cos regions are shown. In λ these are extensions to the 5′ ends of the genome.

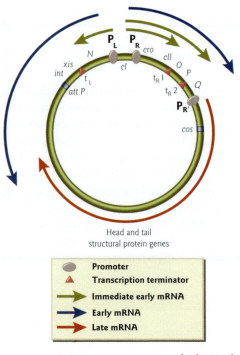

Head and tail
structural protein genes

⬭	**Promoter**
▲	**Transcription terminator**
→	**Immediate early mRNA**
→	**Early mRNA**
→	**Late mRNA**

Fig. 15.2 The circular map of the bacteriophage λ genome. The genes involved in the lytic cycle are indicated. The positions of the phage promoters and transcriptional terminators involved in the lytic cycle are shown together with the mRNAs produced at immediate early, early, and late times. The sequence attP is relevant to lysogenic replication (Section 15.3).

The phage λ genome contains two principal promoters which are recognized by the host cell DNA-dependent RNA polymerase which immediately begins transcription of mRNA. One of the promoters (P_L, Fig. 15.2) directs transcription in a leftwards direction and generates an mRNA that terminates at the end of the gene encoding the N protein. The other promoter (P_R) directs transcription in a rightwards direction to encode the protein Cro. However, termination of transcription at the end of the cro gene is not absolute and some mRNAs extend through the cII, O, and P genes. The proteins are then translated from the polycistronic mRNA. The N protein causes the RNA polymerase to transcribe through the regions of DNA at the ends of the N, cro, and P genes, where it had previously stopped, to generate polycistronic mRNAs encoding several proteins. The mRNA from P_L extends through the N, cIII, xis, and int genes, and mRNA from P_R extends through the cro, cII, O, P, and Q genes. Thus, N protein acts as a transcriptional anti-terminator and allows expression of additional genes. These events occur before phage λ DNA synthesis and are referred to as immediate early (N and cro) and early (cIII, xis, and int from P_L, and Q gene from P_R) gene expression.

The Cro and Q proteins are important for the next phase of gene expression. This occurs after phage λ DNA synthesis and is therefore, by definition, a late event in the replication cycle. Immediately on infection, and at the same time as P_L and P_R are utilized, host cell RNA polymerase recognizes a third promoter region in phage λ DNA, P'_R located immediately after the Q gene. However, transcription is terminated just downstream to synthesize a very short mRNA, even in the presence of the N protein. This short mRNA does not encode a protein. The Q protein is an anti-terminator which causes the RNA polymerase molecules initiating transcription at P'_R to ignore the termination signal and to continue mRNA synthesis. The resulting mRNA is extremely long and extends through the genes encoding the structural proteins which make up the phage head and tail (Fig. 15.2). At the same time as the Q protein is exerting its activity then Cro protein is also at work. The Cro protein binds to the phage λ DNA at the operator elements (O_L and O_R, respectively) of the promoters P_L and P_R. By doing this Cro inhibits transcription from these promoters, stopping production of the early mRNAs. Sufficient Q protein is present to ensure that P'_R which is not affected by Cro protein, remains active. The result is that the only mRNA found at late times encodes the structural proteins which can package the newly synthesized phage λ DNA into progeny virions utilizing the cos site (see Section 11.4).

15.3 ESTABLISHMENT AND MAINTENANCE BACTERIOPHAGE λ LYSOGENY

Establishment of lysogeny

The initial events in the process of lysogeny are identical to those seen in a lytic infection. The circularized phage DNA is transcribed from the two major promoters, P_L and P_R, and also from P'_R. Since transcription from P'_R makes only a small mRNA which does not encode a protein and it is not involved in the lysogeny process, it will not be considered further. Transcription from P_L and P_R makes the mRNAs encoding the N and Cro proteins, with small amounts of the cII, O, and P proteins. Subsequently, the anti-terminator action of the N protein results in the synthesis of cIII, Xis, Int, and Q proteins and larger quantities of cII, O, and P, as described above (Section 15.1). The cII protein acts as a gene activator, directing the host RNA polymerase to begin transcription at two promoters which would otherwise be inactive. These are P_{RE} (promoter for repressor expression) and P_{int} (Fig. 15.3a). The mRNA initiated at P_{int} directs the synthesis of the Int protein which is responsible for integrating the phage λ DNA into the host chromosome (see below). The use of the P_{int} promoter ensures expression of larger quantities of Int protein than can be generated from the polycistronic mRNA transcribed from P_L.

P_{RE} directs transcription of a single gene, *cI*. The cI protein, usually referred to as the cI, or lambda, repressor, is critical in lysogeny and phage with mutations in the *cI* gene can only replicate lytically. The cI protein binds to O_L and O_R, inhibiting transcription from P_L and P_R and preventing production of the early proteins. By inhibiting synthesis of the early proteins, the cI protein prevents the subsequent appearance of the late, structural, proteins and, consequently, of infectious particles. In order to ensure that its own synthesis is not prevented by an absence of cII protein, the cI protein also directs RNA polymerase to an additional promoter, P_{RM} (promoter for repressor maintenance), which initiates transcription of the *cI* gene alone (Fig. 15.3b).

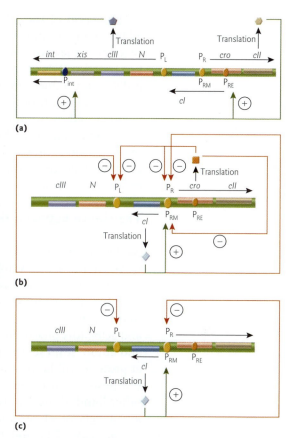

Fig. 15.3 The region of the bacteriophage λ genome encoding the genes responsible for lysogeny. (a) The cII protein activates the P_{RE} and P_{int} promoters to produce mRNAs for the cI and Int proteins, respectively. Note the mRNA from P_{RE} contains sequences antisense to the *cro* gene. (b) The cI protein binds O_L and O_R, inactivating P_L and P_R, respectively, while activating P_{RM}, ensuring its own continued synthesis throughout lysogeny. The Cro protein also inhibits P_L and P_R activity. (c) The steady state of *cI* gene expression during lysogeny.

The action of the cII protein in activating expression from P_{RE} has an additional consequence. The mRNA transcribed from P_{RE} contains some sequences that are antisense to those of the mRNA from P_R encoding the Cro protein. Hybridization of these will prevent translation of the *cro* gene mRNA and reduce the level of Cro protein in the bacterium.

The final result of the process described above is that the phage λ DNA is integrated in the host chromosome and the only gene which is active is that encoding the cI protein. The cI protein, by inhibiting P_L and P_R, prevents the expression of any other phage genes. By activating P_{RM}, the cI protein ensures transcription of its own gene and therefore its own continued synthesis, so stabilizing the lysogenic state (Fig. 15.3c).

The choice between the lytic and lysogenic pathways

As described above, the initial steps in both the lytic and lysogenic pathways are the same. While the lytic process occurs much more frequently, it is clear that some circumstances must favor lysogeny. Several aspects of what determines the choice between lysis and lysogeny remain unclear, but a key factor is the balance between the various repressors and activators produced early in the phage's interaction with the bacterium, and amongst these, the role of the cII protein is critical. If cII is very active cI protein will be produced in large amounts. This will efficiently inhibit synthesis of all genes except its own, and lysogeny will result. On the other hand, if the cII protein is poorly expressed or has low activity, very low levels of cI protein will be present. The Cro protein will inhibit the activity of P_L and P_R and the synthesis of the cII protein, amongst others, will be significantly reduced. Without sufficient cII protein P_{RE} will not be strongly activated and synthesis of the cI protein will, in turn, be further reduced. In this case the lytic pathway will follow.

The activity of the cII protein is determined, at least in part, by environmental factors. The cII protein is susceptible to proteolytic degradation by host enzymes. The levels of host proteases are affected by many factors, especially growth conditions. Healthy bacteria grown in rich medium contain high levels of proteases and when infected are more likely to support lytic replication. By contrast, poorly growing bacteria have lower levels of proteases and so will encourage establishment of lysogeny. It should be clear that it is to the advantage of the phage to undertake a lytic infection in healthy bacteria which will contain high levels of ATP and are equipped to synthesize the virus proteins. When bacteria are nutritionally deficient it is to the advantage of the phage to establish itself as a lysogen and wait until conditions improve.

The cIII protein also plays a role in the choice between lysis and lysogeny. The function of cIII is to protect cII from proteolytic degradation. This protection is not complete since cII can be destroyed even in the presence of cIII. However, if cIII is absent or not functional cII is almost always degraded and the phage can only undergo a lytic lifestyle.

Integration of bacteriophage λ DNA during lysogeny

During lysogeny the phage λ DNA integrates into the genome of the host. This contrasts with the situation during a lytic infection when the phage DNA replicates independently. The integration of phage λ DNA into the host chromosome is carried out by the Int protein. The integrated DNA is referred to as a prophage and genetic mapping studies showed that there is only one site of integration, adjacent to the galactose (*gal*) operon in the *E. coli* genome at a position referred to as the lambda attachment (*att λ*) locus. In 1962, Campbell suggested that crossing-over between the circularized phage DNA and the host genome results in insertion of the entire phage genome as a linear structure into that of the bacterium by reciprocal recombination (Fig. 15.4a). The position of the recombination site in the phage DNA lies downstream of the *int* gene and the region is referred to as the *att P* site (Fig. 15.2). The bacterial and phage attachment sites, *att λ* and *att P*, contain identical tracts of 15 base-pairs indicating that homologous recombination occurs during the integration event. The actual site of the crossover for both integration and excision must take place within, or at the boundaries of, this common sequence (Fig. 15.4b).

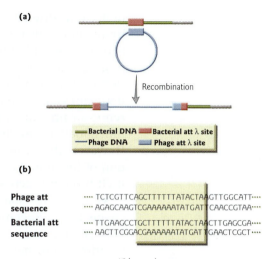

(a)

Recombination

■ Bacterial DNA ■ Bacterial att λ site
─ Phage DNA   ■ Phage att λ site

(b)

Phage att sequence
···· TCTCGTTCAGCTTTTTTATACTAAGTTGGCATT····
···· AGAGCAAGTCGAAAAAATATGATTCAACCGTAA····

Bacterial att sequence
···· TTGAAGCCTGCTTTTTTATACTAACTTGAGCGA····
···· AACTTCGGACGAAAAAATATGATTGAACTCGCT····

15 base-pair common sequence

Fig. 15.4 (a) Proposed model for the insertion of phage λ DNA into the bacterial genome by reciprocal recombination between phage and host DNA. (b) The common sequence in the bacterial and λ attachment sites.

15.4 INDUCTION AND EXCISION OF THE BACTERIOPHAGE λ LYSOGEN DNA

Induction and excision of integrated DNA

Early observations showed that ultraviolet irradiation of a lysogenic culture of *E. coli* stimulates the phage to relinquish the lysogenic state and to enter a lytic mode of replication. This is termed induction of the lysogen. Several other stimuli, such as treatment with potent mutagens, also induce the phage and a common feature of these treatments is that they cause significant changes in the host cell, and in particular expression from an array of genes not previously active. This is referred to as the *E. coli* SOS response and the role of the new gene products is to protect the bacterium against the effects of the stimulus. Just as lysogeny may allow the phage to wait for the host to provide an optimum environment for replication, induction may be seen as a means by which such a phage can escape from a host whose survival is threatened.

A critical component in the SOS response is the Rec A protein. The function of the Rec A protein is to mediate recombination between DNA molecules. However, when the bacterium is subjected to stress, such as irradiation with ultraviolet light, the Rec A protein alters its activity to become a specific protease. The primary target of the proteolytic Rec A protein is a repressor called Lex A. The Lex A protein represses expression from a range of genes and its cleavage removes this repression giving expression of the genes. Rec A protein also cleaves the phage λ cI repressor protein, and when this occurs the promoters P_L and P_R become functional, transcribing the *N*, *cro*, and other genes described above. However, the activity of the altered Rec A protein prevents accumulation of functional cI protein and so only lytic replication can occur even if cII protein is active.

As well as re-establishing the lytic pattern of phage λ gene expression, escape from lysogeny also requires that the λ genome is excised from the host genome and becomes independent again. Reversal of the recombination event between the phage and host genomes which generated the prophage results in the regeneration of the circular phage genome. However, while the integration event requires only the Int protein, for excision to occur the *xis* and *int* genes must both be transcribed as both proteins are required to act together. The mRNA encoding the Xis and Int proteins is transcribed from the reactivated P_L promoter. The excision process is very precise, generating an exact copy of the lambda genome. However, on occasion the excision process can go wrong resulting in the

---

Box 15.2

Specialized transduction

The process of excision is usually very precise but occasional errors can occur, such that the region excised includes a small amount of bacterial DNA, with a corresponding piece of phage DNA deleted. If the portion of bacterial DNA is located to the left of the prophage, it may include some or all of the genes of the *gal* operon, with a compensatory loss of some of the genes at the right end of the prophage. This incomplete phage chromosome then serves as the template for replication, such that essentially all of the phage progeny issuing from that bacterium carry the genes for galactose utilization and have lost some of the phage genes. Infection of a Gal⁻ bacterium with such a phage, under conditions allowing lysogenization, can confer the ability to ferment galactose, since the necessary genes become inserted into the bacterial chromosome. The process of transferring genes from one host to another is called transduction, and since phage λ is able to transfer only the genes immediately adjacent to the *att* λ region of the bacterial chromosome it is referred to as a specialized transducing phage. Many transducing phages are defective in replication, since they may lack some essential phage genes, and these can only be propagated if a normal ("helper") phage is present to supply the missing gene functions.

virus DNA removing a portion of the adjacent cellular DNA. This cell DNA can then be transmitted to other *E. coli* by infection (Box 15.2).

15.5 IMMUNITY TO SUPERINFECTION

An unusual feature of lysogeny is that a bacterial lysogen is immune to superinfection by a second phage of the same type that it carries, but it is usually not immune to other, independently isolated temperate phages. Phages that induce immunity to one another are termed homoimmune while those that do not are called heteroimmune. The immunity that the lysogens display is very different to the immunity seen in animals. Analysis of the genetic factors involved in the specificity of immunity showed that the *cI* gene is responsible.

Lysogens continuously express the *cI* gene and contain significant levels of functional cI repressor until induced (Section 15.4). When the DNA of a superinfecting, homoimmune, phage enters the lysogenic bacterium the cI repressor in the cell binds to the O_L and O_R sequences on the incoming phage genome and prevents phage gene expression, aborting the infection. If the superinfecting DNA is from a heteroimmune phage, the cI repressor of the lysogen cannot bind the incoming DNA and the second phage is able to initiate an infection.

15.6 THE BENEFITS OF LYSOGENY

The fact that temperate phages carry so many genes involved in lysogeny indicates that lysogeny is likely to have some advantages for them. A significant advantage would be to provide a method for a phage which has infected a host low in energy and synthetic capacity to persist in the bacterium until conditions improve. The phage can then "reappear" when conditions have improved.

It is important that in the persistent, lysogenic, state the phage does not represent a significant disadvantage for the survival of the host. In fact, in contrast to the situation with latent animal viruses where no advantage for the host is known, lysogeny often carries a strong selective value for the bacterium, since temperate phages frequently confer new characteristics on the host. This phenomenon manifests itself in many ways and is referred to as lysogenic conversion.

An interesting example of lysogenic conversion is that observed in *Corynebacterium diphtheriae*. Strains of this bacterium which cause the serious childhood disease diphtheria carry the diphtheria toxin (and are called toxigenic). Infection of *Corynebacterium* with phage β isolated from virulent, toxigenic bacilli of the same species, produces lysogens which acquire the ability to synthesize toxin. Loss of the prophage results in loss of toxin production as the structural gene for toxin is carried by the phage itself.

15.7 HERPES SIMPLEX VIRUS LATENCY

HSV-1 establishes latency in the cell bodies of sensory neurons that innervate the area of the epithelium where the primary infection is occurring (Box 15.3). An important question is whether latent HSV-1 DNA is integrated into the host genome. This question was technically difficult to answer because only a small proportion of neurons in the dorsal root ganglion are infected. Estimates indicate that between 1 in 6 to 1 in 60 neurons in a ganglion contain an HSV-1 genome. The DNA in the latently infected neurons differs from the linear DNA in the virus particle. It is not integrated but is circularized and exists as a free, circular, molecule in the cell nucleus.

Gene expression during HSV-1 latency and reactivation

In contrast to the extensive and well-characterized pattern of HSV-1 lytic gene expression described in Section 9.5, the activity of the genome in a latently infected neuron is very limited. It is generally agreed that the primary reason why the lytic program does not occur in these cells is a failure to express the α (immediate early) proteins, but it is not clear why this failure occurs. In neural cells infected with HSV-1 the only viral RNAs which can be detected are the latency-associated transcripts (LATs). These are actually stable introns, produced by splicing of longer RNAs, and they do not encode proteins. The precise role of the LATs in latency is

Box 15.3

Evidence for HSV-1 latency in sensory nerve ganglia

- In mice experimentally infected with HSV-1 via the eye, explanted trigeminal ganglia contain no detectable infectious virus but, if placed in culture together with cells susceptible to HSV-1 productive infection, virus re-emerges.
- In trigeminal ganglia from corpses of people who have died of causes unrelated to HSV-1 infection, HSV-1 DNA is detectable by Southern blotting in samples from those who are seropositive for HSV-1 infection but not those who are seronegative.
- When Southern blots of explanted ganglion DNA are probed for sequences at the termini of the linear genome, these are found in a single "junction" fragment with the individual terminal fragments being undetectable, indicating that the genome has circularized.

> ## Box 15.4
>
> ### Evidence suggesting the involvement of LAT RNA in HSV latency
>
> - LAT$^-$ mutant viruses induce a greater level of apoptosis in rabbit ganglia than LAT$^+$ virus, suggesting that LATs help to prevent neuronal cell death following infection.
> - Infection of mice with a LAT$^-$ virus results in higher levels of neuronal death during establishment of latency than is seen with LAT$^+$, virus indicating a role for LATs in maintaining neuronal cell survival.
> - The number of mouse sensory gangliar neurons expressing virus antigens is higher in mice infected with a LAT$^-$ mutant virus than when infected with a LAT$^+$ virus, suggesting that LATs inhibit expression of virus genes involved in lytic infections.
> - Production of immediate early and early virus mRNA is increased in mouse neurons following infection with a LAT$^-$ mutant virus when compared to infection with a LAT$^+$ virus, suggesting a role for LATs in suppressing virus gene expression.
> - Cells expressing LATs suppress the production of immediate early mRNAs.

not known and a number of possibilities have been proposed (Box 15.4). LATs are transcribed from the opposite strand of the genome to the α gene, ICP0, leading to the suggestion that they interfere with ICP0 synthesis by an antisense mechanism and so prevent the normal pattern of gene expression from proceeding. Certainly, ICP0 mutants reactivate poorly from latency. Alternatively, it has been proposed that the protein activator of the α genes, VP16, which is carried in the virion, fails to enter the nucleus in infected neurons, resulting in a general failure of α gene expression and the establishment of latency by default. One possible explanation for this failure may lie in the localization of VP16's cellular cofactor, HCF. HCF is a nuclear protein in most cell types, but is cytoplasmic in sensory neurons. If HCF and VP16 are restricted to the cytoplasm there is no activation of HSV-1 genes. Stimuli that cause HSV-1 reactivation from latency in neurons also cause relocalization of HCF to the nucleus, supporting this view. A model for the gene expression patterns in HSV-1 latency is shown in Fig. 15.5.

HSV-1 latency and the immune system

An acute HSV-1 infection is resolved by the immune system, and the question which arises is how does the virus establish a latent infection despite this response. The main answer is one of timing. By the time the immune system has succeeded in clearing the acute infection the virus has already travelled up the sensory nerve supplying the infected area and become latent in the cell body of one or more of the sensory neurons

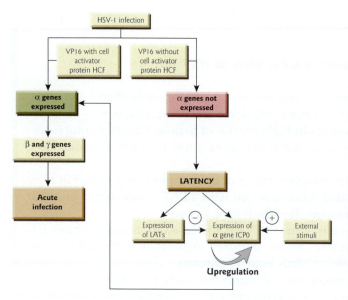

Fig. 15.5 Molecular control of latency in herpes simplex virus type 1 infection. Of 72 viral genes, only the latency-associated transcripts (LATs) are synthesized during latency. External factors that can break the latent state upregulate the synthesis of the viral nonstructural protein, ICP0, which then enhances its own expression. This activates expression of the α genes and unlocks the pathway to acute infection (see text).

that form the dorsal root ganglia. Since each dorsal root ganglion supplies innervation to a specific area, the latent infection is established only in the ganglion supplying the original area of infection. For HSV-1 infecting the oral region this is the trigeminal cranial nerve. All acute HSV-1 infections become latent.

Reactivation of HSV-1 occurs regularly several times each year in about 10% of infected people. It usually results in the familiar pathology of a cold sore in the epithelium around the mouth, the blisters of which contain infectious virus that can be transmitted to any non-immune person, usually an infant. However, up to 40% of reactivations are subclinical and thus virus can be transmitted inadvertently. When latency is broken the virus multiplies in one of the neurons in which it is latent, and infectious virus particles descend the sensory nerve to infect the epithelium that the nerve supplies. Thus, because the virus is confined by the sensory nerve, the reactivated acute infection reoccurs each time in one of the areas of the skin supplied by this nerve. In addition, infection is usually monolateral as the nerves are paired with each supplying only the right or the left side of the body, and virus does not move from one side to the other.

An important question is why people with a normal, functional, immune system suffer repeated acute infections after virus reactivation. Although the immune system does not prevent the cold sores, they are short lived, indicating that the immune system does respond and clear them. The importance of T cell involvement in control of HSV-1 infection is clearly seen when studying individuals who have a T cell deficiency for whom the virus is life-threatening. During reactivation there are two stages of immune involvement (Fig. 15.6). Firstly there is a short term, virus-mediated evasion of the immune response that allows the reactivation to get started and, secondly, there is immune activation which allows the infection to be resolved. Initially in the early stages of reactivation, expression of MHC I proteins in infected cells is inhibited. The main way in which this is achieved is that the immediate early viral protein (ICP47) prevents the translocation of processed peptides into the endoplasmic reticulum, and hence the formation of MHC–peptide complexes. As the virus begins to express its genes another mechanism comes into play and the virion protein that shuts-off host protein synthesis prevents the synthesis of MHC polypeptides. Thus, virus peptides are not presented and virus-specific CD8[+] T memory cells are not activated. In due course, CD4[+] cells and NK cells move into the cold sore lesion where lysis of infected cells is taking place. Viral antigen is processed and presented as MHC class II protein–virus peptide complexes on the surface of antigen presenting cells (local dendritic cells and macrophages). The CD4[+] T cells are activated and, in turn, through the cytokines that T cells secrete, NK cells are also activated. By day 2, both CD4[+] T cells and NK cells are producing γ-interferon. This increases the expression of MHC I proteins in infected epithelial cells. CD8[+] T memory cells are then activated, virus-infected cells are killed, and the infection resolves. This very delicate relationship between HSV-1 and the host allows both to prosper.

15.8 EPSTEIN–BARR VIRUS LATENCY

Epstein–Barr virus (EBV) normally causes a sub-clinical infection in infants. However if infection is delayed until adulthood, infectious mononucleosis (IM or glandular fever) results. The symptoms of glandular fever take several weeks after infection to appear and last for only 2 or 3 weeks. After the symptoms disappear the virus remains in the body, establishing a latent infection in B lympho-cytes. During the acute infection EBV expresses its genes in a controlled, coordinated pattern similar to the situation with HSV-1, though the nature of

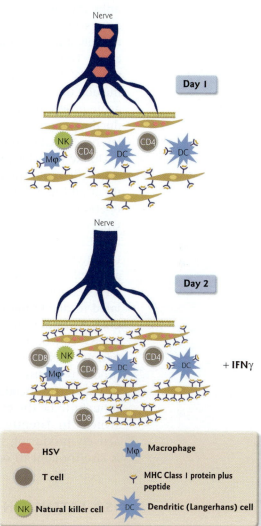

Fig. 15.6 Development and resolution of a cold sore from reactivated HSV-1. On day 1, reactivated virus is released from cells of the dorsal root ganglion and transported down the sensory neuron. It then infects adjacent epithelial cells. Expression of major histocompatibility complex (MHC) I protein complexed with viral peptide on the cell surface is inhibited (see text) and virus-specific memory CD8[+] T cells are not activated. NK cells and virus-specific CD4[+] T cells move gradually into the lesion and release cytokines. By day 2, they are releasing enough γ-interferon to upregulate the expression of MHC class I protein. The CD8[+] memory T cells are then activated and virus-infected cells are cleared. (From Posavad *et al.*, 1998. *Nature Medicine* **4**, 381.)

the genes is different. During latency, the genome is circularized and it is maintained in this state. Between one and nine proteins are expressed, depending on the growth state of the cells and whether or not latency is fully established. One of these genes is formed through the ligation of the two ends of the linear genome to form a circle (see Section 6.3), which links the promoter with the coding regions of the gene. Transcription covers more than half of the 172-kbp genome, with extensive differential splicing to produce alternative mature mRNAs from the primary transcripts. This fact emphasizes the issues of control that arise in maintaining the EBV latent state. The primary transcripts produced during latency include the sequences of a great many of the genes expressed only in the acute program, but these are discarded by splicing and so no protein production occurs; instead, expression is focused on the small number of latency-specific genes.

Unlike HSV-1 latency, where DNA replication does not occur, EBV has a specific replication mechanism which is used during latency establishment and maintenance; this mechanism is completely distinct from that operating during a productive EBV or HSV-1 infection (see Section 6.3). Only one of the latency proteins, an origin-recognition DNA-binding protein EBNA-1, is required directly for DNA replication. The remainder of the latency proteins appear to be involved in modulating cell signalling and cell cycle control. The role of latent replication is to maintain the viral genome in an expanding pool of B lymphocytes which are stimulated to divide upon infection. Subsequently, latency is maintained in a pool of resting, memory B cells.

EBV can be reactivated from the latent state in individual cells by a variety of stimuli. The reactivation is associated with a switch into a productive pattern of gene expression and DNA replication. Expression of two virus-encoded transcriptional activators, BZLF-1 and BRLF-1, is thought to be the event that commits a latently infected cell to make this switch. In latently infected B lymphocyte cell lines, a few cells are actively producing virus at any given time. Similar sporadic reactivation is also thought to occur *in vivo*.

15.9 LATENCY IN OTHER HERPESVIRUSES

Other important examples of human herpes viruses are shown in Table 15.1. Herpes simplex virus type 2 can be sexually transmitted, and its cycle of acute→latent→acute infection is very similar to that of HSV-1, but centers on the genital area. In fact the location of infection is interchangeable and which virus is resident depends on the site of the initial acute infection. This in turn determines which dorsal root ganglion becomes the site of the latent virus. With a genital primary infection, the virus becomes latent in the sacral dorsal root ganglion that innervates that area.

Table 15.1

The establishment of latent infections in man with herpesviruses and the breakdown of latency (reactivation).

Virus	Primary acute infection	Site of latency	Stimulus for reactivation	Reactivated acute infection
HSV-1	Stomatitis: infection of the mouth and tongue	Dorsal root ganglion of the trigeminal (cranial) nerve	E.g. strong sunlight, menstruation, stress	Cold sore*
HSV-2	Genital lesions	Dorsal root ganglion of the sacral region of the spinal cord	Not known, though probably similar to HSV-1	Genital lesions (and infection of neonates)
VZV	Chicken-pox: generalized infection with fluid-filled vesicles over the body surface	Any dorsal root ganglion of the central nervous system	Release of immune control e.g. in the elderly	Zoster (shingles);
EBV	Child: subclinical Adult: glandular fever (IM: acute)‡	B cells or possibly throat epithelium	Not known; frequent	Subclinical†
CMV	Prenatal§ Child: subclinical Adult: subclinical	Salivary glands and probably other sites	Frequent	In all body fluids, especially during the immunosuppression that accompanies pregnancy. A major cause of death in AIDS and transplantation surgery

*10–15% of reactivations are subclinical, but in some individuals this reaches 40%.
†EBV also causes cancer: nasopharyngeal carcinoma and Burkitt's lymphoma.
‡IM: infectious mononucleosis. An example of a more severe clinical disease that occurs when primary infection takes place after childhood – a common microbiological problem of the (over-) sanitized world. "Mononucleosis" refers to the uncommonly large numbers of mononuclear (compare polynuclear) cells (mainly lymphocytes) that are found in the blood.
§The fetus is only at risk when its mother gets a primary infection. AIDS, acquired immune deficiency syndrome; CMV, cytomegalovirus ("cytomegalo" refers to the characteristic swollen cell cytopathology caused by CMV to cells of the kidney, lungs and liver); HSV, herpes simplex virus; IM, infectious mononucleosis; VZV, varicella-zoster virus.

   Varicella-zoster virus (VZV) causes acute chicken pox of children and shingles (zoster) on reactivation. Apart from chicken pox being generalized over the entire surface of the body, the transition from acute to latent infection occurs exactly as with HSV-1. Because of the generalized nature of the initial infection, almost any of the brain and spinal dorsal root ganglia can be latently infected. Reactivation is a rare event which is normally restricted to a single ganglion, and the lesions formed by the reactivated virus are restricted to the precise segment of the body that is innervated by the nerve in which virus was latent. Unlike HSV latency there are specific virus proteins made in latently infected neurons. Latent VZV is maintained in the presence of antibody and a T cell response, both of which may contribute to maintaining virus in its latent state. Immunosuppression (including the natural waning of immunity that accompanies aging) results in virus reactivation.

Another ubiquitous infection is cytomegalovirus (CMV). This is normally a subclinical childhood infection that becomes latent in the salivary glands. It reactivates frequently but reactivations are controlled by the immune system and are completely subclinical. However, the infection can become generalized and life-threatening when there is some immuno-suppression, such as that prescribed to avoid organ rejection during transplantation surgery, and as occurs naturally during HIV-1 infection. In addition although adult infection is subclinical, the virus can pass through the placenta and infect the fetus. As a result, some fetuses suffer severe and permanent brain damage (including microcephaly). This infection, and its severity is also thought to be the result of the immunosuppres-sion that occurs naturally during pregnancy and helps to prevent the fetus (a foreign graft) from being rejected by the maternal immune system.

15.10 HIV-I LATENCY

Following an initial infection with HIV-1, affected individuals carry the virus for the remainder of their life. Unlike the situation with the her-pesviruses during the latent state, most of the HIV-1 genomes continue to express the full range of genes and to produce infectious virus, with approximately 10^{10} virions produced per day, and to be dealt with by the immune response. However, there is an underlying minority of HIV-1 genomes that become truly latent in resting CD4$^+$ T lymphocytes, pro-ducing no progeny virus until activated. In the latent state no virus genes are expressed and the virus genome, which is of course integrated within the genome of the host cell, is replicated along with the cellular DNA. Estimates suggest that approximately 10^6 to 10^7 CD4$^+$ T lymphocytes per person may be infected in this way. The latent virus can reactivate at any time though the stimuli for this are not well understood. This is important as most active virus multiplication, but not the latent virus infection, can be eliminated by chemotherapy (see Section 19.8). Thus, the latent infection acts as a reservoir from which virus can reappear if chemotherapy is stopped. When this occurs the levels of infectious virus in the body often exceed those prior to the treatment and there is a greatly enhanced likelihood of drug-resistant viruses appearing due to selection pressures. This means that the current chemotherapy regime has to be maintained for life, with the attendant difficulties of expense, toxicity, noncompliance, or the eventual breakthrough of resistant mutants.

KEY POINTS

- Latency is an outcome of infection for certain viruses. Herpesviruses have the ability to establish latent infections.
- Following reactivation, infectious virus is produced from latently infected cells.
- In animals, the immune system is key in dealing with reactivation of infection.
- The pattern of gene expression during latency is different to that seen during an acute infection: fewer virus genes, and sometimes no genes, are expressed during latency.
- In phage λ the establishment of latency is determined by a balance between the action of gene activators and gene repressors.
- In HSV-1 only RNA from the LATs is seen during latency.
- Latency requires suppression of the early steps in the cascade of lytic gene expression.

QUESTIONS

- Consider the control of lysogeny in bacteriophage lambda and compare this with the gene expression of herpesviruses during the establishment and maintenance of latency.
- Discuss the potential outcomes of infection of humans with herpesviruses and describe the processes which control and determine the various stages of the infection.

FURTHER READING

Amon, W., Farrell, P. J. 2005. Reactivation of Epstein–Barr virus from latency. *Reviews in Medical Virology* **15**, 149–156.

Efstathiou, S., Preston, C. M. 2005. Towards an understanding of the molecular basis of herpes simplex virus latency. *Virus Research* **111**, 108–119.

Johnson, A. D., Poteete, A. R., Lauer, G., Sauer, R. T., Ackers, G. K., Ptashne, M. 1981. λ repressor and *cro*-components of an efficient molecular switch. *Nature (London)* **294**, 217–223.

Mitchell, B. M., Bloom, D. C., Cohrs, R. J., Gilden, D. H., Kennedy, P. G. 2003. Herpes simplex virus-1 and varicella-zoster virus latency in ganglia. *Journal of Neurovirology* **9**, 194–204.

Ptashne, M. 1986. *A Genetic Switch*. Blackwell Scientific Publications & Cell Press, Palo Alto.

Ptashne, M., Jeffrey, A., Johnson, A. D., Maurer, R., Meyer, B. J., Pabo, C. O., Roberts, T. M., Sauer, R. T. 1980. How the λ repressor and *cro* work. *Cell* **19**, 1–11.

Tsurumi, T., Fujita, M., Kudoh, A. 2005. Latent and lytic Epstein–Barr virus replication strategies. *Reviews in Medical Virology* **15**, 3–15.

Young, L. S., Rickinson, A. B. 2004. Epstein–Barr virus: 40 years on. *Nature Reviews Cancer* **4**, 757–768.

16

Transmission of viruses

Viruses are intracellular parasites and have to find a new host before the original host mounts an effective immune response or dies. However virus infectivity is intrinsically unstable and transmission has to be achieved usually within a few hours. All successful viruses have solved the transmission problem.

Outside the host, the infectivity of most viruses is inherently unstable. After being shed from a host animal it is imperative that viruses quickly and efficiently encounter a new host to initiate a fresh infection. The process of transfer between hosts is referred to as transmission and this is an important step in the life cycle of viruses. Knowledge of how viruses are transmitted may enable the cycle to be broken at this stage, preventing further infections.

Since outside the host, the infectivity of most viruses is inherently unstable, this is an important stage in their life cycle. Knowledge of how viruses are transmitted may enable us to break the cycle at this stage and thus prevent further infections. While the routes by which viruses are transmitted between humans are known and can be investigated in the laboratory, the elements that are important in natural transmission are often difficult to investigate and details are poorly understood. All transmission strategies adopted by viruses have in common the ability to circumvent the outer layer of the skin which is impermeable to viruses, and to bring virus into contact with the naked cells. Person-to-person infections are said to take place by *horizontal* transmission, while those from mother to baby are put in a separate category and described as *vertical* transmission. It is no coincidence that most viruses are spread by the respiratory and fecal–oral routes (both examples of horizontal transmission), since, in order to carry out their normal physiological functions, the lungs and small intestine each have a surface of living cells with an area approximately equivalent to that of a tennis-court (400 m²). It should be remembered that viruses may be usually transmitted by one particular route, but that other routes may also be important under certain specific circumstances. In Section 16.3, we shall discuss *zoonoses* – infections where the virus is transmitted from animals other than humans, to humans.

16.1 HORIZONTAL TRANSMISSION

Summary of horizontal transmission routes

- *Respiratory route: occurs commonly, e.g. rhinoviruses, influenza viruses*
- *Conjunctival route: occurrence rates are not known, e.g. respiratory viruses*
- *Fecal route: occurs commonly, e.g. poliovirus*
- *Sexual route: used by specific viruses, e.g. HIV-1, HBV, papillomaviruses*
- *Via urine: used by specific viruses, e.g. Lassa fever virus, cytomegalovirus*
- *Mechanical route: common with tropical arthropods that feed on humans, e.g. arboviruses, and with high risk behaviors, e.g. HIV-1, HBV*

Transmission via the respiratory tract

Many virus infections are contracted via the respiratory tract. While many of these cause respiratory infections, other viruses causing generalized infections, such as measles, chicken pox, and smallpox, are contracted by this route (Box 16.1). After virus has multiplied, it either infects additional cells within the same host or escapes from the respiratory tract in liquid droplets that result from our normal activities, such as talking, coughing, and sneezing (and particularly singing). These aerosols are inhaled by others and give rise to a "droplet infection" of a susceptible individual. The size of droplets is important, as those of very large (>10 μm) diameter rapidly fall to the ground, while the very small droplets (<0.3 μm) dry very quickly, and virus contained within them is inactivated. Thus, the middle-sized range of droplets are those that transmit infection more efficiently. The precise size of these will determine where the droplets are entrapped by the respiratory system of the recipient, since there are baffles lining the nasal cavities that remove the larger airborne particles; the medium sized droplets get as far as the trachea (throat), while the smallest droplets penetrate deep into the lung (Fig. 16.1). However, these particles may be trapped on mucus that is driven upwards by the cilia lining the tubes of the respiratory tract, counter to the incoming air. This is the mucociliary flow. The mucus also serves as a physical barrier that prevents a virus from attaching to its receptors on the surface of cells.

Box 16.1

Nursery rhymes can tell us about virus infection and transmission – small pox, in this case

Ring-a-ring of roses	[= symptomatic blemishing of the skin]
A pocket full of posies	[= a posy of flowers thought to ward off infections]
Atichoo, atichoo	[= respiratory spread]
We all fall down	[= dead]

The increase in nasal secretion that accompanies many respiratory infections favors the dispersal of the viruses responsible, and the increase in coughing and sneezing increases the production of infectious aerosols. However, transmission experiments from people infected with a rhinovirus to susceptibles sitting opposite at a table proved singularly unsuccessful. Similar data were obtained with influenza virus in London. Here, chosen families were closely monitored for the transmission of naturally acquired infections. Even in the proximity of a family, an infected spouse (from whom influenza virus could be isolated) frequently did not transmit it to their nonimmune partner. This has led to the suggestions that only particular individuals may shed sufficient virus, produce excess nasal secretion, and/or aerosols containing optimum-sized particles and so act as efficient spreaders of infection, or that, in order to be infected, one has to be in a certain physiological state. Apart from the traditional stories of wet feet predisposing you to catch a cold (not true!), it has been shown that recently bereaved people are particularly susceptible to infectious diseases. Thus, the immune system is influenced by one's state of mind. Neuroimmunology is a developing field of exciting study.

In addition to aerosol transmission via the respiratory tract, it is now believed that some respiratory viruses are spread by contact of droplets with the conjunctiva, the layer of cells covering the eye. Natural drainage to the throat would then carry progeny virus to the respiratory tract. Little information is available, but there is a report of an infection that resulted when a frozen chicken was being dismembered and a particle flew into the cook's eye and resulted in infection. Another variation on standard respiratory transmission is the suspicion that large aerosol droplets that land on solid surfaces can then be transferred from fingers to the conjunctiva or directly into the nose. The importance of airborne aerosols relative to physically borne aerosols in regard to virus transmission remains to be determined.

The dynamics of aerosol infections have been studied in detail by the military, from a biological warfare standpoint, and are known to be highly complex. There are many germane factors, including the size of the virus-carrying droplets, which determine how long droplets stay airborne and how quickly they dry up and inactivate the infectivity of the virus particles contained therein. The composition of the liquid component of the droplet too affects its size, desiccation, and virus stability. It has been postulated that some infections may not be spread by inhaled droplets, but through large droplets that fall rapidly from the air onto solid surfaces and are then transferred on fingers to the mouth or nose, or to the conjunctiva. There are other possibilities – as some people appear to produce more infected aerosol than others, or more aerosol droplets of a size that is able to spread that virus efficiently, and that this minority of the human population may be responsible for spreading most of the virus. Key points are summarized in Box 16.2.

Environmental factors which result in seasonal variation in the amount of illness or frequency of isolation of a virus are linked with transmission of respiratory infections. Influenza, for example, is a winter disease, occurring around January in the Northern Hemisphere, June in the Southern Hemisphere, but nearly all the year round close to the Equator. Variations both in the environment (e.g. temperature and humidity) and in social behavior (crowding together in winter with poor ventilation) have been identified as factors which could affect virus stability and transmission and hence the seasonal incidence of virus diseases, but a full explanation of these complex phenomena is not yet available. Since viruses can survive for only a limited time outside the cell, it is fair to assume that infected individuals are present continuously somewhere in the population. Interestingly these seasonal patterns for influenza have not been altered by modern mass air travel, which transports thousands of people, many of whom will be incubating the virus, from winter to summer at the other end of the world. However, there is concern that the spread of a new pandemic influenza virus (see Section 17.5) may be enhanced by air travel.

Transmission via the fecal–oral route

Many viruses are ingested with food or water that is contaminated with feces, and infect and multiply in cells of the alimentary tract. This is the fecal–oral route of infection. The surface of the small intestine normally functions to adsorb nutrients and water, and can potentially be infected by viruses. Two well known picornaviruses that invade by this route are poliovirus and hepatitis A virus. Poliovirus infections traditionally result from ingesting sewage-contaminated drinking water, but more recently have come from ingesting the water of swimming pools that were inadequately disinfected. Hepatitis A virus is commonly experienced by travellers as a result of encounters with

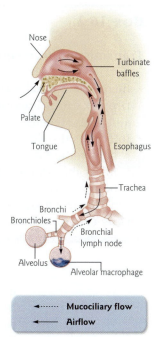

Fig. 16.1 Route taken by aerosol droplets containing virus particles that are breathed into the respiratory tract. Note the defences designed to trap and remove all sorts of small particles: turbinate baffles, the mucociliary flow that runs counter to inspired air, and macrophages. The airways become progressively narrower (here much foreshortened) and end in alveoli, each of which is formed by a single cell. An alveolar macrophage can be seen in one of the alveoli. (Adapted from Mims, C. A., White, D. O. 1984. *Viral Pathogenesis and Immunology.* Blackwell Scientific Publications, Oxford.)

Box 16.2

Transmission of respiratory viruses

- Increases in nasal secretion, coughing, and sneezing produce more infectious aerosols.
- Droplet size is important for droplet aerodynamics, penetration of the respiratory tract, and virus stability.
- Infectious droplets from an infected respiratory tract may:
  - be breathed directly into another respiratory tract
  - infect the conjunctiva and then move to the respiratory tract
  - be transferred by fingers from solid surfaces to the respiratory tract.
- Environmental factors that affect virus stability and human behavior may affect transmission.
- Only certain individuals may be able to spread infection efficiently.

---

drinking sewage-contaminated water, or more subtly when such water has been used to wash fruit or salad vegetables, or used to prepare ice-cubes for drinks. All viruses transmitted via the fecal–oral route are excreted in feces, so their spread is favored by poor sanitation and poor personal hygiene. Many of these viruses grow to very high titers (10^9 infectious units (IU)/ml), but would probably transmit even if there were only 10^6 IU/ml of feces – as just 1 µl of feces contains 10^3 IU. Many of these infections occur in early childhood – not surprising, when it is seen how small children investigate strange objects by putting them in their mouth, and that their personal hygiene is not well developed. Some virus infections (rotavirus) are associated with diarrhea which, though not proven, could be an adaptation to improving virus transmission.

With improvements in sanitation it would be expected that there would be a reduction of disease caused by enteric viruses, and this is the experience. However, there have been some unpleasant surprises. In conditions of poor sanitation, poliovirus infects young children as soon as they can crawl outside the house, but results usually in a subclinical gut infection. With improved sanitation, poliovirus is not contracted until adolescence, and is then associated with an increase in the incidence of paralytic poliomyelitis. This appears to be another example of increased severity of a disease seen when a virus infects an "unnatural" host – in this case the age of the individual is the key difference.

Sexual transmission

Only a few viruses are spread by sexual transmission but, for them, it is a very successful route indeed. The main examples are HIV-1 and hepatitis B virus (HBV), but herpes simplex viruses (HSV) type 2 predominantly but also type 1, and some types of human papilloma virus, are also spread sexually. HIV-1 has only been recognized as the causative agent of AIDS since 1983 and is currently responsible for more than 40 million infections worldwide. AIDS predisposes the infected individuals to both viral and nonviral infections of exaggerated severity, and has a mortality rate approaching 100% (see Chapter 19). HBV is responsible for primary hepatocellular carcinoma in 0.1% of infected individuals, and even with this low incidence rate it the commonest human cancer because the virus is endemic in the very large population of China and the Far East (see Section 14.5). These viruses are transmitted by heterosexual intercourse, both from male to female, and from female to male, and homosexual intercourse. For HIV-1 the risk of infection is marginally less (by two- to threefold) if the female is the carrier. This may be because ejaculate contains not only virus but also around 10^6 lymphocytes, some of which may be infected and producing virus. Virus is also spread by homosexual behavior and the risk for receptive anal intercourse (which is 1/300 to 1/1000) is about the same as that in male–female intercourse when the male is infected. Sexual transmission of infection is affected by sociosexual behavior and spread of these viruses is increased by promiscuity. Rates of HIV-1 infection are stable in some countries, but the worldwide pandemic is moving eastwards and shows no signs of slowing down.

Transmission via urine

Transmission of viruses in urine is rare as urine is usually sterile. However, a few viruses, such as Lassa fever (an African arenavirus) and sin nombre (a North American bunyavirus), are excreted in the urine of their hosts, and are thought to be transmitted in this way. These viruses infect wild mice, but in rural areas at certain times of the year mice enter houses. Infected urine can then contaminate surfaces where food is prepared and virus is ingested. Alternatively, virus-contaminated dust particles may be breathed in. Both viruses cause rare human fatalities. In addition, cytomegalovirus and polyomaviruses are excreted in the urine of small children in large amounts, and may be transmitted by this route.

Transmission by mechanical means

Animal viruses transmitted by mechanical means include those which infect their hosts by various methods that directly puncture the normally impermeable skin layer. This route involves the role of *virus vectors* – animals that transmit the virus to man. Such a relationship is a zoonosis (see Section 16.3). The vectors of virus transmission are usually biting arthropods (mosquitoes which are insects, and ticks which are arachnids) found in tropical parts, which feed on human blood by piercing the skin with their mouth parts. Preparatory to the next meal, the arthropod injects saliva

Box 16.3

Evidence that yellow fever virus is transmitted by a mosquito

In the late 1920s definitive evidence that yellow fever virus is transmitted by mosquitoes was presented. The evidence was:

- The infection was transmitted between monkeys or from infected patients to monkeys by injection of blood.
- The mosquito *Aedes aegypti* was able to transmit the infection between monkeys.
- Infected mosquitos were able to transmit the virus throughout their lives, confirming that the virus was able to replicate in the insect.
- A single bite from an infected mosquito was sufficient to transmit the infection to monkeys.

as an anticoagulant, and in so doing introduces the virus that was picked up from the last blood meal. Viruses spread by arthropods (*arthropod-borne viruses*) are known collectively as *arboviruses* (Box 16.3).

The main groups of animal viruses that are spread by mosquitoes belong to the alpha-, flavi-, reo-, rhabdo-, and bunyavirus groups. Transmission of a virus is specific to a particular mosquito species. Such viruses have a complex life cycle, and are adapted to multiply in the tissues of the invertebrate vector as well as those of the vertebrate host. Although arboviruses are found mostly in the tropics, examples from more temperate regions are the flaviviruses, tick-borne encephalitis virus, that is found in Europe from Austria eastwards, and louping-ill virus of grouse that is spread by *Ixodes* ticks and is found in the northern UK, Ireland, and Norway. Sheep and humans in those regions are occasionally infected with louping-ill virus through bites from infected ticks. Other arthropod-borne animal viruses do not multiply in their vector but are carried passively. For example, myxomavirus (a poxvirus) that causes myxomatosis is spread between rabbit hosts on the contaminated mouth parts of infected mosquitoes in Australia and rabbit fleas in the UK.

The serious blood-borne infections of humans (HBV and HIV) can be transmitted by mechanical means between intravenous drug abusers. Virus can be carried from one person to another by sharing nonsterilized hypodermic syringes and/or needles that are contaminated with infected blood. The practice of drawing blood into the syringe increases the risk. For HIV the risk of infection is 1 in 160, higher than for sexual activity. However, there is also risk of transmission through any shared device which can cause minor cuts or abrasions and transfer blood, such as combs, razors, and needles used for acupuncture, tattooing or body-piercing. Transmission has also occurred in dental surgeries. Infection is easily avoided by using good hygiene and sterile equipment. HBV can also be spread through infected saliva by biting. However, there is no evidence that these viruses are transmitted by biting insects.

Most plant viruses are spread by mechanical means. This reflects that plant cells, having a robust cell wall structure, are less readily infected. The viruses are introduced into the cells of plants by animal vectors, such as plant-feeding aphids, leafhoppers, beetles, and nematodes, by fungi, or by nonspecific abrasions made to the plant tissue by the wind, etc. which expose cells sufficiently for them to be infected. Transmission by plant-feeding animals is a specific process and only certain species are implicated in the transmission of a particular virus. Like animal viruses, some plant viruses multiply in the vector, while others are passively transmitted.

16.2 VERTICAL TRANSMISSION

Occurs between mother and fetus/baby: common route for transmission of specific viruses, e.g. rubella virus

Vertical transmission is the transfer of virus from mother to fetus/baby, in contrast to the horizontal transmission between other individuals. Rubella virus (a togavirus) is the classic example of a vertically transmitted virus, although it is normally spread by the respiratory route. In an adult the infection (German measles) is manifest as a mild skin rash or is subclinical. However, the virus can cross the placenta and multiply in the fetus. As a result the fetus can die or can be born with serious congenital malformations that affect the cardiovascular and central nervous systems, the eyes, and hearing. The risk of malformations arising from rubella virus infection is high in the early stages of pregnancy (up to 80%) and decreases as fetal development proceeds, to become almost no risk by the fifth month of pregnancy or later. Fortunately, there is an excellent live attenuated vaccine (MMR, measles virus, mumps virus and rubella virus combined) that is given to young children and protects them through adulthood. Other viruses which infect the mother and can be passed vertically to her fetus/baby are listed in Table 16.1 together with the outcomes affecting the fetus or newborn infant.

Analysis of the exact route of vertical transmission is not possible as the virus can be transmitted to the zygote from an infected oocyte or sperm, or the zygote can be infected from virus present in cells of the uterus, or via the bloodstream in placental mammals. Any virus that infects the mother while she is producing eggs or offspring can, in theory, be vertically transmitted, while those whose genomes are integrated with that of the reproductive cells of the host cannot help but be transmitted by

Table 16.1

Examples of vertical transmission of viruses from mother to baby and their consequences.

Virus	Possible modes of infection			Possible adverse outcomes		
	Transplacental	During birth*	After birth	Death of fetus	Clinical disease after birth	Persistent infection
Rubella	+	−	−	+	Congenital abnormality	+
CMV (primary)	+	−	−	?	Congenital abnormality	+
CMV (reactivated)	−	+ (2% of all babies)	+ (bm)	na	−	+
HIV-1	?	+	+ (bm)	−	Up to 15% of babies born to infected mothers: AIDS at age 2	+
HBV	+	+	+	−	−	+
HSV (genital)	+	+	+	+	Herpes lesions	+
HPV (various types)	−	+	−	−	−	+

*A caesarean delivery can help avoid infection.
bm, breast milk; CMV, cytomegalovirus; HBV, hepatitis B virus; HIV-1, human ??? type 1; HPV, human papillomaviruses; HSV, ??? na, not applicable; ???

this route. The genomes of mammals, including humans, contain large numbers of *endogenous retroviruses* in their germline cells (see Section 23.3). These are remnants of infections which occurred long ago in evolutionary time and they generally do not produce infectious virus. In addition, as shown in Table 16.1, infection can also occur during the birth process in *perinatal* transmission. This risk can be minimized by Caesarean section. Also included here, as they cannot be distinguished, is infection of the baby by the mother after birth as a result of infected breast milk, or possibly via saliva. Accordingly, HIV-1-infected mothers are advised not to breast-feed their babies. The use of anti-HIV drugs is also beneficial, and the combined precautions reduce the risk of vertical transmission of HIV-1 from around 15% to less than 2%.

16.3 ZOONOSES

A zoonosis is an infection that occurs when the infectious agent is transmitted from an animal other than a human, to a human. Usually, the

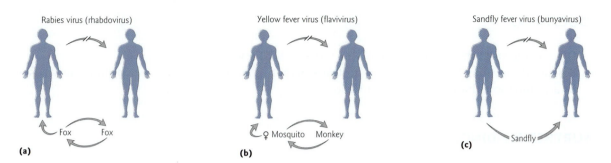

Fig. 16.2 Examples of zoonoses. In (a) the fox vector is the primary host. Rabies can be passed on infected saliva by biting and also, but more rarely, by the respiratory route. Biting arthropods are the vectors in (b) and (c), and the virus multiplies in both the vertebrate and the arthropod. In (b) monkeys are the primary host and yellow fever virus is spread by the female mosquito (the male mosquito does not feed). The primary host of sandfly fever virus (c) may be the gerbil, but as shown here humans can be the reservoir when there is a high incidence of infection. In these examples, virus does not pass from person to person.

main host is another vertebrate and the human is only infected accidentally. Biting arthropods are often the vectors responsible for zoonotic transmission. Rabies is a well known zoonosis, although rabies virus (a rhabdovirus) is transmitted directly by the bite of an infected vertebrate. Rabies is an unusual virus as it infects so many different species. There are animal reservoirs for rabies all over the world, e.g. the red fox in Europe, the skunk, racoon and other animals in North America, the vampire bat in South America, the yellow mongoose in Southern Africa, and the jackal in India. Three types of zoonosis are shown in Fig. 16.2.

KEY POINTS

- Most infections are spread horizontally (person–person), but vertical spread from mother to fetus or baby is an important route for some viruses.
- Most viruses are spread by the respiratory or fecal–oral routes of infection.
- Viruses are also spread via urine, by sexual behavior, or mechanically; a few are contracted by infection of the eye.
- A virus may be usually transmitted by one particular route, but it may spread by other routes under certain circumstances.
- Much of our ignorance is due to the difficulty of investigating the natural transmission of human infections.
- Some infections are zoonoses in which a virus is spread from a non-human animal to humans; many are transmitted no further.

QUESTIONS

• Discuss the role of arthropod vectors in virus transmission.
• Discuss the proposition that horizontal transmission is more important in maintaining viruses in a population than vertical transmission.

FURTHER READING

Alexander, D. J., Brown, I. H. 2000. Recent zoonoses caused by influenza A viruses. *Review of Science and Technology of the International Office of Epizootics* **19**, 197–225.

Beaty, B. J., Marquardt, W. C. 1995. *The Biology of Disease Vectors*. ASM Press, Washington.

Belshe, R. B. 1998. Influenza as a zoonosis: how likely is a pandemic? *Lancet* **351**, 460–461.

Cleaveland, S. 2003. Emerging infectious diseases of wildlife. *Microbiology Today* **30**, 155.

Glezen, W. P. 1999. Maternal immunization for the prevention of infections in late pregnancy and early infancy. *Infections in Medicine* **16**, 585–595.

Kovarik, J., Siegrist, C.-A. 1998. Immunity in early life. *Immunology Today* **19**, 150–152.

Krauss, H., Weber, A., Appel, M., Enders, B., Isenberg, H. D., Schiefer, H. G., Slenczka, W., von Graevenitz, A., Zahner, H. 2003. *Zoonoses: infectious diseases transmissible from animals to humans*, 3rd edn. ASM Press, Herndon, VA.

Kruse, H., Kirkemo, A-M., Handelande, K. *et al.* 2004. Wildlife as a source of zoonotic infection. *Emerging Infectious Diseases* **10**, 2067–2072.

McIntyre, J. 2003. Mothers with HIV. *British Medical Bulletin* **67**, 127–135.

Mims, C. A., Nash, A., Stephen, J. 2000. *Mims' Pathogenesis of Infectious Disease*, 5th edn. Academic Press, London.

Also check Appendix 7 for references specific to each family of viruses.

17

The evolution of viruses

Like all living organisms, viruses are not static entities but are subject to evolutionary pressures and undergo evolutionary change.

Viral genomes accumulate mutations in the same way as all other nucleic acids and, where conditions enable a mutant to multiply at a rate faster than its fellows, that mutant virus will have an advantage and will succeed the parental type. Where viruses have an evolutionary advantage is that many polymerases (notably in the RNA viruses) lack proofreading mechanisms, so that mistakes in replication are not corrected. Thus, mutants accumulate more rapidly than seen with DNA-based organisms where DNA replication has proofreading. Viruses also have a shorter generation time relative to their host. In any discussion of evolution, it is pertinent to ask where viruses first came from (see Section 1.9). The absence of any fossil records and the scarcity of other evidence have not, of course, prevented scientists from speculating about their origins. The two prevailing opinions are that viruses have either arisen (i) from degenerate cells that have lost the ability for independent life, and/or (ii) from escaped fragments of cellular nucleic acid.

The history of viruses extends as far back as man's historical written or pictorial records. Amongst the earliest references is a 3500-year-old bas relief sculpture from Egypt depicting a man supporting himself with a crutch. Careful examination shows that the calf muscle of his right leg is withered in the characteristic aftermath of paralytic poliomyelitis. A slightly later excavation of the same civilization found the mummified body of the pharaoh, Rameses V, whose face bears the characteristic pock marks of smallpox. Such evidence tells us of the long association of these viruses with mankind.

Whatever their origins, viruses have been a great biological success, for no group of organisms has escaped their attentions as potential hosts. Viruses are largely species-specific with respect to their host and usually do not cross species boundaries. It seems likely that every animal species has the same range of viruses as occurs in humans. In Chapters 13–16, the various ways in which viruses interact with their hosts are discussed, illustrating how viruses cause a variety of changes, ranging from death to the imperceptible. Evolution of any successful parasite has to ensure that the host also survives. The various possible virus–host interactions can be thought of as different ways in which viruses have solved this problem.

17.1 THE POTENTIAL FOR RAPID EVOLUTION IN RNA VIRUSES: QUASISPECIES AND RAPID EVOLUTION

The lack of a proofreading function in the polymerases of RNA viruses means that base substitutions (mutations) occur at the rate of between 10^{-3} and 10^{-5} per base per genome replication event (i.e. one mutation for every 1000–100,000 newly synthesized bases). Not all of the resulting mutants will be viable but many are, resulting in an extremely heterogeneous population of viruses. This is called a quasispecies. This has a number of implications. For example, it is not possible to define the genome sequence of that virus population precisely, and any sequence in a database will represent only one member of the quasispecies. The quasispecies phenomenon also endows RNA viruses with the ability to adapt rapidly to and to exploit any environment they occupy. This is seen on both a micro and a macro level, and arises because one or more members of the quasispecies will inevitably have a selective advantage over others. The micro level is infection of the individual, and virus evolution during the life time of the individual is seen, particularly in life-long infections with HIV-1 (see Chapter 19) and hepatitis C virus (see Section 20.8). It is likely that in these infections, the immune system is a major selective force, and that the quasispecies phenomenon allows these viruses to persist in the face of that immune response. The macro level of evolution is seen in viruses with a worldwide distribution. This could impinge upon people with different genetic backgrounds, especially those that affect the MHC (HLA) haplotype that controls T cell immunity. This leads to the establishment of different specific genetic sequences, known as genotypes, or clades of a virus in different parts of the world, and a classification that is based on sequence rather than serology (e.g. HIV-1, Chapter 19).

Influenza A virus does not cause persistent infections but virus variants evolve under the selection pressure of antibody stimulated during an acute infection. These variants are in turn passed on to susceptible individuals in a country or a continent. Thus, the virus evolves with the effect that, in approximately 4 years, it is not recognized by the immune memory established by the original host, and he or she can be reinfected (Section 17.5). This is called *antigenic drift*, and essentially results in the formation of new serotypes of virus.

Not all RNA viruses exhibit such obvious variation, even though they have the same potential for change as HIV-1 and influenza virus. For example, the overall antigenicity of polioviruses types 1, 2, and 3, and measles virus has not changed over their known history of approximately 50 years, and the original vaccines still provide the same level of protection. Thus, by comparison with HIV-1 and influenza virus, the polioviruses and measles virus are stable. However, their genomes do accumulate mutations, and minor changes in antigenicity (in individual epitopes) of the virion can be detected by probing with monoclonal antibodies. In addition, viruses from different global areas can be distinguished by their genotype. It is not understood why, for example, measles virus is antigenically stable and influenza virus is not, as in many ways these two infections are similar (both viruses are highly infectious, spread by the respiratory route, and cause acute infections). Presumably measles virus is under a less effective or more constrained evolutionary selection pressure than influenza virus, but exactly what these pressures are is not known.

17.2 RAPID EVOLUTION: RECOMBINATION

Recombination is the other major force in virus evolution and takes place in a cell that is simultaneously infected by two viruses. Usually, the two genomes are highly related with regions of homology between their genomes that permit the replicating enzyme to move from one strand to another. Thus at a stroke, the daughter molecule has some of the properties of both parents. However, both parts of the resulting genome have to be sufficiently compatible for the progeny to be functional. Both DNA and RNA genomes undergo recombination. Detailed monitoring of the HIV-1 pandemic shows new recombinant viruses arising between different clades, and such plasticity makes the prospect of control by vaccine problematic (see Chapter 19). Viruses with segmented genomes can also undergo recombination by acquiring an entire genome segment from another virus, and this occurs at higher frequency. This form of recombination is known as *reassortment*. The effect on antigenicity is enormous as a virus can acquire an entirely new coat protein in a single step. The prime example is influenza A virus where this process, called *antigenic shift*, is responsible for pandemic influenza in man (Section 17.5). However, a viable recombinant requires compatibility between all eight genome segments, and this puts some limitation on the creation of new viruses.

17.3 EVOLUTION OF MEASLES VIRUS

Measles virus

- Member of the paramyxovirus family.
- Infects only humans.
- Unusual in that every infection causes disease.
- Infection results in lifelong immunity.
- Measles can be countered by use of a live vaccine.

F. L. Black studied the occurrence of the measles in island populations (Table 17.1), and found a good correlation between the size of the population and the number of cases of measles recorded on the island throughout the year. A population of at least 500,000 is required to provide sufficient susceptible individuals (i.e. new births) to maintain the virus in the population. Below that level, the virus eventually dies out, until it is reintroduced from an outside source.

On the geological time-scale, humans have appeared only recently and have only existed in population groups of over 500,000 for a few thousand years. In the days of very small population groups, measles virus could not have existed in its present form. It may have had another strategy of infection, such as persistence, which would have allowed it to infect the occasional susceptible passer-by, but there is no evidence of this.

Table 17.1

Correlation of the occurrence of measles on islands with the size of the island population. (From Black, F. L. 1966. *Journal of Theoretical Biology* **11**, 207–211).

Island group	Population ($\times 10^{-3}$)	New births per year ($\times 10^{-3}$)	Months of the year in which measles occurred (%)
Hawaii	550	16.7	100
Fiji	346	13.4	64
Solomon	110	4.1	32
Tonga	57	2.0	12
Cook	16	0.7	6
Nauru	3.5	0.17	5
Falkland	2.5	0.04	0

However, Black has speculated upon the antigenic similarity of measles, canine distemper, and rinderpest viruses. The latter infect dogs and cattle respectively, animals that have been commensal with humans since their nomadic days. Black suggests that these three viruses have a common ancestor which infected prehistoric dogs or cattle. The ancestral virus evolved to the modern measles virus when changes in the social behavior of humans gave rise to populations large enough to maintain the infection. The first such population occurred 6000 years ago when the river valley civilizations of the Tigris and Euphrates were established.

17.4 EVOLUTION OF MYXOMA VIRUS

Myxoma virus

- *Member of the poxvirus family.*
- *Natural host is the South American rabbit in which it causes only minor skin outgrowths.*
- *Also infects the European rabbit causing myxomatosis, with lesions over the head and body surface; infection is 99% lethal.*
- *Is spread by arthropod vectors that passively carry virus on their mouth parts: mosquitoes in Australia and rabbit fleas in the UK.*

Myxoma virus was released in England and Australia upon wholly susceptible host populations of the European rabbit in an attempt to control this serious agricultural pest. This experiment in nature was carefully studied with respect to the changes occurring in the virus and the host populations and provides an object lesson in the problems of biological control.

In the first attempts to spread the disease in Australia, myxoma virus-infected rabbits were released in the wild but, despite the virulence of the virus and the presence of susceptible hosts, the virus died out. It was then realized that this was due to the scarcity of mosquito vectors, whose incidence is seasonal. When infected animals were released at the peak of the mosquito season, an epidemic of myxomatosis followed. Over the next 2 years, the virus spread 3000 miles across Australia and across the sea to Tasmania. However, during this period it became apparent that fewer rabbits were dying from the disease than at the start of the epidemic. The investigators then compared the virulence of the original virus with virus newly isolated from wild rabbits by inoculating standard laboratory rabbits. Two significant facts emerged: (i) rabbits infected with new virus isolates took longer to die, and (ii) a greater number of rabbits recovered from infection. From this it was inferred that the virus had evolved to a less virulent form (Box 17.1). The explanation was simple:

Box 17.1

Evidence for the evolution of myxoma virus to avirulence in the European rabbit after introduction of virulent virus into Australia in 1950

After release into Australia viruses were isolated from wild rabbits and their virulence tested by infecting laboratory rabbits. The percentage of survial in the laboratory rabbits and the mean survival times were calculated.

Mean rabbit survival time (days)	Mortality (%)	Year of isolation			
		1950–1951	1952–1953	1955–1956	1963–1964
<13	>99	100	4	0	0
14–16	95–99		13	3	0
17–28	70–95		74	55	59
29–50	50–70		9	25	31
>50	<50		0	17	9

mutation produced virus variants which did not kill the rabbit as quickly as the parental virus. This meant that the rabbits infected with the mutant virus were available to be bitten by mosquitoes for a longer period of time than rabbits infected with the original virulent strain. Hence the mutants could be transmitted to a greater number of rabbits. In other words, there was a strong selection pressure in favor of less virulent mutants which survived in the host in a transmissible form for as long as possible.

The second finding concerned the rabbits themselves, and the possibility that rabbits were evolving resistance to myxomatosis. To test this hypothesis, a breeding program was set up in the laboratory. Rabbits were infected and survivors were mated and bred with other survivors. Offspring were then infected, the survivors mated and so on. Part of each litter was tested for its ability to resist infection with a standard strain of myxoma virus. The result confirmed that the survivors of each generation progressively increased in resistance. However the genetic and immunological basis for this is not understood.

This work shows (i) how a virus which is avirulent and well adapted to peaceful coexistence with its host can cause lethal infection in a new host, and (ii) how evolutionary pressures rapidly set up a balance between the virus and its new host which ensures that both continue to flourish. The latter remains a stumbling-block for biological control of pests that attack animals or plants. Today, rabbits are still a serious problem to agriculture in the UK and Australia.

17.5 EVOLUTION OF INFLUENZA VIRUS

Influenza A viruses

- *Members of the orthomyxovirus family.*
- *Comprise 144 viruses (subtypes) that naturally infect aquatic birds; cause subclinical gut infections and virus is spread by the oral–fecal route.*
- *Can evolve to infect man, via domesticated poultry; causes respiratory disease that is spread by infected droplets; serious morbidity and mortality, although can be subclinical.*
- *Initial introduction of a new influenza virus in man is called an* antigenic shift; *virus then undergoes gradual change or* antigenic drift – *Darwinian evolution in response to positive pressure from virus-specific antibody. Shift and drift refer to changes in the two major surface proteins of the virus particle.*
- *Influenza can be countered by use of a killed vaccine, and recently by a live vaccine, and by antiviral drugs.*

Background

There are three groups of influenza virus, types A, B, and C, that all cause respiratory disease in man and can be distinguished by the antigenicity of their internal virion nucleoproteins. Type A viruses cause the worldwide epidemics (pandemics) of influenza, and both type A and B viruses cause epidemics. Type C viruses cause only minor upper respiratory illness and will not be discussed further. In terms of natural history, the primary hosts of influenza A viruses are wild aquatic birds (such as ducks, terns, and shore birds). Influenza B viruses infect only humans. In man A and B viruses cause disease only in the winter, usually January and February in the Northern hemisphere, and June and July in the Southern hemisphere. At the equator, virus is present at a low level throughout the year. The cause of this periodicity is not known. In addition, a limited number of other species are naturally infected with type A influenza viruses. These and the directions of transmission are shown in Fig. 17.1. More recently (2005), cats and dogs were found for the first time to be infected.

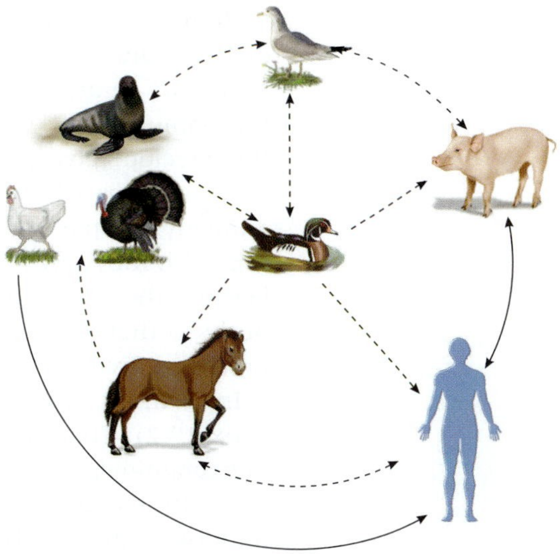

Fig. 17.1 Animal species that are naturally infected with influenza A viruses. Wild birds of the sea and shore form the natural reservoir (top). Known routes of transmission are indicated by continuous arrows and probable routes of transmission by broken arrows.

Typical influenza in man is a lower respiratory infection with fever and muscular aches and pains, but can range from the subclinical to pneumonia (where the lungs fill with fluid). In the elderly and people

Box 17.2

Influenza viruses have a formal descriptive nomenclature

Influenza virus type   A

Subtype (example)   H3N2
Where H is the hemagglutinin and N is the neuraminidase. There may be any permutation of H subtypes 1–16 and N subtypes 1–9.

Strain (example)   A/HongKong/1/68(H3N2)
The strain designation indicates (in order): the type, where the strain was isolated, the isolate number, the year of isolation, the subtype.

The nomenclature of nonhuman strains also includes the host species, e.g. A/chicken/Rostock/1/34(H7N1).

with underlying clinical problems of the heart, lungs, and kidneys, and in diabetics and the immunosuppressed, influenza can be life-threatening. Immunity to an influenza virus is effective at preventing reinfection by that same strain. The viral antigens relevant to protective immunity are the envelope glycoproteins, the hemagglutinin (HA) and neuraminidase (NA) (see Figs 3.14 and 3.15), and immunity is mediated by virus-specific antibody. Despite the induction of the immune response repeated infection with influenza virus is common. The repeated infections are possible because the HAs and NAs of influenza A and B viruses evolve continuously so that previously acquired immunity is rendered ineffective. The processes involved in the evolution are discussed below.

Influenza A viruses are classified formally as *types*, *subtypes*, and *strains* (Box 17.2). There are 16 subtypes of hemagglutinin and nine subtypes of neuraminidase. All 144 permutations are found in nature in wild aquatic birds. By comparison, influenza in man is a side show, with currently only two subtypes, H1N1 and H3N2, in circulation.

Two mechanisms of evolution

Influenza A viruses undergo two types of change affecting their major surface glycoproteins called *antigenic shift* and *antigenic drift* (Box 17.3). Since the start of modern virology there have been four shifts that occurred in 1889, 1918, 1957, and 1968. A shift results in a pandemic (a worldwide epidemic), is always preceded by an abrupt change in hemagglutinin subtype and, apart from the 1968 virus, also by a change in neuraminidase subtype. Drift results in epidemics, and is caused by gradual evolution under the positive selection pressure of neutralizing antibody. A new shift virus immediately starts to undergo continuous antigenic drift.

Box 17.3

Evidence for antigenic drift and shift in influenza virus

- Antigenic drift causes epidemics of influenza A and B in man. It results from a gradual change in antigenicity of the influenza virus particle hemagglutinin (HA) and neuraminidase (NA) proteins, through amino acid changes that affect antibody epitopes (see Figs 3.14 and 3.15). It takes ≥4 amino acid substitutions in ≥2 antigenic sites, and approximately 4 years, to evolve a drift variant that can escape immunity previously acquired to the parent virus and so cause significant disease.
- Antigenic shift causes influenza pandemics in man. Only influenza A viruses are involved. Shift results from an influenza virus acquiring a new HA and usually new NA. This can arise by genetic reassortment of an existing human virus with a bird virus that provides the new HA and NA genes. It may also arise by a bird virus adapting/evolving to grow in man, and may involve intermediary mammalian hosts. Once formed, a shift virus causes a sudden pandemic that spreads around the world in 1–2 years, but it takes an unknown time (years?) to evolve to this point. A new shift virus starts drifting immediately.

Influenza B viruses only undergo drift and this is believed to be because they infect only man and have no other animal reservoir.

The complex evolutionary processes that influenza viruses undergo can be better understood against a background of their biology. These viruses are highly infectious and cause an acutely cytopathogenic infection. Thus, they are a victim of their own success which results in a near universal immunity among hosts who, given adequate conditions, can live for around 80 years. Apparently, the production of immunologically naïve individuals by new births is not enough to allow influenza A and B viruses to survive. Thus, they have evolved strategies for changing their antigens in order to increase the proportion of susceptible individuals in the population.

The phenomena of antigenic shift and drift in influenza have significant implications for human infection and they are monitored by a worldwide network of laboratories, coordinated by the World Health Organization (WHO), that isolate and classify currently circulating influenza viruses. In this way new strains can be quickly spotted and the vaccine changed appropriately.

Antigenic shift

The appearance of shift viruses in man is shown chronologically in Fig. 17.2. Virus was first cultivated in the laboratory in 1933 by intranasal inoculation of ferrets with some nasal secretion from a virologist (Wilson

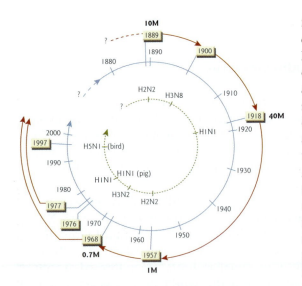

Fig. 17.2 History of antigenic shifts of influenza A viruses in humans. The outer circle denotes the year of emergence of a new subtype that is shown in the inner circle, the duration of the reign of that subtype, and when it is replaced by another subtype. A time scale is shown in the middle circle. Approximate worldwide mortality figures for each shift are indicated in millions. The 1900 shift did not cause a serious pandemic. Currently (2006) H3N2 and H1N1 subtypes coexist. Some occasional infections of people with bird or pig viruses that are not transmitted person to person are noted in the inner circle (but see text).

Smith) who had influenza, and was then successfully passaged in embryonated chickens eggs. However, retrospective information has also been obtained by studying influenza virus-specific antibodies in human sera that had been kept stored in hospital freezers for other purposes. As can be seen, antigenic shift has occurred sporadically from the first recorded shift of 1889 at intervals of 11, 18, 39, 11, and more than 32 years. A brief history of the terrible 1918 pandemic is shown in Box 17.4. Shift, by definition, introduces a virus into a population that has no pre-existing immunity, so it is always associated with an explosive pandemic with high morbidity and mortality, although these vary between different shifts. Since 2001, a new H1N2 reassortant virus has been isolated in many countries in mainly young people. However, this virus has not spread widely, presumably because it is sensitive to existing immunity to its parent human H1N1 and human H3N2 viruses. Approximate mortality figures are included in Fig. 17.2.

Until 1977, only one virus subtype was in circulation at any one time, and this virus was replaced completely when a new subtype was introduced. What causes the original subtype to disappear is not known. In 1977, during the reign of the H3N2 subtype, an H1N1 virus appeared that was identical to the H1N1 virus that had been in circulation in 1950. Thus this was not strictly speaking a shift, but a reintroduction. Where the 1977 virus came from is not known. It had not been infecting the human population during its "absence" as it would have undergone 27 years of antigenic drift (see below). It is as if it had been frozen, and some say that this is literally what happened. However, this is conjecture. At the time of writing, the drifted descendants of the 1968 H3N2 and the 1977 H1N1 cocirculate and continue to undergo antigenic drift. However a new shift could occur at any time.

The mechanism of antigenic shift

As indicated, the natural reservoir of influenza A viruses comprises wild aquatic birds, such as ducks, terns, and shore birds. These avian viruses are tropic for the gut (not the respiratory tract), cause a subclinical infection, and are evolutionarily stable. Many of these birds migrate enormous

Box 17.4

1918 influenza

The 1918 shift virus was unusual in several ways. It is estimated that in just 1 year it killed around 40 million people worldwide. (The enormity of this can be appreciated when compared to the 34 million that have died in 25 years of the AIDS pandemic.) The highest mortality was in young adults and not the elderly and other high risk groups. The overall mortality rate was 25 times greater than other pandemics (2.5% compared with 0.1%). Neither the virulence of the 1918 virus nor its age tropism is fully understood, and it is feared another high mortality virus might arise at any time.

The seemingly intractable problem of analyzing the virus from 1918 that had never been cultivated was solved by extracting fragments of virus genome from the preserved tissue of victims of that pandemic. These are either pieces of lung preserved in formalin in the US army pathology archives or from the exhumed bodies of people who died and were buried in the Arctic permafrost. Virus RNA has been analyzed by RT-PCR and and the complete genome has now been sequenced. Phylogenetics allows hypothetical family trees of influenza A viruses to be constructed, but the resulting relationship of the 1918 virus with other viruses is perplexing. The analysis of the sequence suggests that the virus adapted rapidly from birds to man and was not a reassortant.

Reverse genetics has allowed the functions of 1918 RNAs to be investigated by making infectious virus which has one or more of its RNAs replaced by a 1918 virus RNA. The main finding is that the 1918 HA converts an avirulent virus into one that is pathogenic for mice. This reassortant virus stimulates macrophages to make cytokines and chemokines that attract inflammatory cells into the lung and causes local hemorrhage – as did the virus in people in 1918. Exactly how the 1918 HA differs in this regard from other HAs is not clear.

A complete infectious 1918 virus, A/Brevig Mission/1918, has now been recreated in the USA using reverse genetics and viral RNA from a body preserved in the permafrost. This has caused immense concern worldwide as nobody in the world has immunity to the 1918 virus, and the consequences of an escape would be horrific. However, it was considered that the need to understand what made the virus so terrible made the risk worth bearing. A raft of safety measures including a high level containment laboratory were used for this experiment. Compared with an H3N2 virus, also derived by reverse genetics, the 1918 virus needed no added protease to cleave the HA and form infectious virus, grew to higher titer in mouse lung, caused more lung pathology, and killed chicken embryos. The H3N2 virus with the 1918 HA had increased virulence, but was still less virulent than the complete 1918 virus, showing that other 1918 genes also contribute to virulence. In short, the reconstructed 1918 virus is highly virulent. However, what makes it virulent is still not understood.

distances (e.g. from Siberia to Australia) and spread virus as they go via infected feces. In essence, antigenic shift in the circulation of human influenza virus occurs when virus from a wild bird infects a human being (Fig. 17.3). However, like most viruses, avian influenza virus is species

Box 17.5

Some major evolutionary groupings of vertebrates

Class *Reptilia*: reptiles
Class *Aves*: birds
Class *Mammalia*: monotremes (e.g. platypus)
                   marsupials (e.g. kangaroo)
                   placentals (e.g. man)

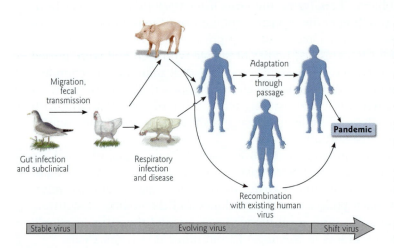

Fig. 17.3 Summary of the events leading up to an antigenic shift of human influenza virus. The evolutionary time scale is not known but probably takes several years.

specific and does not readily infect other bird species let alone mammals which are members of another major taxonomic group (Box 17.5). Thus, the virus has to adapt by progressive mutation before the jump to humans can occur.

It is thought that the first link in the chain is the infection of freerange domestic poultry (mainly chickens, ducks, turkeys, and quail) by migrating wild birds. This is an avirulent respiratory infection, although it is not understood how the switch from a gut to respiratory infection comes about. The infection passes from bird to bird until mutations take place in the protease cleavage site of the hemagglutinin precursor polypeptide. Cleavage is necessary for virus infectivity. Normally, the HA can only be cleaved by a protease that is present solely in the respiratory tract, but after an increase in the number of basic amino acid residues in the HA cleavage site it can be cleaved by another protease that is widely distributed throughout tissues, and this allows the virus to grow throughout the body and cause serious disease. At this stage the virus is able to jump from Class *Aves* to Class *Mammalia*, and multiply (but not yet spread) in humans (Fig. 17.3). One of the major gaps in modern virology is the understanding of animal virus transmission in general, and what gene or genes are responsible for its control. Another variation in the evolutionary progression of virus from wild birds to humans may involve domestic pigs.

In rural areas poultry and pigs are often kept together, giving ample opportunity for the crucial bird-to-mammal adaptation. At this point the virus has two evolutionary options. It can continue to accumulate mutations and become better adapted to humans, or it can recombine with a human influenza virus strain and so acquire genes that are already fully adapted. In fact, it may do both. One essential adaptation is a change to the viral attachment site on the HA protein so that the virus can use as a receptor an *N*-acetyl neuraminic acid that has an α2,6 linkage to the carbohydrate moiety on a glycoprotein or glycolipid (as is found in cells of the human respiratory tract), rather than the α2,3 linkage that is most common in birds. The advantage to the virus of achieving this by recombination is that at its simplest, there need only be a recombination of the RNA segment that encodes the new hemagglutinin with the seven existing segments from the human strain that control person-to-person spread. This scenario would account for the 1968 shift where an H3N2 virus replaced the H2N2 virus.

Recombination in influenza viruses is very efficient as the genome is segmented – it comprises eight single-stranded, negative-sense RNAs. The hemagglutinin and neuraminidase proteins are encoded by distinct RNA segments. When a cell is infected simultaneously with more than one strain of virus, newly synthesized RNA segments reassort at random to the progeny (Fig. 17.4). Many, though not necessarily all, of the 2^8 (256) possible genetic permutations that can be formed between two viruses are genetically stable. Reassortment occurs readily in cell culture, in experimental animals, and in natural human infections between all type A influenza viruses, but not between A and B viruses. Even after a reassortment event, more mutational adaptation may be needed before the virus is able to cause a recognized pandemic. Thus it may be that a shift virus is present in the human population, undergoing mutational adaptations for a few years before it causes a pandemic.

Tracking influenza viruses in domesticated animals

Two instances of abortive antigenic shifts took place in the eastern USA in 1976 and in Hong Kong in 1997. These viruses were noted and notable because they caused death in a small number of young people, and it was feared that they signalled the start of a new lethal pandemic. Both had their origins in nonhuman animals. The 1976 virus was also isolated from local domestic pigs, and the 1997 H5N1 virus from chickens that had been brought as live birds to the local poultry market. Of 18 people infected with H5N1 virus, six (33%) died, but these represent deaths among hospitalized, seriously ill cases, and it is not known

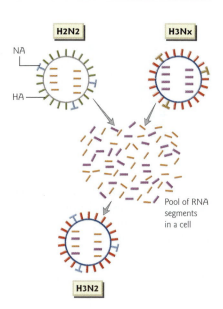

Fig. 17.4 Recombination (reassortment) between an existing human influenza A virus (H2N2) and a new virus from the wild bird reservoir (H3N*x*, where *x* represents an unknown neuraminidase subtype; see text) that gives rise to antigenic shift. The two viruses simultaneously infect a cell in the respiratory tract, and the eight genome segments from each parent assort independently to progeny virions. The example shows a novel progeny virion (H3N2) that comprises the RNA segment encoding the H3 avian hemagglutinin and the seven remaining segments from the existing human virus.

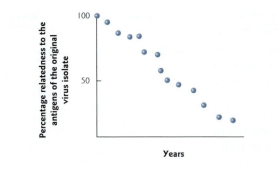

Fig. 17.5 Diagram showing antigenic drift of type A influenza virus in humans. This could represent either the HA or NA. Each point is a virus strain isolated in a different year.

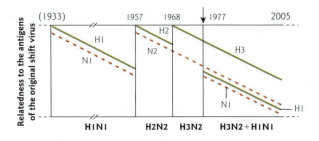

Fig. 17.6 Diagram showing the course of antigenic shift and drift of influenza A viruses in man. The first virus, isolated in 1933, was H1N1. This arose by antigenic drift from the 1918 virus. Other shift viruses appeared in 1957 (H2N2) and 1968 (H3N2). A 1950 H1N1 virus reappeared in 1977. Drift is shown schematically. The 1957 N2 was acquired by the H3N2 shift virus, and has drifted from 1957 to the present day.

how many mild cases there were. However, neither virus had the ability to spread from person to person, and no more infections were seen. Since that time, surveillance has increased, particularly in southeast Asia, of domestic poultry and infected people in order to get an early warning of any impending pandemic. The H5N1 virus was isolated only from poultry in 2001 and 2002, but then caused some cases of fatal disease in humans in 2003, 2004, and 2005. It has been isolated from countries in southeast Asia including Indonesia, Thailand, Vietnam, Japan, South Korea, Cambodia, Laos, and China. Human infection rates are low but mortality has been put as high as 67%. Usually there is H5N1 infection of people's own chickens, and presumptive bird-to-human transmission. There may have been very occasional human-to-human transmission but, in the absence of a pandemic, this is clearly (and fortunately) still an inefficient process. Outbreaks in poultry cause alarm and have provoked culls of 20 million or so birds by the authorities. Other economic costs are also huge as trading partners ban imports of poultry products. The known species range of influenza A viruses was extended in 2003 when H5N1 virus spread with some fatalities to tigers and leopards in a Thai zoo, as a result of animals being inadvertently fed infected chicken carcasses, and domestic cats have been infected experimentally. In the last few years various subtypes have been isolated from poultry and humans, including H9N2 virus which causes only mild respiratory disease in man in China and Hong Kong, H7N7 virus in Holland and Belgium (with one human fatality), H7N1 in Italy, H7N2 virus in the USA, and H7N3 in Canada; viruses isolated from pigs in Europe and North America include H1N1, H1N2, and H3N2 subtypes. In 2005, H3N8 virus was found for the first time in dogs in the USA. This overall increase in virus isolation may partly be the result of better surveillance rather than any rise in virus dissemination. To avoid the possibility of serious infection, workers handling infected birds wear safety suits and are offered prophylactic anti-influenza drugs.

Antigenic drift

Influenza A viruses have been isolated every year from humans around the world since their discovery in 1933. Each new isolate is tested serologically with antibody to all other influenza strains. It soon became apparent that the hemagglutinin and neuraminidase of new isolates were antigenically slightly different from those of the previous year, and that the difference increased incrementally year by year. This difference is reflected in one or more amino acid substitutions in one or more of the antigenic sites on the HA or NA. This phenomenon is explained by the assumption that influenza virus mutants carrying modified antigenic determinants have an evolutionary advantage over the parent virus in the face of the existing immune response. However, although both the HA and NA drift, the HA is the major neutralization antigen and therefore appears to be evolutionarily the more significant. Thus year-by-year new amino acid substitutions – and new epitopes – appear. Epitopes can also disappear if a mutation creates an attachment site for a carbohydrate moiety that then sterically prevents antibodies from reacting with those epitopes. The name "antigenic drift" is very apt (Fig. 17.5). The nature of the selective force which drives this process is discussed below. In practice a drift variant that can cause significant disease arises about every 4 years, and this has on average four or more amino acid substitutions in two or more antigenic sites. At the same time the "old" strain is completely replaced by the "new" strain. Influenza B viruses also undergo antigenic drift, but this is a slower process and also several different strains cocirculate. The reason for this difference is not known.

Antigenic drift is at least as important in causing human influenza as antigenic shift. In the UK there are 4000–14,000 deaths associated with epidemics of influenza every year, and this extrapolates to 400,000 to 1,400,000 deaths per year worldwide. Thus, in the twentieth century, drift viruses were responsible for approximately 40 to 140 million deaths worldwide. (In the UK an influenza epidemic is formally defined as 400 cases of influenza per 100,000 people.)

The mechanism of antigenic drift

As soon as a new shift virus appears and infects people, it begins to drift (Fig. 17.6). Drift of influenza A viruses is linear due to the dominating effects of favorable mutations (Fig. 17.7). Influenza B viruses undergo less drift than type A viruses, and multiple virus lineages coexist. Drift happens on a global scale. It can only be theorized how drift takes place as it is an assumption that drift variants arise from virus circulating in the previous year. It is generally believed that variants are selected by neutralizing antibody (Box 17.6).

Type A

Type B

Fig. 17.7 Model of antigenic drift of influenza type A and type B viruses. Points on the same level represent drift variants that arise in the same year. The branch length indicates the relative change in antigenicity from virus in the preceding year. Drift is shown for an arbitrary 7-year period. See text for further discussion.

Box 17.6

Testing antigenic drift in the laboratory

- Antigenic drift has been modeled in the laboratory using a neutralizing monoclonal anti-body (Mab) specific for the HA. Virus and MAb are mixed together and inoculated into an embryonated chicken's egg. This is a surprisingly efficient process and, after just one passage, the progeny virus is no longer neutralized by the selecting MAb. This "drift" virus is also known as an *escape mutant*, and represents the growth to population dominance of an antigenic variant virus that already existed in the inoculum. Sequencing shows there is a single amino acid substitution in the expected antigenic site. If the drift virus is subjected to a MAb to a different epitope, then the process repeats and the progeny virus now has two amino acid changes compared with the original (Fig. 17.8). However if two or more MAbs are mixed together to resemble an antiserum, no progeny virus – mutant or wild-type – is produced at all. Thus drift can only take place if an antiserum effectively contains antibody to only a single epitope.

- Influenza virus HA molecules have five antigenic sites (although sometimes one is hidden by a carbohydrate group). In H3 viruses these are labeled A–E. Each site comprises around 10 epitopes, thus there are 50 epitopes in all, and in theory during infection the immune system should make antibodies to all of these. However when tested by immunizing rabbits with virus and measuring the amount of antibody made to individual epitopes, it was found that one epitope in site B was dominant, and there were only traces of antibody to two other epitopes (Fig. 17.9). The amounts varied between animals. In mice the results were similar but the dominant epitope was in site A. Thus for unknown reasons the immune system responds selectively to foreign epitopes.

- While some antisera were completely neutralizing, others were shown to select escape mutants like a MAb. Thus it would seem that drift operates because certain individuals have an antibody response that is highly biased to one epitope. The derivation of the ≥4 amino acid substitutions in ≥2 sites, referred to above, could therefore be achieved as shown in Fig. 17.10. The biased antibody response may be genetically controlled, and drift variants may arise in a subpopulation with the relevant antibody response. Further, since adults suffer repeated influenza infections, they are likely to make a complex antibody response that cannot give rise to drift variants. The simplest antibody response is likely to occur in children after their first infection, and this may be responsible for the selection of drift variants.

Unsolved problems relating to influenza virus

Despite considerable knowledge of influenza virus (the complete genome sequence, the atomic structure of the HA protein, infectious nucleic acid) there are major areas of ignorance. Most of these concern viral epidemiology

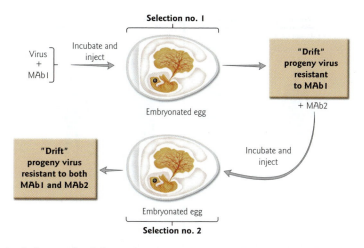

Fig. 17.8 "Antigenic drift" in the laboratory: a single neutralizing monoclonal antibody can select a population of influenza virus escape mutants that is no longer neutralized by the selecting MAb. Another round of selection with a second MAb produces virus that now carries two amino acid substitutions, but selection with the two MAbs simultaneously is completely neutralizing (not shown).

which is immensely difficult to study. Space precludes discussion but the following problems still require an answer about an important though common virus:

- Why is influenza a winter disease?
- How does the same virus appear simultaneously in different places around the globe at the start of an epidemic (i.e. it does not appear to follow any apparent transmission chain)?
- Why is the seasonal restriction of influenza apparently unaffected by the year-round introduction of virus by air travellers incubating influenza? Does this virus not infect people in summer or does it infect them without causing disease?
- How does a clinically significant drift variant travel from northern to southern latitudes in 6 months (and did so prior to frequent air travel)?
- How does a shift virus (other than the reoccurring H1N1 of 1977) replace the existing virus?
- How does a drift virus replace the existing virus?
- The hypothesis of antigenic drift outlined above needs verifying.
- Why did the 1918 shift virus have a higher mortality than other shift viruses?

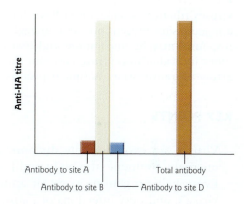

Fig. 17.9 An epitope-biased serum antibody response in a rabbit injected with influenza A virus. Nearly all the HA-specific antibody is accounted for by the response to a single epitope in antigenic site B. (From Lambkin, R. and Dimmock, N. J. 1995. *Journal of General Virology* **76**, 889–897.)

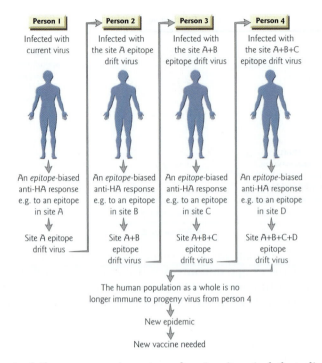

Fig. 17.10 How antigenic drift may occur in nature, bearing in mind that clinically significant drift viruses (that cause epidemics) have four or more amino acid residues changed in two or more antigenic sites present on the hemagglutinin protein. Note that people with biased antibody responses may be uncommon and that there may be any number of nonselective infections of other individuals occurring between the four persons shown here, as indicated by the broken arrow of transmission. A similar process could cause drift of the NA protein.

KEY POINTS

- Viruses are evolutionarily dynamic and respond to changes in the macroenvironment and microenvironment (the body and immune system).
- Every living organism, plant or animal, amoeba or elephant, has its own extensive range of viruses, and new infections of man can arise by nonhuman viruses crossing species, e.g. HIV, influenza A virus. Cross-over is accelerated by loss of habitat and increased proximity of man and animals.
- All established human viruses undergo Darwinian evolution as a result of evolutionary pressure acting on random variants arising by mutation, although the scale of change ranges from, say, measles virus (low) to influenza A virus (high).
- Many new infections in a host species result in high morbidity and mortality, but evolution selects hosts that are genetically more resistant to infection and less virulent progeny virus (e.g. myxoma virus in the European rabbit).
- A new shift variant of influenza A virus could arise at any time and because of its efficient transmission such virus is a major threat to human health. Better antiviral measures than are currently available are needed to combat a virus with the potential of the 1918 virus.

QUESTIONS

- Describe the processes of antigenic drift and antigenic shift in influenza virus and indicate, with reasons, which is likely to generate a more pathogenic virus.
- Using examples, discuss the proposition that viruses with RNA genomes evolve more rapidly than those with DNA genomes.

FURTHER READING

Black, F. L., 1966. Measles endemicity in insular populations: critical community size and its evolutionary implications. *Journal of Theoretical Biology* **11**, 207–211.

Domingo, E., Holland, J. J., 1997. RNA virus mutations and fitness for survival. *Annual Review of Microbiology* **51**, 151–178.

Domingo, E., Webster, R. G., Holland, J. J., 1999. *Origin and Evolution of Viruses*. Academic Press, New York.

Fenner, F., Ratcliffe, F. N., 1965. *Myxomatosis*. Cambridge University Press, London.

Griffiths, P. D., 2004. Should vaccines and antiviral therapy for influenza now be deployed strategically? *Reviews in Medical Virology* **14**, 137–139.

Guan, Y., Poon, L. L. M., Cheung, C. Y. *et al.* 2004. H5N1 influenza: a protean threat. *Proceedings of the National Academy of Sciences of the United States of America* **101**, 8156–8161.

Kobasa, D., Takada, A. E. A., Shinya, K. *et al.* 2004. Enhanced virulence of influenza A viruses with the haemagglutinin of the 1918 pandemic virus. *Nature (London)* **431**, 703–707.

Lamb, R. A., Krug, R. M. 1996. Orthomyxoviridae: the viruses and their replication. In *Fields Virology* (B. N. Fields, D. M. Knipe, and P. M. Howley, eds), Vol. 1, 3rd edn, pp. 1353–1395. Lippincott-Raven, Philadelphia.

Lipatov, A. S., Govorkova, E. A., Webby, R. J. *et al.* 2004. Influenza: emergence and control. *Journal of Virology* **78**, 8951–8959.

Nicholson, K. G., Wood, J. N., Zambon, M. 2003. Influenza. *Lancet* **362**, 1733–1745.

Oxford, J. S. 2000. Influenza A pandemics of the 20th century with special reference to 1918: virology, pathology and epidemiology. *Reviews in Medical Virology* **10**, 119–133.

Reid, A. H., Taubenberger, J. K., Fanning, T. G., 2004. Evidence of an absence: the genetic origins of the 1918 pandemic influenza virus. *Nature Reviews Microbiology* **2**, 909–914.

Webby, R. J., Webster, R. G., 2003. Are we ready for pandemic influenza? *Science* **302**, 1519–1522.

Also check Appendix 7 for references specific to each family of viruses.

Part IV

Viruses and disease

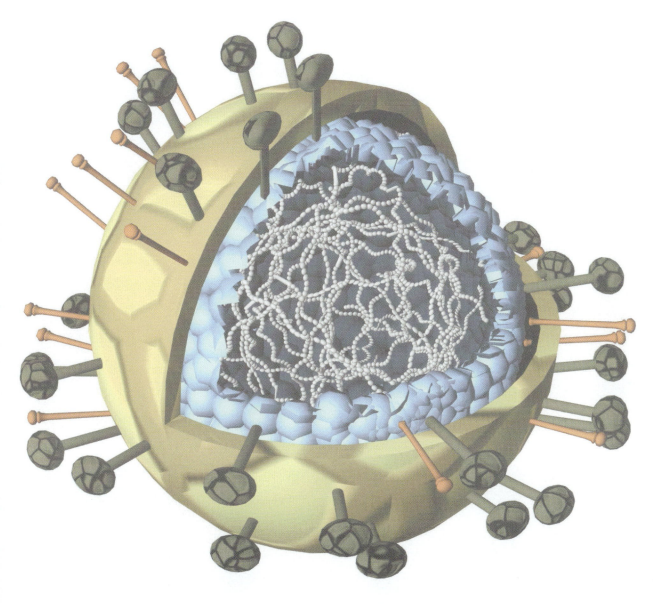

18

Human viral disease: an overview

Between them, the viruses that infect humans can target, and cause disease in, essentially every tissue in the body. Even closely related viruses can cause markedly different diseases, both in severity and type. Thus, an understanding of viral disease processes must be developed for each virus individually.

Since the discovery of viruses and their role in human disease, there have been many significant advances in our understanding of these agents and progress in both the prevention and treatment of infection (see Chapter 21). However, despite this progress, viral diseases continue to be a major component of the morbidity and premature mortality burden in the human population world-wide. There is also the constant threat of the emergence of new viruses to afflict us (see Section 23.4). Human immunodeficiency virus, for example, represents one of the greatest viral threats to our species that we have seen; it is considered in Chapter 19. Another, less familiar, aspect of viral disease, the role of viruses as causes of cancer, is considered in Chapter 20. This chapter gives an overview of human viral disease, focusing principally on acute viral infections (those from which we suffer transiently and then recover), which are our most common experience of viral disease. It explores the factors that govern the nature, severity, and geographical distribution of viral disease and some aspects of viral pathogenesis in specific examples.

Table 18.1

The principal disease-causing viruses of humans.

Major target organ/system*	Virus (family; genus; species)
Blood and lymphoid system	Herpesvirus; Lymphocryptovirus; Epstein–Barr virus (HHV4) Paramyxovirus; Rubulavirus; mumps virus Parvovirus; Erythrovirus; human parvovirus B19 Retrovirus; Deltaretrovirus; human T-lymphotropic virus 1 Retrovirus; Lentivirus; human immunodeficiency virus 1, 2
Eye	Adenovirus; Mastadenovirus; human adenovirus B Herpesvirus; Simplexvirus; herpes simplex virus 1 (HHV1)
Fetus (infection *in utero*)	Herpesvirus; Cytomegalovirus; human cytomegalovirus (HHV5) Togavirus; Rubivirus; rubella virus (German measles)
Gastrointestinal tract	Adenovirus; Mastadenovirus; human adenovirus A, F Astrovirus; Mammastrovirus; human astrovirus Calicivirus; Norovirus; Norwalk virus Reovirus; Rotavirus; rotavirus A
Genital tract	Herpesvirus; Simplexvirus; herpes simplex virus 1, 2 (HHV1, 2) Papillomavirus; Papillomavirus; human papillomaviruses (especially 6, 11, 16, 18, 31)
Heart	Picornavirus; Enterovirus; Coxsackie B viruses
Liver	Picornavirus; Hepatovirus; hepatitis A virus Hepadnavirus; Orthohepadnavirus; hepatitis B virus Flavivirus; Hepacivirus; hepatitis C virus
Multisystem/hemorrhagic fevers	Arenavirus; Arenavirus; Lassa fever virus Bunyavirus; Nairovirus; Crimean–Congo hemorrhagic fever virus Bunyavirus; Hantavirus; Hantaan virus Bunyavirus; Hantavirus; sin nombre virus Filovirus; Ebolavirus; Ebola virus Filovirus; Marburgvirus; Marburg virus Flavivirus; Flavivirus; dengue virus Flavivirus; Flavivirus; yellow fever virus
Nervous system	Bunyavirus; Orthobunyavirus; California encephalitis virus Bunyavirus; Phlebovirus; Sandfly fever virus Flavivirus; Flavivirus; Japanese encephalitis virus Flavivirus; Flavivirus; tick-borne encephalitis virus Flavivirus; Flavivirus; West Nile virus Herpesvirus; Simplexvirus; herpes simplex virus 1, 2 (HHV1, 2) Herpesvirus; Varicellovirus; varicella-zoster virus (HHV3) Paramyxovirus; Morbillivirus; measles virus Picornavirus; Enterovirus; polio virus Rhabdovirus; Lyssavirus; rabies virus

Table 18.1 (*Cont'd*)

Major target organ/system*	Virus (family; genus; species)
Respiratory tract	Adenovirus; Mastadenovirus; human adenovirus B, C, D, E
	Coronavirus; Coronavirus; human coronavirus, SARS coronavirus
	Orthomyxovirus; Influenzavirus A, B; influenza A, B viruses
	Paramyxovirus; Morbillivirus; measles virus
	Paramyxovirus; Pneumovirus; respiratory syncytial virus
	Paramyxovirus; Respirovirus; parainfluenza 1, 2, 3
	Picornavirus; Rhinovirus; human rhinoviruses
Skin	Herpesvirus; Varicellovirus; varicella-zoster virus (HHV3)
	Papillomavirus; Papillomavirus; human papillomaviruses (especially 1, 2, 4)
Testes	Paramyxovirus; Rubulavirus; mumps virus

*Viruses can cause disease in more than one system and therefore can appear more than once in the table. In some cases, viruses listed as causing disease in a system do so in only a small proportion of infected people.

18.1 A BRIEF SURVEY OF HUMAN VIRAL PATHOGENS

We now know that viruses are capable of affecting all the major tissues and systems in the human body. The diseases they cause range from trivial to life-threatening. The outcome of infection by a virus is not absolute but can vary considerably in nature and severity between individuals. Table 18.1 contains a brief survey of this spectrum of pathogens, classified by their principal clinical manifestation. A comprehensive review of the disease processes for each of these viruses is beyond the scope of this text. Examples of viral pathogenesis in the most common target organs/systems for acute virus infection are considered in Sections 18.5 to 18.8.

18.2 FACTORS AFFECTING THE RELATIVE INCIDENCE OF VIRAL DISEASE

The likelihood of an individual suffering from a particular viral disease depends on a number of interrelated factors. These include the following: geographical distribution of the virus and any necessary vector species; climate; human and vector migration patterns; socioeconomic conditions, particularly the quality of nutrition and sanitation; individual genotype, especially the MHC haplotype since it determines the immune response; and the age at infection.

Viruses cannot be avoided; virus infections are a fact of life. All individuals inevitably encounter viruses that infect them and make them unwell. However, the extent to which the threat of viral disease should be a cause for concern to an individual will vary greatly depending on where in the world they live, the nature of their community, their socio-economic circumstances, and their genetics.

The effect of location on viral disease incidence

Many viruses are found throughout the world, often with high rates of seroprevalence (i.e. most people in the community have been infected). However, some are much more restricted in their range and, in many cases, this reflects a restricted range for an obligatory vector species. For example infection by the hantavirus, sin nombre, is found in only the southwestern USA, reflecting the distribution of its vector, the deermouse. Other hantaviruses occupy other geographical niches based on the location of their specific rodent host. Viruses can also be constrained in range by geographical isolation. Historically, it is documented that the exploration of the globe by Europeans spread measles and variola (smallpox) viruses into communities that had apparently never experienced them before and, as a result, epidemics with high mortality were initiated in indigenous populations. In retrospect, the same form of geographical constraint can now be seen to apply to the mosquito-borne West Nile virus (WNV; see Section 23.4). This virus was long known in southern Europe and Africa but unknown in the Americas before 1999; it has now spread across North America, having been introduced from its traditional range by an unknown route. The mosquitoes of North America are clearly quite capable of transmitting WNV, they just had not been exposed to it prior to this introduction event.

The prevailing climate also affects the probability of acquiring particular viral diseases. Arbovirus infections (viruses transmitted by biting arthropods, see Chapter 16) are essentially limited to the less seasonal areas of the globe because the colder winters experienced elsewhere restrict the ability of the vector species to survive. Climate also affects the survival of virus while it is outside a host, since particle stability can depend on hydration/humidity (see Section 16.1). Viral genomes may also be vulnerable to inactivation by ultraviolet radiation in sunlight. Finally, climate influences human behavior and physiology. Several respiratory virus infections, such as influenza A and respiratory syncytial viruses, show annual winter peaks of incidence (Fig. 18.1). Maybe this is because we tend to spend more time in confined spaces with poor air circulation at this time of the year, although there are also respiratory virus infections with summer peaks of infection. Perhaps also, the reduced sunlight intensity affects our ability to counter infections. The light-sensitive pineal gland in the center of the brain secretes melatonin and controls our circadian rhythms; low light is recognized to cause seasonal affective disorder (SAD) in some people. This is a complex syndrome of depression and malaise, and mental state is now thought to affect the functioning of the immune system. So, what the weather is doing can have an impact on the efficiency of infectious virus spread between individuals, and hence on the propagation (or not) of an epidemic.

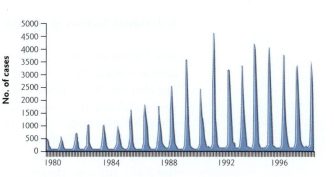

Fig. 18.1 Monthly incidence of respiratory syncytial virus infection in the UK showing pronounced annual epidemics peaking in November–January.

The effect of environment on viral disease incidence

Any virus that is cleared from a host, with the host becoming immune to subsequent infection, requires access to a sufficient number of new hosts in order to perpetuate itself in a community, as already discussed for measles virus (see Section 17.3). Below this critical host community size, the virus will die out in that community until reintroduced from outside. Thus, the size of a community and its population density – whether a person lives in an urban or rural area – can determine whether or not they will be exposed to a virus. This of course assumes little movement of people in and out of the community, which is no longer true for much of the world, but in isolated areas this factor is still relevant. However, even with extensive population mixing bringing viruses into communities regularly, the greater population density in towns and cities still increases an individual's exposure to some viruses. Conversely, people in rural areas may be more vulnerable to zoonotic infections, because of an increased likelihood of encountering other species and the viruses that they carry.

Beyond the size of the community, another important factor is its socio-economic status. Many of the principal viral pathogens, such as rotavirus, poliovirus, and hepatitis A virus, are spread by the fecal–oral route. Thus, rates of infection are determined by the quality of sanitation and the availability of clean drinking water. Nutritional status also affects the severity of viral disease. For example, the possibility of a fatal outcome of measles virus infection is increased by poor nutrition. Indirectly, socioeconomic factors also influence another parameter: the age at which you become infected. Living in a poorer household results in earlier exposure to a range of different viruses, such as Epstein–Barr virus, poliovirus, hepatitis A virus, and hepatitis B virus. Intriguingly, in each of these cases the acute diseases are less severe and/or a greater proportion of infections are subclinical with early exposure to the virus. Finally, deprivation and poverty tend to increase the level of prostitution, which exacerbates the spread of sexually transmitted infections, such as HIV, that are therefore more prevalent in such communities (see Chapter 19).

Individual factors in viral disease susceptibility: genetics

In addition to all the above external factors, which can create differences between individuals in their probability of suffering viral disease, there are also intrinsic individual characteristics that are important. Most obviously, the ability of the immune system to respond to a virus is not the same for everyone. Apart from the case of people who have recognizable problems with their immune systems, each of us has a different ability to respond to specific antigens, determined by the precise nature of the MHC antigens that our genomes encode (see Chapter 12). MHC antigens are produced from highly polymorphic gene loci, so the combination of MHC antigens varies greatly between individuals. There are also differences between communities, with particular MHC alleles being common in one part of the world and rare in another. Natural selection produces MHC antigens in a population that are suited to respond to the prevailing antigens. When a new antigen (e.g. a new virus) enters a population, perhaps only a small proportion of people will be able to mount a strong immune response to it. If the new virus kills those less able to mount a response, then the remaining individuals, and the MHC alleles they carry, expand within the population. This phenomenon may partly explain the severity of disease that typically arises from newly introduced viruses.

Specific alleles at other loci can also affect viral infection. HIV requires a coreceptor known as CCR5 for successful infection (see Section 19.3). Some people are relatively resistant to HIV infection because they are homozygous for an allele of the CCR5 gene that cannot express a functional protein. The frequency of this allele is highest in North European and West Asian populations. As the HIV pandemic continues, natural selection can be expected to increase the frequency of this CCR5 allele, and of alleles at any other loci that favor the survival of HIV-infected individuals.

18.3 FACTORS DETERMINING THE NATURE AND SEVERITY OF VIRAL DISEASE

Why a virus causes the disease that it does is one of the most important questions in virology, and one of the most difficult to answer. Individual virus-specific features are likely to be crucial in each case. Factors that are thought generally to be relevant include the cell-type tropism of the virus, its portal of entry into the body, and the relationship that the virus establishes with the host immune system.

One does not have to look far into the diversity of human viral pathogens to see the difficulty in making general or predictive statements about why viruses cause the specific diseases that they do. Within a single virus family, such as the adenoviruses, there are viruses causing respiratory tract infections ranging from mild to severe, and other viruses that cause infections of the eye, gut, or urinary tract. Equally, viruses from at least three distinct families with very different molecular biology and particle structures can all cause the acute symptoms of hepatitis (liver infection). Thus there is no feature of a novel virus that can be observed outside of a human host and then related to knowledge about other viruses to predict the nature and/or severity of the disease that the new virus causes; each virus must be studied independently (see Chapter 4).

Starting from first principles, one plausible determinant of the disease-causing capacity of a virus is its interaction with cell-surface receptors. As discussed in detail in Chapter 5, a virus generally cannot infect a cell successfully in the absence of its specific receptor so the distribution of the receptor around the tissues of the body will act as a restriction on the range of tissues that can be infected and hence on the number of systems in the body where signs

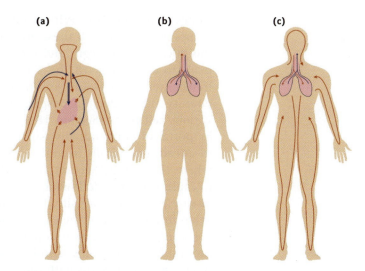

Fig. 18.2 Three scenarios for the relationship between the route of entry of a virus, its circulation in the body, site of replication, and sites of disease manifestation. (a) Virus enters and circulates widely but can only infect cells in one location, where it replicates and causes disease, e.g hepatitis B virus, with entry directly into the blood or by sexual contact and replication in the liver. (b) Virus infects cells at the portal of entry, where it replicates causing signs and symptoms, but cannot spread elsewhere despite its use of a cell receptor that is widely distributed, e.g. rhinovirus, respiratory syncytial virus with entry and exit via the respiratory tract. (c) Virus infects cells at the portal of entry, where it replicates and then spreads systemically to give a variety of signs and symptoms, e.g. measles virus, which again enters and exits via the respiratory tract. Blue: movement of infecting virus; red: movement of progeny virus.

and symptoms of infection might be experienced. In fact, many of the receptors used by viruses (see Table 5.1) are ubiquitous molecules present on many different cell types. However, there are some examples of receptor tropism correlating with disease, such as hepatitis B virus, which appears only able to enter and to infect primary hepatocytes (Fig. 18.2a). In a variation of this concept, influenza A virus can only bind its receptor and enter cells following proteolytic activation (see Section 10.7); the limited tissue distribution of suitable activating proteases is important in restricting most strains of influenza virus to the respiratory tract.

Another potential determinant of the disease profile of a virus is restriction by portal of entry. Viruses enter our bodies by one of a few routes (Chapter 16). When they do so, they will encounter certain tissues before others. If these are receptive to infection then this is where the infection will begin and where signs and symptoms might be most likely to occur (Fig. 18.2b). Thus, viruses such as rhinovirus, which are transmitted via aerosols, are inhaled and cause symptoms of upper respiratory tract infection despite having a receptor that is displayed on various cell types (Section 18.6). Equally, viruses that may be restricted by portal of entry initially, such as measles virus, may have the ability to spread systemically from this site of infection and cause symptoms elsewhere (Fig. 18.2c; Section 18.8). However, it is also possible for a virus to cause no effects following infection at its portal of entry, but to cause disease at a secondary site. For example, birds infected intranasally with avian influenza nevertheless excrete virus from the gut.

Having gained access to a particular tissue in the body, the severity of the resulting symptoms is still unpredictable; it depends crucially on the interaction of the virus with the host immune system. At one level, there is clear evidence that the immune response serves to limit the spread of virus in the body and hence the severity of disease. Herpes simplex virus, for example, whilst generally giving mild symptoms, causes a systemic and life-threatening infection when acquired pre- or neonatally, when the immune system is not mature. However, it is also the case that the immune response causes many of the general signs and symptoms of a virus infection (Section 18.4) and can also be responsible for more specific aspects of pathogenesis, e.g. hepatitis (Section 18.7). Many viruses encode proteins whose function is to reduce the impact of specific aspects of the host immune response. For example, human adenovirus 5 makes a protein that blocks MHC class I antigen maturation, and other proteins that block apoptosis, part of the innate response to infection. When adenovirus pathogenesis is studied in an animal model, removal of these functions from the virus exacerbates disease. In other words, having a stronger immune response to the infection makes the disease worse in this case.

18.4 COMMON SIGNS AND SYMPTOMS OF VIRAL INFECTION

The classic signs and symptoms of a virus infection, fever and malaise, are caused by the host response to infection, not by the virus damaging cells of the host.

Fever (elevated body temperature), aches and pains, and general malaise are features of a variety of different viral infections. Often these occur in advance of more virus-specific signs and symptoms, and they are all associated with activation of the innate immune response. A key element in the innate response is the production of interferon α/β and the subsequent generation of an antiviral state (see Section 12.2). When the antiviral properties of interferon were recognized, attempts were made to use it as a general treatment for viral infections but with limited success. In fact it was found that administering interferon generated many of the same signs and symptoms as are caused by viruses themselves. It is our body responding to a virus infection that usually generates the first symptoms we are aware of, not virus-mediated destruction of cells.

There are close links between the innate and adaptive arms of the immune response (see Chapter 12). One of these is the production of the cytokine interleukin 1β (IL-1β) by macrophages and other cells in response to interferon α/β which then stimulates T cell activation by antigen, potentiating the response. Studies with the poxvirus, vaccinia, in an animal model have demonstrated that the release of this cytokine is also responsible for the fever (pyrexia) associated with infection. Mice infected with wild-type vaccinia actually show a reduction in body temperature during the infection but a mutant that is unable to make a protein called B15R instead causes an increase in temperature. In other words, the B15R protein is used by vaccinia virus to block the fever response of the host. B15R is now known to bind and inactivate IL-1β, implicating this cytokine as a key mediator of temperature elevation. The downstream effects of IL-1β also include heightened sensitivity to inflammatory pain, which may be linked to the generalized aching associated with many viral infections.

18.5 ACUTE VIRAL INFECTION 1: GASTROINTESTINAL INFECTIONS

The classic specific symptoms of viral infection of the gastrointestinal (GI) tract are vomiting and/or diarrhea. These symptoms are both manifestations of dysfunction in the GI tract that results from the virus infection.

Whilst there are a number of different viruses that can infect the gastrointestinal (GI) tract, two of the most significant agents are rotaviruses and noroviruses. Rotaviruses pass through the stomach following ingestion and infect the cells at the tips of the villi of the epithelium lining the upper part of the small intestine. This leads to loss of these cells and so-called blunting of the villi (Fig. 18.3). The intestinal epithelium is responsible for absorption of nutrients from the gut contents. When this process is perturbed, the osmotic balance across the epithelium is lost and there is an outflow of water and electrolytes from the body into the lumen of the gut, resulting in diarrhea. Such an effect could explain the clinical signs of rotavirus infection. However, there is also evidence that active

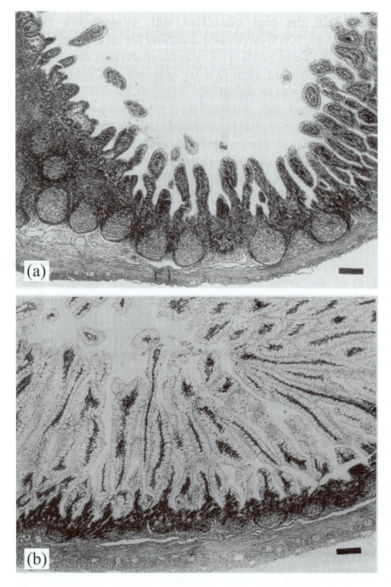

Fig. 18.3 (a) Section from the ileum of a piglet 72 hours post-infection with rotavirus. (b) Section from the ileum of a control, mock-infected, piglet. Note the shortening of the villi and the absence of vacuoles (intracellular white areas) in the villous cells in (a) as compared with (b). These are indicative of normal active absorptive function. The bar represents 50 μm. (Adapted from Ward, L. A., Rosen, B. I., Yuan, L., Saif, L. J. 1996. *Journal of General Virology* **77**, 1431–1441, with permission from the Society for General Microbiology.)

secretion of fluid into the gut lumen, caused by a viral protein acting as a toxin in a manner similar to certain bacterial endotoxins, contributes to rotaviral diarrhea (Box 18.1).

Rotavirus disease is particularly a problem in babies and young children. Signs of rotavirus diarrhea begin within 2–3 days of infection and continue for 4–7 days. Treatment is straightforwardly provided by oral rehydration with an osmotically balanced saline/glucose solution. However, in the absence of such treatment, the loss of fluids can prove life-threatening, especially for young children who are not well nourished. Thus, of the estimated 140 million rotavirus infections per year in children under 5 years old, approximately 400,000 result in death, almost all in developing countries.

Whilst rotavirus infection can cause vomiting as well as diarrhea, usually at the onset of clinical disease, the noroviruses cause vomiting specifically. Noroviruses are commonly known for causing "winter vomiting disease." Outbreaks of infection occur particularly in relatively confined communities, such as on cruise ships. Again, the virus passes through the stomach and infects the epithelium of the small intestine, with damage to the villi and features of malabsorption from the gut. The vomiting is thought to be due to delayed emptying of the stomach, which results from impaired gastric motor function caused by the infection. Signs and

> **Box 18.1**
>
> **Evidence for a secretory component in rotaviral pathogenesis**
>
> - Acute diarrhea is seen at a time when damage to the intestinal villi is minimal, in both babies and experimentally infected animals.
> - Conversely, in experimental animals, recovery from diarrhea is seen at a time when overt damage to the small intestine is still apparent.
> - The rotavirus NSP4 protein (produced in infected cells but not included in virus particles) can cause diarrhea when introduced into the gut of young experimental animals.

symptoms begin within 24 hours after infection and continue for 1–2 days. Severe dehydration is less likely than for rotavirus infection because of the shorter duration of disease, but treatment is similar if required.

18.6 ACUTE VIRAL INFECTION 2: RESPIRATORY INFECTIONS

The classic specific signs of respiratory tract infections are coughing and sneezing. Both reflect the effects of infection on the production and movement of secretions in the airway.

A large number of different viruses cause respiratory tract infections and disease in humans (Section 18.1). The severity of respiratory signs and symptoms in these infections broadly relates to how deep into the respiratory tract the infection penetrates. Thus rhinoviruses normally infect only the nasal epithelium and result in mild upper respiratory tract disease, whilst respiratory syncytial (RS) virus more usually reaches the lung where it can cause bronchiolitis and/or pneumonia (Fig. 18.2b). The severity of nonspecific symptoms also varies considerably, with fever being not normally associated with rhinovirus infections but a significant feature in influenza A virus infections. Various aspects of influenza A virus infection have been considered in other chapters of this book.

Sneezing is a reflex response to irritation and build-up of fluid in the nasal passages. Rhinovirus infection of the nasal epithelium causes enhanced secretion of fluid, also narrowing of the passages, and these promote the sneezing reflex. Rhinovirus infection does not cause extensive cytopathology in the affected tissues. Instead, it is thought that the effects of infection on respiratory tract function are mediated by a host response, possibly involving IL-8 which is a key component of the inflammatory response and which is elevated in nasal secretions during rhinovirus infection.

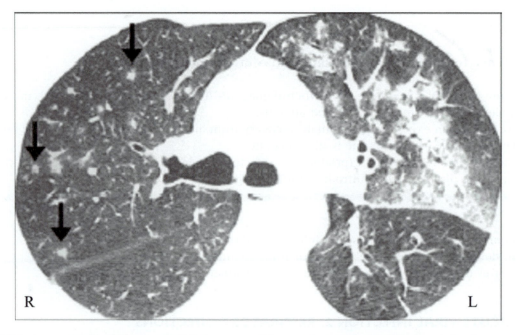

Fig. 18.4 An example of lung consolidation caused by respiratory syncytial virus infection in an immunosuppressed patient. The lungs have been imaged by high resolution CT scan. There are nodular lesions in the right lung (R, arrows) while the left lung (L) shows areas of consolidation (white) where no air is present. From Escuissato, D. L. *et al.* 2005. *American Journal of Roentgenology* **185**, 608–615. (Reprinted with permission from the *American Journal of Roentgenology*.)

Coughing results from a build-up of mucus when viruses infect lower in the respiratory tract. The respiratory epithelium contains ciliated cells that form a conveyor, known as the respiratory escalator, which is constantly moving the layer of mucus that coats and protects the cells upwards towards the nose (see Fig. 16.1). If the ciliated cells become damaged or the quality of the mucus changes making it too thick to move, then the mucus accumulates and has to be removed by coughing. RS virus initially infects the epithelium of the nasopharynx but, within 1–3 days, can move down to the lower respiratory tract where it causes coughing and wheezing. Respiratory distress is seen due to underlying bronchiolitis and/or pneumonia when the virus infects the lungs. The infection causes loss of epithelial cells in the bronchioles, hyperproliferation in compensation for this loss, excess mucus secretion, and a strong inflammatory response with cellular infiltration into the infected tissues. Together, these cause consolidation of portions of the lung with what can be a dramatic loss of air exchange function in those areas (Fig. 18.4). Patients suffering these severe symptoms of RS virus infection require hospitalization for supportive care. Treatment with an aerosol of the antiviral compound, ribavirin, is sometimes used in these cases (see Table 21.6).

18.7 ACUTE VIRAL INFECTION 3: INFECTIONS OF THE LIVER

The signs and symptoms of hepatitis, which include fatigue and jaundice, result from impairment of liver function, particularly bile production and blood detoxification.

Viruses from several different families infect the liver and can cause hepatitis, an inflammatory destruction of liver tissue that results in the classic appearance of jaundice (yellowing of the skin and eyes resulting from impaired disposal of hemoglobin breakdown products by the liver). Some also cause persistent or chronic infections, the long term consequences of which can be severe (see Sections 14.5 and 20.8).

Hepatitis A virus (HAV), a picornavirus, is transmitted by the fecal–oral route and causes acute infections of the liver that are normally cleared rather than becoming persistent. Hepatitis, when it occurs, arises a month or more after infection. The virus is a close relative of the enteroviruses that infect the gut, and HAV can cause gastrointestinal symptoms that generally occur before hepatitis is seen. However, it is not certain whether the virus actually replicates initially in the gut or whether it moves directly to the liver as the primary site of replication. There are close links between the gut and the liver that can facilitate virus movement between them. Nutrients are carried from the gut to the liver via the portal vein, supplying 80% of blood to the liver, while bile produced in the liver is delivered to the gut via the bile duct (Fig. 18.5). Once it has reached the liver, HAV replicates in hepatocytes and is shed into the bile and hence back into the gut, from which it is excreted for onward transmission.

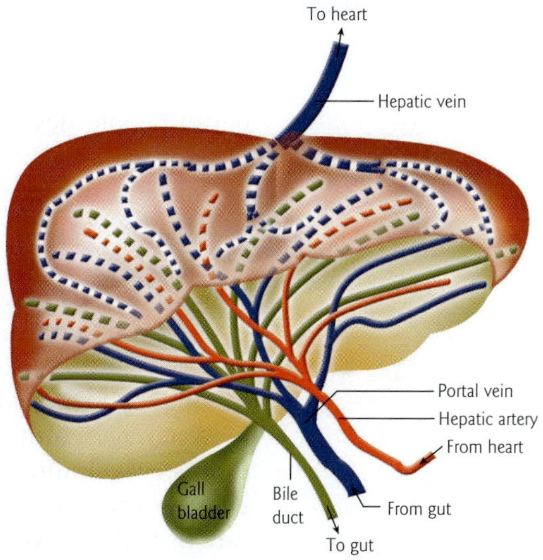

Fig. 18.5 A representation of the flows of blood and bile to and from the human liver. Oxygenated blood arrives via the hepatic artery while the bulk of blood flow into the liver comes from the gut via the portal vein, supplying nutrients for metabolism. Blood returns to the circulation via the hepatic vein. One of the products of the liver is bile, which is stored in the gall bladder and delivered to the gut via the bile duct.

In common with other viruses causing hepatitis, HAV replication does not appear to kill hepatocytes; instead, the damage to the liver that causes disease is mediated by the host immune response, particularly the cytotoxic T cell response (Box 18.2). In most adults, HAV infection is self-limiting and virus is eventually cleared with resulting immunity to reinfection. There is no specific therapy; patients are advised to avoid other

Box 18.2

Evidence for immune-mediated pathogenesis in hepatitis A virus infections

- The phase of infection in which virus is shed from the liver largely precedes the onset of symptoms of hepatitis (a manifestation of liver cell death).
- HAV infections in cell culture are largely noncytopathic.
- Cytotoxic T cells cloned from liver biopsies obtained from patients with active HAV disease can cause lysis of infected cells.

toxic insults to the liver (e.g. alcohol) to aid recovery. In a small number of people (less than 1%), the tissue destruction is extreme, resulting in liver failure which is fatal unless a transplant is available.

Only a proportion of HAV infections result in hepatitis. Interestingly, this is linked to the age at infection. Thus, in young children (most of whom actually become infected very early in life), fewer than 10% of HAV infections result in hepatitis whilst 70–80% of adults develop disease. This difference is thought to reflect the beneficial effect of maternal antibody in protecting against disease following HAV infection acquired in infancy. One consequence of this difference in disease incidence between age groups is that, in countries where HAV is endemic because of poor sanitation, the burden of HAV disease is paradoxically low because the average age at infection is low. HAV-related disease is also rare in highly developed countries because most people do not encounter the virus; in such populations, HAV can cause outbreaks of disease. In between these extremes, as attempts are made to provide clean water and sewage disposal in poorer countries, so the average age at infection rises and so does the incidence of HAV disease. There is an effective killed vaccine available for HAV, which is recommended for people traveling to endemic areas from places where infection is rare.

18.8 ACUTE VIRAL INFECTION 4: SYSTEMIC SPREAD

Some viruses cause little or no disease associated with replication at the primary site of infection, but spread around the body and cause disease elsewhere.

Some viruses are best known for the signs and symptoms that arise from their systemic spread around the body rather than for disease caused by infection at the point of entry. Two good examples are measles virus and varicella-zoster virus (VZV), which causes chicken pox on primary infection. Both these viruses cause skin rashes that can reach all extremities of the body, indicative of systemic spread, but both actually initiate infection in the respiratory tract (Fig. 18.2c). Measles virus can, in fact, cause significant symptoms of respiratory tract infection prior to the rash appearing, especially in immunocompromised children, but VZV does not. Other viruses can spread to specific secondary sites where disease consequences are manifest, e.g. poliovirus neuroinvasion from the gut to cause paralytic disease.

VZV and measles virus both replicate initially in cells of the respiratory epithelium. From there, they gain access to cells of the immune system, particularly monocytes. Since these circulate around the body, they provide the means by which virus can spread away from the respiratory tract. Both viruses then replicate further in lymphoid tissues before spreading to the skin and, for

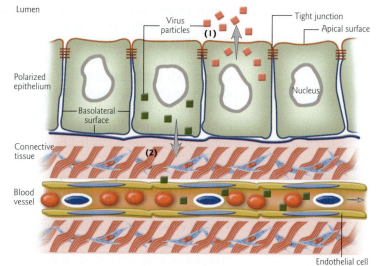

Fig. 18.6 A simplified diagram of a polarized epithelium showing virus exiting an infected epithelial cell either (1) via its apical surface to reach the lumen of the organ, or (2) via the basolateral surface to reach the vascular system. The apical and basolateral surfaces of the epithelial cells are separated by tight junctions that connect the cytoplasm of adjacent cells and prevent movement of membrane proteins between the two surfaces. Not to scale. See Section 18.8 for further details.

measles virus, many other organs. In the skin, VZV replicates in epithelial cells, causing the fluid-filled vesicular lesions which contain infectious virus that are characteristic of chicken pox. VZV then spreads on into sensory neurons where it becomes latent. Measles virus replicates in endothelial cells before spreading to the overlying epidermis. The measles rash, which is a smooth discoloration of the skin, coincides with the onset of an antibody response. It reflects immune complex formation in the skin, and fades as virus is cleared. In the same way, Koplik's spots, a diagnostic feature of measles virus infection, form on mucosal surfaces of the mouth. Thus, while measles virus and VZV both spread systemically from a respiratory infection and cause skin rashes, these rashes are completely different; the VZV lesions contain infectious virus while the measles rash does not.

What determines whether or not a virus can spread away from its site of initial infection? As just discussed, being able to infect cells that move around the body is one factor. Another is exactly how the virus exits the cells it originally infects. Many viruses initially infect epithelial cells because these are the cells that are first encountered at each of the principal routes of entry of virus into the body. Internal epithelia are polarized, which means that the cell sheet has directionality with the two surfaces being functionally distinct. The apical surface faces the lumen of, for example, the gut or airway while the basolateral surface faces into the body, interfacing with connective tissue, the blood, and lymphatic systems. Many viruses exit epithelial cells preferentially to one or other surface (Fig. 18.6). Clearly, a virus that exits apically, as is normally the case with influenza A virus for example, has less opportunity for systemic spread than one exiting basolaterally. Viruses that are released basolaterally into the circulation generate what is known as a viremia.

18.9 ACUTE VIRAL DISEASE: CONCLUSIONS

Viruses establish infection initially at sites in the body that are determined by characteristics of the virus and its route of entry. Disease signs and symptoms may arise from this infection and/or from the results of subsequent virus spread to other parts of the body. Wherever in the body specific indicators of virus infection occur, it is reasonable to infer that the virus is present in that tissue. However, it is not always the case that these signs and symptoms are caused by the virus damaging cells of the tissue. Often, the infection is not itself grossly cytolytic but instead, the host immune response to infection leads to tissue damage and consequent dysfunction and overt symptoms.

KEY POINTS

- Many different viruses are significant human pathogens.
- All parts of the body can be affected by viral pathogens.
- Similar viruses can cause different diseases and similar diseases can be caused by very different viruses.
- The nature and severity of signs and symptoms caused by a virus can vary greatly between infected individuals.
- Whilst everyone will suffer from many virus infections during their lifetime, the chance of being infected by specific viruses depends on your genetics, which part of the world you live in, and the nature of your community.
- The nature of the disease caused by a virus is determined by a series of factors, including the distribution of its cell receptor, its portal of entry into the body, and whether or not it has the ability to spread systemically.
- Viruses can achieve systemic spread either by infecting circulating cells, or by achieving release of particles into the blood from a site of infection.
- Symptoms of acute viral infection are frequently due to the host immune response to infection rather than to a cytopathic effect of the virus.
- Specific symptoms of acute viral infection result from transient dysfunction in the affected organs or systems of the body.

QUESTIONS

- Discuss the factors that affect the incidence of viral disease in developed versus developing countries.
- Discuss the factors that influence the nature and severity of disease during an acute virus infection.

FURTHER READING

Collier, L., Oxford, J. 2006. *Human Virology*, 3rd edn. Oxford University Press, Oxford.

Mims, C. A., Nash, A., Stephen, J. 2000. *Mims' Pathogenesis of Infectious Disease*, 5th edn. Academic Press, London.

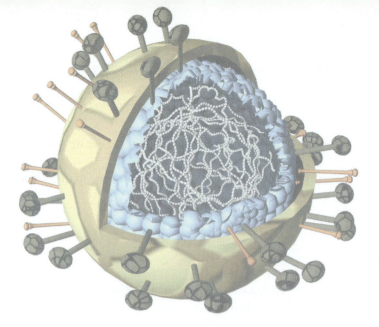

19

HIV and AIDS

HIV is a relatively new, sexually transmitted, human virus. There is treatment that suppresses but does not eliminate the virus, but this is too expensive for 90% of the world's population. In the absence of treatment, and normally after many years of infection, the virus damages the immune system to the point where the body cannot combat common infections and AIDS (acquired immune deficiency syndrome) results.

During the past 35 years human immunodeficiency virus (HIV) type 1 has become a common infection of mankind. There is also a second, less common, less virulent, but closely related virus, HIV-2. Both infect mainly helper T lymphocytes and macrophages (cells of the immune system, Chapter 12) using the cell surface CD4 protein as the primary receptor. If there is no treatment, there is a long, essentially symptomless, incubation period of, on average 10 years. By that time CD4$^+$ lymphocytes have declined to such a low level that the immune system can no longer function efficiently, resulting in the immunodeficiency which gives the virus its name. The consequence of this is that the affected person can no longer restrain certain normally harmless passenger microorganisms (viruses, bacteria, or fungi) which then cause overt clinical disease. A collection of diseases such as this, unrelated except for a common underlying cause, is called a syndrome – hence the name "acquired immune deficiency syndrome" or AIDS. HIV infection is almost always lethal and this is typical of viruses that infect a new host species (Chapter 18).

Immense progress has been made in understanding HIV through intensive research over the past 20 years although, at the time of writing (June 2006), no curative treatment or effective vaccine has yet been achieved. However, HIV research is a rapidly moving field, so contemporary literature should be consulted to update the knowledge detailed here.

19.1 THE BIOLOGY OF HIV INFECTION

Milestones for AIDS and HIV

1970s – Silent pandemic
1981 – AIDS is recognized for the first time
1983 – AIDS is linked to infection with a new virus
1985 – Initial characterization of HIV-1
1995 – Triple therapy with inhibitors of reverse transcriptase and viral protease
2006 – Still no vaccine or universally affordable treatment available

The discovery of HIV

Historically, AIDS diseases were recognized before the virus that was responsible for them. In 1981, astute clinicians noted an unusual pneumonia caused by *Pneumocystis carinii*, a yeast-like organism, and identified a rare cancer (Kaposi's sarcoma) in homosexual young men in New York and California. Some of those with the sarcoma also had pneumocystis pneumonia. These conditions were then linked to immune deficiency. Later an association with a variety of common infections, leading to death, was recognized, and AIDS was defined. (Kaposi's sarcoma is actually caused by a ubiquitous human herpes virus (HHV type 8) under conditions of immunodeficiency; Section 20.5). Various causes were considered but it was not until 1984 that it was thought that a virus was involved. Then in 1985 a virus new to science, HIV-1 (then known as human T cell lymphotropic virus type 3 (HTLV-3) or lymphocyte-associated virus (LAV)), was isolated. Once viral antigens were available as diagnostic reagents, stored sera could be tested retrospectively for the presence of antibody to HIV and few HIV infections were found in humans prior to 1970. However, since we now know that the incubation period in developed countries averages 10 years (shorter in the developing world), it is clear that HIV had spread in a silent but explosive pandemic throughout that decade.

The spread of HIV

The HIV/AIDS scenario is tragically grim. The United Nations Programme on AIDS (UNAIDS) estimates that nearly 40 million people are infected

Table 19.1

Estimates by the UNAIDS of the number of HIV-1-infected adults and children at the end of 2004.

Developed-world countries	4.3 million
South and southeast Asia	7.1 million
Sub-Saharan Africa	25.4 million
Worldwide	39.4 million

with HIV worldwide, of which 45% are women. More than 95% of all new infections arise in the developing world (Table 19.1). During 2004, 4.9 million people became infected, equating to 13,000 new infections every day or nine new infections every minute, and 3.1 million died of AIDS. The epicentre for HIV is sub-Saharan Africa with 3.1 million new infections in 2004, with the highest incidence in Botswana which is estimated to have 50% of its population infected. The predicted expansion of the pandemic into south and southeast Asia (approximately 1 million new infections in 2004), east Asia including China (290,000), and eastern Europe and central Asia (210,000) has sadly been fulfilled. There is still no vaccine and no treatment for the immunodeficiency, but a combination of two nucleoside analogs and a protease inhibitor (triple therapy; Section 19.8) is remarkably successful in halting virus replication and the progression to AIDS. This is also known as HAART (highly active anti-retrovirus therapy), and has been in use since 1994. However, although virus in plasma (blood from which cells have been removed) is undetectable under successful triple therapy, the treatment does not clear virus completely. There is latent virus present in resting $CD4^+$ memory T cells and other cell reservoirs. Treatment must be continued without a break for the foreseeable future, or the virus rebounds within a few days to its original level, and is often then drug-resistant. HAART is expensive and this has restricted it largely to developed countries. Overall less than 12% of HIV-infected people receive HAART.

Where did it all start?

The history of HIV is surprisingly sketchy, despite much effort to trace its origins. The earliest HIV-1 infections can be traced by serology to people in central parts of sub-Saharan Africa. Either it is a long-established but rare infection in these relatively isolated populations or a new infection by a virus which has jumped from another species. The latter fits with the generalization that only new infections, not yet in balance with their hosts, cause serious disease. Also, the closest relative of HIV-1 so far identified is a simian immunodeficiency virus from chimpanzees (SIV_{cpz})

(Fig. 19.1). There may have been several separate introductions into man. Chimpanzees are a food source and young animals are kept as pets, so infection may have been contracted through cuts and abrasions during butchery or by close contact. In any event, the factors which caused the virus to spread, firstly into urban areas in Africa (the earliest reported human virus is 1959), and from there to the USA and the rest of the world, are not known. There are many gaps in the story. Similarly, HIV-2 is most closely related to SIV$_{sm}$ and is endemic in West Africa (Fig. 19.1). There have been at least six independent introductions into man, again this may have been from sooty mangabeys used for food, or kept as pets.

What sort of virus is HIV-1?

HIV-1 is a typical member of the lentivirus genus of the retrovirus family (Baltimore class 6). The name is derived from the Latin *lente*, which refers to the *slow* onset of the disease. However, there is nothing slow about the rate of virus multiplication. There are several well-characterized lentiviruses (Table 19.2) that infect a number of different vertebrate species. They have a number of common features (Box 19.1). HIV-1 is globally dispersed, causes over 99% of HIV infections, and is usually fatal. The serologically distinct HIV-2 is much less pathogenic and is largely restricted to West Africa.

In fact there is not one HIV-1, but a collection of different evolutionarily linked groups or *clades*. The classification of clades (A, B, C, D, F, G, H, J, and K) is based solely on sequence. A to D are the

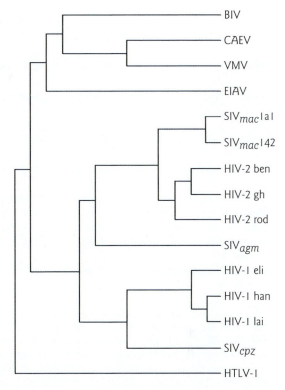

Fig. 19.1 Dendrogram showing the relatedness of some major lentiviruses. BIV, bovine immunodeficiency virus; CAEV, caprine arthritis-encephalitis virus; VMV, visna-maedi virus; EIAV, equine infectious anemia virus; SIV, various strains of simian immunodeficiency virus (SIV$_{cpz}$ was isolated from a chimpanzee); HIV, various strains of human immunodeficiency virus type 1 and 2; HTLV-1, human T cell lymphotropic virus type 1. (Courtesy of Georg Weiller.)

major clades; clades are defined as having an Env protein that differs in sequence from other viruses by at least 20%, and a Gag protein that differs by 15%. Clades differ enormously in their geographical distribution (Fig. 19.2), and the picture is complicated by their genetic interactions that result in various recombinants, called *circulating recombinant forms* (CRFs). Most HIV-1 in the developed world is clade B, virus in Africa is clade A, C, D or a CRF, and in India is clade C. Compare this distribution with the frequency of the virus in Table 19.1; as clades also differ in antigenicity of the envelope gp120–gp41 protein each area will probably need a different vaccine. Racial and ethnic differences in MHC haplotypes which affect the nature of peptides presented to T cells adds to this problem.

Table 19.2

Some typical members of the lentivirus subfamily, their hosts, and diseases.

Virus	Host	Main target cell	Clinical outcome
Visna-maedi	Sheep	Macrophage	Pneumonia (= maedi) or chronic demyelinating paralysis (= visna); little immunosuppression*
Visna-maedi	Goats	Mammary macrophage	Arthritis; rarely encephalitis; little immunosuppression
Equine infectious anemia	Horses	Macrophage	Recurrent fever; anemia; weight loss, little immunosuppression
Bovine immunodeficiency	Cattle	Not known	Weakness; poor health
Feline immunodeficiency	Cats	CD4$^+$ T cell	AIDS
Simian immunodeficiency†	Monkeys	CD4$^+$ T cell	Subclinical or AIDS
Human immunodeficiency type 1	Humans	CD4$^+$ T cell	AIDS
Human immunodeficiency type 2	Humans	CD4$^+$ T cell	AIDS

*The same virus can cause two different diseases in sheep.
†Different SIV strains are named after the species of monkey from which they were first isolated, e.g. SIV$_{man}$ from the mandrill, SIV$_{agm}$ from the African green monkey, SIV$_{sm}$ from the sooty mangabey. These are all Old World African monkeys, and SIV strains cause no disease in their natural host. However, although it does not naturally infect Asian macaque monkeys, SIV$_{sm}$ does do so under experimental conditions and causes AIDS. SIV$_{mac}$ is thought to be SIV$_{sm}$ which accidentally infected a laboratory macaque. (This is a further example of a virus interaction with its natural host being relatively benign compared to that with a new host.)

Box 19.1

Common features of lentivirus infections

- Infection of bone marrow-derived cells.
- Integration of DNA copy of the viral genome into host DNA.
- Persistent viremia.
- Life-long infection.
- Prolonged subclinical infection.
- Weak neutralizing antibody responses.
- Continuous virus mutation and antigenic drift.
- Neuropathology.

B	CRF02_AG, other recombinants	A	D	CRF01_AE,B	Insufficient data
B, BF recombinant	F, G, H, J, K, CRF01 other recombinants	C	A, B, AB recombinant	B, C, BC recombinant	

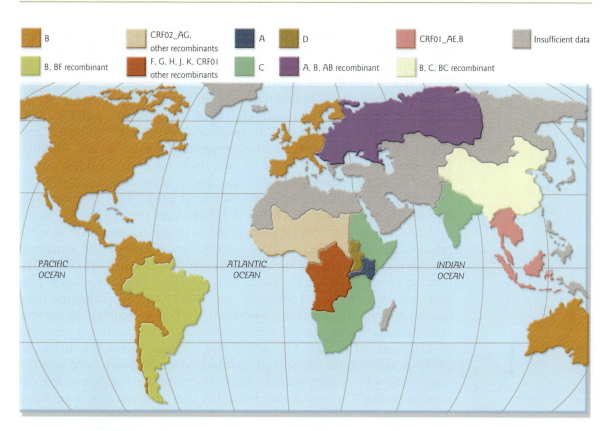

Fig. 19.2 Global distribution of HIV-1 clades, recombinants, and circulating recombinant forms (CRF). The 10 different epidemic patterns are indicated by different colours. HIV-2 accounts for only a minority of infections, mainly in West Africa but now also in India. (From F. E. McCutcheon and IAVI, 2003. http://www.iavi.org/iavireport/0803/clades-vax.htm)

19.2 MOLECULAR BIOLOGY OF HIV-1

Gene expression

Like all retroviruses, HIV-1 is diploid and contains two identical molecules of positive sense ssRNA. Its genome conforms to the basic oncovirus plan with the standard gene order 5'–*gag*–*pol*–*env*–3', described in Fig. 9.8, but is complicated by a number of additional regulatory genes (Fig. 19.3). One dsDNA provirus molecule with flanking long terminal repeats (LTRs) is produced from the RNA template using the virion reverse transcriptase, and the DNA is integrated into the host genome (see Chapter 8). Reverse transcriptase was the first target for antiviral chemotherapy against HIV-1. Simpler oncornaviruses synthesize only two RNAs, the viral genome, which doubles as mRNA for the Gag (**g**roup **a**nti**g**en) and Gag-Pol proteins (see Section 9.9), and an mRNA with a single splice, which encodes the envelope protein. By contrast, HIV-1 imposes upon this basic plan a number of further mRNAs and proteins (Tat, Rev, Vif, Vpr, Vpu, and Nef) with positive or negative regulatory activity (Figs 19.3 and 19.4). In all, HIV-1 encodes 16 proteins.

Initially after infection, cell transcription factors result in the synthesis of a small amount of full-length viral RNA from integrated HIV-1 DNA, which is then multiply spliced to form small mRNAs encoding Tat and Rev. Tat and Rev are RNA-binding proteins. Tat binds to the so-called TAR element, a sequence in the nascent RNA that is encoded just downstream from the transcription start site, and upregulates transcription by increasing processivity of RNA polymerase II (Box 19.2). This increases Rev to a certain critical level that downregulates the cytoplasmic accumulation of multiply spliced RNAs and favors the accumulation of nonspliced and several singly spliced mRNA, which encode the structural proteins Gag and Pol, and Env, and Vif, Vpu and Vpr respectively (Fig. 19.4). Rev achieves this by binding to an RNA sequence (the RRE) that is present only in nonspliced and singly spliced mRNAs and facilitating their transport from the nucleus (Box 19.3). Nef is a complex pleiotropic (multifunctional) regulator. It down-modulates cell surface expression of CD4 and MHC class I proteins, and increases virus infectivity. In addition, it enhances the spread of infection in the body from infected macrophages by causing them to firstly chemoattract T cells, and then to activate resting T cells with the result that they are then able to replicate

Box 19.2

Evidence for the mechanism of action of Tat protein

- Unlike other DNA sequences that upregulate transcription by binding transcription factors, the TAR element has to be positioned downstream of the transcription start site and oriented as found in HIV in order to act. This suggests that the TAR element might be working as RNA.
- Purified Tat protein binds the RNA form of TAR, not the DNA form, confirming that Tat works through RNA.
- In nuclear run-on experiments (where ongoing transcription is labeled through a brief incorporation of radioactive precursors in isolated nuclei in the test tube), Tat does not increase the rate of synthesis of RNA from DNA close to the HIV promoter (the LTR) but does massively increase the rate of synthesis of RNA from distal parts of the HIV genome. The first result indicates no change occurs in the rate at which new RNA polymerase II molecules begin transcription at the promoter when Tat is present, while the second result indicates that the processivity of the enzyme (how efficiently it holds onto its template and synthesizes a long RNA) is increased.
- Tat–TAR complex binds a host cell protein kinase that phosphorylates RNA polymerase II, so increasing its processivity.

Box 19.3

Evidence for the mechanism of action of the Rev protein

- Comparing cytoplasmic mRNA from wild-type and Rev-defective mutant HIV infected cells, the full-length and singly spliced mRNA that encode Gag, Gag-Pol, and Env are selectively depressed by the absence of Rev while multiply spliced mRNAs are not affected (Fig. 19.4). Since all the mRNAs come from the same promoter, this Rev effect must be achieved post-transcriptionally.
- Rev has high affinity for a specific RNA sequence in the Env reading frame of HIV mRNA. This sequence can confer Rev-dependence for cytoplasmic accumulation on a heterologous mRNA produced from a suitable reporter gene construct. Thus Rev binding to RNA is required for it to enhance cytoplasmic mRNA accumulation.
- Rev is located in the nucleus as assessed by immunofluorescence. If two cells, one expressing Rev and the other not, are fused together to form a heterokaryon (one cell with two nuclei), and further protein synthesis is blocked with a chemical inhibitor, then Rev can be seen to accumulate in the second nucleus. Thus Rev can move from one nucleus to the other via the cytoplasm, reflecting an ability to move reversibly between nucleus and cytoplasm in a normal cell, a process known as shuttling.
- The sequence in Rev that is required for shuttling, the nuclear export signal (NES), is essential for Rev-dependent mRNA accumulation. It binds to components of a host cell nuclear export pathway.

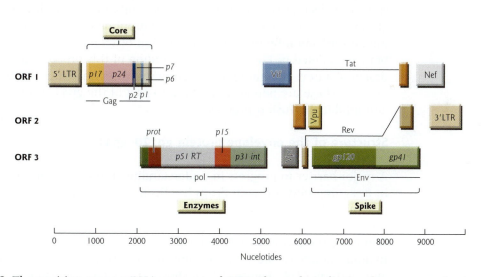

Fig. 19.3 The positive sense ssRNA genome of HIV. The scale indicates the genome size in nucleotides. Note that all three open reading frames (ORFs) are used, but only simultaneously in the region of nucleotide 8500. *Gag, pol,* and *env* are transcribed and translated to give polyproteins which are then cleaved by proteases to form the virion core proteins, enzymes, and spike proteins as shown. The nomenclature "p17" indicates a polypeptide having an M_r of 17,000, etc. Genes not aligned require a frame shift for expression. Tat and Rev proteins are expressed from spliced RNAs.

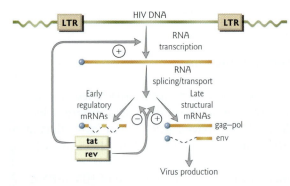

Fig. 19.4 Regulation of HIV-1 gene expression from provirus DNA integrated into the cellular genome. This occurs in two phases: early expression of regulatory genes and late expression of structural genes. (From Cullen, B. R. 1991. *Journal of Virology* **65**, 1053.)

the virus. Vpr and Vpu are concerned in morphogenesis of virions and their release from the cell. Vif is associated with virus infectivity. Tat and Rev are essential for replication *in vitro*, while the others are not and are consequently called accessory proteins. However they are presumably needed for the success of the HIV *in vivo*. Much more is still to be learned about all of them.

Virion proteins

Production of structural proteins from the *gag*, *pol*, and *env* genes and their location in the virion is shown in Figs 19.3, 19.4, and 19.5a. In brief *gag* encodes the virion core proteins, *pol* encodes virion enzymes, and *env* encodes the virion envelope protein. All are derived from polyprotein precursors. Polyproteins from *gag* and *pol* are cleaved by the viral aspartyl protease (Pro), itself a product of *pol*, and that from *env* by a cellular protease. Hence the importance of aspartyl protease inhibitors in the chemotherapy of HIV. The distal part of the envelope or spike protein, glycoprotein 120 (gp120), comprises five variable (V) loops that tend to vary in sequence. An outline of a monomer is shown in Fig. 19.5b, but the spike is actually a trimer. There are approximately 72 spikes per virion (far sparser than influenza virus which has ten times more spikes on a similar surface area). gp41 is the transmembrane anchoring part of the glycoprotein and has an N-terminal hydrophobic fusion sequence that initiates infection by promoting the fusion of viral and cell plasma membranes (detailed in Fig. 5.3). Understanding of this process has led to the design of a new generation of experimental antiviral drugs (T-20 or enfuvirtide). These are peptides that bind to part of the gp41 fusion sequence and inhibit the fusion process.

Structure of the envelope protein gp120–gp41

A major barrier to progress in understanding antibody neutralization and in designing new antiviral drugs has been the lack of structural information on gp120. For over 10 years it resisted all attempts at crystallization, probably because the molecule was too flexible. This was solved by removing the more flexible parts (loops V1, V2, and V3, and the carbohydrate moieties) and immobilizing the structure by binding it to a FAb and part of CD4. Figure 19.6a shows a skeleton version of a gp120 monomer. This forms a U-shaped hairpin with an inner and an outer domain, and the CD4-binding site, the coreceptor binding sites (Section 19.3) and a bridging sheet at the "U." The CD4-binding site is a conformational region that allows the virus to attach to its target cell. The fleshed-

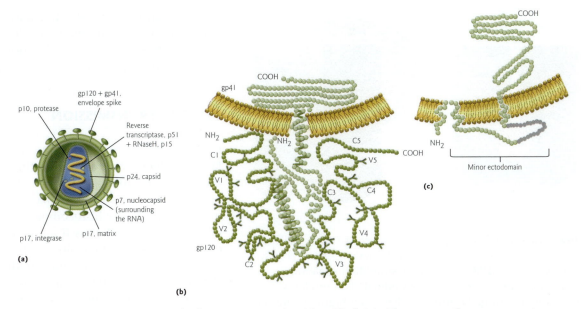

Fig. 19.5 An HIV-1 virion with the major proteins identified (a). There are others present (see text). (b) A monomer of the trimeric envelope spike protein adapted from the structures suggested by Leonard *et al.* 1990. *Journal of Biological Chemistry* **265**, 10373) and Gallagher (1987, *Cell* **50**, 327). Each circle represents an amino acid residue. The gp160 has been cleaved to form gp41 (open circles), the COOH-terminal transmembrane anchor, and the distal gp120 (solid circles). gp120 contains five antigenically variable (V) regions and five intervening conserved (C) regions; it has several intramolecular disulfide bonds, and is heavily glycosylated. gp41 is also glycosylated (not shown here). (c) An alternative scheme for the C-terminal transmembrane and tail regions of gp41. This has three transmembrane (tm) domains and a 100-residue intravirion tail; the N-terminal tm domains are formed by the same residues that comprise the single tm region in (b); tm 2 and tm 3 support 41 residues, the minor ectodomain, which expresses several epitopes (gray), including one that is neutralizing (adapted from Hollier, M. J., Dimmock, N. J. 2005. *Virology* **337**, 284). The major N-terminal gp41 ectodomain and gp120 are not shown.

out version (Fig. 19.6b) shows how the virus evades antibody. Loops V1–V3 lie on top of, and partly obscure, the two receptor binding sites and the bridging sheet, while most of the inner face is obscured by juxtaposition with the other two monomers. Much of the outer face is obscured by carbohydrate. In all, very little of gp120 and gp41 are exposed to B cells so that antibodies are not made to neutralization epitopes, and the few antibodies that are made have a very small target to aim at. The V loops that are exposed to antibody are able to sustain amino acid changes that enable virus to evade antibodies that are made against them. The structure of the main external region of gp41 (the ectodomain) has also been solved. The hydrophobic N-terminus of gp41 is released after gp120 interacts with cell receptors and inserts into the cell membrane as the fusion entry process proceeds (see Fig. 5.3). HIV-1 can also be neutralized

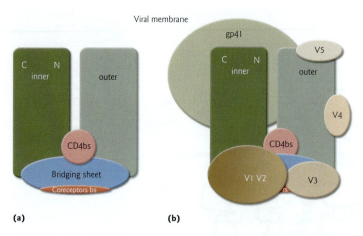

(a)   **(b)**

Fig. 19.6 Model of a gp120 monomer of HIV-1 based on the atomic structure determined by Kwong *et al.* (1998, *Nature, London* **398**, 648). (a) A simplified skeleton structure, and (b) with V loops added. The model is orientated with gp41 and the virus membrane at the top of the page. The monomers meet in a trimer and obscure the inner face. Much of the outer face is obscured with carbohydrate. Only a small proportion of the gp120 surface has epitopes that interact with neutralizing antibody (see text), although there are also neutralizing epitopes on gp41. There is also an atomic structure of an isolated gp41 ectodomain, but it is not known how it interacts with gp120. (From Burton, B. R., Parren, P. W. H. I. 2000. *Nature Medicine* **6**, 123.)

through epitopes on gp41. These include epitopes on the main ectodomain and the newly described minor ectodomain (Fig. 19.5c).

19.3 HIV TRANSMISSION

CD4 is a protein present on the surface of helper T cells and cells of the monocyte–macrophage lineage. Its normal function is to recognize and interact with major histocompatibility complex (MHC) class II antigens on target cells. It is the main (but not sole) receptor through which both HIV-1 and HIV-2 initiate infection. Successful HIV infection also requires the presence of the chemokine receptors, CCR5 or CXCR4, which have been hijacked by the virus as coreceptors. These also bind to specific regions on gp120 (Fig. 19.6). People can contract an HIV infection in three different ways.

Sexual transmission

The main route of infection is through the transmission of infected CD4[+] T lymphocytes or free virus during sexual activity. HIV is spread equally well by heterosexual and by male homosexual activity, but is better spread by infected males than infected females. In Africa, HIV is spread predominantly by heterosexual contact. In Western countries, male homosexual contact has been responsible for the majority of infections, but now the greatest percentage increase in new infections is due to heterosexual activity. Bisexual activity provides a conduit between hetero- and homosexual people.

Transmission through infected blood

Injection using hypodermic needles contaminated with HIV can cause infection. The virus spreads quickly between injecting drug abusers who share unsterile equipment, and there is a similar risk with clinical practitioners who do not have effective sterilization available for hypodermic syringes and needles. Before the virus was recognized, some hemophiliacs were accidentally infected by being given clotting factor VIII prepared from blood contaminated with HIV. Heat sterilization and screening of blood donors have now eliminated this risk.

Vertical transmission

Babies born to HIV-positive mothers may be infected. Fortunately, there is only about a 20% chance of infection, but at this age AIDS develops with a much shorter incubation period of around 2 years. The risk is reduced by giving the mother antiviral therapy (Section 19.8). The exact route of transmission is not known but breastfeeding increases the risk of infection, and is therefore not recommended.

19.4 COURSE OF HIV INFECTION AND DISEASE

This section relates to people from the developed world. Although infection in developing countries is essentially similar, the disease course may be hastened by other factors such as nutritional status and burden of infection. However data are less complete. The course of infection in Box 19.4

Box 19.4

Classification of stages of progression of HIV infection in an adult to AIDS, where antiviral chemotherapy has not been used

Initially HIV-1 causes a short self-limiting flu-like illness that starts a few weeks after infection and lasts only a few days; there is then a period of some years during which there are high levels of multiplication and infectiousness. Eventually immunosuppression commences:

- **Category A:** ranges from primary HIV infection to asymptomatic subclinical infection, to persistent generalized lymphadenopathy (swollen lymph nodes), which indicates that immune suppression has started and is progressing. Without treatment this stage lasts on average 10 years. Formally defined as having <500 CD4$^+$ T cells/µl or /mm^3 of blood (normally 800–1200 cells/µl). Infectious.
- **Category B:** symptomatic people, presenting with a selection of conditions not found in category C. These include weight loss, opportunistic infections (e.g. candidiasis, fever, diarrhea lasting more than 1 month, and more than one episode of shingles). These conditions suggest a defect in cell-mediated immunity. Most category B people have 200–500 CD4$^+$ T cells/µl blood. Infectious.
- **Category C:** AIDS and includes people with more severe infections or cancers. Ranges from *mild* with constant infections to *severe* with severe infections, major weight loss, myopathy, peripheral nervous system disease, and central nervous system disease (dementia). Most category C people have <200 CD4$^+$ T cells/µl blood. Infectious.

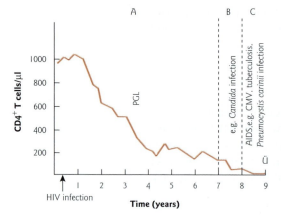

Fig. 19.7 A case study of AIDS, which ended in death in year 9. The CD4+ T cell loss was exceptionally rapid. This case occurred before anti-HIV therapy was available. PGL, persistent generalized lymphadenopathy; CMV, cytomegalovirus. A, B, and C are defined in Box 19.4.

refers to HIV antibody-positive people aged 13 or more years (as opposed to babies where death ensues in around 2 years). Primary infection is a brief influenza-like illness and proceeds from category A to full-blown AIDS (category C). An average incubation period to AIDS of 10 years means that some will develop disease earlier and some will remain healthy for much longer. The stages are formally defined by the number of CD4+ T cells in the circulation (Box 19.4). If an AIDS patient does not die from opportunistic infection with adventitious microorganisms, HIV progresses to infect CD4– T cells and causes disease in the muscles and the peripheral and central nervous systems. "AIDS dementia" or collapse of brain function is the final stage. The time taken to progress to AIDS, and the symptoms involved, vary greatly between individuals. An example of one case study and the progressive decline in CD4+ T cells is shown in Fig. 19.7. With successful antiviral chemotherapy a person remains subclinically infected with HIV, with low-to-undetectable levels of virus. The virus is never eliminated but the person may not be infectious.

What cells are infected?

HIV-1 isolates are classified as M-tropic or T-tropic. M-tropic viruses infect both CD4+ **m**acrophages and T cells, and use the CCR5 coreceptor. T-tropic viruses preferentially infect CD4+ **T** cells and use the CXCR4 coreceptor. These properties depend on the gp120 sequence. Most infections take place through the mucosal surface of the genital tract. Virus binds to the surface of a dendritic cell through the surface protein, DC-SIGN, a C-type lectin (a protein that specifically binds carbohydrate moieties). HIV may be taken up into an intracellular vesicle, but does not enter the cytoplasm or cause an infection. The dendritic cell then acts as a transporter and migrates carrying the virus to lymph nodes, where the virus is transferred to, and infects, a CD4+ T cell. CD4+ macrophages can also be infected but these are found infrequently in lymph nodes and blood. Although CD4+ T cells are the main target, other cells that do not express the CD4 protein may also be infected, which suggests that here the virus can use a different receptor molecule. In the central nervous system, it is thought that HIV infects microglial cells, which belong to the same cell lineage as the macrophage. Infection via the mucosal surface of the rectum, resulting from anal intercourse, may also take place through dendritic cells. Unlike the simple retroviruses, HIV can infect resting cells, in this case resting CD4+ T cells, as its proviral DNA is transported into the nucleus through associated viral proteins that carry nuclear localization signals. There it is integrated, but little transcription takes place until the cell is activated by antigen.

During the initial/primary infection by HIV, virus-specific CD4$^+$ T cells are stimulated to proliferate by viral antigens. As virus infects and replicates in activated CD4$^+$ cells, these same cells are preferentially destroyed. At the same time there is a dramatic expansion of virus-specific CD8$^+$ T cells which coincides with the suppression of viremia. However CD8$^+$ T cell proliferation is dependent on CD4 T cell help. Thus there is a fine balance between virus destroying CD4$^+$ T cells and leaving enough CD4$^+$ T cells to help produce virus-specific activated CD8$^+$ cells. It is suggested that this balance determines the plasma virus load (concentration) at the end of the acute phase, and it is this virus load that determines the rate of progression to AIDS (a low virus load means a longer subclinical period, and a high load means rapid progression to AIDS). Antiviral drug therapy (Section 19.8) during primary infection may help preserve HIV-specific CD4$^+$ T cell function, reduces the ensuing plasma virus load, and prevents the onset of AIDS.

During the intermediate stages of infection, there is relatively little virus and few infected cells in the plasma, and initially this led to the mistaken view that there was little virus replication at this time. Later it was found that most of the infected cells were in lymphoid tissue and large amounts of virus were being continuously released into the circulation. However this was balanced by its efficient removal. It was found that a healthy HIV-infected person produces the colossal total of approximately 10^{10} virions per day. In addition there is a small, but vitally important, reservoir of latently infected resting CD4$^+$ memory T cells (10^3–10^4/person), and possibly other reservoirs, that cannot be shifted by drug therapy (Section 19.8). Latently infected cells that produce no viral proteins on the cell surface are invisible to the immune system and, because the viral genome is integrated, can divide and produce daughter cells that are also infected.

19.5 DEATH AND AIDS

The key failure in immune responsiveness is the loss of the helper function of the CD4$^+$ lymphocyte. This cell is central to immune function, as most antibody responses are helper T cell-dependent and helper cells also assist in the maturation of T cell effectors, such as CD8$^+$ cytotoxic T cells, and can have cytotoxic activity in their own right. One of the functions of T cells is to control a variety of microorganisms (viruses, bacteria, fungi, protozoa; Table 19.3), a selection of which we all carry. The immune system is normally unable to evict these microorganisms but ensures they remain subclinical and harmless. However, when the T cells are compromised by HIV infection, microorganisms can multiply unchecked, and infections are more severe and prolonged than normal. People in different parts of the world carry different microorganisms, so AIDS patients suffer different diseases. For example, *Mycobacterium tuberculosis* is commonly

Table 19.3

Common infections associated with AIDS.

Viral	Cytomegalovirus, Epstein–Barr virus	Generalized infection
Bacterial	*Mycobacterium tuberculosis*	Tuberculosis
Fungal	*Candida albicans*	Thrush
	Pneumocystis carinii (a yeast-like organism)	Pneumonia

carried in some developing countries. The period of chronic adventitious infection may last for several years but eventually becomes overwhelming and death ensues.

19.6 IMMUNOLOGICAL ABNORMALITIES

While loss of CD4$^+$ T cells is the major result of HIV infection, there are many other alterations to immune cells and functions, some of which are listed in Box 19.5. Those in the first category are always found in AIDS patients and are therefore diagnostic. Others are found irregularly.

Box 19.5

Some immunological abnormalities associated with AIDS*

Diagnostic abnormalities
- CD4$^+$ T cell deficiency.
- Reduction in levels of all lymphocytes.
- Lowered cutaneous DTH.
- Nonspecific elevation of immunoglobulin concentration in serum.

Other abnormalities
- Decreased proliferative responses to antigen.
- Decreased cytotoxic responses to all antigens.
- Decreased response to new immunogens.
- Decreased CD8$^+$ T cell cytotoxic response to HIV.
- Decreased macrophage functions.
- Production of autoantibody.
- Decreased NK cell activity.
- Decreased dendritic cell number and activity.
- Loss of lymph node structure.

*The effects of HIV on the immune system extend far beyond the CD4$^+$ T cell and vary greatly between individual patients. DTH, delayed-type hypersensitivity, a T-cell-mediated reaction.

What causes the immunological abnormalities?

Immunological abnormalities result from one or more of the following causes:

- The virus kills the infected cell, and in so doing removes cells that are important for immune function.
- Envelope protein expressed on the surface of an infected cell attaches to CD4 molecules on noninfected CD4$^+$ T cells and causes the cell membranes and cells to fuse and form a single cell (a syncytium). This is a lethal process and fused cells die.
- Infected cells suffer no direct cytopathology but viral proteins expressed on the cell surface are recognized by antibody, and the cells are then attacked by complement or phagocytic cells bearing Fc receptors (Chapter 12).
- Infected cells proteolytically process viral proteins and present peptide–MHC I or II protein complexes on their surface. These cells are then attacked by CD8$^+$ or CD4$^+$ cytotoxic T cells (Chapter 12).
- Noninfectious virus or free envelope protein which is released from infected cells can attach to noninfected CD4$^+$ cells and render those cells liable to immune attack, either by antibody or, if the antigen is processed and presented by MHC proteins, by cytotoxic T cells, as described above.

19.7 WHY IS THE INCUBATION PERIOD OF AIDS SO LONG?

The onset of AIDS is defined by CD4$^+$ T cell numbers dropping below a concentration of 200 cells per microliter of plasma. Thus the question can be rephrased in terms of the time taken (10 years on average in the developed world) to reduce CD4$^+$ cells to this level. It is now clear that this is not due to viral latency, as an HIV-positive person is infectious at all times (see above). Figure 19.8 shows a simplified version of the relationship between the circulating virus, T cells, and overt clinical disease. After the initial infection, the virus-specific CD8$^+$ T cell response increases and virus reaches a stable low level plateau. The CD4$^+$ T cell population falls initially due to the infection, but then recovers. At some point years later, the onset of AIDS is announced by a dramatic rise in circulating virus and fall in circulating T cells. The key event is the reduction in CD4$^+$ T cells to a level that makes them unable to sustain the immune system. This permits the adventitious infections that will eventually kill the patient (Table 19.3). Exactly why this takes several years is not clear (Box 19.6). Data from people who survive far longer than the average 8-year infection (long term nonprogressors), and from those on triple therapy (Section 19.8) indicate that key to survival is the maintenance of a low plasma virus concentration.

Box 19.6

Some hypotheses about why there is an extended lag period before CD4 T cell numbers reduce sufficiently to result in immunodeficiency in HIV-1-infected individuals

Activation of virus and death of CD4⁺ cells is limited by the rate at which CD4⁺ cell clones are stimulated by cognate antigen

HIV only multiplies in dividing CD4⁺ T cells. In nature, T cells only divide when they are stimulated by the cognate antigen which their unique T cell receptor recognizes, and many T cells never divide as they never meet their cognate antigen. Activated infected cells then produce a burst of virus before dying by one of the mechanisms already discussed. More CD4⁺ cells are infected by the progeny HIV and the cycle repeats itself. Thus, the time taken to reduce the number of CD4⁺ cells to the level that defines the onset of AIDS depends on the time it takes to come into contact with enough antigens to react with all those T cells. Taking this hypothesis to logical absurdity, an HIV-positive person would not develop AIDS if he or she met no CD4⁺ cell-specific antigens.

Autoimmunity

The envelope protein of HIV-1 has amino acid sequence homology with the MHC class II protein. Normally, the immune system recognizes MHC antigens as self and mounts no immune response to them or to the epitope which is mimicked by HIV. Here the long incubation period reflects the number of years it takes for the self-tolerance mechanism to break down, and the immune system to respond to the MHC class II-like epitope on the HIV-1 envelope. Once that response is in place, the immune system can then react with the cross-reactive MHC class II epitope, and kill them. CD4⁺ T cells, CD8⁺ T cells, and B cells all express MHC class II proteins constitutively, while other cells only express them after exposure to γ-interferon (Chapter 12).

Evolution of the envelope protein in the infected individual

Sequencing of HIV isolates from infected individuals shows that the envelope protein can be divided into variable and constant regions (Fig. 19.5b). These isolates also differ antigenically. Of particular importance is that the envelope protein evolves in infected individuals, and this is probably driven by the immune response, in a manner similar to antigenic drift in influenza virus (Section 17.5). Evolution of viral proteins during the lifetime of an infected individual is also found in other lentiviruses. Thus infected cells are continuously removed by the immune system, until the virus evolves a variant to which the immune system cannot respond. This variant virus kills the cells and eventually gains the upper hand. This evolutionary plasticity of the envelope protein is also problematic for vaccines, and is discussed later.

19.8 PREVENTION AND CONTROL OF HIV INFECTION

No other virus has been subjected to such intense scrutiny as HIV but, as yet, there is no sign of any effective vaccine. At first sight, this is surprising, since the smallpox vaccine has been in use for over 200 years and in the last 50 years some very effective vaccines have been devised and used against other viruses. The problem with HIV is complex. Unlike measles, HIV-1 infection does not result in an immune response that can elimin-

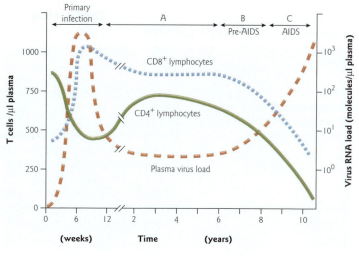

Fig. 19.8 The course of HIV-1 infection, the inverse relationship between infectious virus load and T cell concentrations in plasma. A, B, C refer to the categories of infection (see Box 19.4 and text).

ate the virus. So a vaccine for HIV-1 must do much more that replicate the normal anti-HIV-1 response. The rules governing immunity are not understood. For example, a microorganism is usually sensitive to only one particular arm of the immune system (it might be the CD8$^+$ cytotoxic T cell), but in general it is not known what this is, or how to make a vaccine that stimulates the required response. Thus, vaccines are made empirically. By definition, a successful vaccine stimulates the required immune response, but also raises irrelevant responses. This problem is not unique to HIV-1 – there are a number of other poor vaccines which have defied attempts to improve them (e.g. against cholera and typhoid). We need to understand more about the immune system in general before the rational design of vaccines becomes a reality. That day cannot come soon enough for HIV. Until a vaccine is achieved, our only defences against HIV are a set of antiviral drugs, none of which is curative, in conjunction with modification of behavior to avoid infection.

Prevention of HIV infection

In the absence of any vaccine, avoiding contact with HIV is paramount (Table 19.4). The main risk, already discussed, is through sexual contact. Education programs have been promulgated worldwide, with advice to have as few sexual partners as possible, and to practise safer sex (whether heterosexual or homosexual) by avoiding the exchange of body fluids and by using condoms as a barrier to infection. The possibility of using topical (largely vaginal) virucides to kill virus is being actively explored. However, there are immense social problems in dealing with any sexually transmitted disease, and these vary around the world according to social and religious conventions. Historical parallels are not encouraging, as it is well documented that many were undeterred by the risk of

Table 19.4

Control measures for HIV and AIDS.

Measure	Comment
Avoidance of infection	Available, but not always practised
Diagnosis and detection	Good
Prevention by vaccine	Not available
Chemotherapy	Inhibitors of reverse transcriptase and viral protease reduce virus levels, and retard progression to AIDS; however, they do not eliminate virus from the body and the person remains HIV-positive. A new generation of experimental drugs prevents virus entry into cells by inhibiting fusion of the viral and cell membranes
Directly killing of virus particles	HIV is a labile virus and topical virucides can inactivate infectivity by their detergent-like properties. Used vaginally; still experimental

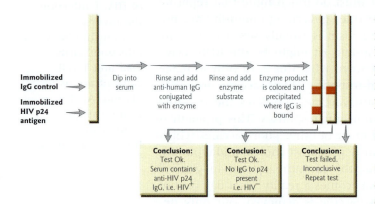

Fig. 19.9 The dip-stick test for antibody to HIV.

contracting syphilis at a time when no treatment was available. The involvement of male and female prostitutes is a particular problem, as, in some areas of the world, a high proportion are infected with HIV.

Diagnosis and detection

Control of any disease needs quick, cheap, and reliable methods for detecting signs of infection. These are in place. Possession of serum *antibody* to the HIV internal antigen, p24, is the main diagnostic criterion, since this protein provokes the strongest and most reliable antibody response. This test decides if a person is "HIV-positive." It is used as a "dip-stick" in an enzyme-linked immunosorbent assay (ELISA) format (Fig. 19.9). The test is not carried out until 2 months after the suspected infection, to allow antibody to form and avoid a falsely negative diagnosis. With babies born to HIV-positive mothers, a period of 3–6 months has to elapse to allow the decay of maternal HIV antibody in the infant's circulation. Positive results are normally confirmed with a second test, e.g. RT-PCR (reverse transcription and polymerase chain reaction) for viral RNA.

It is not easy to isolate HIV-1, although many strains can be grown in CD4$^+$ cells in the laboratory. This is a slow and labor-intensive process and has the intrinsic problem of all such techniques, that of selecting variants from the infected person which happen to grow well in culture but which are unrepresentative of the original virus population. However this can be circumvented by use of RT-PCR directly on viral genomes obtained from a patient. A positive result is the amplification of a DNA molecule of a known size, as seen after gel electrophoresis. This amplified DNA can be sequenced directly to give accurate information on the virus present in the body.

Chemotherapy

Chemotherapy of HIV infections has two mains goals: (i) to prevent the progression of infection to AIDS, and (ii) to clear the infection completely. The first objective has been achieved through the advent of the treatment known as HAART (highly active anti-retrovirus therapy) or just ART, which now usually means triple drug therapy. This commenced in 1994, and under the best conditions can reduce plasma virus to undetectable levels and restore the CD4$^+$ T cell population and function. However it does not clear virus from resting CD4$^+$ memory cells (and possibly other reservoirs) in which the virus is latent, and even a few days without drugs sees a rapid rebound to normal virus levels. Rebound virus may be drug-resistant and can be passed on with catastrophic results. Thus chemotherapy must be continuous. It is calculated that with the natural turnover of the pool of resting CD4$^+$ memory cells it would take 10–60 years for the latently infected cells to disappear. Thus an HIV-positive person will have to have triple therapy for the rest of their life – say 60 years – and it is not known if the drug regimen can be tolerated for this time or if resistant virus will eventually emerge. Triple therapy is prohibitively expensive for all but the richest countries and is still not widely available where it is most needed, in Africa and Asia. The cost is approximately $10,000 per person per year, whereas the *total* health budget in some African countries is less than $10 per person per year. Recent high level political negotiations with the big pharmaceutical companies are beginning to bring down the price and make treatment more widely available.

Triple therapy comprises the simultaneous use of a mixture of three anti-HIV-1 drugs. Individual drugs inhibit synthesis of proviral DNA from viral RNA or inhibit the action of the virus protease (see Fig. 19.3). The former are either nucleoside analogs or non-nucleosides, such as nevirapine. They target different sites on the reverse transcriptase enzyme. The best known of the nucleoside class is the deoxythymidine analog, AZT (zidovudine). This is 3'-azido-2',3'-dideoxythymidine (Fig. 19.10). Its anti-retrovirus properties were discovered before the HIV emergency, but at that time there was no known human retrovirus infection. Nucleoside analogs are incorporated into viral DNA, and because they all lack a 3' hydroxyl group, no further nucleotide can be added and DNA synthesis ceases. Others analogs with similar function (such as stavudine [d4T; deoxythymidine], lamivudine [3TC], zalcitabine [ddC; deoxycytidine], and

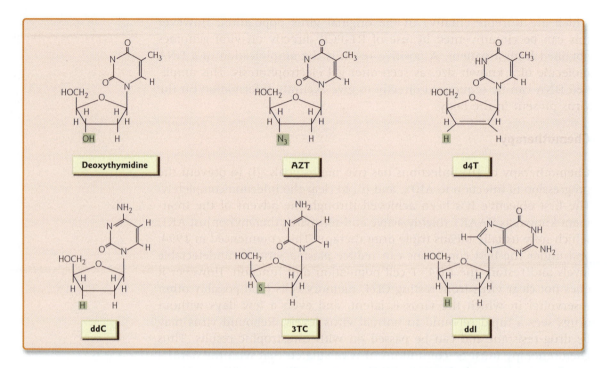

Fig. 19.10 Comparison of the formulae of AZT (3'-azido-2',3'-dideoxythymidine) and deoxythymidine, and the formulae of some other nucleoside analog inhibitors of HIV DNA synthesis. Note that none of the analogs shown have a hydroxyl group in the 3' position (green). Thus they cannot form a bond with the next nucleotide, and act as chain terminators. All these molecules require phosphorylation by cellular enzymes before being incorporated into DNA. See text for commercial names.

didanosine (ddI; deoxyguanosine]) have since been discovered (Fig. 19.10). The protease inhibitors were tailor-made after the atomic structure of the viral protease had been determined. Examples are saquinavir, indinavir, ritonavir, and nelfinavir. Triple therapy usually uses two nucleoside inhibitors and one protease inhibitor. Triple therapy avoids the appearance of the resistant mutants which arise during monotherapy as discussed in Section 21.8. An additional advantage of triple therapy is that the inhibitors are synergistic, meaning that the overall protection is greater than the sum afforded by the drugs individually. Thus the amount of each drug taken can be decreased, reducing the inevitable toxicity and the very high cost. Key points on triple therapy are summarized in Box 19.7.

Vaccines

The discussion so far has focused on people who are already infected. While therapeutic vaccines aim to boost the immune responses of those already infected, the main aim of a vaccine is to put in place immune

Box 19.7

Key points on triple therapy for HIV infection

- Uses three drugs simultaneously; these bind different parts of the viral reverse transcriptase or protease molecules.
- Reduces virus load in plasma.
- Restores lost immune functions.
- Halts progression to AIDS.
- Does not clear latently infected CD4$^+$ memory cells.
- Lapses in therapy result in virus rebound to normal levels.
- Is not tolerated by all people.
- Has to be taken for the foreseeable future.
- Is expensive (approximately $10,000 per person per year).
- Has only been in use since 1994.
- Not known if therapy can be tolerated for a lifetime.
- Not known if resistant virus will eventually break through.

Table 19.5

Some animal models for HIV infection and AIDS, and their limitations.

Virus	Animal	Comment
HIV-1, 2	Chimpanzee	Infection but no AIDS; used to test vaccines
HIV-2	Rhesus monkey	Infection and lymphadenopathy – similar to early stages of AIDS
SIV	Macaque monkey	Infection and AIDS
SHIV	Macaque monkey	Infection and AIDS
FIV	Cats	Infection and AIDS
HIV-1	SCID-hu mouse	Infection, no AIDS

SIV, simian immunodeficiency virus; SHIV, a simian virus which has been genetically engineered to express the envelope protein of HIV-1; FIV, feline immunodeficiency virus; SCID-hu mouse, mice with severe combined immunodeficiency, i.e. in their homozygous form, lack both B and T cell responses, which have been infused with human (hu) lymphocytes that create a human immunity (see text).

responses that prevent HIV from initiating infection. Regrettably, this has not so far proved to be possible.

Animal model systems

Since it is ethically not possible to test the efficacy of a vaccine or a treatment by deliberately infecting people with a lethal agent such as HIV, vaccine trials can only be made in people whose lifestyle places them at high risk from infection (see below) or in model animal systems. The latter have limitations, as only primates have a CD4 protein sufficiently closely related to the human CD4 protein to permit infection with HIV, but no primate infected with HIV-1 progresses to AIDS (Table 19.5). The use of higher primates, such as chimpanzees, is fraught with problems, and, aside from ethics, chimpanzees are expensive to maintain and cannot be used in sufficient number for statistically meaningful experiments. Many experimenters use SIV and New World monkeys as a model of the human situation, but closer to the human situation is SHIV, an SIV that retains the capacity to infect and cause disease in monkeys but whose envelope gene has been replaced by that of HIV. This allows HIV envelope-specific immunity to be assessed.

A major breakthrough came when immunologists took mice, congenitally deficient in both B and T cells (known as severe combined immunodeficient (SCID) mice), and reconstituted them with human lymphocytes. By conventional immunological wisdom, this experiment was a waste of time, as the transplanted cells should have mounted an immune response against mouse antigens in a classic "graft-versus-host" reaction. For some reason, not understood, this did not happen and it is now possible to study infection of human lymphocytes in these animals, and their reaction to experimental vaccines. However, human T cells do not repopulate the mice in their normal proportions and the mice do not develop AIDS. Another approach is to study other nonhuman immunodeficiency lentivirus–host systems (Table 19.2) and hope that there are lessons to be learned that are relevant to the human situation.

Immunity to HIV

Infected people and infected animals mount antiviral antibody and T cell responses, and the appropriate laboratory assays can demonstrate the presence of anti-HIV neutralizing antibody and cytotoxic T cells. Why, then, do HIV-positive people (i) not clear the virus and (ii) go on to develop AIDS? Virus surviving in a host in the face of an active immune response has many precedents, e.g. hepatitis B and herpes simplex viruses (discussed in Sections 14.5, 15.7, and 18.7), and viruses have a variety of ways of evading immune responses. It needs to be remembered that antibody and T cells cannot get inside cells, and act only on antigens that are exposed on the cell membrane. Also while antibody can recognize almost any molecule, T cells recognize only peptide fragments that are complexed with MHC proteins and expressed on the cell surface. Antibodies that bind to epitopes that mediate neutralization can render the virus noninfectious. Both antibodies and activated cytotoxic T cells can kill infected cells bearing their cognate antigens (see Chapter 12). However the HIV-infected person presumably does not make antibody to the right epitope, or enough of this antibody, or antibody of high enough affinity to deal with

Box 19.8

Some ways in which the HIV-1 virion avoids stimulating the synthesis of neutralizing antibody and/or its action

- Both gp120 and gp41 are extensively glycosylated, and this masks much of the viral protein.
- The gp120 has hypervariable (V) loops that vary in sequence under immunological pressure without losing essential functions.
- The variable loops, V1 and V2, partly mask the binding site for the primary virus receptor, CD4, shielding it from antibody.
- As well as varying, the V3 loop may attract the immune response away from other epitopes that are less able to vary, and has been described as a "decoytope."
- The CD4 binding site is a recessed cavity which makes it difficult for an antibody paratope to contact it.
- Infected cells release large amounts of monomeric gp120 that has highly immunogenic surfaces that are not exposed in the mature trimer due to monomer–monomer interactions. Antibodies made to these surfaces are non-neutralizing, and monomeric gp120 is regarded as an immunological decoy.

See Fig. 19.6 for details of the gp120-gp41 envelope protein.

the infection. HIV-1 has a large number of ways in which it evades the host's immune responses, and some of these that relate to antibody are listed in Box 19.8. In essence the mature HIV-1 envelope protein is poorly or nonimmunogenic, and stimulates antibodies of a specificity that poses no danger to the virus. There is the same sort of problem with T cells.

Failure of experimental vaccines to protect against HIV infection

All vaccines to date have been produced empirically; that is, they have been made without knowledge of the exact immune response(s) needed to control a particular infectious disease. In the case of HIV, many of the types of vaccine discussed in Sections 21.1–21.7 have been investigated: inactivated whole virus, purified viral protein, recombinant protein, recombinant protein expressed in a vaccinia virus vector, peptides, and DNA immunization; there has also been intensive work on attenuated vaccines but so far only with SIV (Table 19.6). All stimulate some part of the immune system, but none appear to reliably stimulate an immunity – or enough of it – which gives protection. Responses are often transient. There is individual variation in immune responses, and antibody responses are rarely cross-reactive with other HIV-1 strains. It is putting it mildly to say that the scientific world is surprised by this universal lack of success – it was expected that at least one of the preparations would

Table 19.6

Different types of experimental HIV vaccine.

Type of vaccine	Comment
Killed (inactivated) vaccine	Work ongoing on the problem of totally inactivating virulent virus; more work on immune responses is needed. Problems are the expense and the lack of production of, and hence immunity to, nonstructural proteins. Elicits antibody
Live attenuated vaccine (mostly with SIV with deletions in *nef*)	Attenuated for adult monkeys but not for neonates. Induces variable protection in adults against infection and/or disease after challenge by virulent virus. Infected animals have a persistent infection, which eventually reverts to virulence, usually by repair/partial repair of the deletion. Elicits both antibody and T cell responses
Soluble recombinant gp120	Simulates antibodies that are poorly neutralizing and not cross-reactive; non-neutralizing epitopes are dominant. Not an option
Soluble recombinant gp120 trimers	Genetically engineered to possess disulfide bonds that stably hold the trimer in its native conformation. Express the desired external neutralizing epitopes, but not yet tested for induction of immune responses
Expression of gp120 by recombinant vaccinia virus	Often used in prime-boost mode (infect with recombinant virus and boost with viral protein); has given protection of primates. Elicits both antibody and T cell responses
Peptides and various protein expression systems	Give poor neutralizing antibody and T cell responses
DNA vaccines: immunization with plasmids encoding gp120 and other proteins	The DNA is transcribed and protein made at the site of injection. Systemic and mucosal immune responses are being evaluated. Gives weak immune responses but has protected chimpanzees. May be best used to prime the immune system in prime-boost vaccination

have been protective. Vaccinologists are now, for the first time ever, forced to follow a logical approach, to determine what types of immune response(s) protect against HIV and how these can be stimulated, but it is taking a long time to accomplish this.

Production of a protective vaccine is one of the major goals of the worldwide campaign to prevent AIDS, and an international collaborative scientific effort, with large amounts of scientific time and money, is bent on achieving this goal. It is generally agreed that the aim is to stimulate both virus-specific antibody and T cell responses. Antibody work has concentrated on the envelope protein that carries the only neutralization epitopes, and the best way of presenting it to the immune system. T cell immunity, however, can be directed against peptides from any HIV protein, but it is not yet clear which is the best candidate. One experimental vaccine consists of a number of known T cell peptide epitopes from different HIV proteins joined together by chemical synthesis.

> **Box 19.9**
>
> **Key points on clinical trials with HIV vaccines or drugs**
>
> - Phase I: small trials testing safety and/or immunogenicity; 8–12 months' duration; 10–30 people with a low risk of contracting HIV infection.
> - Phase II: mid-sized trials for further safety and immunogenicity testing, with variation of dose, route of administration, and sample population; 18–24 months' duration; 50–500 people with low to high risk of contracting HIV infection.
> - Phase III: a large scale efficacy trial for the prevention of infection in high risk volunteers (protective vaccine) in which incidence of infection is compared with people who receive a placebo vaccine; the end-point is difficult to determine; takes a minimum of 3 years and requires several thousand people; cost of one trial about $25 million.

Early on it was realized that systemic immunity (immunity in the blood and the body core) would not protect against HIV infection of the genital or rectal mucosae. These naked epithelial surfaces, which are the route of entry for many different pathogens, have a local mucosal immune system that has to be stimulated by the direct application of the vaccine (Chapter 12). Local cognate B and T lymphocytes then migrate to the draining lymph node and, after undergoing activation, return to mucosal surfaces, and form a barrier to infection. Mucosal vaccines can often stimulate systemic immunity, but not vice versa. There is increasing evidence that mucosal immunity is crucial to preventing HIV infection.

Various human HIV vaccine trials have been conducted in healthy HIV-negative people, and are continuing to be undertaken (Box 19.9). By early 2005, there were 27 phase I trials, seven phase II trials, and one phase III trial in various countries including some in Africa. The cost and logistic problems are enormous, and many doubt that there is sufficient evidence to suggest that the vaccine candidates in use are likely to be successful in phase III trials. However the political pressure to get "real vaccine work" underway cannot be underestimated. The results of the first phase III trial in Thailand to produce neutralizing antibody using soluble monomeric clade B and clade E gp120 was not successful. This was perhaps not greatly surprising as the native conformation of the HIV envelope protein is a trimer. More worrying is the poor initial results of a phase II trial in Kenya designed to stimulate HIV-specific T cells using a synthetic peptide composed of major epitopes from several different HIV proteins.

HIV is susceptible to neutralizing antibody

Many hundred HIV-1-specific monoclonal antibodies have been produced and although many do not neutralize infectivity, some neutralize quite well, and a few neutralize very well indeed and protect primates from infection. The reader will be familiar with the fact that the gp120 monomer has a preponderance of non-neutralizing epitopes, but there are other complications lying in wait in the neutralization of HIV. Virus that has been adapted to grow in laboratory T cell lines is relatively easy to neutralize but primary virus (virus isolated from an infected person in peripheral blood mononuclear cells) requires 100-fold higher concentration of the same antibody for significant neutralization. It is well known that adaptation of any virus to growth in cells in the laboratory is associated with selection of virus variants, and this is well illustrated here. The other problem is clade variation (Section 19.1), as it would be useful if the immunity stimulated by any future vaccine protected against as many HIV-1 clades as possible. A small number of MAbs are known that neutralize primary viruses from a range of different clades. These include the gp120-specific MAbs b12 and 2G12, and the gp41-specific MAb 2F5. All are human antibodies, which proves that our immune system has the ability to produce such antibodies. Now we only have to learn how to stimulate the immune system to do this. However, as yet, none of the three antibodies can be raised by immunization of man or any animal, even by preparations of envelope protein that carry the cognate epitope in a recognizable form. It is not known why.

Neutralization, while important, is a laboratory phenomenon, and what is vital is the ability of antibody to protect the whole animal or person from infection. This can be tested by injecting preformed antibody into an animal or a person. Here HIV and chimpanzees or SCID-hu mice and SHIV or SIV and monkeys have been used (see Table 19.5). (This is called *passive transfer* of antibody or *passive immunization* as the animal does not actually make the antibody.) The animals are then challenged with infectious virus by the intravenous or preferably the more natural genital route. In this way it was shown that some MAbs were capable of protecting the animal against HIV-1 infection. This was a vitally important experiment as it showed in principle that vaccines that stimulated the *right sort of immunity* could prevent HIV infection, and that this could be achieved by antibody alone. Key points are summarized in Box 19.10.

Any successful vaccine would need to stimulate antibodies to two or three different epitopes, or neutralization escape mutants would soon be selected (see also Section 17.5). Escape mutants arise with the same frequency as any other mutant, and the arguments above for using triple therapy apply equally to antibodies. One advantage of having more than one MAb present is that they may act synergistically, so that less of each is needed, again as described for chemical inhibitors. Presumably the binding of one antibody alters the conformation of the protein so that a second antibody can bind more easily.

> **Box 19.10**
>
> **HIV-specific antibodies**
>
> Primary virus isolates are intrinsically difficult to neutralize.
>
> There are nine virus clades and a variety of recombinants that differ in sequence and antigenicity of their gp120-gp41 envelope protein, hence:
> - Different clades have very few neutralizing epitopes in common.
> - Only a few MAbs are known that neutralize primary viruses from different clades.
> - The same MAbs protect against infection in model systems *in vivo*.
> - Locally produced mucosal antibody may be essential for preventing infection.

Future vaccines

Today, the vaccine problem is being attacked on a number of different fronts. Empirical work is still in progress and is investigating the almost infinite permutation of immunization protocols in terms of type of vaccine, amount inoculated, site(s) of inoculation, type of adjuvant, number of doses, and the interval between them. The main aim, though, is directed at achieving a fundamental understanding of the immune system. This is a huge and inevitably slow endeavor, but will in the end provide the information necessary to devise an anti-HIV vaccine as well as establishing principles that will enable vaccines to be made to many other problem diseases. A future vaccine, whether protein or nucleic acid or a combination of both, will probably consist of a mixture of different immunogens, each of which stimulates a different facet of the immune response, and this will surely include both antibody and T cell responses. A major question is whether or not vaccine-induced T cells will be more effective at controlling HIV infection than are T cells that arise during natural infections. New adjuvants are being used and the ability of cytokines to enhance immune responses is being evaluated. Mucosal responses that provide local immunity are likely to be essential in preventing the establishment of infection. Whatever the form of any new vaccine, it will need to be inexpensive so that, unlike current drug therapy, it is available to developing countries where it is so desperately needed.

19.9 THE COST OF THE HIV PANDEMIC

Every country has a finite amount to spend on health care, and AIDS is making substantial inroads into this budget at a time when health care costs are generally rising. Cost is a major issue in any chronic disease where treatment has to be provided over a number of years, and the cost in human terms of any AIDS case is incalculable. Because of the high incidence of infection, the immense loss of life in Central and East Africa is devastating the countries of those regions. Demographic changes are predicted in a manner not seen since the time of the Black Death in Europe in the fourteenth century. There is every indication now that HIV is advancing through Asia on a scale similar to that in Africa, and will result in social and economic disaster unless preventative measures are rapidly put in place. This is not helped when some countries will not face the fact that they have an HIV problem.

19.10 UNRESOLVED ISSUES

We do not understand why HIV envelope antigens that express known key epitopes do not stimulate cognate antibodies. This occurs in non-infected people, so the phenomenon appears to be a property of the immunogen, rather than of virus-induced immunosuppression. Although inactivated (noninfectious) virus particles that have native envelope trimers on their surface are also poor immunogens, there is much interest in the immunogenicity of gp120 trimers that have been genetically engineered to remain permanently bound together. A small proportion of HIV-infected people do not progress to AIDS, probably because they can control their virus load. It would be valuable to understand which part of the immune response of these long term nonprogressors is responsible, and what makes them different from the majority who develop AIDS within the normal timescale. Finally, it would benefit education programs enormously to understand why people are prepared to risk contracting HIV and their lives for momentary sexual gratification.

KEY POINTS

- HIV is a typical member of the Lentivirus genus of the retrovirus family.
- HIV is a newly emerged human disease, resulting from contact with other animals (primates); it is still rapidly spreading throughout the world.
- Emerging diseases are often virulent as they are not adapted to their new host species.
- HIV is primarily a sexually transmitted virus.
- HIV infects cells that carry the CD4 protein on their surface, notably CD4$^+$ T cells which play a central role in many immune responses.
- HIV rarely causes disease directly but over many years leads to a decline in the immune system (AIDS) that allows a variety of microorganisms to cause serious infections; it is these that eventually cause death. The time to AIDS is proportional to the virus plasma load after the initial acute infection.

- Antiviral drugs are available that inhibit virus multiplication; these do not clear virus from the body and so have to be taken for a lifetime; they act by reducing virus load; because of cost drugs are widely available only in developed countries.
- Most HIV proteins have T cell epitopes, and both parts of the envelope protein (gp120-gp41) have neutralizing epitopes.
- A handful of antibodies specific to certain HIV envelope protein epitopes on gp120 and gp41 neutralize primary virus strains and protect primates from infection.
- No HIV antigen preparation yet tested stimulates effective neutralizing antibody or antiviral T cells; hence there is no HIV vaccine.
- A worldwide network of scientists is actively collaborating to find a vaccine for HIV.
- Unlike, say, influenza virus, infection with HIV is easily avoided as infection requires a risky behavior; avoiding such behavior is the only effective way to remain HIV-negative.

QUESTIONS

- Discuss the origins of HIV-1, its emergence as a significant infectious agent in humans, and the factors that have led to a global pandemic.
- Virus variation is an important property of HIV-1. What role does this feature of the virus play in the pathogenesis of AIDS and how does it influence the strategies used to treat HIV-1 infection?

FURTHER READING

Burton, D. R., Desrosiers, R. C., Doms, R. W. *et al.* 2004. HIV vaccine design and the neutralizing antibody problem. *Nature Immunology* **5**, 233–236.

Butera, S. T., ed. 2005. *HIV Chemotherapy*. Caister Academic Press, Portland, OR.

Desrosiers, R. C. 2004. Prospects for an AIDS vaccine. *Nature Medicine* **10**, 221–223.

Doan, L. X., Li, M., Chen, C., Yao, Q. 2005. Virus-like particles as HIV-1 vaccines. *Reviews in Medical Virology* **15**, 75–88.

Douek, D. C., Picker, L. J., Koup, R. A. 2003. T cell dynamics in HIV-1 infection. *Annual Review of Immunology* **21**, 265–304.

Escourt, M. J., McMichael, A. J., Hanke, T. 2004. DNA vaccines against human immunodeficiency virus type 1. *Immunological Reviews* **199**, 144–155.

Fauci, A., Touchette, N. A., Folkers, G. K. 2005. Emerging infectious diseases: a 10-year perspective from the National Institute of Allergy and Infectious Diseases. *Emerging Infectious Diseases* **11**, 519–525.

Garber, D. A., Silvestri, G., Feinberg, M. B. 2004. Prospects for an AIDS vaccine: three big questions,

no easy answers. *Lancet Infectious Diseases* **4**, 397–413.

Greene, W. C., Peterlin, B. M. 2002. Charting HIV's remarkable voyage through the cell: basic science as a passport to future therapy. *Nature Medicine* **8**, 673–680.

Johnson, W. E., Desrosiers, R. C. 2002. Viral persistence: HIV's strategies of immune system evasion. *Annual Review of Medicine* **53**, 499–518.

Klein, M. 2003. Prospects and challenges for prophylactic and therapeutic HIV vaccines. *Vaccine* **21**, 616–619.

Levy, J. A. 1998. *HIV and the Pathogenesis of AIDS*, 2nd ed. ASM Press, Herndon, VA.

Mascola, J. R. 2003. Defining the protective antibody response. *Current Molecular Medicine* **3**, 209–216.

Mascola, J. R., Lewis, M. G., VanCott, T. C. *et al.* 2003. Cellular immunity elicited by human immunodeficiency virus type 1/simian immunodeficiency virus DNA vaccination does not augment the sterile protection afforded by passive infusion of neutralizing antibodies. *Journal of Virology* **77**, 10348–10356.

Mascola, J. R., Montefiori, D. C. 2003. HIV-1: nature's master of disguise. *Nature Medicine* **9**, 393–395.

McIntyre, J. 2003. Mothers with HIV. *British Medical Bulletin* **67**, 127–135.

Moore, J. P., Kitchen, S. G., Pugach, D., Zack, J. A. 2004. The CCR5 and CXCR4 coreceptors: central to understanding the transmission and pathogenesis of human immunodeficiency virus type 1 infection. *AIDS Research and Human Retroviruses* **20**, 111–126.

Pantaleo, G., Koup, R. A. 2004. Correlates of immune protection in HIV-1 infection: what we know, what we don't know, what we should know. *Nature Medicine* **10**, 806–810.

Pillay, D., Taylor, S., Richman, D. D. 2000. Incidence and impact of resistance against approved antiretroviral drugs. *Reviews in Medical Virology* **10**, 231–253.

Rambaut, A., Posada, D., Crandall, K. A., Holmes, E. C. 2004. The causes and consequences of HIV evolution. *Nature Reviews Genetics* **5**, 52–61.

Rinaldo, C. R., Piazza, P. 2004. Virus infection of dendritic cells: portal for host invasion and host defense. *Trends in Microbiology* **12**, 337–345.

Sharland, M., Gibb, D. M., Tudor-Williams, G. 2003. Advances in the prevention and treatment of paediatric HIV infection in the United Kingdom. *Sexually Transmitted Infections* **79**, 53–55.

Shattock, R. J., Moore, L. P. 2003. Inhibiting sexual transmission of HIV-1 infection. *Nature Reviews Microbiology* **1**, 25–34.

Whitney, J. B., Ruprecht, R. M. 2004. Live attenuated HIV vaccines: pitfalls and prospects. *Current Opinion in Infectious Disease* **17**, 17–26.

Zolla-Pazner, S. 2004. Identifying epitopes of HIV-1 that induce protecting antibodies. *Nature Reviews Immunology* **4**, 199–210.

The IAVI website: http://www.iavi.org/iavireport/0803/clades-vax.htm

The United Nations AIDS website: http://www.UNAIDS.org

Also check Appendix 7 for references specific to each family of viruses.

20

Carcinogenesis and tumor viruses

It is estimated that viruses are a contributory cause of 20% of all human cancers but, in each case, infection is just one factor that contributes to disease development. As a result, these viruses cause cancer in only a small percentage of the people who become infected. Studying viruses that can cause cancer in people or in laboratory animals has revealed a great deal about the processes that underlie the formation of all cancers, not just those that are caused by viruses.

The great majority of viruses of vertebrates are not oncogenic, i.e. they do not have the ability to initiate a cancer (for the meaning of "oncogenic" and other cancer-related terms, see Box 20.1). However, for many years it has been recognized that certain viruses can induce tumors in appropriate experimental animals. More recently, good evidence has implicated some viruses in the development of specific human cancers or naturally occurring cancers in animals. As a result of the importance of understanding human cancer as a route to finding effective treatments, those viruses that caused such disease experimentally were subjected to intensive study from the 1960s onwards. Our detailed understanding of these viruses today owes much to this driving motivation for research.

For each of the viruses associated with human malignancies, infection does not lead inevitably to the disease. Carcinogenesis is multifactorial and hence a rare event, even though the associated

Box 20.1

Definitions of cancer-related terms

Adenocarcinoma: A carcinoma developing from cells of a gland.
Benign: An adjective used to describe growths which do not infiltrate into surrounding tissues. Opposite of malignant.
Cancer: Malignant tumor; a growth which is not encapsulated and which infiltrates into surrounding tissues, the cells of which it replaces by its own. Its cells are spread by the lymphatic vessels to other parts of the body (metastasis). Death is caused by destruction of organs to a degree incompatible with life, extreme debility, and anemia or by hemorrhage.
Carcinogenesis: Complex multistage process by which a cancer is formed.
Carcinoma: A cancer of epithelial tissue.
Fibroadenoma: Tumors of mixed cell type.
Fibroepithelioma: Tumors of mixed cell type.
Fibroblast: A cell derived from connective tissue.
Leukemia: A cancer of white blood cells.
Lymphoma: A cancer of lymphoid tissue.
Malignant: A term applied to any disease of a progressive and fatal nature. In the context of carcinogenesis, an adjective describing a tumor which grows progressively and invades other tissues. Opposite of benign.
Mesothelioma: A tumor of the cells lining the body cavity, surrounding the lungs.
Neoplasm: An abnormal new growth, i.e. a cancer.
Oncogenic: Tumor causing.
Sarcoma: A cancer developing from fibroblasts.
Telomerase: The enzyme that maintains the length of the telomeres at the ends of the chromosomes; it is normally inactive in somatic cells.
Transformation: A constellation of phenotypic changes of cells in culture.
Tumor: A swelling, due to abnormal growth of tissue, not resulting from inflammation. May be benign or malignant.

virus infection may be common. Various factors, such as host genotype, age at infection, diet, environmental carcinogens (other than viruses), and other invading organisms, may all contribute to the process of virus-associated oncogenesis. Thus to call a virus a "tumor virus" is rather misleading, as the name refers to one relatively infrequent aspect of its life cycle. A more accurate name would be "tumor-associated viruses." A tumorigenic interaction with a host offers essentially no advantage to the virus. Usually, it loses the ability to produce progeny and to transmit to other hosts in the course of such an interaction, whilst threatening the life of its existing host.

20.1 IMMORTALIZATION, TRANSFORMATION, AND TUMORIGENESIS

Normal cells, i.e. cultures established from healthy tissue and then passaged, senesce and die after a fixed number of divisions (about 50). No matter how carefully they are looked after, something within the cells limits their lifespan. Immortalized cells, by contrast, are able to be passaged indefinitely in cell culture, allowing them to form permanent cell lines. A key event in immortalization is the reactivation of telomerase, the enzyme that maintains the specialized DNA sequences at the ends of the chromosomes; it is the reduction in length of these sequences that leads to senescence. Transformed cell lines are recognized by further changes in phenotype as compared with immortalized cells (Box 20.2). These usually result in visible changes in appearance, a process known as morphological transformation. Members of the same families of virus that contain tumor-associated viruses are able to immortalize and transform primary mammalian cells in culture and/or to transform already immortalized cells. The process of transformation in cell culture is closely linked with tumorigenesis *in vivo* (Box 20.3).

There is now overwhelming evidence that the mechanisms underlying immortalization, transformation, and tumorigenesis involve alteration to a cell's DNA, i.e. mutation. In principle, mutation could lead to these changes through either the loss of functions that are required for normal cell behavior or the acquisition of functions that disrupt normal behavior, mediated either by changes to existing genes or the arrival of new genes. Genes whose loss is tumorigenic are known as tumor-suppressor genes, whilst genes that can promote tumor formation if they sustain mutations that either alter the protein product in specific ways or increase its level of expression are known as oncogenes. Oncogenes are involved in controlling cell signalling and regulating the cell division cycle. Applying the term "oncogene" to a normal gene that has such activity only when

Box 20.2

Some altered properties of transformed cells

- Multiply for ever (immortalization).
- Grow to higher saturation density.
- Have reduced requirement for serum growth factors.
- Grow in suspension in soft agar.
- Form different cell colony patterns.
- Grow on top of normal cell monolayers.
- Readily agglutinated by lectins.

Box 20.3

Evidence for a link between transformation and tumorigenesis

- Transformed cells, but not normal or simply immortalized cells, can often form tumors in suitable animal hosts.
- The same genetic changes that have been found to underlie transformation of cells in culture are also found frequently in naturally occurring tumors.
- Several of the defining properties of transformed cells reflect decreased cell adhesion and increased mobility, which are essential features of invasive malignant cells *in vivo*.
- Cell lines established from tumors have similar properties to transformed cell lines.
- Where a virus can both transform cells in culture and cause tumors *in vivo*, the viral genes that are required for the two processes are often the same.

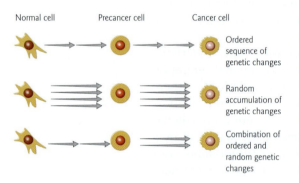

Fig. 20.1 Alternative scenarios for multistep progression from a normal to a cancer cell occurring by the acquisition of genetic alterations, sometimes including the effect of specific virus genes. Genetic alterations are symbolized by arrows, though the number of arrows shown is not intended to indicate the actual number of events involved in converting a normal cell into a fully malignant cell. The specific nature of early events in the pathways probably predisposes cells to suffer further mutations.

mutated is rather confusing; therefore the non-mutated forms of these genes are sometimes called proto-oncogenes, to distinguish them from their mutated, oncogenic forms. The involvement of viruses in oncogenesis is through altering the expression of proto-oncogenes, tumor suppressor genes, and other host cell genes and/or the provision of new virus encoded functions. Remember, though, that transformation and tumorigenesis are multistep processes, in which the input of viral genes is just one possible event; it is the accumulation of changes that leads to the tumor cell phenotype. These changes may have to occur in a defined order or may accumulate randomly (Fig. 20.1). Thus there may be multiple routes by which a cell can reach a malignant state.

As well as the acquisition by cells of an array of genetic alterations, tumor formation involves complex interactions with the immune system of the host animal. The immune system is important as it provides a check on the spread of cells bearing viral or nonself tumor antigens (even tumor cells arising without virus involvement often display novel antigens on their surface). Hence transformation in culture is not an infallible indicator of malignancy *in vivo*. However, studying transformation has been very revealing of the events required for the development of malignant potential by cells.

20.2 ONCOGENIC VIRUSES

A diverse collection of viruses is associated with naturally occurring or experimental tumors. To have the chance to be a cause of a tumor, a virus must be capable of a noncytolytic interaction with cells and, if genes from the virus are to be able to contribute to the ongoing maintenance of a tumor cell phenotype during rapid cell division, they must be stably inherited by daughter cells; this means either becoming integrated into the cell genome or being replicated episomally.

All the viruses listed in Table 20.1 are known or suspected to be onco-genic in humans or animals. Particularly good evidence links hepatitis B and C viruses (HBV, HCV) with primary hepatocellular carcinoma (HCC), Epstein–Barr virus (EBV) with nasopharyngeal carcinoma and Burkitt's lymphoma, and human papillomaviruses (HPV) with cervical cancer. Complete observation of Koch's postulates (see Section 14.1) is out of the question, so other criteria, such as epidemiological data, are also used to make these associations.

Representatives of most major families of DNA viruses are associated with cancer, with the exception of the poxviruses. As poxviruses replicate in the cytoplasm, they do not have the same opportunity to affect the function of the nucleus as do the other DNA viruses; they are also acutely cytopathic. It is important to realize that the genes of a virus can initiate or contribute to a tumorigenesis process only if the virus infection does not kill the cell. For almost all the viruses identified as tumorigenic in Table 20.1, a natural mechanism exists whereby cells can potentially survive an infection, giving the chance that they will develop an altered phenotype subsequently. The herpes- and retroviruses routinely establish latent and noncytopathic infections respectively in their natural hosts (see Chapter 15). The papillomaviruses, with their bi-partite life cycle (see Section 9.3), maintain their genomes in cells which do not support virus production. Adeno- and polyomaviruses normally establish acute or chronic productive infections in their natural hosts and so cause tumors only in species which are nonpermissive for virus production. The exception is polyoma virus in laboratory mouse strains; in this case, tumorigenicity is linked to a defective immune response to the virus. Finally, it is important to consider the genetic status of the virus. Within a population of virus particles of any type, there will typically be many defective particles which, through random mutation, lack some functions essential for productive infection. If the genes which are important for transformation or tumor formation remain intact, then such a particle should be capable of transforming what is normally a permissive cell.

Survival of the infected cell is only part of the story. If viral genes are to contribute actively to an altered cell phenotype which includes uncontrolled cell growth, they must be successfully inherited by all the

Table 20.1

Viruses known or suspected to have oncogenic properties in humans, and some selected non-human tumor viruses.

Virus family	Type	Tumor	Animal*	Cofactor/Comment
Adenovirus	9	Benign fibroadenoma, sarcoma of breast tissue	Female rat	Newborn only, not natural host
	12	Sarcoma	Hamster	Newborn only, not natural host
Flavivirus	Hepatitis C	Primary hepatocellular carcinoma	Humans	? Alcohol, smoking
Hepadnavirus	Hepatitis B	Primary hepatocellular carcinoma	Humans	Age at infection, aflatoxin, alcohol, smoking
Herpesvirus	Epstein–Barr	Burkitt's lymphoma	Humans	Malaria, genome rearrangements
		Nasopharyngeal carcinoma	Humans	Dietary nitrosamines, HLA genotype
		Hodgkin's disease	Humans	?
		Immunoblastic lymphoma	Humans	Immunodeficiency
	HHV-8	Kaposi's sarcoma	Humans	Immunodeficiency (e.g. HIV-1 infection), HLA genotype
		Primary effusion lymphoma		?
	Herpesvirus saimiri	Experimentally induced lymphomas and leukemias	Owl monkey	Species not the natural host
	Marek's disease virus	T cell lymphoma	Chicken	Species
Papilloma	Human papilloma types 16, 18, 31	Cervical neoplasia	Humans	Smoking
	Bovine papilloma type 4	Warts, fibroepithelioma	Cattle, hamster, rabbit	Age (newborn), consumption of bracken fern
Polyoma	SV40	Possibly neural tumours	Humans	Not the natural host
		Possibly mesothelioma	Humans	Asbestos, not the natural host
		Gliomas, fibrosarcomas	Hamsters	Newborn, not the natural host
	Polyomavirus	Sarcoma, carcinoma, multiple tissues	Mouse, other rodents	Newborn, mouse strain, HLA genotype
Retrovirus	Human T cell lymphotropic virus type 1	Adult T cell leukemia, lymphoma	Humans	?
	Rous sarcoma virus	Sarcoma	Chicken	
	Mouse mammary tumor virus	Adenocarcinoma	Mouse	Strain, gender, hormone status
	Feline leukemia virus	T cell lymphosarcoma, leukemia, fibrosarcoma	Cat	?

*Tumors are restricted to the species mentioned; this is the natural virus host unless indicated otherwise.
HLA, human leukocyte antigen (i.e. human form of major histocompatibility complex, or MHC antigen).

daughter cells of the originally infected cell. This can be achieved most easily by integration of the viral DNA into the host genome. Of the transforming viruses, only the retroviruses have an integration function as part of their normal life cycle. Alternatively, the viral DNA may carry an origin of replication which allows it to be copied during S phase (DNA synthesis phase) of the cell cycle and partitioned between daughter cells at subsequent mitosis. Both Epstein–Barr virus (and possibly other oncogenic herpesviruses) and the papillomaviruses can do this, although papillomavirus-induced tumors ultimately contain integrated viral DNA, as for adenovirus, etc. Failing this, the relevant viral genes must be inserted into the host genome by random nonhomologous recombination, which is a rare event, occurring experimentally in roughly 1 in 10^5 infected nonpermissive cells. This is how adenovirus, SV40, or polyomavirus DNA is retained in transformed and tumor cells.

Although maintaining key viral genes in the cell to contribute to the altered phenotype on an ongoing basis is the most obvious mechanism by which viruses might cause tumorigenesis, there is an alternative. It is possible to conceive of viruses acting transiently, contributing to tumorigenesis in a "hit-and-run" mechanism, e.g. by promoting genome instability and hence acquisition of mutations by the infected cell, so that viral genes need not be sustained within the cells all the way to the appearance of disease.

20.3 POLYOMAVIRUSES, PAPILLOMAVIRUSES, AND ADENOVIRUSES: THE SMALL DNA TUMOR VIRUSES AS EXPERIMENTAL MODELS

The polyomavirus, papillomavirus, and adenoviruses are completely distinct in the molecular details of their gene expression. They also have completely different disease profiles; adenoviruses cause a variety of respiratory, gastrointestinal tract or eye infections, polyomaviruses cause urinary tract infections and papillomaviruses cause warts on various epithelial surfaces. It is particularly interesting, therefore, that the mechanisms by which these viruses can immortalize and transform appropriate cells in culture should be so similar.

Genetics of viral transformation

SV40, polyomavirus, human adenoviruses, and bovine papillomavirus type 1 (BPV1) can transform appropriate nonpermissive rodent cell types in culture (human papillomavirus effects on human cells are considered further in Section 20.4). While SV40, polyomavirus, and adenovirus DNA becomes integrated during this process, BPV1 DNA is maintained as an episome. Studies of these transformed cells revealed which were

Box 20.4

Evidence that defined the transforming gene(s) of BPV1, adeno- and polyomaviruses

- Adenovirus, SV40, and polyomavirus:
  - In cells transformed by these viruses, viral DNA is stably integrated into the host genome, as detected by Southern blotting, without any apparent preference as to the integration site.
  - Typically, only part of the genome is integrated; comparing the retained sequences between multiple cell lines to find the regions in common identified the crucial genome regions for transformation, i.e. the transforming genes.
  - Cloned genomic DNA from these viruses can transform cells; by cloning progressively smaller genome fragments and testing their transforming capacity, the individual transforming genes were identified.
- BPV1:
  - Transformation is achieved with cloned genome in a plasmid, which is maintained episomally.
  - BPV1 transforming genes were identified by testing the transforming capacity of cloned subfragments.
  - Once genes essential for BPV1 replication were deleted, transformation required random integration of the transforming genes as for SV40, etc.

Table 20.2

The oncogene products of the small DNA tumor viruses.

Virus	Protein(s) implicated in cell transformation
SV40	Early region, large T antigen
Polyoma	Early region, large T antigen plus middle T antigen
Human papilloma	E6 and E7 proteins
Bovine papilloma type 1	E5, E6, and E7 proteins (depending on cell type)
Human adenovirus	E1A and E1B proteins

the important viral genes for transformation in each case (Box 20.4); these genes are summarized in Table 20.2 and are the same genes that are required for tumorigenesis. All of these genes are essential for normal virus replication in their natural target cells.

Studies of the adenovirus transforming genes were important in developing the idea that multiple genetic changes are needed to observe transformation or the appearance of a tumor. Neither the E1A nor E1B genes alone were sufficient to produce permanently transformed cell lines from baby rat kidney (BRK) cells, but both genes together could efficiently transform these cells (Fig. 20.2). E1A alone produced abortive transformants, cells which appeared transformed but which could not be stably maintained. Thus, at least two separate functions are needed to fully transform BRK cells. Only the fully transformed cells were tumorigenic in congenitally T-cell immunodeficient mice (known as athymic nude mice). Similarly, polyoma virus large T and middle T genes together are needed for full transformation by this virus.

By contrast, SV40 large T antigen alone is able to fully transform BRK cells. Further experiments showed that it contains three separate transformation functions, one or more of which is needed to achieve cell transformation depending on which cells are used for the experiment. Thus, some T antigen mutants are defective for transformation of one rodent cell line but not another. This result is taken as a reflection of the different level of mutation already present in the genomes of these different cell lines before introducing the T antigen gene. Even the most "normal" cell line in culture has undergone some genetic changes during its establishment and so the further changes that are needed before a transformed phenotype is observed will vary.

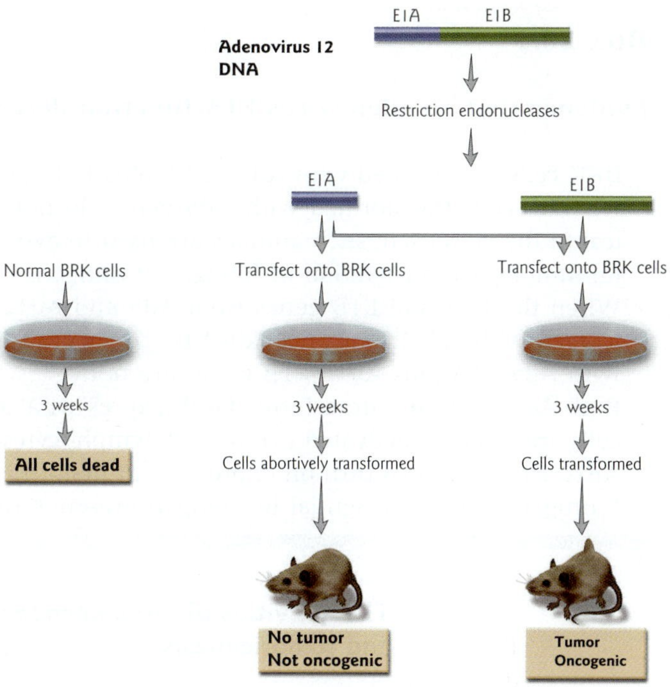

Fig. 20.2 Demonstration of cooperation between gene functions to achieve full oncogenic transformation. While the E1A region alone of adenovirus type 12 is transforming, the cells do not survive indefinitely (abortive transformation). Only cells carrying the adenovirus type 12 E1B gene as well are fully transformed and tumorigenic in syngeneic rats. The E1B gene alone gives no transformation. BRK, baby rat kidney.

The role of the immune response in controlling viral oncogenicity

The E1A and E1B genes of a range of different human adenovirus serotypes, e.g. Ad5 and Ad12, can transform rodent cells in culture. However only a few of these serotypes, such as Ad12 but not Ad5, can cause tumors in normal immunocompetent animals. The difference resides in the E1A genes of these viruses (Box 20.5). Ad12 E1A is able to block T cell recognition of transformed cells whereas Ad5 E1A is not. It seems that Ad12 oncogenicity resides in the ability of the virus to "hide" the cells it has transformed from the immune system, a form of specific immunosuppression. However, it is important to remember that this Ad12 E1A function has the primary purpose of optimizing the survival of the virus during natural infections rather than enhancing its tumorigenicity! These studies of adenovirus oncogenicity demonstrate how important the immune system is in preventing the growth of tumors that would otherwise arise.

Box 20.5

Evidence that an adenovirus E1A function determines adenovirus oncogenicity

- BRK cells transformed with Ad12 E1A plus E1B genes generate tumors in syngeneic rats whereas cells transformed with Ad5 genes do not. (Syngeneic means genetically identical to the cells used; such animals are used to avoid rejection of the tumor cells through immunological recognition of foreign rat antigens.)
- When the E1A and E1B genes from Ad5 and Ad12 are mixed and matched, cells transformed with Ad12 E1A plus Ad5 E1B genes are tumorigenic in this assay whereas cells with Ad5 E1A plus Ad12 E1B genes are not.
- Both Ad5- and Ad12-transformed cells express viral antigens but only the Ad5-transformed cells are lysed by activated cytotoxic T lymphocytes (CTL).
- Ad12 E1A functions turn off expression of major histocompatibility complex (MHC) class I antigen, which is essential for antigen presentation to CTL.

The activities of virus-encoded proteins leading to transformation and tumorigenesis

Through the last few years, details of the mechanisms by which SV40 T antigen, adenovirus E1A, E1B proteins, etc. contribute to transformation have become much clearer. In each case, the protein achieves its effect(s) through one or more specific interactions with host cell proteins involved in regulating the mammalian cell cycle. Remarkably, these interactions are very similar for each virus, revealing how fundamental the disruption of the cell cycle is to their normal life cycles. The domain structures of these proteins is summarized in Fig. 20.3. Genetic studies have shown that each of the interactions highlighted is important for transformation by that protein. The following is a simplified account of some of these interactions. For more detail, readers are referred to texts dealing with mammalian cell biology.

One of the targets of the viral transforming proteins is called Rb or p105, a tumor suppressor gene product that is a key cell cycle regulator in mammalian cells. This protein prevents an unscheduled S-phase (DNA synthesis) in the cell by binding and rendering inactive a transcription factor needed to turn on essential S-phase genes. When a cycle of DNA synthesis and division is correctly signalled, Rb protein is specifically phosphorylated, so releasing its bound transcription factor which allows S-phase to proceed. SV40 and polyoma large T antigens, the adenovirus E1A proteins, and papillomavirus E7 protein all inactivate Rb as well. Each one binds to Rb via a protein sequence (Fig. 20.3, red) that contains a short conserved sequence motif, Leu-X-Cys-X-Glu. Binding allows a second region in each protein (Fig. 20.3, yellow) to release the transcription factor and so enable S-phase progression. Two other relatives of Rb are targeted in the same way. Each of the small DNA tumor viruses needs the cell it infects to enter S-phase to support its own replication and so these viral functions are absolutely crucial to infection *in vivo*, where newly infected cells typically will not be dividing rapidly or even at all. However, when allowed to act outside of the normal context of infection, these activities promote transformation because they remove a key negative regulation on continuous cell division. The fact that the Rb gene is a tumor-suppressor means that its inactivation by direct mutation has been observed to lead to tumors, again demonstrating the mechanistic link between transformation and tumorigenesis.

Another mechanism common to the small DNA tumor viruses involves the very important cell protein p53. This protein acts as a sensor of the replicative health of the cell, preventing the survival of damaged cells that might otherwise threaten the health of the organism. Whenever the cell's DNA is damaged, p53 activity is induced and this leads to

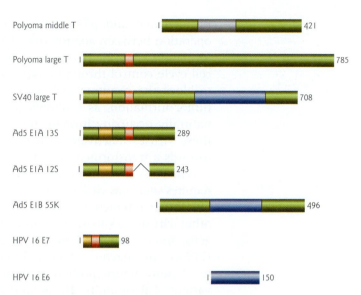

Fig. 20.3 The transforming proteins of SV40, polyoma, papilloma, and adenovirus. Horizontal green bars represent the proteins indicated, not to scale. The adenovirus E1A 13S and 12S polypeptides are highly related, differing only in the removal (by splicing of the mRNA) of 46 amino acid residues from the shorter protein. For expression and sequence relatedness of the SV40 and polyoma proteins, see Section 9.2. The various colored blocks indicate functional rather than sequence relatedness, although there is sequence similarity between the indicated Rb-binding sequences. Red: Rb-binding; blue: p53 binding; yellow: accessory region required to release transcription factors from bound Rb; grey: Src protein binding. For definitions of Rb, p53, and Src, see text.

the cell cycle being halted, to allow DNA repair systems a chance to fix the damage and, if repair fails, to the induction of apoptosis. When the restraint on cell proliferation normally imposed by Rb protein is inactivated, p53 activity is induced. Having inactivated Rb, these viruses therefore need a mechanism for blocking the p53 response. SV40 large T antigen (but not polyoma large T), papillomavirus E6, and adenovirus E1B 55K proteins all bind to p53 (Fig. 20.3, blue), inactivating and/or targeting it for proteolytic degradation. Also, a second E1B protein (19K) acts independently to block apoptosis. The polyoma middle T antigen interacts with the Src protein (Fig. 20.3, grey). The Src protein is encoded by the gene acquired as cDNA by Rous sarcoma virus (see Section 20.6) and its interaction with polyoma middle T antigen leads indirectly to a block on the p53 response. Thus, it appears that all the small DNA viruses target both

Rb-mediated and p53-mediated regulation of the cell cycle. The co-operation between adenovirus E1A and E1B genes to achieve full transformation (Fig. 20.2) can now be understood; E1A proteins overcome cell cycle control through effects on Rb, and E1B 55K protein is needed to block the resulting p53 response. p53 is a crucial protection against tumor formation, as is shown by the very high frequency with which naturally occurring human tumors contain a mutated p53 gene. When these viruses inactivate p53, the infected cell takes an important step towards a transformed or tumorigenic state.

When SV40 transforms growth-arrested rodent cells, the process requires small t as well as large T antigen. This discovery led to the definition of another interaction with host cell proteins which has counterparts in other viruses. SV40 small t antigen, polyoma small and middle T antigens, and adenovirus E4 Orf4 protein all bind to protein phosphatase 2A (PP2A), an enzyme which is important in cell cycle regulation. PP2A is a heterotrimeric protein, with active A and C subunits regulated by a variety of B subunits. These viral proteins act as alternative B subunits, displacing existing B subunits to alter enzyme activity and hence disrupt regulation of cell division.

20.4 PAPILLOMAVIRUSES, SV40, AND HUMAN CANCER

Human papillomaviruses and carcinogenesis

Certain human papillomaviruses play a causative role in the development of cervical cancer, one of the most common cancers of women world-wide. They are also associated, more rarely, with other cancers of the genitalia in men and women. Vaccines are now in trials that it is hoped will protect against the most common of the cancer-causing HPVs and thus also protect people from the tumors that these viruses can cause.

There are over 100 human papillomavirus (HPV) types, as defined by a minimum level of sequence difference between viruses since no system for serotyping exists. These viruses cause warts (which are benign tumors) and can be subdivided into those infecting the skin and those infecting internal epithelial cell layers. Among the latter, the two most prevalent are types 6 and 11, which are associated mainly with genital warts, but also with papillomas of the vocal chords and respiratory tract. The occurrence of genital warts correlates with sexual activity from an early age and a high number of sexual partners, reflecting the transmission of HPV by sexual contact. The incidence of cervical cancer also correlates with these indicators and analysis of cervical carcinoma biopsies routinely reveals HPV DNA of one of a small group of types, most commonly HPV16, 18, and 31, termed high risk HPV. These are only minor

Box 20.6

Features unique to E6 and/or E7 from high risk HPV

- E7 is more effective at releasing transcription factor from Rb than is E7 from low risk types.
- E6 binds to a class of host proteins known as PDZ proteins as well as to p53.
- E6 re-activates the expression of the host telomerase gene.
- E6 and E7 together promote mitotic instability (chromosome segregation errors at mitosis).

components of the total load of genital HPV infections but seem selectively to be involved with this serious disease. Since high risk HPVs are now believed to be the trigger for cervical carcinoma in almost all cases, a vaccine that would protect against infection would be of great importance in preventing this disease. At the time of writing (June 2006), a multivalent vaccine, based on self-assembled particles formed of the L1 major capsid protein of each virus, is giving promising results in clinical trials.

Progression from wart to carcinoma involves successive changes in the properties of the HPV-infected cells and is invariably associated with a change from maintenance of the HPV DNA as a free episome to its integration into the host genome and loss of expression of all but the E6 and E7 genes. The selective retention of these genes indicates the importance of E6 and E7 proteins in the development of the tumor. As discussed above (Section 20.3), these two proteins affect key regulators of the cell cycle, but these activities are broadly common to all papillomaviruses. E6 and E7 from high risk HPVs must have some distinct molecular characteristics that promote progression from a benign wart to an invasive carcinoma. This question is still being studied, aided by the observation that only high risk E6 plus E7 can transform primary human keratinocytes (the natural target for these viruses), a property that correlates with their tumorigenic potential (Box 20.6).

SV40 as a possible human tumor virus

SV40 is not naturally an infectious agent of humans. However, in cell culture, human cells are semi-permissive for this virus, indicating a low level of replication. During the 1950s, when the mass vaccination campaign for polio was beginning, the live Sabin vaccine was grown in cultures of African green monkey kidney cells that, it was subsequently realized, contained SV40 as an inapparent infection. Thus, many millions of people were exposed to SV40 during this period when they received the Sabin vaccine. (Current vaccine production methods avoid the risk of such contamination.) More recent epidemiological studies have shown ongoing transmission of SV40 within the human population.

Clearly, there has not been a massive epidemic of SV40-induced human tumors following this mass exposure to the virus. However, there have been sporadic reports of SV40 sequences being isolated from choroid plexus tumors and ependymomas (brain), and more recently mesothelioma (a tumor of the cells lining the body cavity). The latter tumor has long been believed to be strongly associated with exposure to asbestos. It has been suggested that SV40 might act with asbestos to promote this rare tumor since, in cell culture, SV40 acts synergistically with asbestos to transform mesothelial cells. Whenever virus is found in tumor tissue, a crucial question is whether the virus is actively contributing to the disease or merely taking advantage of the unusual cell environment to grow more readily. The data on this point are not conclusive and so a true causative role for SV40 in any of these diseases has yet to be proven.

20.5 HERPESVIRUS INVOLVEMENT IN HUMAN CANCERS

Two human herpesviruses are associated with human cancers. Epstein–Barr virus infection is ubiquitous but the disease outcomes vary greatly. Several types of cancer are apparently caused by EBV in association with different cofactors. The existence and disease-causing potential of the second virus, HHV8 or KSHV, was revealed by the epidemic of immunodeficiency due to HIV-1 infection.

Epstein–Barr virus

EBV infects everyone. Childhood infection is subclinical, but infection of young adults can cause prolonged debilitation (glandular fever or infectious mononucleosis, named after the extensive proliferation of host T lymphocytes in the blood in response to the expanding pool of infected B cells). Whether or not the primary infection is symptomatic, EBV establishes a lifelong latent infection of circulating B cells (see Chapter 15). It has been suggested that primary infection occurs in epithelial cells, but this is now doubted. However, in both these cell populations, EBV infection is associated with specific tumors (Table 20.1).

The stages by which EBV causes cancer are only now beginning to be unravelled. Natural infection results in a permanent pool of infected B cells, proliferation of which is only held in check through the action of T cells. This explains the incidence of EBV-associated B cell proliferative disease after organ transplantation, since such organ recipients have to receive some immunosuppressive therapy to prevent rejection of their grafts. Inadvertent immunosuppression may also play a part in the development of Burkitt's lymphoma, another B cell tumor. This tumor forms in the lymphoid tissue of the jaw, and is found in high incidence in children, particularly boys, aged 6–9 in tropical Africa and New Guinea where the tumors almost universally contain EBV DNA. This geographical distribution is markedly coincident with the distribution of the malaria parasite, *Plasmodium falciparum*, which Dennis Burkitt (who first described the tumor) suggested as a possible disease cofactor. It has been demonstrated that malaria drastically lowers the control of EBV-infected cells by EBV-specific T lymphocytes, by decreasing the proportion of helper T cells in relation to suppressor T cells. Repeated attacks by the parasite are thought to reduce the immune response to the level needed to allow a tumor to establish. Exactly how malaria achieves this immunosuppression is not known.

Besides EBV infection, a feature of Burkitt's lymphoma cells is that they characteristically contain one of three chromosome translocations, which join a specific region of chromosome 8, usually to a site in chromosome 14 or more rarely to sites in chromosomes 2 or 22. These chromosomal translocations place a cellular proto-oncogene, *myc* (see Section 20.6), on chromosome 8 under the transcriptional control of the chromosome regions encoding the immunoglobulin heavy and light chains respectively (chromosomes 14, 2, 22). The latter are of course highly active in B cells, so the effect of the translocations is to deregulate *myc* expression and cause inappropriate production of its protein product, a DNA-binding protein responsible for aspects of transcriptional control in the cell. This then impairs the cell's ability to control its division. Nasopharyngeal carcinoma involves epithelial cells and again shows a marked prevalence in a restricted geographical area, in this case southeast Asia. Here, consumption of salted fish from an early age, which is thought to expose the infected cells to various carcinogens, is thought to be a cofactor in disease development. Possession of certain MHC haplotypes also acts as a risk factor.

Human herpesvirus 8

HHV-8 (also known as KSHV) is the most recently recognized human herpesvirus. It was identified through the cloning of sequences found consistently in a tumor, Kaposi's sarcoma, which has become a defining feature of AIDS, particularly in homosexual men (see Chapter 19). HHV-8 is now thought to be significantly involved in the development of this tumor, which appears only very rarely in the population in the absence of HIV-1-induced immunodeficiency. Another rare tumor associated with HHV-8 is primary effusion lymphoma, a body cavity tumor of B cell origin. The molecular features of the virus that are responsible for its tumorigenic activity are still being defined. HHV-8 is now known to be present at varying levels of prevalence in different populations, but always at a frequency greater than the malignancies believed to be associated with infection. Thus, as with the other examples already described, HHV-8, although suggested to be a cause of certain human cancers, is not sufficient of itself to elicit a tumor.

20.6 RETROVIRUSES AS EXPERIMENTAL MODEL TUMOR VIRUSES

The first retroviruses to be discovered were oncogenic and these viruses were much studied subsequently for this reason. It is now clear that few retroviruses are oncogenic in the context of natural infection. However, the study of oncogenic retroviruses led to the discovery of oncogenes in the host genome and these genes are now understood to be of great importance in the development of cancer.

Highly oncogenic retroviruses that also transform cells in culture

Rous sarcoma virus was isolated in 1911 as a filterable agent capable of causing sarcomas at the site of injection in chickens. It was named after its discover, Peyton Rous, who later received the Nobel prize. This virus is an example of a subset of the oncogenic retroviruses which transforms immortalized cells in culture and causes tumors in animals with high efficiency and rapid onset. Each of these viruses has, in its genome, sequences that have been acquired from the host and genetic analysis reveals that these are responsible for the cell transformation and tumorigenic phenotypes. The acquired genes represent cDNA copies (reverse-transcribed mRNA) from cellular proto-oncogenes (Section 20.1); indeed, analysis of such viruses was the way in which the oncogenic potential of these normal components of our genomes was first identified. The study of these oncogenes has been hugely informative as to the mechanisms which control the growth of mammalian cells and how these can go wrong, leading to tumor development.

What causes a normal cell gene to take on the character of an oncogene when it is captured as cDNA by a retrovirus? Deregulated expression from the powerful viral LTR promoter is at least partly responsible for these genes acquiring oncogenic properties (Fig. 20.4a). Another factor is that, when compared to their cellular counterparts, the acquired oncogenes often have mutations which affect protein function; classic examples are the altered *ras* genes of Kirsten and Harvey murine sarcoma viruses. The same *ras* mutations have also been found in the DNA of human bladder carcinomas which have no viral involvement in their development.

With the exception of Rous sarcoma virus, which acquired its oncogene, termed *src*, 3′ to its *env* gene, all highly oncogenic retroviruses have lost essential viral genes (*gag, pol*, or *env*, see Section 9.8) during the acquisition of their oncogenes. They are therefore defective and can grow only with support from a standard retrovirus, such as the avian (ALV) or murine (MLV) leukemia viruses, from which they derived. The oncogenes play no part in the virus life cycle. The detailed mechanism by which these viruses arose is unclear. Models proposed involve integration of an intact provirus adjacent to a cellular gene, subsequent production of a hybrid mRNA containing viral and cell sequences, and then recombination during reverse transcription to create a provirus with its essential flanking LTRs (see Chapter 8). There is no obvious evolutionary advantage to the virus in acquiring cellular sequences at the expense of its own essential genes, so these events may best be viewed as rare accidents of the retroviral life cycle.

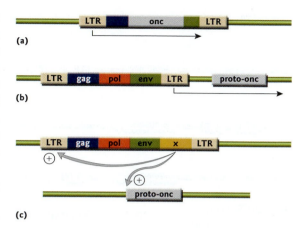

Fig. 20.4 Models of retrovirus oncogenesis involving integration of viral genomes into cellular DNA. (a) Integration of a defective viral genome carrying an oncogene whose transcription is under the control of the strong promoter in the viral LTR. (b) Integration of a virus that lacks an oncogene, so that its right-hand LTR drives expression of a cellular proto-oncogene (proto-onc). (c) Expression of a transcription-enhancing product from viral gene X which affects both viral and cellular transcriptional control, upregulating expression of a proto-oncogene.

Other oncogenic retroviruses

If immortalized fibroblastic cells in culture are infected with standard retrovirus, such as ALV or MLV, very little morphological change occurs. The cells are not transformed nor do they show cytopathology. Despite this, these retroviruses can still produce malignant disease, in this case leukemia, when introduced into a suitable host, but unlike the highly oncogenic retroviruses, the efficiency is low (only a few of the infected animals show disease) and the time taken to observe disease is much greater. Genetic analysis does not reveal any viral oncogene to which the oncogenic properties of these viruses can be allocated. Thus the abilities of the virus to go through a productive cycle and to cause leukemia cannot be separated by mutation.

The clue to understanding how ALV, MLV, etc. cause disease came from an analysis of DNA from the tumors they induced. Although the sites of provirus insertion were not identical, in each case insertion had occurred in the same region of the host genome. Characterization of the surrounding DNA led to the discovery of the *myc* gene, already known from its presence in cDNA form in a highly oncogenic retrovirus. This result led to the concept of retroviral insertional mutagenesis (Fig. 20.4b). By inserting upstream of a proto-oncogene, such as the *myc* gene, the right viral LTR, which contains a powerful promoter identical to that in the left LTR (see Sections 8.6 and 9.8), drives its high level expression. This alters the timing and level of expression of the gene, so causing aberrant cell behavior (activation of *myc* expression is also involved in Burkitt's lymphoma, Section 20.4). In fact, insertional mutagenesis can work even when the insertion is downstream of the target gene, or in the wrong orientation to drive transcription directly via the LTR promoter. This is because the

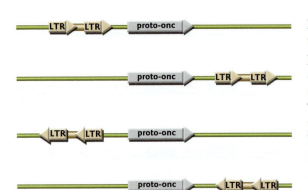

LTRs also contain enhancer elements that will upregulate expression from a gene's endogenous promoter whenever they are placed in reasonable proximity to it (Fig. 20.5). Why does disease caused by these viruses have a long latency? This is because retroviral integration is a random process. Many rounds of infection will typically be needed to give any probability of a provirus inserting in a region of the genome that can have an oncogenic effect.

Fig. 20.5 Four variations on the retroviral insertional mutagenesis model shown in Fig. 20.4b. Integrated proviruses are indicated in brown. Transcriptional regulation of a cellular gene will be altered by the provirus in all four scenarios, because the enhancers in each LTR operate in a position- and orientation-independent manner to upregulate transcription from the gene's endogenous promoter.

20.7 RETROVIRUSES AND NATURALLY OCCURRING TUMORS

Mammary tumors in mice

Given the random nature of retroviral integration, natural retroviral infections might be expected to confer a risk of mutational activation of cellular proto-oncogenes, as is seen experimentally with the avian and murine leukemia viruses. One example of this is the disease observed with mouse mammary tumor virus (MMTV), which is endemic in certain mouse strains and is spread from mother to pup in the milk. When mammary tumors are examined for MMTV DNA, the proviral integration sites are in reasonable proximity to one of a small number of genes, known as the *int* genes. As a result, the expression of the *int* gene will come under the influence of the viral LTR (Figs. 20.4b, 20.5). The MMTV LTR is sensitive to steroid hormone regulation, perhaps explaining why the onset of disease is linked to pregnancy in infected female animals. The fact that specific targeting of these regions by provirus is seen in tumor cells is evidence that the effect on neighboring genes is relevant to the disease process, although the precise pathway by which each of these *int* genes affects cell behavior is not fully understood.

Adult T cell leukemia in humans

The first human retrovirus, T cell lymphotropic virus type 1 (HTLV-1), was not recognized until 1980, followed soon after by HIV-1. HTLV-1 is associated with adult T cell leukemia in a minority of those infected.

Adult T cell leukemia (ATL) is a tumor of CD4+ T lymphocytes. The geographical distribution of ATL cases across the globe is markedly nonuniform, with hot-spots in Japan and the Caribbean Islands. This distribution closely matches that of HTLV-1 infection (seroprevalence), suggesting an involvement of the virus in this disease. Confirming this,

ATL cells always contain integrated HTLV-1 DNA. However, other factors are also involved because, as with all the other examples of human viruses associated with cancer, only a minority (about 1%) of those infected show the disease and there is a 20–40 year interval between becoming infected and developing ATL. One of these factors is the nature of the immune response that is mounted to HTLV-1, and in particular to the Tax protein that it encodes, during the initial infection. If immunological tolerance to Tax is established in an individual, due to infection occurring when very young or because their particular MHC antigens cannot adequately present Tax peptides to cytotoxic T cells, then the chance of subsequently developing ATL is increased.

Analysis of the HTLV-1 genome shows that no acquired oncogene is present; all the genes are required for virus replication and none of them is related to a host cell gene. Also, there is no common region of integration of HTLV-1 in the DNA from different tumors, suggesting that insertional mutagenesis is not involved. A clue to the mechanism is the fact that, despite apparently lacking an oncogene, HTLV-1 can transform T lymphocytes in culture. This activity depends on the viral Tax protein, which is encoded by an additional viral gene to the minimal retroviral *gag*, *pol*, and *env* genes. The probable mechanism of Tax action is shown in Fig. 20.4c, and is somewhat similar to that of the adenovirus E1A protein. Tax is a transcriptional activator that interacts with and alters the activity of a number of host cell transcription factors. Together with these factors, Tax acts on the viral LTR and also on host cell genes such as those for interleukin 2 (IL-2) and its receptor; IL-2 is a key growth factor for T cells. Tax also interferes with Rb and p53 functions (Section 20.3), again promoting loss of control of cell division. These changes alter the cell's growth properties so as to predispose it to accumulate further mutations as it divides, a process which can eventually lead to the cell initiating a tumor.

20.8 HEPATITIS VIRUSES AND LIVER CANCER

Hepatitis B virus chronic infection is strongly associated with liver cancer in later life. With 300 million carriers, this makes HBV probably the most potent human carcinogen after cigarette/tobacco smoking. A second hepatitis virus, HCV, is also strongly associated with liver cancer.

Infection by hepatitis B virus (HBV) may either be resolved by the immune response or become chronic (see Section 14.5), the latter being most frequent in people infected in infancy. Those who become chronically infected have a 200-fold increased risk of developing primary (i.e. derived from liver cells; not metastatic from other locations) hepatocellular carcinoma (HCC) over the rest of the population. In most cases of HBV-positive liver cancer, the tumor cells contain integrated HBV sequences. The integration site is the same in all cells of a tumor, indicating that integration occurred in a single cell which subsequently divided to form the tumor. There is however no consistent integration site among different patients. These observations are suggestive of a direct role of virus function(s) in the development of disease. As the HBV X protein can both transactivate transcription and disrupt p53 function (Section 20.3), it has been suggested as the likely culprit. However, its expression is usually undetectable in tumor cells and the integrated sequences are usually "scrambled" by rearrangements so as to prevent most viral gene expression. Also, any proposed mechanism must take account of the long time lag (perhaps 40 years) between primary infection and the appearance of cancer. Therefore, an alternative theory is that the virus exerts an indirect effect, through the long term damage to the liver which it inflicts. In a chronic infection, cells are constantly being lost through immune-mediated destruction of infected cells. The liver responds by regeneration and this persistently increased rate of division among hepatocytes may then predispose the cells to accumulate mutations, ultimately leading by chance to the emergence of a cell with a set of mutations which gives it cancerous properties.

A persuasive piece of evidence for the theory that HBV can cause liver cancer through the indirect effects of chronic liver damage has come from the more recent discovery and characterization of hepatitis C virus (HCV), infection by which also predisposes to liver cancer after a long delay. HCV is the only tumor virus listed in Table 20.1 which does not have any DNA involvement in its replication cycle. Because its genome is always in the form of RNA, none of the mechanisms by which genetic material might persist in a dividing population of pre-cancerous cells (Section 20.2) are available to it. Therefore "hit-and-run" or indirect mechanisms like that proposed for HBV are the only options. Given the long delay between infection and disease, the latter is the more likely model. This conclusion is also supported by the fact that other, nonvirological, agents which cause chronic liver damage, such as alcohol, also give rise to an increased risk of hepatocellular carcinoma. The extent of HCV infection in developed countries is only now becoming apparent and promises to be a major public health problem.

20.9 PROSPECTS FOR THE CONTROL OF VIRUS-ASSOCIATED CANCERS

Vaccination against tumor-causing viruses has the potential to reduce substantially the incidence of virus-associated cancers. A vaccine for hepatitis B virus already exists and is in widespread use, and vaccines for the tumor-causing papillomaviruses are in an advanced stage of development and testing.

The power of vaccines against tumor viruses to prevent the tumors that these viruses cause has been amply demonstrated with Marek's disease virus of chickens, a herpes virus, which causes a cancer of cells of the feather follicles, and with successful experimental vaccines against herpesvirus saimiri and EBV. Immunization against tumor viruses is as easy (or as difficult) as immunization against any other virus. An excellent vaccine for humans against HBV is already on the market, and is now used in nationwide vaccination programs in most countries. In the long term, reduction in the numbers of people becoming infected with HBV should be reflected in declining incidence of HCC.

Since vaccines rarely succeed in completely blocking a virus infection, only in rendering it subclinical, and virus-initiated tumors may arise from a single infected cell, it can be argued that vaccines that prevent standard disease caused by a virus may be less effective at preventing any long term oncogenic consequences of infection. However, the probability of an individual infected cell later giving rise to a tumor is very low — the oncogenic potential of the virus comes from the large number of such cells produced during the initial infection, each of which is then at risk of progressing into a tumor. Thus, by reducing the extent of the initial virus infection, a vaccine should substantially reduce the number of infected cells at risk of malignant progression and hence the probability of cancer caused by that virus arising.

To combat other known virus-induced cancers, several new vaccines are needed, e.g. to EBV, HCV, and HPV-16 and 18. Indeed, a vaccine for these papillomaviruses is in an advanced stage of development and excellent results from clinical trials have been recently reported. One problem in applying such vaccines may be that each of these viruses only rarely causes cancer. Even with a disease as emotive as cancer, will there be sufficient interest to persuade people to avail themselves of immunization, or to have their children vaccinated, given that the disease consequences are so uncertain and distant in time, and that any vaccination carries some small risk? Will potential recipients understand that, at best, such vaccines can only protect against virally induced tumors, when many still think of cancer as a single disease? If not, when many vaccinees still experience cancer the public may feel that such vaccines have failed and acceptance may decline. Development of cancer vaccines must therefore proceed in tandem with the more traditional anticancer approaches involving the elimination of cofactors, education about the risk of cofactors to inform people's lifestyle choices, improved early diagnosis and better treatment. These are as applicable to virus-induced cancers as to other cancers.

KEY POINTS

- Normal cells cannot grow and divide indefinitely in culture whereas immortalized, transformed, or tumor-derived cells can.
- The underlying basis of the changes in cell behavior that lead to tumor formation is the acquisition of mutations.
- Some viruses carry genes that have the effect of substituting for one or more of the mutations that would be needed to convert a normal cell into a tumor cell.
- For a virus to contribute to tumorigenesis, it must carry one or more relevant genes that become inherited by daughter cells at division and the cell must not be killed by the infection.
- Key elements of the growth control pathways of cells are targeted by many viruses in order to favor their own growth.
- Diverse viruses are associated with specific human cancers, always in conjunction with one or more cofactors. No virus infection is known that leads inevitably to cancer development in humans.
- Collectively, viruses are causally involved in the development of about 20% of human cancers.
- Vaccines against specific viruses may protect against the development of the cancers with which they are associated.

QUESTIONS

- Discuss the mechanisms by which certain viruses are thought to subvert the normal growth controls of cells in culture and to contribute to tumor formation *in vivo*.
- Write an essay on the significance of virus infections as a cause of human cancer.

FURTHER READING

Barmak, K., Harhaj, E., Grant, C., Alefantis, T., Wigdahl, B. 2003. Human T cell leukemia virus type 1–induced disease: pathways to cancer and neuro-degeneration. *Virology* **308**, 1–12.

Block, T. M., Mehta, A. S., Fimmel, C. J., Jordan, R. 2003. Molecular viral oncology of hepatocellular carcinoma. *Oncogene* **22**, 5093–5107.

Bouchard, M. J., Schneider, R. J. 2004. The enigmatic X gene of hepatitis B virus. *Journal of Virology* **78**, 12725–12734.

Butel, J. S. 2000. Viral carcinogenesis: revelation of molecular mechanisms and etiology of human disease. *Carcinogenesis* **21**, 405–426.

Classon, M., Harlow, E. 2002. The retinoblastoma tumor suppressor in development and cancer. *Nature Reviews: Cancer* **2**, 910–917.

Dilworth, S. M. 2002. Polyoma virus middle T antigen and its role in identifying cancer-related molecules. *Nature Reviews: Cancer* **2**, 1–6.

Eiben, G. L., Da Silva, D. M., Fausch, S. C., Le Poole, I. C., Nishimura, M. I., Kast, W. M. 2003. Cervical cancer vaccines: recent advances in HPV research. *Viral Immunology* **16**, 111–121.

Endter, C., Dobner, T. 2004. Cell transformation by human adenoviruses. *Current Topics in Microbiology and Immunology* **273**, 163–214.

Gazdar, A. F., Butel, J. S., Carbone, M. 2002. SV40 and human tumors: myth, association or causality? *Nature Reviews: Cancer* **2**, 957–964.

Harris, S. L., Levine, A. J. 2005. The p53 pathway: positive and negative feedback loops. *Oncogene* **24**, 2899–2908.

zur Hausen, H. 2002. Papillomaviruses and cancer: from basic studies to clinical application. *Nature Reviews: Cancer* **2**, 342–350.

Schultz, T. F., Sheldon, J., Greensill, J. 2002. Kaposi's sarcoma associated herpesvirus (KSHV) or human herpesvirus 8 (HHV8). *Virus Research* **82**, 115–126.

Young, L. S., Rickinson, A. B. 2004. Epstein–Barr virus: 40 years on. *Nature Reviews: Cancer* **4**, 757–767.

Also check Appendix 7 for references specific to each family of viruses.

21

Vaccines and antivirals: the prevention and treatment of virus diseases

Most vaccines prevent infections (prophylaxis); most drugs treat infections (therapy). Since prevention is always better than cure, and because really effective antiviral therapy has been difficult to achieve, there is great importance attached to producing safe and effective vaccines against viruses.

People are concerned to protect themselves, their plants and animals against death, disease and economic loss caused by virus infections. Historically, immunization with vaccines (vaccination) has been far more effective than antiviral chemotherapy, but this is changing as more antiviral drugs come on stream. Nonetheless the better option is to prevent infection with a vaccine, than to treat it with drugs. Not all vaccines work, or work as well as they should, and for some viruses (notably HIV-1) there is still no vaccine available despite years of work. These problems will be discussed below. The rationale of vaccination is to raise virus-specific immunity without the individual having to experience the disease. Since plants do not have an inducible immune system, prevention of plant virus diseases has relied upon other means, such as selective breeding of plants that are genetically resistant to the virus or its vector, or by control of the vector. Control of animal virus diseases through improvements in nutrition, management of transmission, and education is also a vitally important aspect of prevention, especially in the underdeveloped world.

21.1 PRINCIPAL REQUIREMENTS OF A VACCINE

Properties of an ideal vaccine

- *Cause less disease than the natural infection; ideally no clinical side effects.*
- *Be genetically stable.*
- *Stimulate effective and long-lasting immunity in the individual and/or stimulate sufficient herd immunity to deprive virus of the susceptibles needed for its survival.*
- *Have public support.*
- *Be affordable.*

A vaccine should cause less disease than the natural infection

Conventional vaccines comprise either infectious ("live") or noninfectious ("killed") virus particles (Table 21.1). The process of producing a virus which causes a reduced amount of disease for use as a live vaccine is called *attenuation*. The disease-causing virus is referred to as *virulent* and the attenuated strain as *avirulent*. Note that these are only relative terms – there are no absolutes. Avirulent vaccines have been obtained by selecting for naturally occurring avirulent variants. This is achieved empirically, and experience has shown that it can be helped by multiplication in cells unrelated to those of the normal host, by multiplication at a subphysiological temperature, or by recombination with an avirulent laboratory strain. An alternative is to use a natural strain that is antigenically related to the virulent strain and which causes less disease in that host. This was the very first approach to vaccination, known since 1798 when Edward Jenner used cowpox virus to vaccinate against smallpox. Vaccinia virus, a related virus of unknown origin, is now used for the same purpose, and has given rise to the term "vaccine" which is now synonymous with any immunogen used against an infectious disease.

Naturally, killed vaccines should cause no disease at all. However, since the vaccine is made from virulent virus, it is essential to kill every infectious particle present (Fig. 21.1). An advantage of killed preparations is that any other unknown, contaminating viruses will probably also be killed in the same process. However, a major disadvantage is that the killed virus does not multiply, so that an immunizing dose has to contain far more virus than a dose of live vaccine, and repeated doses may be required to induce adequate levels of immunity. This increases both the cost and the amount of nonviral material injected; the latter may result in hypersensitivity reactions when these substances are experienced again. Another problem is that killed vaccine does not reach and stimulate mucosal immunity in the intestinal and respiratory tracts where virus normally gains entry to the body (see Fig. 12.3), and may not stimulate the type of immunity needed for protection. However, providing

Fig. 21.1 Exponential kinetics of virus inactivation. Note the "resistant" fraction which is inactivated more slowly.

Table 21.1

Some human and veterinary virus vaccines.

Infecting virus	Family	Live or killed vaccine	Disease (if named differently from the virus) and other comments
Human viruses			
Hepatitis A	Picornavirus	Killed	Hepatitis
Hepatitis B	Hepadnavirus	Killed*	Hepatitis
Influenza A and B	Orthomyxovirus	Killed*, live	
Measles	Paramyxovirus	Live[†]	
Mumps	Paramyxovirus	Live	
Polio	Picornavirus	Killed, live[‡]	Poliomyelitis
Rabies	Rhabdovirus	Killed	
Rubella	Togavirus	Live	German measles
Variola	Poxvirus	Live vaccinia virus[§]	Smallpox
Yellow fever	Flavivirus	Live	
Varicella-zoster	Paramyxovirus	Live	Chicken pox
Veterinary viruses			
Canine distemper	Paramyxovirus	Live	Affects dogs
Equine influenza	Orthomyxovirus	Killed	Affects horses
Foot-and-mouth	Picornavirus	Killed, live	Affects cattle and pigs
Newcastle disease	Paramyxovirus	Live	Fowl pest; affects poultry
Parvo	Parvovirus	Killed, live	Causes death in young dogs and cats
Pseudorabies	Herpesvirus	Killed, live	Aujeszky's disease of pigs
Marek's disease virus	Herpesvirus	Live turkey herpesvirus[§‖]	Lymphoma in chickens

*The HBV vaccine is the surface (S) antigen protein encoded by recombinant DNA – a subunit vaccine; the killed influenza vaccine is a natural subunit vaccine comprising the virion hemagglutinin and neuraminidase surface antigens.
[†]Vaccines may be combined and administered together, such as MMR – measles, mumps and rubella viruses.
[‡]The killed vaccine was used successfully in Scandinavian countries, and is now being used instead of the live attenuated vaccine in countries where the virus is no longer endemic.
[§]A nonidentical virus that provides protection.
[‖]This is the first vaccine-induced protection from cancer. Likewise the hepatitis B vaccine is the first vaccine against a human (liver) cancer.

that the killed vaccine is a potent immunogen, the high levels of serum IgG which result from injection can often serve in place of local immunity, presumably because there is a sufficient concentration of IgG to diffuse to the extremities.

Genetic stability

An attenuated live virus vaccine must not revert to virulence when it multiplies in the immunized individual, but even the excellent poliovirus vaccine is not perfect in this regard (Box 21.1). Vaccines which are formed

Box 21.1

Evidence for the basis of poliovirus vaccine attenuation and its reversion to virulence

There are three poliovirus serotypes and immunity to each is required for protection against poliomyelitis. Accordingly the vaccine is trivalent. The avirulent type 3 component of poliovirus live oral vaccine is known to spontaneously revert to virulence, with the result that there is a very, very low incidence of vaccine-associated paralytic poliomyelitis (approximately one case per 10^6 doses of vaccine for the first dose and diminishing to zero by the third dose). Viruses isolated from such patients are neurovirulent when inoculated into monkeys in the test normally applied to ensure the safety of new batches of vaccine.

Long after the vaccine had been made, sequencing techniques were developed, and Jeffrey Almond and Philip Minor in the UK compared the wild-type Leon strain of type 3 poliovirus with the vaccine strain that was derived by Albert Sabin. They found that only 10 of the 7431 nt that comprise the poliovirus genome had mutated during the attenuation process (which, incidentally, was achieved empirically by adaptation of the virus to growth in cell culture at $31°C$). There were two mutations in the 5′ noncoding region, one in the 3′ non-coding region, three leading to amino acid changes and four silent mutations in the coding regions (Fig. 21.3a). Next they compared the vaccine strain with the virulent revertant, and found that revertants have seven nucleotide changes, one in each of the terminal noncoding regions and five in the coding regions. Of the latter, one is silent (at position 6034) and four result in amino acid changes in structural proteins (Fig. 21.3b), but it is not possible to conclude which of the mutations is associated with the return to virulence. However, comparison of revertant genome with the attenuated genome shows that they have one nucleotide change in common, at residue 472 in the 5′ noncoding region from cytosine to uridine and back to cytosine. The implication that this change was responsible for reacquisition of virulence was strongly backed up by the finding that the 472U→C occurred in all viruses responsible for cases of vaccine-associated illness.

This is exciting stuff, but the final story has not yet been told, as it is not obvious how mutation of residue 472 in the 5′ noncoding region affects virus function. Molecular modeling shows that the 472U→C greatly alters the secondary structure of the viral RNA, but exactly how this affects virulence is not known. Virus with 472U has reduced replication and translation and decreased interaction of its RNA with a cellular RNA-binding protein in neuronal cells, but not in cells of non-neuronal origin. Secondly, the mutation at nucleotide 472 increases but does not fully restore neurovirulence to that of the original Leon strain. Thus one or more of the other residues may contribute to virulence. In anatomical terms the avirulent vaccine virus multiplies solely in the epithelial cells of the small intestine, while virulent virus invades the central nervous system. It remains to be explained how the 472 mutation is responsible for this difference in cell tropism.

by inactivating virulent strains of viruses must not be restored to infectivity by genetic interactions with each other or with viruses occurring naturally in the recipient.

Effective and long-lasting immunity

A vaccine should produce both effective and long-lasting immunity. The requirement for these properties is both scientific and sociological. As explained in Chapter 14, different viruses are susceptible to different parts of the immune system; hence it is necessary for the vaccine to stimulate immunity of the correct type, in the correct location, and in sufficient magnitude for it to be effective. This is most readily achieved with a live vaccine as its interaction with the immune system will most closely mimic that of the virulent virus. Killed vaccines struggle to achieve this, for example the killed influenza virus vaccines currently in use, which are administered by injection and do not raise virus-specific IgA antibody at the mucosal (respiratory epithelial) surface where it is needed, or virus-specific T cell-mediated immunity. However, sufficient IgG to be protective is thought to diffuse from the bloodstream to the respiratory surface in about 70% of individuals immunized. If effective immunity is not achieved, the virus being immunized against may still be able to multiply. This can occur with foot-and-mouth disease virus of cattle (a picornavirus). Worse still, partial immunity provides a selection pressure which favors the multiplication of naturally occurring antigenic variants, which in time replace the parental virus strain. Even animals with effective immunity against the original strain are not protected from the new strain, and the long, expensive process of vaccine development and vaccination has to begin again. Immunogenicity can be enhanced by mixing virus with an *adjuvant*, which stimulates the immune system in ways that are poorly understood. The virus binds to the adjuvant or, with oil-based adjuvants, is emulsified. The latter acts as a depot and releases the immunogen over a long period which can result in a greater immune response. However, many adjuvants cause local irritation and currently only alum (aluminum phosphate and aluminum hydroxide) and MF59 (a squalene oil in water emulsion) are licensed for use in humans.

Individual versus herd immunity

It might seem that every individual should be immunized, but a 100% immune population is not usually needed to eliminate a virus (which is as well, as it can never be achieved). What is required is a sufficient percentage of immune individuals to break the chain of transmission of the virus. Thus *herd immunity*, rather than individual immunity, is the key issue. A natural experiment that illustrates this point is measles virus in island populations. Natural measles infection leaves people immune to reinfection, and a population of at least 500,000 is needed to generate a sufficient supply of susceptible children to prevent the virus from dying out (Section 18.3). The exact percentage of immune individuals required for herd immunity is different for each infectious agent, and can be calculated.

The public perception of vaccines

Vaccines are only beneficial if people are persuaded to take them. Problems arise if by chance a person becomes ill at the time he or she is given a vaccine, as the vaccine may be blamed for the illness even though there is no connection. Another problem is the unrealistic expectation that vaccines are totally harmless; there is always a risk of side effects, however small, that has to be balanced against the risks of not vaccinating. The public seems to find difficulty with the concept of risk. For example, in countries where poliovirus was endemic and every child was exposed, there was a 1 in 100 chance of contracting paralytic poliomyelitis, while the risk of poliomyelitis from the live vaccine due to it reverting to virulence is around 1 in 1000,000 (Box 21.1). This seems like a good deal – unless it is you or your child who develops the vaccine-associated disease. Parents in particular feel that they are to blame, and this dissuades other parents from allowing the vaccine to be given to their children. What is worse is that perception of an adverse event associated with one vaccine may persuade them from taking up other vaccines. The problem is exacerbated by heavy media publicity. The real problem is that as the disease becomes rarer due to the success of the vaccine, the risk from vaccine use becomes closer to the risk of natural infection. Both are extremely low. It is for this reason that the attenuated poliovirus vaccine has been replaced with the lower risk inactivated vaccine. Vaccine manufacturers strive to make vaccines safer but there is risk in any medical intervention, and a vaccine can never be absolutely safe. Finally, during a mass vaccination campaign, it is inevitable that a few vaccinated people will become ill, and even die, as a result of events that are completely unconnected with the vaccine, and this would have happened at exactly the same rate had the vaccination not taken place. Thus it is erroneous to say, as some do, that the vaccine caused those illnesses. It is important that governments minimize bad publicity for vaccines, and compensate all who have been damaged as the result of vaccination.

Public faith in any medicine is rapidly lost if it is not effective in the majority of cases (see above). Consequently, to avoid bad publicity, a vaccine must protect the majority of individuals who have received it. Vaccination should result in effective immunity, and this should also last for years, preferably a lifetime. However, it is surprisingly difficult to persuade people to come forward to be immunized. A single-shot vaccine

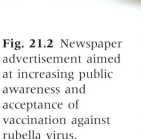

SALLY'S
MUM GOT
GERMAN
MEASLES.

LOOK WHAT
SALLY GOT

Fig. 21.2 Newspaper advertisement aimed at increasing public awareness and acceptance of vaccination against rubella virus.

requiring only one visit to the clinic is ideal, but has been realized only for measles, mumps, rubella (MMR) and yellow fever viruses. All other vaccines require at least two doses. This is an even greater problem in the developing world, where there is little healthcare infrastructure to implement a multidose vaccination regimen. Some employers have circumvented this problem by offering vaccination at the place of work. Vigorous advertising campaigns are widely used also (Fig. 21.2) to overcome public apathy. In tropical countries, where vaccination is often most needed, there is the added problem of ensuring that a vaccine has a good shelf-life (i.e. that its infectivity or immunogenicity is stable for periods when refrigeration may not be available), or vaccination will be ineffective.

Another hope for the improved efficacy of killed vaccines is their administration by mouth or intranasally. This both avoids "jabs" (Section 21.2) and stimulates mucosal immunity (see Fig. 12.3). Intranasal vaccines look the more promising (Box 21.3), as oral vaccines suffer from problems of digestive enzymes and immunological tolerance of foreign (food) antigens in the gut.

21.2 ADVANTAGES, DISADVANTAGES, AND DIFFICULTIES ASSOCIATED WITH LIVE AND KILLED VACCINES

Inactivation of infectivity

A killed vaccine must not be infectious, yet it should be sufficiently immunogenic to stimulate protective immunity. Inactivating agents should target the viral nucleic acid and not just attack virion proteins, since nucleic acid alone can be infectious and could be released by the action of the cell on virion coat proteins. Secondly, inactivating agents must lend themselves to the industrial scale of manufacturing processes. Formaldehyde and β-propiolactone are two inactivating agents which have been used. The former reacts with the amino groups of nucleotides and cross-links proteins through ε-amino groups of lysine residues. β-propiolactone inactivates viruses by alkylation of nucleic acids and proteins. The basic problem of preparing a killed vaccine is that every one of the 1,000,000,000 (10^9) or so infectious particles which are contained in an immunizing dose of virus must be rendered noninfectious, and this was the problem which faced Jonas Salk before the first poliovirus vaccine could be presented to the general public in 1953. Inactivation of the infectivity of polio- and other viruses is exponential and has the kinetics shown in Fig. 21.1. Frequently, it is found that a small fraction of the population is inactivated far more slowly than the majority, so that the whole virus population has to be kept in contact with the inactivating agent for much longer than is predicted by the initial rate of inactivation. This has the concomitant risk that immunogenicity will be altered or destroyed. Examination of the "resistant" fraction shows that it results from inefficient inactivation of particles trapped in aggregates or clumps. An added problem is the risk associated with growing large quantities of a virulent strain of a pathogenic virus. Despite these difficulties, the inactivated Salk vaccine has been used very successfully. For live vaccines, the problem is ensuring that it is genetically stable and remains attenuated. Albert Sabin overcame these problems for poliovirus, and attenuated all three serotypes to produce the successful live vaccine, which has been in use since 1957. Nonetheless, the type 3 vaccine reverts to virulence at the rate of one case of poliomyelitis per 10^6 first doses of vaccine. The molecular explanation of attenuation and reversion of this vaccine is discussed in Box 21.1 and illustrated in Fig. 21.3.

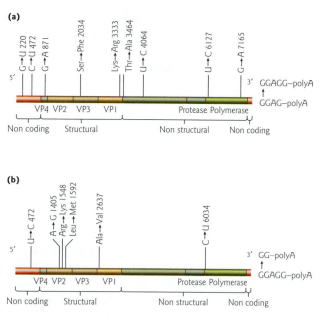

Fig. 21.3 Sequence changes occurring (a) when the original type 3 poliovirus wild-type neurovirulent Leon strain was attenuated to form the Sabin vaccine, and (b) when the vaccine strain reverted to the neurovirulent 119 isolate. (Drawn from information in Almond, J. W. 1987 *Annual Review of Microbiology* **41**, 153–180.)

Routes of administration, and a public relations problem

The means of administration is an important factor in persuading people to accept a vaccine. Injections via the traditional "jab" (hypodermic syringe and needle) are painful and unpopular, particularly with small children and their mothers. A compressed-air device, which also impels vaccine through the skin, was developed for the military in the USA, and is coming into more general use. A painless alternative for respiratory viruses is inhalation of an aerosolized form of the vaccine, and this has the additional advantage of getting it to the appropriate local site (Box 21.3). A disadvantage of killed viruses is that they do not multiply, and thus require at least two injections to provide a primary and then a secondary immune response, while live vaccines like the natural infection should take only one. However, in practice most vaccines require more than one dose.

Cost

The production and use of vaccines in humans is subject to legislation by the Food and Drug Administration in the USA, by the Department of Health in the UK, and similar agencies elsewhere. The precautions laid down for the production of a safe vaccine are necessarily stringent. However, this means that the cost of the final product will be correspondingly high. Precautions increase as knowledge is obtained about hitherto undetected viruses in cells used in vaccine manufacture, or other potential hazards, and as they do so the costs increase yet again. Yet another problem, particularly in the USA, is the cost to manufacturers of defending lawsuits where vaccine-induced damage is alleged; this trend is now spreading to other countries. Regulations governing the use of veterinary vaccines are not so demanding and the products are correspondingly cheaper. The other major aspect which directly affects cost is the quantity of vaccine which has to be produced. Live vaccines are inexpensive as only a small dose is needed to initiate an infection, whereas an immunizing dose of killed, nonmultiplying vaccine costs many times more. For example live and killed poliovirus vaccines cost around US 7 cents and US$3 per dose respectively. Finally, the costs involved with packaging, distribution, administration, and the syringe and needle if injection is needed, are considerable. These all add up to make vaccination a luxury which poor countries, usually those in the greatest need, cannot afford. Provision of vaccines is one of the ways in which the richer countries can, and often do, provide valuable aid to such countries.

Age of vaccinees

All vaccines are tested exhaustively for safety and efficacy in adults before they are adopted for general use. However, the immune responses and susceptibility to some infections of infants and the elderly differ significantly from this norm, and these groups require separate testing. The susceptibility of infants to infection/disease is different from that of mature adults in that they are less susceptible to, for example, poliovirus, cytomegalovirus, and Epstein–Barr virus, but more susceptible to, for example, rotavirus and human respiratory syncytial virus, so new live vaccines are introduced with caution. In addition, the immune system of infants is not completely mature until 2 years of age, with different parts developing at different rates, and may not be able to give the response needed to resist a particular infection. The immune response of the elderly (like all body systems) gradually declines with age, and may not respond as vigorously to the same vaccine dose as that of younger adults.

Thus a more aggressive vaccination regimen may be required to achieve the same level of immunity. In addition, the elderly are more susceptible to certain infections (e.g. rotavirus, human respiratory syncytial virus, influenza A virus), which again has implications for safety of live vaccines. For these reasons the new live attenuated influenza A vaccine is licensed only for those aged 5–49 years.

A vaccine case study: human influenza

Influenza has major health and economic implications, even in nonpandemic years (Box 21.2). There is continuous global monitoring for new shift and drift viruses that is overseen by the World Health Organization (WHO). Shift viruses are the most serious because of the explosive outbreaks and high mortality they can cause (Section 17.5). So far, these have all first appeared in winter (June, July) in the Southern hemisphere. This gives the vaccine manufacturers just 6 months to produce a new vaccine in time to protect people in the Northern hemisphere (January, February). The correct choice of vaccine strains is vitally important.

Until recently only a killed vaccine was available for protection against influenza. This is prepared by growing virus in embryonated chicken's eggs, although it is hoped to move this to cell culture. Purified virus is then treated ("split") with detergent to release the HA and NA proteins from the lipid membrane of the virus, and then centrifuged again to separate the HA and NA from the other viral proteins which are pyrogenic (cause fever). The vaccine is trivalent, and comprises the HA and NA proteins from the currently circulating H1N1 and H3N2 viruses and the current influenza B virus. A single subcutaneous injection is given. The vaccine is recommended only for the elderly, people with chronic clinical problems (mainly of the heart, lungs, and kidneys), diabetics, the immunocompromised, and key healthcare workers. It is not an ideal vaccine as it produces serum antibody but no T cell-mediated or mucosal immunity, and has to be repeated annually, even if the vaccine components have not been changed. Its efficacy in preventing influenza is about 70%.

Box 21.2

Every year in the USA influenza causes

- 70 million lost working days.
- 38 million lost school days.
- 17–50 million infections.
- 36,000 deaths.
- 3–15 billion lost dollars in associated costs.

Box 21.3

FluMist™ – a new live attenuated flu vaccine that is inhaled

For years a killed vaccine has been used to combat the flu, but this vaccine is injected and does not stimulate local respiratory tract mucosal or T cell-mediated immunity. The new vaccine is a *cold-adapted* (*ca*) mutant virus, so called as it was selected to grow at subphysiological temperatures. This procedure empirically selects virus mutants that are attenuated for humans, and was pioneered by Dr Hunein Maassab of the University of Michigan, USA. There are master mutated A and B viruses, both of which have *ca* mutations in the genes encoding nonenvelope proteins. The mutated genes responsible are then used in reverse genetics to construct an attenuated virus that has the envelope proteins of currently circulating virus. MedImmune Vaccines Inc. in conjunction with Wyeth Vaccines Inc. produce the vaccine, which was licensed in 2003 for those aged 5–49 years. It is not deemed safe for those of more advanced years, with asthma, or the immunocompromised. The vaccine is given as an aerosol, which gets it into the respiratory system, where it causes a subclinical respiratory infection that gives the desired local immunity which the killed vaccine fails to do. Children aged 5–8 receive 2 doses (priming and boost), and those aged 9–49 years 1 dose, on the premise that their immune systems have already been primed by a natural influenza infection. Like all flu vaccines, FluMist™ comprises strains matching the envelope proteins of currently circulating influenza A H3N2, H1N1 and influenza B viruses. Vaccination costs US$46 – three times as much as the killed vaccine. In trials it protected 85% of 18–41 year olds from experimental virus challenge, and 87% of 5–7 year olds from natural infection. As antigenic drift proceeds new antigenically updated viruses will have to be made, and people re-vaccinated.

Recently after much preliminary testing, a live attenuated influenza vaccine was licensed in the USA for use in healthy people (Box 21.3). Flu vaccine measures are now supported with specific anti-influenza antivirals (Section 21.8).

Multivalent vaccines

The immune system can respond to more than one antigen at a time; hence it is possible to immunize with a vaccine "cocktail." Killed multivalent vaccines have advantages in minimizing the number of injections and the resulting inconvenience to vaccinees and the healthcare system, and in practice, this gives good immunity – a UK vaccine comprises five elements that protect against diphtheria, tetanus, whooping cough, *Hemophilus influenzae* B, and poliovirus). Multicomponent live vaccines are more problematic as there may be mutual interference in multiplication, possibly as a result of the induction of interferon (see below). However, the live and killed poliovirus vaccine contains the three serotypes, and a triple live vaccine of measles, mumps and rubella viruses (MMR) is in routine use.

Postexposure vaccines and rabies

Once signs and symptoms of an infection are apparent, it is usually too late to immunize. It might appear that the very effective postexposure vaccination against rabies is an exception, but this is done at the time of the suspected *inoculation* of virus by the bite of a rabid animal. It works because transport of virus along peripheral nerves to its target cells in the spinal cord and brain takes several weeks, and the disease does not start until it gets there, by which time the vaccine has induced immune responses. Incidentally, the modern rabies vaccine is an inactivated vaccine produced in tissue culture. This has none of the problems of Louis Pasteur's infected rabbit brain suspension which required multiple injections into the abdomen and was prone to cause life-threatening hypersensitivity reactions.

21.3 PEPTIDE VACCINES

From the foregoing pages, the reader will appreciate that the manufacture of vaccines is a difficult process. The central problem is that of safety – making sure that a killed vaccine contains no residual virulent particles or that a live attenuated vaccine does not revert to virulence. Neither can be guaranteed absolutely. Safety and the attendant problems of expense and efficacy have encouraged research into alternative ways of producing vaccines.

Understanding antigenic determinants

Most killed vaccines (Table 21.1) consist of whole virus particles that in the main stimulate an antibody-based protective immunity, which is directed to only a minor part of the surface structure of the virus. The rest is either nonimmunogenic or stimulates an immune response which is not protective. (This observation does not discount the importance of cell-mediated immunity against virion or nonvirion proteins and this is actively being sought for future vaccines.) However as a result of the limited immunogenicity of the virion surface, research has pursued a reductive approach to determine if a vaccine can be made that consists only of those vital immunogenic regions of the virus particle.

X-ray crystallographic analysis of the influenza virus hemagglutinin (HA) protein, and of natural or laboratory-derived antigenic mutants, has shown that only five regions of the molecule (antigenic sites) bind neutralizing antibody (Fig. 21.4), and antibodies that bind other regions of the HA are not neutralizing. Thus knowing the sequence of the HA

Fig. 21.4 Diagram of an influenza A virus hemagglutinin spike deduced from X-ray crystallographic analysis. The spike is composed of three identical monomers. Only when antibody binds to the red areas is the virus neutralized. The attachment site that binds to *N*-acetyl neuraminic acid on the host cell is in black.

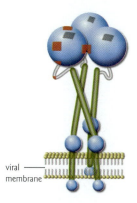

viral membrane

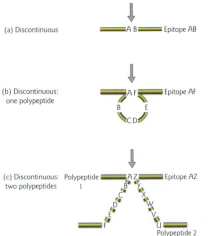

Fig. 21.5 Antibody epitopes constructed in various ways from segments of polypeptide sequence represented by an arbitrary two-letter code and viewed by the immune system in the direction arrowed.

protein, it is possible to chemically synthesize a permuted series of overlapping peptides that cover the antigenic sites and, by immunizing animals with them, discover peptides that can stimulate immunity to the virus. However, such peptides only mimic continuous epitopes (Fig. 21.5), and also take no account of the three-dimensional structure that the antigenic site possesses. Even more complex are discontinuous epitopes (Fig. 21.5b,c) as they cannot be mimicked by linear sequence. However, study of the protein–antibody crystal structure can reveal which residues of the protein are in contact with the antibody para-tope, and it may eventually be possible to synthesize a peptide (e.g. -AF- in Fig. 21.5b) that mimics these contacts.

From the discussion so far, the reader might conclude that a vaccine could, in theory, consist solely of peptides in the correct conformation. However, there is more to the story and we should now take a step back to consider how the immune system is stimulated to synthesize antibodies. Analysis of the immune system (see Chapter 12) has shown that CD4[+] helper T cells are needed to assist B lymphocytes to become plasma cells, and that T cell epitopes usually differ from those that bind neutralizing antibodies. Thus a minimal peptide vaccine has to have two epitopes, one to stimulate the helper T cell and the other to stimulate the B lymphocyte.

Prediction of epitopes

Although epitopes on the influenza virus HA protein have been located, as described above, such work is laborious and expensive. One short-cut is based on the understanding that epitopes are likely to be parts of the molecule which are rich in hydrophilic amino acids. These will extend into the aqueous environment and be easily "seen" by cells of the immune system. Regions of the protein that are rich in hydrophobic amino acids are generally folded into the internal parts of the molecule, where they are hidden from sight. All one needs to determine likely antigenic sites in a protein is its amino acid sequence and a computer program to search for hydrophilic regions, such as that devised by Kyte and Doolittle (Fig. 21.6). Hydrophilic peptides are then synthesized chemically, covalently linked to a protein carrier molecule to provide the necessary CD4[+] T helper cell epitope, and used to stimulate antibody. The resulting antiserum almost invariably reacts with the stimulating peptide, but the key tests of the antiserum to determine if the computer has guessed right are reactivity with the *native protein*, the ability to neutralize virus *in vitro*, and above all the ability to protect against infection *in vivo*. The procedure has been used successfully with hepatitis B surface antigen (HBSag). Recent refinements have suggested that antigenic determinants are likely to be not only hydrophilic but also those parts of a protein with the highest mobility. Even though epitopes can be predicted, it is still not possible to make a peptide with a desired conformation (e.g. as in Fig. 21.5). A free peptide will take up all possible conformations at random, and antibodies will be made to all of these. It is a matter of luck if there is enough antibody present to react with the conformation that the peptide takes up in the native protein.

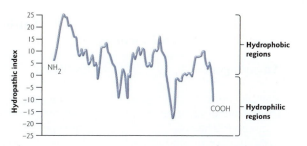

Fig. 21.6
Identification of hydrophilic regions (putative epitopes) of a protein by calculating the hydropathic value of each segment of amino acid sequence across a moving window of seven residues, i.e. the amino acid in question and the three residues on either side.

21.4 GENETICALLY ENGINEERED VACCINES

Genetic engineering offers the chance to create vaccines for viruses that are not amenable to the classical techniques of attenuation or killed whole virus preparation. Recombinant vaccines may be either noninfectious protein, produced from cloned genes, or infectious recombinant viruses, where a safe vaccine strain is used to carry genes from a second virus, against which immunity is desired.

Heterologous protein expression

Killed vaccines require a considerable amount of virus for their production and it is a major problem to produce this cost-effectively, to ensure that no infectious virus survives the inactivation procedure, and that the cells used contain no contaminating infectious agents. However DNA technology offers a solution as it can be used to identify the part of a viral genome that encodes the virus protein against which protective immunity (usually antibody) is directed. Thus DNA technology can express the entire protein or the fragment of the protein that contains the antigenic site, either alone or fused with a gene encoding a carrier protein. Viral DNA, or a DNA copy if the virus has an RNA genome, has to be excised and inserted into an appropriate eukaryotic expression vector, together with control (promoter, stop, and polyadenylation) signals. Thus, we have a small part of the viral genome, by definition noninfectious, which can be introduced into cells, which are then grown on an industrial scale to produce very large amounts of protein very cheaply. Hopefully the protein or protein fragment is large enough to spontaneously form the natural three-dimensional structure of the protein found in the virus particle. Of course, with this approach, it is not necessary to grow the virus in culture, a great advantage for viruses like HBV.

   Does it work? Table 21.2 summarizes the state of what is still very much an experimental development. Genetic engineers would very much like to use bacterial expression systems which are well understood from the

Table 21.2

Genetically engineered vaccines.

Advantages	Problems
Noninfectious	Identification of neutralization antigen or epitope
Large-scale production methods available	Need for proper co- and post-translational modifications of viral polypeptide
Cost-effective to produce	Need for the expressed protein to achieve proper conformation to avoid poor immunogenicity
Can use genes from noncultivable viruses	Need to purify the viral protein away from cell constituents

viewpoints of both their genetics and large-scale industrial production. However, human and veterinary diseases are caused by viruses whose newly synthesized polypeptides undergo eukaryotic-type cotranslational and post-translational modifications, such as glycosylation and proteolytic cleavage, which prokaryotic cells cannot accomplish. Thus eukaryotic cells are essential. However, in addition to cultured cells from higher animals (for which technology has now been intensively developed), there is several thousand years' experience in the bulk culture of yeasts, also eukaryotes, for brewing and baking. Yeast offers an expression system with many of the advantages of a prokaryote in its handling, but with the essential eukaryotic features.

The method works well if a linear epitope gives protective immunity, but antibody is usually directed to conformational determinants (Fig. 21.5). At the moment, we do now know how to instruct a peptide to fold in a particular way. Thus success is only achieved by chance. The alternative is to synthesize the whole protein which will probably fold correctly.

A chemically synthesized experimental peptide vaccine against foot-and-mouth disease vaccine also proved very successful, and both raised neutralizing antibody and protected cattle, the main species at risk. Good luck and good science came together here, as it was only realized after the event that the peptide used, residues 146–161 of the coat protein VP1, was a linear B cell epitope that also contained a helper T cell epitope (Section 21.3).

There are many problems still to be overcome before conventional vaccines can be replaced by recombinant ones. Some have already been attended to (HBV: see above). Others require a more profound understanding of the immune response that is effective in a particular disease situation and how to stimulate it. One intrinsic problem is that subviral entities are generally poorly immunogenic.

Yet another recent development is the *DNA vaccine*, where DNA encoding an immunogen is injected rather than the immunogen itself. This type of vaccine is still experimental. The use of cytokines (soluble protein mediators of the immune system) as adjuvants is another area of vaccine research.

Hepatitis B virus vaccine

There are estimated to be more than 350 million carriers of hepatitis B virus (HBV; hepadnavirus) worldwide. The virus infects people of all ethnic groups, but is endemic in the People's Republic of China and the Far East due to perinatal infection. In the developed world, HBV is transmitted mechanically or sexually through contact with infected body fluids. As its name indicates, HBV causes acute liver disease, a debilitating infection that can progress to a chronic lifelong carrier state which may prove fatal. In addition, chronic infection may lead to one of the commonest types of human cancer, primary hepatocellular carcinoma (see Sections 14.5 and 18.7).

The problem with making an HBV vaccine was that the virus could not be grown in cell culture (and today still grows only poorly), and infects only humans and higher primates, such as chimpanzees. This was solved by cloning and expressing HBSag, which elicits protective immunity, in yeast. This was the first *recombinant vaccine*, and it has proved to be very effective in clinical trials and came to the commercial market in 1986.

Importantly, children born to infected mothers can be immunized at birth and protected. The immunology is complicated and not well understood, but evidently the infant's immune system is sufficiently developed, has not been tolerized to HBSag, and there is no interfering maternal antibody. Today, this is the standard commercially available vaccine. Once it was one of the more expensive vaccines, now a course of three injections in developing countries costs US$1.50.

Genetically engineering a virus as a vaccine

Live vaccines evoke the most effective immunity and are the cheapest to produce but stable avirulence is difficult to obtain in practice. Hence, the idea of inserting the gene for the desired neutralization antigen of a pathogenic virus into a pre-existing live vaccine, so that it is expressed naturally as the virus multiplies, is particularly attractive. This has already been achieved experimentally for antigens of numerous viruses, including influenza, rabies, herpes simplex type 1, and hepatitis B viruses, using vaccinia virus as the live vaccine.

Vaccinia virus was once used universally as a vaccine when smallpox was endemic, but today would only be employed in emergency if variola virus was used in instances of terrorism. Animals infected with recombinant vaccinia carrying genes for HBSag or influenza virus HA respond with excellent antibody-mediated and T cell-mediated immune responses, so vaccines of this type are an exciting prospect. An experimental HIV vaccine in which a series of T cell peptide epitopes has been inserted into the vaccinia virus genome is currently in clinical trials in Africa. Vaccinia

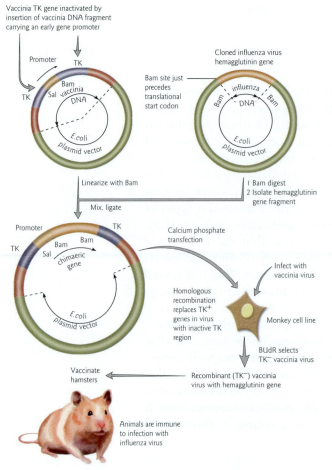

Fig. 21.7
Construction of an infectious vaccinia virus recombinant expressing the influenza virus hemagglutinin. BUdR (bromodeoxoyuridine) inhibits DNA synthesis.

virus is produced in cell culture and costs approximately 7 US cents per dose, although modification of the virus would inflate this price.

The mechanics of producing a recombinant vaccinia virus vaccine are outlined in Fig. 21.7. This starts by inserting a cloned gene into a plasmid under the control of the promoter of the vaccinia thymidine kinase (TK) gene. Upon transfection into vaccinia virus-infected cells, recombination takes place, because of the homologous TK sequences which both virus and plasmid possess. The TK gene is inactivated by insertion of the foreign sequence, which means that recombinants can be selected, because they are unable to utilize (and hence be inhibited by) the DNA synthesis inhibitor bromodeoxyuridine. Consequently, only recombinant virus produces plaques in the presence of the agent.

One problem is that vaccinia virus is not sufficiently attenuated for use today, since it sometimes results in a generalized, fatal infection, with an incidence of about 1 in 25,000 doses. While this was acceptable when the alternative was contracting highly lethal smallpox, such a risk could not be tolerated in a vaccine, for example, against influenza. One solution is to use MVA (modified virus Ankara), a vaccinia virus that has undergone spontaneous genomic deletions, or to use other less virulent members of the pox virus family such as canary pox virus. Some insertions of foreign genes have the additional effect of reducing vaccinia virus virulence by an order of magnitude. In addition, coexpression of cytokine genes by vaccinia virus (interleukin-2 and interferon-γ) completely abolishes its lethality even for highly susceptible athymic (nude) mice (these mice lack T cell-mediated immunity and all antibody-mediated immunity except IgM).

Another problem is that immunity directed against the vaccinia virus itself may preclude its use on a subsequent occasion. However, the anti-vaccinial immunity may be relatively short-lived and, in the days of smallpox, vaccination was repeated every 3 years. The vaccinia genome can accommodate around 25 kbp of DNA without losing infectivity, so one way to circumvent the problem of pre-existing immunity to vaccinia would be to make a recombinant virus that expresses 10–20 foreign

Table 21.3

The world's leading infectious diseases.

Disease	Comments
Vaccine-preventable diseases (tetanus, yellow fever, measles, diphtheria, and Hib – *Haemophilus influenzae* B)	Each year 46 million infants are not fully immunized; 1.7 million children die, the greatest number (46%) dying from measles
Chronic hepatitis (hepatitis B virus)	350 million infectious carriers; these have a 200-fold greater risk of developing liver cancer than the population as a whole
Diarrheal disease	750 million children are infected annually and 4 million die; around 50% of these are due to viruses (mainly rotaviruses)
Acute respiratory infections	3.5 million children die annually
Sexually transmitted diseases including HIV-1	1 out of 20 teenagers and young adults contract such diseases each year; 40 million are infected with HIV-1
Tuberculosis	1600 million people carry the bacterium; annually there are 10 million new cases and 3 million deaths
Malaria	100 million cases annually; almost half the world's population lives in malarial areas
Schistosomiasis	200 million cases annually

antigens. This would be a one-off polyvalent vaccine that would immunize against a broad spectrum of virus diseases.

21.5 INFECTIOUS DISEASE WORLDWIDE

Table 21.3 puts in context the impact of major virus diseases (measles, rotavirus-induced diarrhea, some respiratory viruses, human immunodeficiency virus (HIV), and hepatitis B virus (HBV)) relative to bacterial diseases (tetanus, diphtheria, tuberculosis, and some sexually transmitted organisms), protozoal diseases (malaria), and the minute snail-borne worm which causes schistosomiasis.

21.6 ELIMINATION OF VIRUS DISEASES BY VACCINATION

The World Health Organization (WHO) is responsible for the worldwide control of virus diseases. Table 21.4 shows some of the achievements and current goals, made possible through the development of suitable vaccines.

Eradication of smallpox

Smallpox, which is caused by variola virus, has been totally eradicated by vaccination. The live vaccine was vaccinia virus which was administered

Table 21.4

Goals, past and future, for the control of some virus diseases.

Virus	Commencement of control program	Goal	Result
Variola (smallpox)	1966	Worldwide eradication	Last case in Somalia in 1977
Polio types 1, 2, and 3	1988	Eradication worldwide by 2002, amended to 2005	Not achieved
Measles	1991	Eradication from the Americas by 2000	Not achieved
Hepatitis B	1992	Vaccination as part of national immunization programs in countries with a chronic infection of >8%	Adopted by 135 countries

intradermally by scarifying the skin. The virus multiplies at that site, causing a boil-like reaction, and stimulates both antibody-mediated and T cell-mediated protective immunity. The last naturally occurring case was recorded in Somalia in October 1977. Yet less than 200 years ago smallpox was endemic throughout the old world. Child mortality from smallpox in England was up to 25%, and in India alone in 1950 over 41,000 people died. The decline of smallpox in those countries where it was recently endemic has been a triumph for the vaccination program administered by WHO. Smallpox is the only infectious disease that has ever been deliberately eliminated and this experience has served to emphasize that successful elimination is possible only if a disease fulfills certain criteria (Table 21.5). These do not obtain with many infections.

Evidently variola virus fulfilled all of these criteria. The elimination of smallpox took several decades, and one by one countries were declared smallpox-free. The final assault on the virus on the Indian subcontinent and Africa was not done by mass vaccination but by "ring" vaccination. This means that when a new case of smallpox is identified, all people in that location are immunized to form a ring of immunity. The virus cannot break out of the ring to infect new susceptible hosts and dies out. Fortunately, other poxviruses (e.g. whitepox of monkeys, which is antigenically similar to smallpox) are apparently unable to step into the ecological niche vacated by variola virus and replace it as a pathogen of humans.

With the elimination of smallpox new problems arise. With no endemic disease, vaccination was discontinued, so that most people born after *c.* 1970 are susceptible to the virus. Thus it is imperative that the virus is stored securely so that it cannot escape. There was once a move to destroy all stocks of the virus but this meant losing approximately 10^5

Table 21.5
Criteria for the worldwide eradication of human viruses.

Criterion	Variola virus	Measles virus	Poliovirus types 1–3	Comment
There must be overt disease in every instance so that infection can be recognized	+	+	–	99% of poliovirus infections are subclinical
The causal virus must not persist in the body after the initial infection	+	+	+	But people whose B cell system is deficient can have persistent poliovirus infections and excrete virus for years
There must be no animal reservoir from which reinfection of humans can occur	+	+	+	But great apes are susceptible to poliovirus
Vaccination must provide effective and long-lasting immunity	+	+	+	
Viral antigens must not change	+	+	+	

bp of irreplaceable genetic information. We have no way of knowing the value of the smallpox virus genome, but it encodes proteins which are similar to those found in eukaryotic cells and these could conceivably be valuable. The problem was partly resolved by cloning the entire viral genome in fragments, so that virus proteins can be expressed in the absence of infectious virus. However the genome could still be ligated back together. Virus was stored in just two places – in the USA at the Centers for Disease Control, Atlanta, and in the State Research Centre of Virology and Biotechnology at Koltsovo in the old USSR, now Russia. However it is suspected that other stocks of the virus are held illegally and there is a risk that they may be used in acts of bioterrorism. As a result the US and other governments have recently commissioned new stocks of vaccinia vaccine for use in emergencies.

Control of measles

Prior to the development of a vaccine, measles was an inevitable childhood infection, with 130 million cases annually worldwide and a mortality of 2.3% (3 million per year). However in the developing world, child mortality was, and still is, as high as 80%. The difference was not the virus but resulted from other factors, in particular malnutrition that makes the infection more virulent. Measles vaccine is safe, effective, and cheap (US$0.26), and reached nearly all countries by 1982. The occurrence of measles has fallen steadily, but vaccine coverage is still very patchy

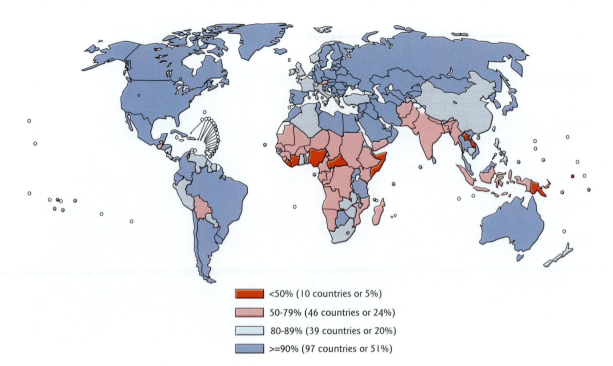

Fig. 21.8 Coverage of measles vaccination of infants around the world, 2004. (From WHO/UNICEF.)

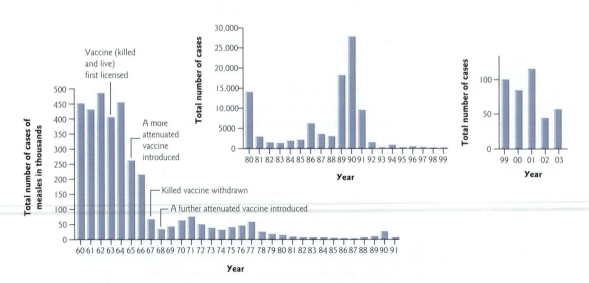

Fig. 21.9 Fall, rise, and fall again in the number of reported cases of measles in the USA following vaccination, 1960–2005. Note change of scale in the insets. About one-third of the cases in the last few years have been contracted outside the USA and imported by visitors or returning residents.

and in many African countries less than 50% of children are immunized (Fig. 21.8). As a result in 2002, there were 30–40 million cases in children worldwide, mostly in developing countries, with 800,000 deaths. Measles is also the major cause of preventable blindness. In developed countries, death from measles results from respiratory and neurological causes in about 1 of every 1000 cases. Encephalitis occurs at the same rate and survivors are often permanently brain damaged. Rare cases of subacute sclerosing panencephalitis (SSPE) are also caused by measles virus (see Section 14.7). The aim is to achieve more than 90% vaccination and a resulting high level herd immunity that will interrupt natural measles transmission. However, this has only been achieved in a few countries; in the others coverage is scantier, or there is no information (Fig. 21.8).

Some way ahead of the world scene, the Department of Health, Education and Welfare announced in 1978 their intention of eliminating measles virus from the continental USA. The criteria for successful elimination were met by measles virus (Table 21.5): it causes an acute infection nearly every time (in SSPE, persistent virus is not infectious); there is no animal reservoir; there is an excellent live vaccine (95% effective), and there is only one serological type of virus. The measles eradication program in the USA started from a recorded annual number of approximately 500,000 cases, although the actual number was estimated at 3–4 million. This was so successful that by 1983, only 1500 cases of measles were reported – and measles became a rare disease in the USA (Fig. 21.9). However, a high vaccination rate was necessary, and to achieve this, legislation was passed prohibiting children from attending school unless they had been immunized or had had a natural infection. (The ethics of this decision make for an interesting debate.) Despite these measures, there was a disturbing increase in the annual number of measles cases in 1989–91 to nearly 28,000 (Fig. 21.9), which is thought to reflect partly the problems of maintaining a high level of vaccination, particularly in socially deprived inner-city areas (around 50% only vaccinated in some areas), but also a failure of some children of younger than 1 year old to be protected by maternal antibody. This was due to the mother's immunity resulting from vaccination rather than infection, with a lower titer of antibody and thus a smaller amount of antibody being transferred across the placenta to the fetus. Hence immunity waned more rapidly than in the prevaccination era and left the infants susceptible to infection at a younger age than in the past. However, more thorough vaccination since 1989 with 2 doses of vaccine has reduced the total number of US cases to around 50–100 per year (Fig. 21.9). At the present time the vaccination program must be continued to prevent the epidemic spread of the virus following its inevitable introduction by an infected visitor to an otherwise susceptible population. Measles in non-immune adults is more severe than in children, so the consequences of ceasing measles vaccination would be far worse than if it had never been started. In 1989 the WHO set targets to eradicate measles from the Americas by 2000 (pretty much achieved), and Europe by 2007 (unlikely to be achieved as it is proving problematic to maintain adequate vaccine coverage), and to reduce mortality in Africa, southeast Asia, and the Western Pacific (again a problem of adequate vaccine coverage but for different reasons).

The planned eradication of poliovirus

In 1985, the Pan-American health authorities announced a plan to eradicate poliovirus from North, Central and South America. In 1988, the policy was extended worldwide by WHO in its Global Polio Eradication Initiative. However Table 21.5 shows that poliovirus falls down on one of the key criteria for eradication as 99% of infections are subclinical. In addition, although there is no natural animal reservoir, Old World primates are susceptible to infection and chimpanzees in the wild have been infected. The live oral vaccine which comprises the three attenuated serotypes is the vaccine of choice. Laboratories have been set up in all countries, good public relations encouraged, with events such as national vaccination days, and intensive vaccination mounted whenever there is an outbreak. The campaign has been very successful, and the last case of natural disease (cf. vaccine-caused disease, Section 21.2) in the Americas was in Peru in 1991, and in southeast Asia in Cambodia in 1997. By 2004, there was endemic transmission of wild polioviruses in only six countries (Afghanistan, Egypt, India, Nigeria, Niger, and Pakistan). A snapshot of the progress resulting from vaccination is summarized in Fig. 21.10.

Box 21.4

Problems with polio

- In 2005, polioviruses were endemic in only six countries: Afghanistan, Egypt, India, Nigeria, Niger, and Pakistan.
- There is mostly type 1 in Egypt and India, and use of a monovalent vaccine has been recommended.
- In 2004, 1265 children worldwide were recorded with poliomyelitis, mostly in Nigeria.
- Vaccination in parts of Nigeria was suspended in 2003 because of anxieties in the Muslim community, which allowed virus to spread.
- Unrest has interrupted vaccination in some areas (e.g. Darfur, Sudan, Chad), and allowed virus transmission to become re-established.
- In 2004/5, vaccination levels were not high enough to prevent cases being imported into Angola, Burkina Faso, Central African Republic, Cameroun, Côte d'Ivoire, Ethiopia, Indonesia, Mali, and the Yemen.

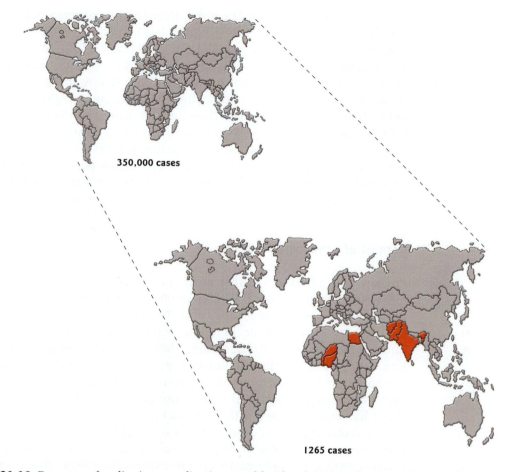

Fig. 21.10 Progress of poliovirus eradication worldwide, showing the countries in which virus was endemic in 1988 (top) and in 2004. (Modified from the WHO Expanded Program on Vaccination.)

However, a further nine countries had re-established transmission following vaccination difficulties (Box 21.4). Eradication measures include "National Vaccination Days" with house-to-house vaccination to help build up adequate levels of protection. In countries where poliomyelitis has been abolished, immunity nevertheless still needs to be maintained as the virus can be reintroduced from elsewhere by infected travelers, as occurred in 2004/5 in the Yemen and in Indonesia. The other danger is reversion to virulence of the oral vaccine, and to obviate this, vaccination is switched to inactivated virus in areas where eradication has been achieved.

There are still other problems to overcome. Careful analysis of sewerage (where all polioviruses end up) has shown that neurovirulent poliovirus has been replicating and circulating in some countries where there has been a highly vaccinated community for years. While this virus is clearly derived from the live vaccine, it has recombined with other enteroviruses to make a brand new chimeric virus. Now there is concern that a virus like this may be able to fill the vacuum left by poliovirus.

As eradication of poliovirus approaches, attention is focused on the day that vaccination will cease, saving the world $1.5 billion – money that can be directed towards other health priorities. Before that can happen, all other sources of the virus from which reintroduction might occur have to be eliminated or contained. These include frozen stocks of poliovirus held in many laboratories where poliovirus was (and still is) studied for research purposes, and frozen fecal samples in hospital pathology laboratory freezers all over the world that may contain poliovirus. All these have to be destroyed or made secure before we can have a world without poliovirus, and where there is no need for poliovirus vaccination.

Control of other virus diseases

Worldwide eradication of other virus diseases is unlikely to be easy, but the WHO is studying the feasibility of eliminating rabies virus, which is responsible for 50,000 deaths per year. The latter presents a new level of difficulty as it is endemic in many common wild animals: foxes in Europe, racoons in the USA, vampire-bats in South America, wolves in Iran, and jackals and mongooses in India. These transmit virus directly to humans by biting or indirectly via domestic animals, notably dogs, but such infections are incidental to the normal lifecycle of rabies virus. The WHO has experimented with immunizing wild animals in inaccessible mountainous regions by using aircraft to drop pieces of meat doctored with live vaccine. Foxes in Belgium were successfully immunized with vaccinia virus genetically engineered to express the rabies virus envelope protein (see Section 21.7 for an account of this type of vaccine). The eradication of hepatitis B virus is being considered by the WHO as part of its cancer prevention program (Section 21.6).

21.7 CLINICAL COMPLICATIONS WITH VACCINES AND IMMUNOTHERAPY

In some circumstances the normal vaccination procedure may be ineffective, inadvisable, or even dangerous (Box 21.5). When in doubt, a killed vaccine is safer than a live vaccine.

Potentiation of disease

Under certain circumstances, immunity acquired from administration of a killed vaccine has potentiated the disease when the wild virus was contracted. This occurred in a clinical trial of a potential vaccine against

> ## Box 21.5
>
> ### Vaccine complications
>
> - Maternal IgG is transmitted to offspring via the placenta and may prevent vaccine from stimulating an active immune response; depending on its concentration, maternal antibody wanes by about 2–6 months of age, and vaccination can then be carried out effectively.
> - A more severe infection with a live vaccine may develop in very young children compared with older children.
> - Certain clinical conditions can make a vaccine infection more severe than normal, e.g. people with eczema are prone to a generalized infection with vaccinia virus instead of a local infection at the site of vaccination.
> - Fetal development can be deranged by virus infection, e.g. by rubella virus, so vaccination should be avoided during pregnancy.
> - An existing virus infection can sometimes interfere with the multiplication of live vaccine, with the result that effective immunity is not established.

respiratory syncytial virus (a paramyxovirus, which causes a lower respiratory tract infection in very young children). The problem revolves around the stimulation of CD4$^+$ T helper cells, which are classified on a functional basis into T helper type 1 (Th-1) cells and Th-2 cells (Box 12.2). Any immunogen stimulates a balance of these two cell types. Th-1 cells stimulate cell-mediated immunity (including CD8$^+$ cytotoxic T cells and IgG2a) that has evolved to deal with intracellular infections), and Th-2 cells stimulate pathological inflammatory responses (including eosinophilia and IgE) that have evolved to deal with parasitic worm infections. It is thought that the batch of killed vaccine used in the trial was altered by the inactivation process in such a way that it produced a response biased to Th-2 cells, rather than the Th-1 cell response seen in natural infection. Thus when the immunized children contracted a natural respiratory syncytial virus infection, only the Th-2 arm had immunological memory, and this response exacerbated the pathology caused by the virus. Unfortunately, this was so severe that some of the children died.

In this example, disease was potentiated only in infants, the group that was in greatest need of protection. This demonstrates the insoluble difficulty that possible untoward effects of a vaccine will not be revealed until it is administered to people, and secondly the need to test vaccines in all age-groups.

Passive vaccination or immunotherapy

One way to treat an already established disease is to administer immunoglobulin with activity against the infecting virus. This was quite widely

Box 21.6

Immunotherapy of human respiratory syncytial virus (HRSV) infection

- There is no vaccine available for HRSV and most young children become infected by the age of 2 years, with a self-limiting respiratory infection.
- A small proportion of children, particularly of a younger age, develop a severe, life-threatening lower respiratory illness and require hospitalization and intensive care.
- One treatment is with intramuscular Synagis™, a humanized neutralizing MAb IgG specific for the virion fusion protein; this has a market value of over US$3 billion per year.
- Another licensed treatment is RespiGam™, IgG from pooled human sera, collected from adults, with high titer HRSV-specific antibody activity.

used in the past but these days its use is not common. Passive vaccination is really chemotherapy with antibody, since no immune response is stimulated and the immunity lasts only as long as the immunoglobulin survives in the body – its half-life is around 30 days. Further complications are the possible generation of an anaphylactic response if successive doses of the same "foreign" immunoglobulin are given. So it is better to use pools of immunoglobulin from humans who are immune to the particular virus rather than serum from an immunized animal, monoclonal antibody from human hybridomas, cloned human antibody, or mouse immunoglobulin that has been humanized by replacement of all but its paratope with human IgG sequences (Section 23.1).

One success story is the treatment of life-threatening, human respiratory syncytial virus infections of infants with a humanized neutralizing monoclonal antibody (Box 21.6).

21.8 PROPHYLAXIS AND THERAPY WITH ANTIVIRAL DRUGS

Ever since the successful introduction of antibiotics in the 1940s to control bacterial infections, there has been the hope that similar treatments for virus infections (antivirals) could be found but only some 50 years later is this hope being realized. About 40 antiviral drugs are approved for human use, half of them specifically for HIV-1. Most antivirals are specific for just one or a few viruses; they are not broad-spectrum treatments like antibiotics.

In general antiviral drugs face two main difficulties. Firstly there is the problem that by the time clinical signs and symptoms appear virus replication has reached such a peak that the antiviral has little *therapeutic* effect.

In some circumstances, such as in the face of an approaching epidemic, it may be best to treat people *before* they are infected, that is to use the antiviral as a *prophylactic* or preventative measure. Prophylactics are not widely used but amantadine and rimantidine, a methylated derivative (Table 21.6), are employed in this way in communal homes for the elderly to combat the spread of influenza. The other problem is that virus multiplication is tied so intimately to cellular processes that most potential antivirals cannot discriminate between them and are thus toxic. However, viruses do have unique features and it is these that antivirals have to target. An antiviral could inhibit any stage of the virus multiplication cycle, i.e. attachment, replication, transcription, translation, assembly, or release of progeny virus particles. Examples of current antivirals are shown in Table 21.6. Another problem is the selection of resistant mutants and this is discussed below. Finally nearly all current antivirals act only against replicating virus, and are ineffective, for example, against latent infections. Thus anti-HIV drugs do not clear virus from the body, and current drugs have to be taken for ever to maintain virus suppression and avoid damage to the immune system. Similarly, anti-herpes virus drugs do not affect the rate of virus reactivation.

Empirical screening to discover drugs is being replaced by rational design

The search for antivirals has traditionally been by high-throughput empirical (trial and error) screening, but this is moving successfully to rational design as understanding of viral processes and properties increases. The new anti-influenza drug, zanamivir (Relenza), was designed *de novo* as a competitive inhibitor for the viral neuraminidase, mimicking the substrate, *N*-acetyl neuraminic acid. After budding, progeny virus attaches to the cell in which it was made through its HA proteins, and is subsequently released by the action of its neuraminidase proteins. Relenza binds to, and inhibits, neuraminidase activity with the result that virus remains sequestered at the cell surface and is unable to infect new cells. However rational design and empiricism will continue side by side for the foreseeable future; there is an upsurge of interest in natural plant products as potential antivirals.

Inhibition of viral DNA synthesis by nucleoside analogs

An important class of antivirals are the nucleoside analogs which are incorporated into DNA. The first versions were used to treat DNA virus infections of the conjunctiva and cornea of the eye, as these cause permanent scarring and hence loss of vision. The nucleoside analog was incorporated into nascent DNA but was unable to base-pair properly and so prevented replication and transcription. However this first generation

Table 21.6

Some antiviral drugs in clinical use.

Antiviral compound	Mode of action	Usage
Inhibitors of infection – of attachment/entry of virus into target cells		
Pleconaril	Binds to a hydrophobic pocket on the virion and blocks attachment/uncoating	Several different picornaviruses
Antibody (passive immunotherapy)	Neutralization	Although antibodies specific for all viruses exist, this therapy tends to be used only for life-threatening infections (e.g. Ebola virus, human respiratory syncytial virus – see below)
Synagis™; RespiGam™	Neutralizing IgG specific for the human respiratory syncytial virus fusion protein	Used to treat life-threatening respiratory infections of infants
Inhibitors of viral fusion (entry into the cell)		
Enfuvirtide (T-20)	A gp41 peptide that inhibits fusion of viral and cell membranes	HIV-1
Inhibitors of virus replication		
Aciclovir (Zovirax)	Nucleoside analog; terminates DNA synthesis as the deoxyribose moiety has no 3′ hydroxyl; requires phosphorylation by viral thymidine kinase (tk)	Herpes simplex viruses 1 and 2; less effective against VZV
Ganciclovir	As above; activated by a viral enzyme other than tk	HCMV infections of the immunocompromised, e.g. during AIDS
Zidovudine (AZT), didanosine (ddI), zalcitabine (ddC), stavudine (d4T), lamivudine (3TC), abacavir (ABC), emtricitabine (FTC)	Nucleoside analogs that inhibit the reverse transcription process; terminate DNA synthesis as the deoxyribose moiety has no 3′ hydroxyl	HIV-1; used in "triple therapy" in combination with protease inhibitors; retards virus replication and progress to AIDS (see Section 19.8)
Nevirapine, delavirdine, efavirenz	Non-nucleoside inhibitors of reverse transcription	HIV-1; used in combination as above
Lamivudine	As above	Hepatitis B virus; leads to high levels of resistance

Table 21.6 (*Cont'd*)

Antiviral compound	Mode of action	Usage
Adefovir (PMEA)	Competitive inhibitor of reverse transcriptase and chain terminator	Hepatitis B virus
Bicyclic furo(2,3-d)pyrimidine nucleoside analogs (BCNAs)	Require phosphorylation specifically by VZV thymidine kinase	Primary VZV (chicken pox) and reactivated VZV (shingles/zoster)
Ribavirin	Inhibits RNA synthesis: blocks inosine 5′-monophosphate dehydrogenase and hence GTP synthesis	Human respiratory syncytial virus (used as an aerosol); Lassa fever
Ribavirin + interferon-α in combination	Ribavirin: as above; interferon-α creates an antiviral state	Hepatitis C virus
Cidofovir	Acyclic nucleoside analog; DNA polymerase inhibitor	Broad spectrum; DNA viruses including papillomavirus, adenovirus, HCMV, and other herpesviruses resistant to primary choice drugs
Ion channel blockers		
Amantadine, Rimantadine	Blocks the M2 proton channel in the virion envelope	Type A influenza viruses
Inhibitors of protease activation of viral proteins		
Includes saquinavir, indivir, ritonivir, nelfinavir, aprenavir, lopinavir, atazanavir	Prevent post-translational cleavage of structural protein precursors	HIV-1; used in "triple therapy" in combination with nucleoside analogs; retard virus replication and progress to AIDS (see Section 19.8)
Inhibitors of virus release from host cells		
Zanamivir (inhaled); oseltamivir (oral)	Analogs of the viral neuraminidase substrate	Influenza A and B viruses; must be taken before or early in infection
Killing the infected cell		
Interferon-α	Upregulates MHC class I and facilitates action of CD8+ cytotoxic T cells	Although generally antiviral, is effective *in vivo* only against selected infections: chronic hepatitis B and C

AIDS, acquired immunodeficiency syndrome; AZT, 3′ azido-2′, 3′-dideoxythymidine; HIV-1 human immunodeficiency virus type 1; HCMV, human cytomegalovirus; MHC, major histocompatibility complex; VZV, varicella-zoster virus.

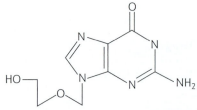

Fig. 21.11 The formula of the chain-terminating nucleoside, aciclovir. Note that most of the cyclic sugar ring is missing – hence the name. The hydroxyl group shown has to be phosphorylated before aciclovir can be incorporated into nascent DNA.

of nucleoside analogs was incorporated into both cell and viral DNA, and so killed all dividing cells – hence its use was restricted to the poorly vascularized surface of the eye. In practice, 0.1% 5'-iodo-2'-deoxyuridine (IUdR), an analog of thymidine, improved healing in 72% of infections with herpes simplex type 1, vaccinia virus, and adenovirus. In extreme circumstances, such as life-threatening cases of encephalitis (infection of the brain) caused by herpes virus type 1 or vaccinia virus, analogs of cytidine (cytosine arabinoside) or adenosine (adenine arabinoside) were used systemically despite their toxicity. However these have been largely superseded by the next generation of nucleoside analogs like aciclovir (see below) and ganciclovir. Another important nucleoside analog is AZT (3-azido-2', 3'-dideoxythymidine). This was the first anti-HIV drug, but the rate of viral mutation and generation of resistant mutants rendered it ineffective when used alone. It is now employed to very good effect in combination with another nucleoside analog and a drug that inhibits the viral protease (see Section 19.8).

Aciclovir: a selective nucleoside analog

Properly known as 9-(2-hydroxyethoxymethyl)guanine, aciclovir (or Zovirax) was the first really effective antiviral to be discovered. This is a purine analog that is particularly effective against acute infection with herpes simplex viruses (Fig. 21.11). It is not incorporated into cellular DNA in uninfected cells, as only the viral thymidine kinase can accomplish the phosphorylation of the hydroxyl group on the sugar ring that is necessary for aciclovir to be incorporated into DNA (Fig. 21.12). Thus, in aciclovir, the ultimate aim of creating a compound which is selectively toxic in infected cells has been achieved. Aciclovir has no activity against latent infections but, when used early on reactivated virus, is effective against cold sores, corneal eye infections, and genital infections. It is also invaluable in treating life-threatening generalized herpes simplex virus infections which can occur in the immunocompromised, particularly people treated with immunosuppressants in the course of transplant surgery or cancer chemotherapy. Aciclovir-resistant mutants do arise but do not seem to be a significant problem.

The direct antiviral action of α- and β-interferons

This antiviral effect is called "direct" to distinguish it from the indirect antiviral effect when interferons-α and -β stimulate the expression of MHC I proteins, described above. Most viruses induce interferons-α and -β and, once induced, they are active against a whole spectrum of viruses, not just the one responsible for interferon induction. The action of interferon can be divided into two stages (Fig. 21.13).

The first stage is *induction* in which the genes coding for interferon are derepressed and the interferon protein is released from the cell. Induction of interferon in human cells is controlled by chromosome 9. All multiplying viruses induce interferon and the specific inducer is double-stranded RNA (dsRNA). This is a unique indication of infection as normal cell metabolism does not produce it. It is not necessary for the inducing virus to have a completely double-stranded genome, as single-stranded cRNA (ssRNA) can at least partially base-pair with its template to form dsRNA. Some DNA viruses transcribe RNA from both genome strands, and these can then form dsRNA and act as an inducer of interferon (see Section 12.2).

The second stage is the creation of an *antiviral state* in neighboring cells when their cell surface interferon-specific receptors bind secreted interferon. In humans, the receptor for interferon-α/β is encoded by chromosome 21. This sets in motion a hugely complex series of events in which the transcription of hundreds of cellular genes is induced. Most of these genes are activated through the **Ja**nus **k**inase-**s**ignal **t**ransducer and **a**ctivator of **t**ranscription (JAK-STAT) pathway. The cell is now prepared to respond immediately should a virus infect it. However the fully active antiviral state is not achieved until the cell is infected, and virally encoded dsRNA is again responsible. Induction of the antiviral state by interferon allows the cell to respond in two ways, both of which curtail virus replication. Firstly, dsRNA stimulates the phosphorylation of certain cellular proteins, notably the eukaryotic initiation factor, eIF2α, which impairs their function in the initiation of protein synthesis. Secondly, dsRNA activates a ribonuclease (RNase L) which degrades mRNA, and also stops protein synthesis. The mechanism is complex (Fig. 21.14): binding of interferon to interferon receptors stimulates the synthesis of 2′,5′-oligoadenylate synthetase, but this is only active in the presence of viral dsRNA, i.e. when the cells are infected. The dsRNA-activated 2′,5′-oligoadenylate

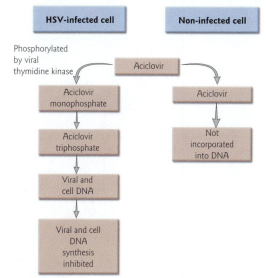

Fig. 21.12 Aciclovir is toxic for herpes simplex virus-infected cells but not noninfected cells

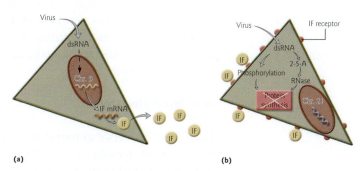

Fig. 21.13 (a) Induction of interferons-α/β (IF) by viral double-stranded (ds) RNA. The dashed arrow indicates that we do not know if dsRNA is required to enter the nucleus to initiate interferon mRNA synthesis or if it acts through an intermediate. (b) Binding interferon by nearby cells creates in them an antiviral state through inhibition of protein synthesis – either from the phosphorylation that results from the induced protein kinase (PKR), or by stimulating 2′,5′-oligoadenylate synthetase and ribonuclease (RNase) L.

Binding of interferons to their receptors stimulates synthesis of:

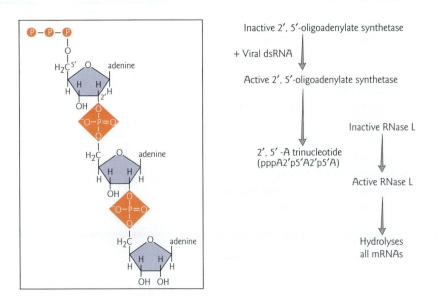

Inactive 2', 5'-oligoadenylate synthetase

+ Viral dsRNA

Active 2', 5'-oligoadenylate synthetase

Inactive RNase L

2', 5' -A trinucleotide
(pppA2'p5'A2'p5'A)

Active RNase L

Hydrolyses
all mRNAs

Fig. 21.14 Action of interferons-α/β: activation of RNase L. Inset: the formula of the 2',5'-A trinucleotide.

synthetase then synthesizes an unusual tri-ribonucleotide 2',5'-A (pppA2'p5'A2'p5'A), which has 2'p5' bonds instead of the normal 3'p5' linkage. This in turn activates RNase L. During the interferon-induced antiviral state all protein synthesis, both viral and cellular, is inhibited, and if continued this would lead to the death of the cell. However, when the virus infection is inhibited, synthesis of viral dsRNA ceases, the inhibitory mechanisms are no longer brought into play, and the cell returns to its normal state.

Interferon therapy

With their universal antiviral activity and high specific activity interferons-α, -β, and -γ should be ideal therapeutic agents, but clinical trials have not been encouraging. In early experiments in the 1960s at the Common Cold Research Unit in Salisbury, UK, it was necessary to give volunteers 14×10^6 units of interferon-α/β as a nasal spray four times per day, including the day before infection, in order to combat a rhinovirus, one of the causative agents of the common cold (Table 21.7). Unfortunately even with high doses, interferon therapy is only effective if started before the virus is inoculated, and is thus not applicable to natural rhinovirus infections. In addition interferons also have side-effects, causing fever, local inflammation, muscular pain, fatigue, and malaise – effects that result from their action as cytokines and regulators of the immune system.

Table 21.7

Treatment of common colds caused by a rhinovirus infection with interferon-α/β.

	Number of people with colds/number of people inoculated	Number of people virus positive/number of people inoculated
Interferon-treated	0/16	3/16
Mock (placebo)-treated	5/16	13/16

Experimental treatment of a variety of human virus infections with interferons has been generally disappointing and it is still not clear why this should be. However, interferon has been of particular use in the treatment of persistent infections caused by hepatitis B virus (hepadnavirus) and hepatitis C virus (flavivirus). In addition it has been used in the treatment of human papillomaviruses. The human papillomaviruses cause benign tumors (warts) on the surface of the body, the larynx and the ano-genital regions. Genital warts are a particular concern, as they may develop into cancer of the cervix (see Section 20.4). Over half the warts injected with interferon regress, but they tend to recur when interferon is stopped. Warts also recur after their surgical removal; combined surgery and interferon treatment is a better option.

After an acute HBV infection, a proportion of patients develop a chronic infection (see Table 14.3 and Fig. 14.3), which can lead to immune complex disease, and destruction of the liver (cirrhosis); these people have a 10% chance of developing liver cancer. Virus-specific activated CD8$^+$ cytotoxic T lymphocytes (CTLs) are present, but are unable to clear the infection, because the virus depresses the expression of MHC class I proteins on infected liver cells. Treatment with α-interferon is beneficial, and acts by enhancing expression of MHC class I proteins (see Section 12.2) and not through its antiviral activity. HBV-specific peptides are then presented in sufficient amount by MHC class I proteins to allow the CD8$^+$ CTLs to lyse the HBV-infected liver cells. The treatment is about 50% successful and, under optimum circumstances, can clear the infection completely (Fig. 21.15). Lamivudine therapy is also used (Table 21.6). Hepatitis C virus, although a member of a different family, causes a progression of infections and liver disease which, if anything, is more severe than that caused by HBV. Worldwide about 170 million people are infected with hepatitis C virus. About 70–90% of acute infections become persistent and 20–30% of these become chronic with a proportion progressing to liver cancer. Combined treatment with ribavirin and α-interferon is expensive ($US12,000–$15,000) and takes 6–18 months; success in clearing virus depends on the virus strain: for genotype 1 there

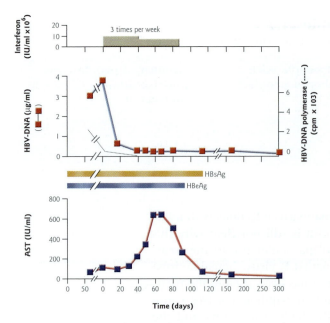

Fig. 21.15 Successful interferon-α treatment of a patient with a chronic hepatitis B virus (HBV) infection. Note the fall in viral DNA, viral DNA polymerase, and the viral antigens HBeAg and HBSAg. These all became undetectable, showing that the infection had been cleared. The transient rise in aspartate aminotransferase (AST), a marker of liver damage, signifies the destruction of HBV-infected liver cells by virus-specific CD8+ CTLs following the upregulation of MHC class I antigens (see text). (From Thomas 1990. *Control of Virus Diseases. Society for General Microbiology Symposium 45*, pp. 243–259. Ed. N. J. Dimmock, P. D. Griffiths, & C. R. Madeley. Cambridge: Cambridge University Press.)

is a 40–50% success rate, and for the less common genotypes 2 and 3 there is an 80% success rate. Virus levels are always reduced but, if not completely cleared, there is a rebound and virus returns rapidly to its original concentration.

Selection of drug-resistant mutants and the advantage of multiple drug regimens

The problem of acquired drug resistance is common in all rapidly multiplying biological entities and is well known in bacteria. It is acute in all viruses and particularly those with RNA genomes, as they have no proofreading mechanism for checking the fidelity of genome replication, and so their populations contain large numbers of mutants. A demonstration of this – and a remedy – has been found during the treatment of HIV-1 infections with the nucleoside analog, AZT. Although a very effective inhibitor *in vitro*, resistance to AZT appeared so rapidly in infected people that the drug was ineffective. The problem is a high mutation rate of approximately 10^{-5} mutations per base-pair per replicated genome. These are mostly G→A transitions. The result is that any mutant that happens to have a selective advantage, e.g. because its reverse transcriptase does not bind the inhibitor, will rapidly outgrow wild-type virus and become dominant. The answer is to use two or more drugs simultaneously, as the chance of two appropriate mutations occurring in the same genome is very low, the product of the mutation rate i.e. 10^{-10}. However as the virus population is intrinsically heterogeneous, one of the mutations may be already present, and the double mutant could be created by a single mutational event. Thus the best option, as now used with HIV-1 (Table 21.6), is a combination of three drugs ("triple therapy") preferably targeting different processes, and this could mean a nucleoside analog, a non-nucleoside inhibitor, and a viral protease inhibitor. This is covered in more detail in Section 19.8. Drug-resistant herpes virus mutants also arise when aciclovir is used, although these do not seem to be replacing wild-type virus. It would be a tragedy if the largely cosmetic use of aciclovir in preventing cold sores resulted in it being no longer of any use for life-threatening herpes virus infections, and for this reason there was concern when aciclovir became a nonprescription drug. Influenza A virus mutants are also seen when amantadine is used.

KEY POINTS

- The purpose of a vaccine is to stimulate immunity without having to experience the infection or the disease.
- Killed or noninfectious vaccines are made by inactivating virulent virus.
- Live vaccines are made by attenuating virulent, wild-type virus; they cause an infection in the vaccinee that is inapparent or less severe than the wild-type virus infection.
- Like all medical procedures, vaccination carries a risk, but this is far less than the risk of contracting the wild-type virus infection.
- Making a vaccine is still largely an empirical procedure; what works for one virus may work poorly, or not at all, for another; despite huge efforts there is no HIV vaccine.
- The key to more and better vaccines is to understand what sort of immunity is needed to prevent a particular infection, and how to stimulate that immunity.
- Since the majority of people require vaccinating, this is an expensive measure, so vaccines should be both affordable and cost-effective.
- Vaccination offers the hope of eradicating some virus diseases, but so far only smallpox has succumbed; poliovirus has been eradicated from most countries but it is proving very difficult to remove the last traces; there is hope for the eradication of measles virus.
- Antiviral drugs have been slow to arrive, but 40 have been approved for human use.
- Most antiviral drugs are now being made by rational design through understanding of the molecular details of viral proteins and replication processes.
- Given their antiviral activity, interferons are surprisingly ineffective at treating infections in general, but they have specific uses.

QUESTIONS

- Compare and contrast the properties of live attenuated and killed whole viral vaccines, commenting on the advantages and limitations of each in controlling human viral infections.
- What factors need to be considered when deciding policy on vaccination against viral infections, and attempts to achieve global eradication of named viral infections?
- Discuss the mode of action of antiviral therapeutics, using selected examples, and explain why broad-spectrum antiviral therapeutics are so difficult to achieve.

FURTHER READING

Afzal, M. A. 2002. MMR vaccine. *Microbiology Today* **29**, 116.

Ahmed, R., ed. 1996. Immunity to viruses. *Seminars in Virology* **7**, 93–155 (several articles).

Anon. 1998. Vaccine Supplement. *Nature Medicine* (Vaccine Supplement 4).

Arvin, A. M. 2000. Measles vaccines – a positive step towards eradicating a negative strand. *Nature Medicine* **6**, 744–745.

Bailey, J., Blankson, J. N., Wind-Rotolo, M., Siliciano, R. F. 2004. Mechanisms of HIV-1 escape from immune responses and antiretroviral drugs. *Current Opinion in Immunology* **16**, 470–476.

Behbehani, A. M. 1991. The smallpox story: historical perspective. *American Society for Microbiology News* **57**, 571–576.

Belshe, R. B., Couch, R. B., Glezen, W. P., Treanor, J. J. 2002. Live attenuated intranasal influenza vaccine. *Vaccine* **20**, 3429–3430.

Ben-Yedidia, T., Arnon, R. 1997. Design of peptide and polypeptide vaccines. *Current Opinion in Biotechnology* **8**, 442–448.

Beverley, P. C. L. 1997. Vaccine immunity. *Immunology Today* **18**, 413–415.

Birmingham, K. 1999. Report calculates value for money of US vaccine R&D. *Nature Medicine* **5**, 469.

de Clercq, E. 2004. Antivirals and antiviral strategies. *Nature Reviews Microbiology* **2**, 704–720.

Cox, J. C., Coulter, A. R. 1997. Adjuvants – a classification and review of their modes of action. *Vaccine* **15**, 248–256.

Dixon, B. 2001. Measles, polio, and conscience. *American Society for Microbiology News* **67**, 124–125.

Glezen, W. P., Alpers, M. 1999. Maternal immunization. *Clinical Infectious Diseases* **28**, 219–224.

Katze, M. G., He, Y., Gale, M. 2002. Viruses and interferon: a fight for supremacy. *Nature Reviews Immunology* **2**, 675–687.

Lehner, T., Anton, P. A. 2003. Mucosal immunity and vaccination against HIV. *AIDS* **16**(Suppl 4), S125–S132.

Lewis, P. J., Babiuk, L. A. 1999. DNA vaccines: a review. *Advances in Virus Research* **54**, 129–188.

Maassab, H. F., Bryant, M. L. 1999. The development of live attenuated cold-adapted influenza virus vaccines for humans. *Reviews in Medical Virology* **9**, 237–244.

Meltzer, M. I. 2003. Risks and benefits of preexposure and postexposure smallpox vaccination. *Emerging Infectious Diseases* **9**, 1363–1370.

Meltzer, M. I., Neuzil, K. M., Griffin, M. R., Fukuda, K. 2005. An economic analysis of annual influenza vaccination of children. *Vaccine* **23**, 1004–1014.

Minor, P. D. 2002. Emerging/disappearing viruses. Future issues concerning polio eradication. *Virus Research* **82**, 33–37.

Pomerantz, R. J., Horn, D. L. 2003. Twenty years of therapy for HIV-1 infection. *Nature Medicine* **9**, 867–873.

Racaniello, V. R. 1988. Poliovirus neurovirulence. *Advances in Virus Research* **34**, 217–246.

Robinson, A., Cranage, M. P., Hudson, M. 2003. *Vaccine Protocols*, 2nd edn. Humana, Totowa, NJ.

Spier, R. E. 1998. Ethical aspects of vaccines and vaccination. *Vaccine* **16**, 1788–1794.

Webster, R. G. 2000. Immunity to influenza in the elderly. *Vaccine* **18**, 1686–1689.

Wild, T. F. 1999. Measles vaccines, new developments and immunization strategies. *Vaccine* **17**, 1726–1729.

Yuki, Y., Kiyono, H. 2003. New generation of mucosal adjuvants for the induction of protective immunity. *Reviews in Medical Virology* **13**, 293–310.

Zambon, M. 1999. Active and passive immunization against respiratory syncytial virus. *Reviews in Medical Virology* **9**, 227–236.

The World Health Organization web pages on vaccines: http://www.who.int/vaccines

Also check Appendix 7 for references specific to each family of viruses.

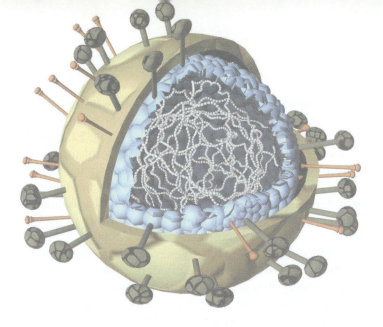

22

Prion diseases

Prion diseases are the subject of widespread interest and concern as threats to public health following an epidemic of bovine spongiform encephalopathy *(BSE) in the United Kingdom in the 1980s and 1990s that spread to other countries, and the associated emergence of a new human disease, variant* Creutzfeld–Jakob disease *(variant CJD).*

This chapter deals with an intriguing group of diseases that affects a variety of animals, including man. These diseases are known collectively as *spongiform encephalopathies* because of the characteristic pathology which they display in regions of the brain. Historically, these were known as slow virus diseases following the demonstration of an infectious basis for the sheep and goat disease, *scrapie*, the first identified disease of this type, and in recognition of the fact that the disease course was prolonged. However, extensive attempts to characterize a classical viral agent associated with these diseases have failed and instead there is a considerable body of evidence which ascribes infectivity in these diseases to a protein. This notion, the *prion hypothesis*, which includes the concept of a replicating protein, has now gained widespread, although not universal, acceptance. Its principal champion, Stanley Prusiner, has received the Nobel Prize for his work on this subject.

22.1 THE SPECTRUM OF PRION DISEASES

Several diseases of animals and humans are now believed to be caused by prions. These are summarized in Table 22.1. All show a characteristic pathology in the central nervous system, with parts of the brain becoming vacuolated or spongey, in many cases also with extensive deposits of extracellular protein fibrils or plaques (Fig. 22.1), but with no sign of inflammation, i.e. no invasion of immune cells. Different brain regions are affected by the various spongiform encephalopathies (Fig. 22.4 shows an example). For most of the diseases listed, the presence of infectivity has been shown by transmission of disease to an experi-mental animal (typically a mouse or hamster) by intracranial injection of a homogenate of diseased tissue and subsequent serial passage to further mice/hamsters. Hence, these diseases have been termed *transmissible spongiform encephalopathies* (TSE).

The symptoms of TSEs vary in detail but generally share the features one would predict for a progressive loss of function in the brain. Early symptoms include personality changes, depression, memory problems, and difficulty in coordinating movements. Later, the deficits become profound, with patients losing the ability to move or to speak. Characteristically, all

Table 22.1

Prion diseases in the date order that transmissibility was demonstrated.

Disease and occurrence	Host species	Date
Scrapie Common in several countries throughout the world	Sheep, goats	1936
Transmissible mink encephalopathy (TME) Very rare, but adult mortality nearly 100% in some outbreaks	Mink	1965
Kuru Once common among the Fore-speaking people of Papua New Guinea, now rare	Humans	1966
Creutzfeld–Jakob disease (CJD) Occurs in iatrogenic*, familial and sporadic forms. The latter has uniform worldwide incidence of 1 per million per annum.	Humans	1968
Gerstmann–Sträussler–Scheinker (GSS) syndrome An inherited disease; less than 0.1 per million per annum	Humans	1981
Chronic wasting disease (CWD) Colorado and Wyoming, USA	Mule-deer, elk	1983
Bovine spongiform encephalopathy (BSE) An epidemic disease in cattle, principally UK and Western Europe, now under control (Section 22.5)	Cattle	1988
Feline spongiform encephalopathy (FSE) BSE disease transmitted into cats	Domestic cat	1991
Fatal familial insomnia (FFI) A very rare inherited disease	Humans	1995
Variant Creutzfeld–Jakob disease (variant CJD) BSE disease transmitted into humans, mainly in the UK (Section 22.6)	Humans	1997

*Relating to medical intervention.

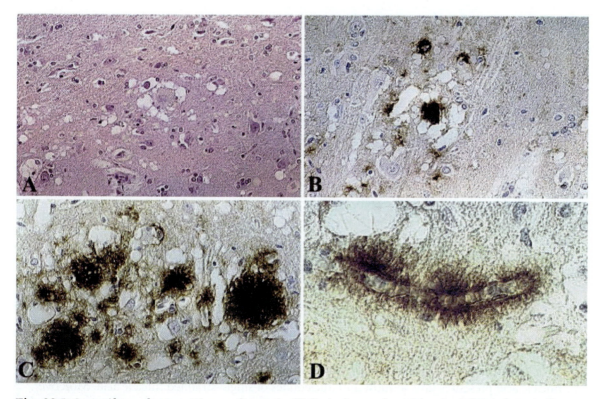

Fig. 22.1 Spongiform degeneration and protein (PrPsc) plaque deposition in CWD-infected elk brain. The images show sections from brain cortex either hematoxylin & eosin stained (A) or with detection of PrPsc deposits by immunohistochemistry (brown stain, B–D). Magnifications: A, ×180; B, C, ×280; D, ×720. (Reproduced with permission from Liberski P. P. *et al.* 2001. *Acta Neuropathologica* **102**, 496–500. © Springer-Verlag.)

the TSEs have long incubation periods (e.g. 1 year for a mouse; 4–6 years for a cow). For the human TSEs, the age at which signs and symptoms first appear depends on the disease type. Sporadic classical CJD symptoms start at a mean age of 60 years while symptoms of inherited forms emerge somewhat earlier (45 years). One of the defining features of variant CJD (Section 22.4) is its early age at onset, below 30 years. To date, all TSEs have been invariably fatal; no effective treatment has been established.

22.2 THE PRION HYPOTHESIS

The infectivity that causes prion diseases is thought to reside in a protein that adopts a stable abnormal conformation rather than in a conventional, nucleic acid-containing, entity.

The essence of the prion hypothesis is that the infectious agent that causes TSEs is a specific structural form of a standard cell protein, PrPc (the cellular form of PrP), a glycoprotein found on the surface of cells in brain and other tissues. In its unusual *prion* form, termed PrPsc (the **sc**rapie form of PrP), the protein has pathogenic properties. For an agent to be infectious, it must be able to replicate itself. An incoming prion is thought to achieve this by driving the conversion of host molecules that are in the standard structural form into the altered, prion form. In short, the infectious aberrant form is thought to act as a template on which existing normal protein molecules are converted into new altered structures that precisely resemble the original infectious agent and are themselves infectious (Fig. 22.2).

Box 22.1

Evidence supporting the prion hypothesis for TSE diseases

- A series of observations argues that the TSE agents cannot be conventional viruses or viroids. Although the scrapie agent passes through filters which admit only viruses and can be titrated in mice (often reaching concentrations of over 10^7 infectious units/ml), it is much more resistant than typical viruses to inactivation by heat, radiation, and chemicals such as formaldehyde. The rate at which infectivity is destroyed by radiation indicates that any nucleic acid present is tiny, at most 250 nucleotides of single-stranded nucleic acid. Moreover, infectivity is not susceptible to nucleases and no unique nucleic acid molecule, even of this minimal size, has been found to copurify with infectivity.
- Several observations directly support the prion hypothesis. Infectivity copurifies with PrPsc, which forms the plaques seen in some TSE-affected brain tissues. The sequence of this protein is identical to the host protein PrPc but its tertiary structure is very different, which renders the protein core insoluble and resistant to protease digestion. PrPsc can catalyze the conversion of PrPc molecules into the PrPsc structure in the test tube. Also, transgenic mice that lack the gene for PrP are completely resistant to experimental transmission of TSEs, as you would expect if this gene provides the substrate for prion replication, while mutations in this gene in humans are associated with inherited forms of TSE (Section 22.3). Until recently, *in vitro* conversion of PrPc to PrPsc was inefficient and did not generate any new infectivity but a much more efficient system has now been described in which new infectivity is generated over many cycles of *in vitro* replication. If this result can be reproduced, it will be the final proof of the hypothesis.
- A prion-type replication mechanism has precedents in yeast. *PSI*+ and *URE3* phenotypes result from the presence of aberrant conformations of two normal proteins, Sup35p and Ure2 respectively. Interestingly, a molecular chaperone (a protein which catalyzes changes in the conformation of other proteins) is involved in generating the *PSI*+ trait in a previously negative (*psi*−) cell population, and in subsequent elimination of the trait from that population. Mutant forms of the normal proteins have also been shown to increase the rate of conversion to give the abnormal phenotypes.

Some of the evidence in favor of the prion hypothesis is summarized in Box 22.1. However, despite the strength of this evidence, the hypothesis is still not universally accepted. This is partly due to natural scepticism about such a novel concept, but there is also the problem of scrapie strain variation (Box 22.2). In essence, when different isolates of sheep scrapie are grown in genetically identical laboratory animals, they show distinct properties that are stable on serial passage. For the prion hypothesis to accommodate these data, PrPc of a specific and defined amino acid sequence must be able to adopt not just one aberrant conformation but several, in each case resembling exactly the conformation of the PrPsc in the infecting scrapie strain, so that its unique pathogenic properties are faithfully reproduced. This idea is difficult to reconcile with our understanding of how stable protein folding is achieved, whereas strain variation is easy to understand if the infectivity has a nucleic acid component.

22.3 THE ETIOLOGY OF PRION DISEASES

The first TSE to be identified was scrapie, a natural infection of sheep, which takes its name from the tendency of diseased animals to scrape themselves against fence posts, presumably to relieve itching of the skin. The disease is usually spread maternally but can also be spread horizontally and its infectious basis was further confirmed by transmission into experimental animals (see Section 22.2). There is also clear evidence for an infectious basis to other prion diseases of animals, such as BSE (see Section 22.5).

Among the human prion diseases listed in Table 22.1, Kuru clearly has an infectious basis. It was spread by ritual cannibalism and, since its identification, its incidence has declined along with this practice. *Iatrogenic* (meaning related to medicine) CJD also shows infectious etiology. It has arisen through various medical treatments involving transfer of certain tissues/tissue extracts between individuals. For example, the treatment of dwarfism, a pituitary growth hormone deficiency, with intramuscular injections of this hormone led to a number of cases of CJD. Disease occurred because the growth hormone was extracted from pituitary glands obtained from cadavers, some of which were harboring inapparent prion disease at the time of death. The risk of transmission was heightened by the need to pool material from a large number of individuals to obtain sufficient hormone for use.

On the other hand, Gerstmann–Sträussler–Scheinker (GSS) syndrome, fatal familial insomnia (FFI), and familial CJD are autosomal dominant inherited genetic disorders while sporadic CJD, as the name suggests, occurs at a random low frequency with no known cause. How

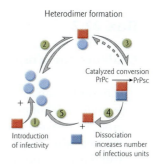

Fig. 22.2 A model for the propagation of TSE infectivity according to the prion hypothesis. PrPc (circles) and PrPsc (squares) represent the normal and an abnormal conformation respectively of the *prnp* gene product. Host PrP in the normal and abnormal conformation are pale and dark blue respectively. The initiating infectivity (red) is proposed to have the same structural features and properties as the progeny PrPsc molecules (although see discussion of the species barrier, Section 22.4). Steps 2 and 3 in the propagation cycle may be reversible. Alternative models differ principally in suggesting that polymerization of the altered structural form is important to its potential for catalyzing further structural conversions.

Box 22.2

Evidence for scrapie strain variation

- Different scrapie isolates are characterized by their very precise, but different, incubation periods in laboratory animals of a given type or inbred strain. A single scrapie strain will produce disease onset at times that vary by only 2 or 3 days among a group of animals whereas the mean incubation period for different scrapie strains ranges from less than 100 days to more than 300 days. Scrapie strains also differ reproducibly in details of their pathogenesis.
- Scrapie strains adapted to one rodent species can be passaged into another species where, after crossing the species barrier when the incubation period is extended and more variable (Section 22.4), they again show reproducible and discrete incubation periods (Fig. 22.3).
- When scrapie strains, serially passaged in mice and then in hamsters, are passaged back to the original mouse strain again they may either display their original properties (incubation time, pathogenic details) or they may "mutate" to a new set of properties that are then stable on further passage.

can these be accounted for within the prion hypothesis? For the inherited diseases, the explanation lies in the discovery of specific mutations in the *prnp* gene, which encodes PrPc, each of which is associated with a distinct disease phenotype (Box 22.3). Studies of these mutations show that subtle differences in the structure of PrP can affect the time course and nature of disease, as the prion hypothesis predicts. Changes in the sequence of PrPc that are linked with disease are proposed to destabilize the protein so that spontaneous conversion into an altered, PrPsc, form inevitably occurs within a normal lifetime. After this, replication of the aberrant structure can take place in the same way as following infection with exogenous prions. Again, the precise nature (shape and structure) of this altered form must determine the disease phenotype. Supporting this idea, mice made transgenic for the human GSS mutant *prnp* gene spontaneously develop disease. Following the same reasoning, sporadic CJD is thought to occur when a molecule

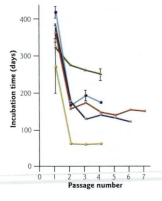

Fig. 22.3 The incubation time to disease of five distinct, mouse-adapted, scrapie strains upon serial passage in hamsters. Each colored line represents data from a different scrapie strain. Passage 1 shows a longer and more variable incubation time for each strain because of the species barrier to transmission (Section 22.4). Subsequent passages show remarkably constant, but strain-specific, incubation times. Standard errors, except where shown, were insignificant. (Data are taken from Kimberlin, R. H. *et al.* 1989. *Journal of General Virology* **70**, 2017–2025.)

Box 22.3

Human *prnp* mutations associated with inherited disease

- Specific mutations in human PrP are associated with either CJD, GSS, or FFI. For example, replacement of glutamic acid by lysine at position 200 results in CJD while replacing proline with leucine at position 102 results in GSS.
- Polymorphisms at other positions in human PrP (not themselves pathogenic) modify the disease pattern associated with pathogenic mutations. A key polymorphism is at position 129, coding either for valine or methionine.
- Almost all disease due to specific CJD and GSS mutations occurs in individuals with methionine at position 129 on both the normal and mutated *prnp* alleles (methionine 129 homozygotes) even though they represent only 40% of the population.
- Replacing aspartic acid at position 178 with asparagine causes FFI when residue 129 in that allele is a methionine but CJD when residue 129 is a valine (Fig. 22.4). Note that these diseases target different regions of the brain. In each case, disease onset and progression are exacerbated when amino acid at position 129 in the normal allele is the same as that at position 129 in the mutated allele, i.e. the person is homozygous at the polymorphic position in *prnp*.

---

of the normal PrPc undergoes this same spontaneous conversion to initiate replication of the relevant PrPsc form. In the absence of a destabilizing mutation, this spontaneous conversion would be a rare event, accounting for the low frequency of sporadic CJD over a normal human lifespan and for the typical late onset of this disease.

22.4 PRION DISEASE PATHOGENESIS

Following infection, PrP expression is required in the central nervous system for brain degeneration to occur. If infection occurs peripherally (e.g. in the gut), cells of the immune system are necessary for neuroinvasion.

If formation of altered conformations of PrP underlies prion diseases of all etiologies, how does infectivity reach the brain in each case? Sporadic and inherited TSEs might arise directly from pathogenic conversion of PrP being initiated within the central nervous system (CNS; comprising the brain and spinal cord) (see Section 22.3). Some cases of iatrogenic CJD, which have arisen through direct introduction of infectivity into the CNS,

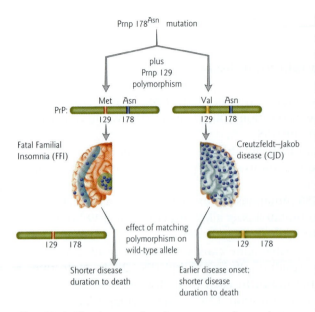

Fatal Familial
Insomnia (FFI)

Creutzfeldt–Jakob
disease (CJD)

Prnp 178^Asn mutation

plus
Prnp 129
polymorphism

PrP:

Met   Asn
129   178

Val   Asn
129   178

129   178

effect of matching
polymorphism on
wild-type allele

129   178

Shorter disease
duration to death

Earlier disease onset;
shorter disease
duration to death

Fig. 22.4 The interaction between pathogenic mutations at position 178 of human PrP and the amino acid present at the polymorphic position 129 on either the mutant or normal allele. Both the nature of the disease pathology, its time of onset, and rate of progression are determined by the position 129 polymorphism. Blue: regions of spongiform degeneration; brown: regions of neuronal loss and replacement with astrocytes and glial cells. See Box 22.3 for further details. (Redrawn, with permission, from Gambetti, P. 1996. *Current Topics in Microbiology and Immunology* **207**, 19–25. With kind permission of Springer Science and Business Media.)

for example by grafting of contaminated dura mater (the covering of the brain) during brain surgery or grafting of contaminated corneas, could similarly arise through propagation of infectivity from the site of exposure through the CNS. Experiments with PrP^o transgenic mice (i.e. they make no endogenous PrP) carrying a graft of PrP$^+$ brain tissue show that expression of PrP in the CNS is essential if pathogenesis is to be seen, leading to the idea that, within the CNS, infectivity is propagated through the tissue from the site of infection by the progressive conversion of normal PrP to a pathogenic form.

Other cases of iatrogenic CJD have come from peripheral exposure to infectivity, for example via injection of contaminated pituitary growth hormone (see Section 22.3), while Kuru, BSE, and variant CJD are, with varying degrees of certainty, known to have been transmitted through ingestion of contaminated food (see Sections 22.5 and 22.6). How, in all of these varying circumstances, does infectivity reach the CNS? Information on this subject has come mostly from studies of infection in various transgenic mouse lines. These show that elements of the immune system are crucially important to amplify the infectivity and transmit it on to the CNS. Indeed the spleen has long been known as a site of replication of these agents and infectivity accumulates there earlier than it does in the brain following peripheral infection (Fig. 22.5). Both B cells and follicular dendritic cells (FDC) have been implicated in TSE agent peripheral replication and, possibly, different types of TSE require different immune cell types for the agent to penetrate the CNS. Amplification of infectivity in FDC seems to be crucial if disease is to be manifested in the brain and the FDC must, as predicted from the prion hypothesis, be expressing PrPc for them to fulfill this role.

A second key aspect of pathogenesis concerns the species barrier to infection. Experimentally, this is observed as a very much greater difficulty (lower frequency of transmission and longer duration to overt disease) in transmitting infection between species than within a species (Fig. 22.3). This has been thought to reflect the need for structural similarity (ideally identity) between the infecting prion and the endogenous PrPc for efficient structural conversion of the latter protein to a prion. Thus human prions transmit disease to standard laboratory mice very poorly, more easily to mice carrying transgenic copies of the human *prnp* gene in addition to the equivalent mouse genes, and most easily to mice carrying the human gene but not the mouse gene. The disease-enhancing effects of homozygosity at position 129 (see Section 22.3) may reflect a similar effect. However, animals which have apparently not been infected in cross-species transfers, in that they remain asymptomatic, can nonetheless harbor considerable infectivity. This calls into question the simple molecular view of the species barrier as a reduced probability of PrP structural conversion.

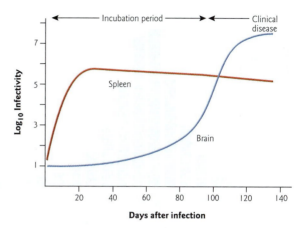

Fig. 22.5 Multiplication of the scrapie agent in the brain and spleen of mice infected peripherally and the time course of disease.

22.5 BOVINE SPONGIFORM ENCEPHALOPATHY (BSE)

BSE (popularly known as mad cow disease) appeared suddenly in dairy cattle in the UK in the 1980s, was formally recognized in 1986, and developed into a large-scale epidemic with serious economic repercussions (Fig. 22.6). Much smaller numbers of cases, in total around 5200 up to the year 2005, have occurred in other European countries, with a very few elsewhere, such as in North America. BSE probably resulted from the scrapie agent of sheep adapting to cattle, an event made possible by the practice of giving cattle artificial food concentrates which contained protein rendered from animal carcasses. It is likely that a single adaptation event originated the epidemic, which was then propagated by the recycling of cattle residues from the abattoir in cattle food concentrates, which gave optimum conditions for a cattle-adapted variant to emerge. Alternatively, a variation of this general hypothesis is that cattle had always harbored the BSE agent at low frequency and that an infectious cattle carcass was recycled into cattle feed, so initiating the epidemic. Whatever the origin, all BSE isolates appear to represent a single prion strain (Box 22.4).

The low number of BSE cases detected so far in North America is surprising given the importation of potentially infected animals into North American herds that took place before the nature of the problem was appreciated, the history of risk-prone cattle feeding practices in the USA and Canada, and the evidence of feed-related transmission in the cases that have been seen. To some extent, the low case count may be

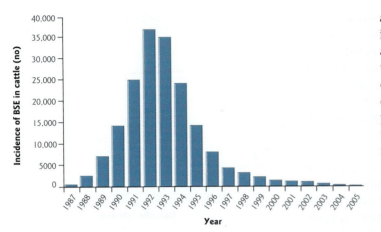

Fig. 22.6 Annual incidence of BSE in the UK cattle herd, 1987–2005. (Graph drawn from data taken from the UK Department for Environment, Food and Rural Affairs (DEFRA) website.)

a reflection of the relatively low intensity of testing of cattle being applied. It may also be the case that the amounts of infectivity in the cattle food chain in these countries did not reach the levels needed to produce an epidemic before measures were put in place to block this route of transmission.

Scrapie has been endemic in sheep in the UK for nearly 300 years, and this feeding practice was in place well before the 1980s, so why did transmission to cattle only occur then? It is likely that this resulted from a change that took place between 1980 and 1983 in the way that meat-and-bone-meal cattle food was prepared from animal carcasses. During this period, the use of organic solvent extraction in the process decreased by nearly 50%. This extraction had involved heating for 8 hours at 70°C and then the application of superheated steam to remove traces of solvent, which would have inactivated scrapie infectivity. Such extreme measures are needed as TSE infectivity is very stable, far more so than typical microorganisms, because of its unique composition (Section 22.2).

Box 22.4

Evidence that BSE isolates are a single strain

- When BSE is transmitted experimentally into mice, different isolates appear to be remarkably homogeneous with regard to incubation period; each of four mouse strains tested shows a characteristic BSE incubation time in Fig. 22.7a. This is a classical definition of strain identity in studies of scrapie.
- BSE isolates share a PrPsc molecular signature on western blots, reflecting the number and location of glycosylation sites and the size of the protease-resistant core of the protein (Fig. 22.7b). By contrast, scrapie strains differ widely in all these properties.
- Different BSE isolates show similar brain pathology upon experimental transmission.

22.6 BSE AND THE EMERGENCE OF VARIANT CJD

Following the BSE epidemic in the UK, an outbreak of a new human TSE, variant CJD, occurred. This represents human infection by the BSE agent, transmitted through the food chain.

Recognition of the BSE epidemic and its likely cause led to a ban on using meat and bone meal in cattle feed to prevent further spread in cattle and, as a precaution, a ban on the human consumption of specified bovine offals (brain, spinal cord, and lymphoid tissues, see Section 22.4) that were considered potentially infectious. However, it was recognized that people had been eating potentially contaminated cattle meat products for a number of years prior to the recognition of BSE as a disease. There were also other risks to be considered, for example calf serum is widely used in the pharmaceutical industry for the growth of cultured cells to make virus vaccines and other cattle byproducts are used in cosmetics.

At the time these measures were taken, the risk of transmission of BSE through the food chain was considered to be low (but not zero) because scrapie had been endemic in the UK sheep population for nearly three centuries and there was no evidence of transmission to humans, despite the large amount of sheep meat consumed. There was also evidence of substantial species barriers to transmission of scrapie infectivity (see Section 22.4). Nonetheless, the disease TME (Table 22.1) had previously emerged in captive mink through feeding them scrapie-contaminated sheep and the subsequent emergence of feline

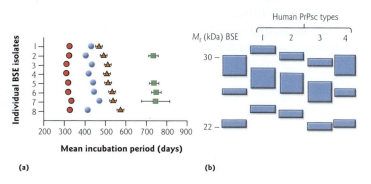

(a)   **(b)**

Fig. 22.7 The biological and molecular properties of different BSE isolates transmitted to mice are very similar to each other, and to those of vCJD isolates. (a) The incubation times of BSE isolates from cattle upon transmission to mice. Each line on the figure represents a different BSE isolate and the four types of symbol show the mean incubation time for that isolate in four different strains of mouse. (Adapted, with permission, from Bruce, M. E. *et al.* 1997. *Nature* **389**, 498–501. Reprinted by permission from Macmillan Publishers Ltd.) (b) Molecular signatures of TSE isolates. A schematic representation of an immunoblot analysis of polypeptides from the brains of mice infected with different TSE isolates that have been separated by SDS–PAGE. The bands represent protease-resistant core fragments of PrPsc, differing in length and glycosylation pattern, that have been detected with antibody to PrPsc. The thickness of each band represents its intensity in the original analysis. Human PrP types 1, 2, and 3 are seen in various sporadic and iatrogenic CJD isolates; the type 4 pattern is seen in variant CJD isolates and is very similar to the pattern consistently generated by BSE isolates in this type of analysis. (Reprinted from *Current Opinion in Genetics and Development* vol. 9, J. D. F. Wadsworth, G. S. Jackson, A. F. Hill & J. Collinge, Molecular biology of prion propagation, 338–345 (1999) with permission from Elsevier, using original data from Collinge, J. *et al.* 1996. *Nature* **383**, 685–690.)

Box 22.5

Evidence that vCJD represents BSE infection in humans

- Infection of mice with material from either BSE or vCJD samples occurs with very similar incubation times and with similar characteristic pathogenic signatures in the brain. Both these features are distinct from classical CJD infections in mice.
- The molecular signatures of the BSE and vCJD agents are similar (Fig. 22.7b).

spongiform encephalopathy (FSE) in domestic cats in the early 1990s, presumably as a result of consuming contaminated food, suggested that this species barrier could also be breached by the BSE or scrapie agent.

This view of the risk to humans from BSE was changed dramatically in 1996 by the recognition of a novel human disease, now known as variant CJD (vCJD). This differed somewhat from previously described CJD (however acquired) in its clinical characteristics, but strikingly also in the age of its victims. vCJD patients are typically much younger (less than 30 versus average 60 years) than are sporadic CJD patients. Experiments have shown that vCJD represents human infection by the BSE agent (Box 22.5).

So far, at least 150 people have been diagnosed with vCJD in the UK and several more in other European countries. Isolated cases in other parts of the world have been linked with probable exposure to infection in Western Europe. All the cases so far have been in people homozygous for methionine at the polymorphic position 129 in their *prnp* genes. This polymorphism is already known to influence disease in the context of pathogenic *prnp* mutations (see Section 22.3). The annual UK death rate from vCJD peaked in 2000 with 28 deaths and has now begun to show a gradual decrease. The number of deaths from sporadic CJD has also increased over the period of the vCJD outbreak in the UK, from a base of around 30 per year in the 1980s to a peak of 77 in 2003. In part, this may be due to increased surveillance for this type of disease, but it is also possible that human BSE infection can be manifest as classical CJD, as well as vCJD, depending on individual patient characteristics.

So far, around 170 variant CJD cases have occurred globally, over 150 of them in the UK. The eventual size of the outbreak is hard to estimate and depends greatly on whether those so far affected are atypical of the rest of the population in their disease susceptibility or incubation period.

For obvious reasons, there is intense interest in predicting the eventual size of the vCJD problem. All the while the annual UK vCJD death rate was holding steady, uncertainty about the incubation time from infection to disease made such prediction very difficult. Was this a plateau, or just the front end of a very gradually increasing curve? Now the firm downturn in the death rate over the past few years has led to estimates that the final number of people affected in the UK will be only a few hundred. However, the methionine homozygotes so far found to be susceptible to vCJD are only about 40% of the population. It is not known whether other genotypes are completely resistant to infection by BSE or simply show a longer incubation time to disease. Recently, a clinically inapparent vCJD infection was detected at death in someone heterozygous at position 129 in PrP, suggesting that, as in mice, infectivity can propagate in such people. If they can also progress to disease within a natural lifespan, the eventual size of the outbreak will be significantly greater.

Another uncertainty in predicting the future trend of vCJD cases is whether or not there will be significant human-to-human transmission. Although the outbreak originated from consuming BSE-infected meat, iatrogenic transmission could amplify the number of cases considerably. This is because people harbor significant prion infectivity long before they show disease symptoms. Infection may be passed on from such people if they donate blood or organs, or during routine surgical or dental procedures. To minimize this risk, blood donations in the UK are no longer being accepted from people who have themselves received a donation in the past, and the way in which certain surgical instruments are reused is being scrutinized, since standard cleaning and sterilizing procedures are not sufficient to eliminate infectivity. In the USA, blood donations are not accepted from people who have lived for extended periods in the UK.

Finally, it is worth noting that infected cattle are not the only source of TSE infectivity in human food. Scrapie-infected sheep meat has been consumed for years, apparently without human disease consequences since the frequency of sporadic CJD is similar in countries with and without endemic sheep scrapie. However there is concern that, during the BSE epidemic, this agent could have infected sheep, its presence perhaps being masked by scrapie in the flocks, and that this might then be infectious for humans consuming sheep meat. There is also the condition chronic wasting disease (CWD) that affects elk and mule deer in an area of the western USA (Table 22.1). The possibility has been considered that hunters and others who eat meat from these animals might also be at risk of TSE infection. However, in a number of cases of human TSE-like disease investigated for a possible connection to CWD, no convincing evidence of a link has so far been obtained.

22.7 UNRESOLVED ISSUES

Despite the explosion of research interest in the TSEs, there remain several significant unanswered questions about them. Firstly, it is unclear why the accumulation of a protein in an aberrant conformation should cause the mass death of neurons and spongiform degeneration of the brain, though it is notable that other neurodegenerative conditions, such as Alzheimer's disease, are also characterized by insoluble protein deposition in the brain. Secondly, it is unclear how the different pathologies associated with each human prion disease, which include different effects on specific brain areas in each case, can be generated through the structural conversion of the same normal protein. In other words, why does conversion of PrP to one abnormal form damage area A of the brain and spare area B, while conversion to another form damages area B but not A when the substrate for conversion, normal PrPc, is present in both areas? Finally, what is the normal function of PrP, aberrations in which seem to have such disastrous consequences? Prevention and cure for TSEs will surely depend on a much fuller understanding of these diseases.

KEY POINTS

- The prion diseases are a collection of delayed-onset neurodegenerative diseases that are mostly transmissible – the transmissible spongiform encephalopathies.
- The infectious entity is very likely to be a stable aberrant conformer of a cell protein called PrP.
- According to the prion hypothesis, protein molecules in the aberrant form catalyze the conversion of further molecules into that form, so replicating the infectious agent.
- The principal human TSE is Creutzfeld–Jakob disease. CJD can arise spontaneously, through inheritance of a mutation in PrP, through infected tissue transplant/injection, or through eating infected material.
- A major epidemic of BSE occurred in cattle in the UK through the feeding of material from recycled animal carcasses.
- BSE infectivity in meat entering the human food chain has transmitted into humans, causing variant CJD in around 170 people so far.

QUESTIONS

- Discuss the evidence for and against the hypothesis that all spongiform encephalopathy diseases are caused by the propagation of an altered conformation of the prion protein.
- Discuss the mechanisms by which transmissible spongiform encephalopathy infectivity is believed to be spread within an individual, between individuals, and between species.
- Describe the prion hypothesis and discuss how it can account for the existence of human diseases with both an infectious and an inherited basis.

FURTHER READING

Aguzzi, A., Polymenidou, M. 2004. Mammalian prion biology: one century of evolving concepts. *Cell* **116**, 313–327.

Bruce, M. E. 2003. TSE strain variation. *British Medical Bulletin* **66**, 99–108.

Chien, P., Weissman, J. S., DePace, A. H. 2004. Emerging principles of conformation based prion inheritance. *Annual Review of Biochemistry* **73**, 617–656.

Glatzel, M., Giger, O., Seeger, H., Aguzzi, A. 2004. Variant Creutzfeldt–Jakob disease: between lymphoid organs and brain. *Trends in Microbiology* **12**, 51–53.

Hill, A. F., Collinge, J. 2003. Subclinical prion infection. *Trends in Microbiology* **11**, 578–584.

Ironside, J. W., Head, M. W. 2004. Neuropathology and molecular biology of variant Creutzfeldt–Jakob disease. *Current Topics in Microbiology and Immunology* **284**, 133–159.

Pattison, J. 1998. The emergence of bovine spongiform encephalopathy. *Emerging Infectious Diseases* **4**, 390–394.

Ricketts, M. N. 2004. Public health and the BSE epidemic. *Current Topics in Microbiology and Immunology* **284**, 99–119.

Silveira, J. R., Caughey, B., Baron, G. S. 2004. Prion protein and the molecular features of transmissible spongiform encephalopathy agents. *Current Topics in Microbiology and Immunology* **284**, 1–50.

Soto, C., Castilla, J. 2004. The controversial protein-only hypothesis of prion propagation. *Nature Reviews Neuroscience* (Suppl. July 2004), S63–S67.

Weissmann, C. 2004. The state of the prion. *Nature Reviews Microbiology* **2**, 861–871.

Weissmann, C., Aguzzi, A. 2005. Approaches to therapy of prion diseases. *Annual Reviews of Medicine* **56**, 321–344.

The UK Creutzfeldt–Jakob Disease Surveillance Unit web site. http://www.cjd.ed.ac.uk/

The UK Government BSE web site. http://www.defra.gov.uk/animalh/bse/index.html

The World Health Organization TSE web pages. http://www.who.int/topics/spongiform_encephalopathies_transmissible/en/

23

Horizons in human virology

Virology is a fast moving field – and much remains to be understood. Techniques are constantly developing and this is opening up new avenues for therapy and also the possibility of using viruses in the treatment of other diseases. Equally, viruses are also far from static – new viruses are continually emerging to affect the human population and this creates a further imperative for virology research.

Where is virology going? Clearly, prevention of virus disease is a major aim of virologists. For virus diseases of humans and other animals, the priorities are improvements in vaccines and the search for antivirals. The relatively successful treatment of HIV in the developed world has underlined the need for affordable treatment to combat the far greater number of cases in the developing world; treatment is still reaching only a few percent of those infected. There is still no sign of an HIV vaccine, despite major clinical trials of three vaccine candidates, as discussed in Chapter 19.

Much of the progress in virology is technique driven, and huge strides were made with the advent of recombinant DNA technology, the polymerase chain reaction (PCR), and monoclonal antibodies. These truly changed the face of virology, as well as those of other branches of biological sciences and medicine, and continue to do so. In this chapter we discuss microarray analysis, which similarly promises much, as does the application of recombinant DNA technology to the derivation of virus-based gene therapy vectors. One area where progress remains slow is the study of virus–host interactions, largely because of the complexity of a whole organism, and because understanding of the immune system is far from perfect, and we pick out some interesting areas where progress is being made. Lastly, we address the concerns that new viruses continue to assail the human population, largely by managing to make the leap from their native host species. There is much to do for those with an interest in viruses.

23.1 TECHNICAL ADVANCES

Using microarrays in virology

Over the past few years, the emphasis in all areas of biological research has been shifting more and more towards a systems-based approach. This means attempting a comprehensive analysis of all the features of a particular situation, rather than targeted analysis of single parameters. The change is being driven by the emergence of technologies capable of achieving such results; virology is no exception to this trend.

Microarray analysis, also known as gene array analysis, is one of the new technologies that underpins a systems-based approach to studying biological processes. It is ideally suited to analyzing differences between two closely matched samples. Such sample pairs arise very readily in virology, e.g. cells that either are or are not infected with a virus, cells infected with two different viruses, or cells infected with wild-type virus and a specific mutant of that virus. The essence of the technique is the use of a very large array of oligonucleotide or cDNA probes, immobilized on a solid support, to measure the level of expression of multiple genes in a tissue or cell sample simultaneously. mRNA is prepared from the sample and amplified *in vitro* by reverse transcription and PCR, with the incorporation of fluorescently labeled nucleotides. This material is then allowed to hybridize to the array so that each probe in the array gets the chance to base-pair with any labeled complementary sequences that may be present. After washing, an image of the array is taken by digital capture and then the levels of fluorescence at each location in the array are calculated by appropriate computer software. Arrays can be purchased that contain probes to all of the genes in the genome of humans or several other species, attached to a silicon chip. Alternatively it is possible to build your own arrays with collections of probes of your choice spotted onto glass slides.

Microarrays are starting to make an impact in the study of viruses. One crucial area is the study of the interaction of viruses with their hosts. It has become clear that most viruses are much more subtle in their effects on the host than a simple view of their replication cycle might suggest. They can manipulate the expression of host genes selectively, increasing the amounts of proteins that are beneficial to them and reducing the expression of others that are not. While gene expression was analyzed at the single gene level, the full extent of these effects could not be determined. Moreover, the analysis was inherently biased towards examining effects known or suspected from other data, as investigators would only analyze the expression of genes that they had some basis for expecting would be altered by the infection. Array analysis addresses these issues, but is not without its own problems. Chief among these is that the experiments, at least those using whole genome arrays, are too costly at present to permit a systematic analysis of an infection at multiple time points and in different cell types, with all the necessary biological replicates. Thus, more limited experiments to suggest new lines of enquiry have to be combined with traditional analysis of the expression of selected genes.

A second area where microarrays may impact on virology is in diagnosis. Earlier (see Section 2.3), the detection of viruses through the use of the polymerase chain reaction to amplify fragments of their nucleic acid was considered. It is clear that such nucleic acid hybridization-based approaches offer quick and highly specific means to test for the presence of virus, and can distinguish even between very closely related viruses or strains of the same virus. The only drawback is that, for each test, primers or probes must be selected to test for a specific virus. In other words, an informed choice has to be made as to which virus(es) to test for, based upon the clinical details in each case. Using a custom array approach would avoid this step, allowing a sample to be tested simultaneously for nucleic acids from a wide variety of viruses. However, issues of robustness and cost would need to be addressed before this approach could become routine.

Antibodies: monoclonal antibodies, humanized monoclonal antibodies, human monoclonal antibodies from DNA libraries, and microantibodies

Since the initial technology for the production of monoclonal antibodies (MAbs) was devised in 1975 by George Köhler and César Milstein at Cambridge (UK), the face of virology has been transformed by their multifarious uses. Monoclonal antibodies continue to be in the forefront of analytical, preparative, and diagnostic technologies, and nowhere have they had more impact than in the creation of a new sphere of employment in the commercial sector. Now, it is difficult to imagine how virology functioned without them.

Humanizing mouse MAbs

Despite their huge impact on virology, there is one important area where the use of MAbs has been frustrated by technical limitations and this is in their therapeutic use. The problem is that MAbs are made by immunizing an animal, extracting primed B cells from the spleen, and then immortalizing the B cell by fusion with a myeloma (B cancer) cell. While the system works for mice, it is not applicable to humans and no human myeloma cell has proved effective. However there has been some success by immortalizing human B cells with Epstein–Barr virus. The problem of treating people with a mouse MAb is that their immune system very rapidly reacts to the mouse antigens. At best, this destroys the antibody and, at worst, initiates a potentially harmful, or even lethal, anaphylactic response. One answer is to humanize a mouse MAb by using recombinant DNA technology to replace the mouse sequences with sequences from a human antibody, except those mouse sequences responsible for binding to antigen (the paratope that is composed of complementarity determining regions (CDRs) of the hypervariable parts of the variable (V) regions of the light (L) and heavy (H) chain (see Fig. 12.4)). By minimizing the amount of foreign (mouse) protein, the problem of rejection is overcome. One of the first humanized MAbs is licensed for use against respiratory syncytial virus.

Isolation of human antibodies from random combinatorial libraries

An alternative procedure does not obtain MAbs by immortalizing antibody synthesizing B cells but, instead, uses molecular biological technology. mRNA is isolated from a pool of antibody-synthesizing cells obtained from a person and subjected to reverse transcription and PCR to make cDNA (Fig. 23.1). The cells are obtained, using a syringe, either from the bone marrow or from blood. In the first PCR reaction, specific primers are used to amplify all V_L regions present and, in the second reaction, all V_H regions are similarly amplified. One V_L DNA and one V_H DNA are then linked together at random, using an oligonucleotide encoding a flexible 15-residue peptide linker $(Gly_4–Ser)_3$, so that, when expressed, the resulting polypeptide folds up to form an antigen-binding site or single-chain Fv fragment.

Any antibody of interest will be present at a very low frequency in the population of expressed hybrid molecules, so a large number of cloned Fv molecules need to be tested to find the desired specificity. This screening problem has been solved in a novel fashion by expressing the V_L–linker–V_H polypeptide on the surface of a filamentous bacteriophage (fd or M13). These phages encode an attachment protein, present in low number – about four copies per virion. This binds to the F pilus of E. coli. The V_L–linker–V_H is expressed as a fusion protein with the phage attachment protein. The phage is still infectious and the V_L–V_H is expressed in a form which is able to bind to antigen, and is called a "phage antibody." Selection of the required antibody activity is achieved by panning a phage population expressing a myriad of random antibody specificities onto the desired antigen immobilized onto a plastic well. Nonbound phage is washed away and bound phage eluted and grown up on a lawn of E. coli. Agar plugs are picked from the resulting phage plaques, and phages then amplified and retested. DNA encoding the V_L–V_H gene can be excised from the cloned phage, and V_L joined to C_L and V_H joined to C_H to create a FAb which is synthesized in bacteria, or a complete antibody molecule which is synthesized in human cell line.

The phage panning technique allows antibody specificities present at a frequency of 1 in 10^6 in the original mix to be selected. This type of procedure has been used recently to obtain powerful neutralizing antibodies from HIV-positive individuals. The system is known as the random combinatorial method of making antibody, since it allows the formation of antibody from a library of V_L and V_H genes. It also permits the pairing of V_L and V_H genes to form specificities that may not exist in the donor.

Immunotherapy with humanized mouse MAbs or human MAbs is not intended to replace the stimulation of immunity with vaccines but, rather, to provide immunity in circumstances where no other treatment is possible or effective. It is used routinely to protect infants who have underlying (e.g. heart) conditions that render them seriously at risk from infection, or otherwise healthy infants who have contracted a life-threatening infection (e.g. human respiratory syncytial virus for which there is no vaccine; see Box 21.6). Immunotherapy could also be considered for the congenitally immunodeficient, or those whose immune systems are compromised by accident or deliberately in the course of transplant surgery or cancer chemotherapy. In addition, it could be of use for protecting the newborn from virus infections carried by its mother, and is being considered as a step to prevent mother–baby transmission of HIV-1 infection.

Microantibodies

While the main aim of a vaccine is to prevent infection by stimulating the body to make its own immunity, the passive administration of antibodies is an invaluable branch of immunotherapy. However, antibodies used as antiviral agents pose many problems. In the first place they can only be produced biologically from cells (hybridomas), and this is an expensive process that is difficult to control on an industrial scale. Antibodies are most easily administered by injection into the blood stream, but being large molecules (M_r, 155,000), they have difficulty

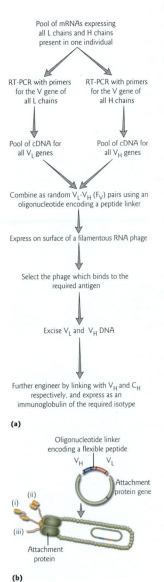

Fig. 23.1 (a) Scheme outlining the random combinatorial method for the cloning of human antibody single-chain Fv fragments. Its success lies in the ability to select rare recombinant phages expressing the required antibody activity (about 1 in 10^6 phages). (b) Construction of the fd phagemid containing the fused V_L–linker–V_H gene. This is transformed into *E. coli*. The bacterium is then infected with phage (fd) and progeny phages are produced, each having a single recombinant V_L–linker–V_H polypeptide displayed on its surface (i). This polypeptide folds to form a functional antibody-binding site (ii). The "phage antibody" requires just one molecule of nonmodified attachment protein (iii) to be infectious.

getting out of the blood system into the tissues where they are needed, and this can take several days. Again, their size can prevent them from penetrating deeply into the tissue where the antibody is needed. Another problem is that the Fc region of an antibody is a powerful activator of complement and other host immune mediators, and when these are stimulated in large quantities they can cause tissue damage, and even be life-threatening. Inevitably, passive antibody administration means that the body is being exposed to large amounts of foreign protein, although this can be minimized by humanizing mouse MAbs, as above, or using human MAbs. However, the Fc region has allotypic regions (amino acid residues that vary from person to person), and the paratope itself is foreign, so both can provoke antibody responses. The inherent multivalency of antibodies may lead to problems of antigen aggregation and deposition in kidneys and other tissues, with the resulting risk of tissue damage from complement activation and attack by host cells. Finally, it is difficult to ensure that antibody is free of adventitious infectious agents that can be covertly produced by the antibody hybridoma.

One way of avoiding many of these difficulties is to use antibody fragments which no longer possess the Fc region. The $F(Ab')_2$ antibody fragment is made by proteolytically removing its Fc region, but it is still bivalent and can cause antibody–antigen complexes and aggregation problems, and is still a large molecule. Smaller antibody fragments have the Fc region removed and are monovalent. These, in order of decreasing size, are FAb (a single antigen-binding arm), single-chain Fv (the heavy and light chain variable domains (V_H and V_L) joined by a flexible peptide linker), and individual V_H and V_L chains. However single-chain Fv and V_H and V_L chains are still relatively large and may be immunogenic.

One solution that solves all these problems is to use a microantibody. This is a peptide with a sequence derived from just one of the six CDRs (complementarity-determining regions) of an antibody. These are highly variable sequences of approximately 8–23 amino acid residues, three on each of the V_L and V_H, that are folded together to form the paratope or epitope/antigen-binding site. The surprise is that such a peptide has sufficient of the sequence and conformation of the paratope to act like an antibody. The best example is a microantibody derived from a monoclonal mouse IgG that neutralizes HIV-1. It comprises 17 amino acid residues from the CDR-H3. Work in the laboratory of one of the authors showed that the microantibody recognizes exactly the same epitope as the IgG from which it derives, and binds to the gp120 envelope protein and neutralizes infectious HIV-1. Such a small structure can be chemically synthesized and so obviates the problems of contamination by infectious agents such as endogenous viruses that can exist in hybridoma cell lines. Its small size permits the study of the poorly understood neutralization process through alteration of individual amino acid residues. The still unsolved problem is to create more microantibodies from other virus-specific antibodies. So far, only three antiviral microantibodies have been made, two that neutralize viruses (HIV-1, human respiratory syncytial virus) directly, and one that recognizes the CD4 receptor and inhibits the production of HIV-1 by infected cells.

In terms of immunotherapy, microantibodies have many advantages over antibodies; their small size allows them to leave the circulation quickly, and to penetrate deeply and rapidly into tissues. Peptides are poorly immunogenic, and as they consist of less than a single paratope, micro-antibodies are unlikely to stimulate even an anti-idiotypic response, thus enabling them to be used in repeat regimens that would normally stimulate serious anti-antibody responses. For similar reasons, labeled microantibody can be used to advantage for other purposes such as in the diagnosis of tumors, and in immunoassays or immunodiagnosis in research or industry where antibodies are now used. Chemical synthesis also means that production of microantibody is easily regulated, and is cost-effective; further the resulting product is sterile. If the technical problems to their production can be overcome, microantibodies have potential applications in all those areas of research, diagnostics, and the clinic where monoclonal antibodies are currently used.

23.2 RECOMBINANT VIRUSES AS GENE THERAPY VECTORS

As well as providing potent and constantly evolving threats to human health, and to animals and plants, viruses also offer the possibility of being used for human benefit through the construction and appropriate use of recombinant viruses. These are viruses where the genome has been altered in a planned way by experimental manipulation. There is intense interest in the potential application of such viruses as gene therapy agents, carriers of therapeutic nucleic acid sequences into our cells, in order to cure specific diseases.

Making and using recombinant viruses

The starting point for any program to generate a recombinant virus is to clone its genome. After genetic material has been isolated from virus particles, it can be manipulated in exactly the same way as any other RNA or DNA molecule. Thus DNA virus genomes may be cloned directly while RNA virus genomes may be cloned as cDNA. These cloned molecules can then be modified by site-specific alteration, or more drastically, segments may be removed and replaced with foreign DNA sequences. In this way, it is quite straightforward to generate modified forms of either an intact viral genome or, if the genome is large, a segment of it, cloned in a bacterial plasmid. What is far more difficult is to complete the process by using these cloned sequences to recreate infectious virus particles. This requires techniques specific to each type of virus and is not yet possible for all virus types; some of the double-stranded RNA viruses remain refractory to all attempts at generating recombinant virus from cloned cDNA.

Recombinant viruses have been invaluable in the study of virus replication cycles. They are also powerful tools in the study of fundamental cell processes in the laboratory, where they can be used to carry cDNA for specific host genes into cells in culture and hence to cause expression of the protein products. Another use is as viral vaccines. As discussed in Section 21.4, recombinant viruses can be created with the specific aim of creating a vaccine for a virus that is not amenable to conventional means of vaccine development. Finally, recombinant viruses have been of great interest over the past decade as potential gene therapy agents. This application is discussed in the following sections.

What is gene therapy?

Gene therapy is simply the introduction of DNA sequences into the cells of a patient with the aim of achieving a clinical benefit. It was originally conceived as a way of restoring normal function in patients with specific inherited single gene defects, such as cystic fibrosis. In such conditions, absence of a normal copy of the gene means that a specific protein function is missing, with severe physiological consequences. By putting back a "good" copy of the gene (normally as cDNA) into the patient's cells, all these consequences should, in theory, be corrected. From this beginning, the gene therapy concept has now grown to encompass a variety of possible applications, with the greatest number of clinical trials being in the field of cancer therapy. When it comes to treating cancer by gene therapy, the goal is totally different from gene therapy for inherited conditions. Rather than trying to restore normal function to the tumor cells, the objective is to kill them, or to cause them to undergo apoptosis – but to do so specifically, so that normal cells are not damaged.

All gene therapy experimentation is carefully regulated. In concept, gene therapy could be applied to either the germ cells or the somatic cells of an individual, but, for ethical reasons, only the latter is allowed. Any attempt to modify the DNA that is passed on to subsequent generations is against the law. Thus, gene therapy is restricted in scope to the treatment of the health problems of the individual. It cannot be used to attempt to eliminate "bad" genes from the human gene pool.

Recombinant viruses for gene therapy

For all the exciting possibilities of gene therapy to become reality, DNA has to be carried across the membranes of cells and ultimately to reach the cell nucleus. This is something that naked DNA is very poorly equipped to achieve. The cell membrane is a huge barrier to direct entry of such a long, highly charged, molecule and, if it is endocytosed into a cell, it is then vulnerable to degradation in lysosomes. By contrast, nucleic acid that is inside an infectious virus particle avoids these problems. First, viruses have evolved specific interactions with cell surface molecules that lead to their efficient entry into the cell and, second, if that entry involves arrival in the cytoplasm within an endocytic vesicle, then viruses have mechanisms to allow efficient escape into the cytoplasm (see Chapter 5). These properties make viruses important tools for gene therapy. A recombinant virus (the vector) carrying a foreign gene (often referred to as a transgene) is rather like a Trojan horse, being able to penetrate the defences of the cell before unloading its cargo. What that cargo does to the cell subsequently is totally dependent on the gene(s) chosen; having carried the gene into the cell, the role of the virus is over. This process of virus-mediated delivery of a gene into a cell is known as transduction.

To be potentially useful as a gene therapy vector, a virus should have a number of features (Box 23.1). Bearing these factors in mind, the viruses that have seen extensive development as vectors so far include mouse retroviruses, human adenovirus, and the parvovirus, AAV. However, no virus meets all of the criteria for an ideal gene therapy vector and, as considered below, there are some significant drawbacks to the vectors that have received extensive attention so far. Thus, there is no such thing as

Box 23.1

Necessary or desirable features in a virus to be used as a gene therapy vector

- The virus must be able to enter the desired human cell (i.e. receptors for the virus must be present on the cell surface, see Chapter 5).
- It must be possible to generate recombinant forms of the virus that are deleted for one or more essential functions and so do not replicate in the patient. Equally, it must be possible to grow these replication-defective recombinants in cell cultures in the laboratory.
- In its deleted, replication-defective form, the virus must be safe to use in people. In particular, it should not be able to recover replication-competence by recombining with other viruses that might be present *in vivo*.
- The virus should deliver its genetic material to the nucleus in the form of DNA.
- It should be possible to grow recombinant forms of the virus to high concentrations.
- The replication-defective form of the virus should be able to accommodate a sufficient length of foreign DNA to be useful (most viruses have quite stringent upper limits on the amount of nucleic acid that they can include in a particle).
- If therapy is to be "permanent," the transgene needs to persist in the cells indefinitely and to be inherited by daughter cells if the cell divides. To achieve this, a viral vector needs to achieve either integration into a chromosome of the cell or else autonomous replication of its genome and partition during cell division.

the perfect viral gene therapy vector and each gene therapy application is likely to need its own vector, chosen and then tailored to meet the precise circumstances.

Retroviral vectors for gene therapy

Promising results were obtained using retroviral vectors in clinical trials of gene therapy for inherited immune deficiencies, but serious safety issues emerged during patient follow-up.

Retrovirus vectors have all their normal coding sequences removed and replaced with a cDNA of interest. The recombinant virus is then grown in a special cell line (packaging cells) which provides the viral proteins that are needed to form particles containing the recombinant genome (Fig. 23.2). When used therapeutically, the virus particles reverse-transcribe their genome and integrate the DNA copy, including the transgene, randomly into the host cell genome using enzymes included in the particle (see Chapter 8), but the absence of all viral genes from the particle means

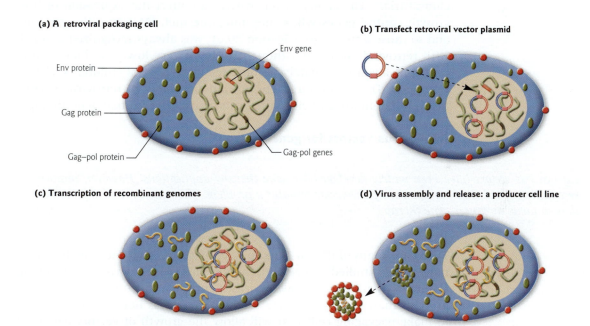

Fig. 23.2 (a–d) Production of recombinant retroviruses by transfection of cloned recombinant genome plasmid into a packaging cell line. Cells may be cultured on as a producer cell line, with recombinant virus being harvested from the growth medium. The essential retroviral protein coding genes, *gag*, *pol*, and *env* (see Section 9.8) are integrated in the packing cell line in two distinct segments, to minimize the risk that recombination between these sequences and the vector could produce a viable retrovirus again.

that no further events of the infectious cycle can occur. Hence the transgene that the vector has carried into the cell is present for the life of the cell and will be passed on if the cell divides.

Retroviral vectors grow only to relatively low concentrations, which means that they have typically been used only to transduce cells taken from a patient; these must then be reimplanted later (*ex vivo* gene therapy). These vectors are thus well suited to treating cells of the lymphoid system, because firstly the cells can be easily explanted and reimplanted, and secondly transgene integration is needed since lymphoid cells are being constantly replaced from progenitor cells. Two inherited immunodeficiencies have been the target of clinical trials of retrovirus-mediated gene therapy, adenosine deaminase (ADA) deficiency and X-linked severe combined immunodeficiency disease (X-SCID). Both conditions are characterized by a profound lack of T lymphocytes and hence a failure of both humoral and cell-mediated arms of the adaptive immune response (see Chapter 12). In both cases, there was evidence of clinical improvement in small numbers of young children treated in this way, and in the case of X-SCID, there was early optimism that the treatment had cured the disease. However, a serious difficulty has emerged from these clinical trials. The ability of retroviruses to affect the expression of host growth control genes when they integrate their DNA into the genome, and so cause a tumor (see Section 20.6), was always recognized as a risk with retroviral vectors but thought to be of very low probability. However, this has unfortunately become reality in several children suffering from X-SCID. It is now recognized that the safety of retrovirus vectors needs to be improved before they can be used for human gene therapy.

Adenovirus vectors for gene therapy

Adenovirus vectors have been widely developed for gene therapy applications. However, immune recognition of cells that have taken up the vector remains a problem. Their most promising application so far is in cancer gene therapy.

There are many different human adenoviruses. The one that has been most widely studied as a gene therapy vector is adenovirus type 5, a mild respiratory pathogen. As adenoviruses encode a much larger number of proteins than retroviruses (see Section 9.4), it has not been possible yet to make packaging cells that will allow the growth of vectors where all the viral genes have been deleted (gutless vectors, Fig. 23.3b). Such vectors can be grown using an intact virus (helper virus) to provide the necessary viral proteins, but most applications to date have used vectors where most of the viral genes are retained so they can be grown in cells that complement their gene deficiencies. A widely tested example is a vector lacking the E1A and E1B genes, which encode proteins required for the expression of the other viral genes, and the E3 gene, which encodes several proteins that are not needed for growth in cell culture but which modulate the host response to infection *in vivo* (Fig. 23.3a). Very high concentrations of vector particles can be obtained, suitable for direct administration to patients for *in vivo* therapy.

The major problem that has been encountered with adenovirus vectors deleted for E1 and E3 genes has been the immune response that develops to vector proteins, which effectively eliminates cells that have acquired the transgene. This response arises in part from some residual expression of the viral genes that remain on the vector and can be reduced by using gutless vectors. However, such responses may be an advantage in the context of cancer gene therapy if the virus can be targeted specifically to tumor cells. A further difficulty is that receptors for adenovirus 5 are not as widely distributed as first thought. Initially, cystic fibrosis was considered to be a promising target for therapy with adenovirus 5 vectors, since this gene defect affects respiratory epithelial cells, which are the natural target of the virus. However, the virus receptors are on the basolateral membranes of these cells, not the apical surface, and so infection is not efficient. Work is continuing to modify adenovirus 5 receptor specificity and to use other adenoviruses as vectors. Using adenovirus vectors for cancer therapy, where the aim of transgene delivery is

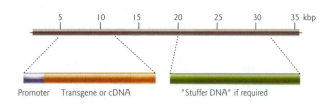

(a) E1,E3 deleted, replication-defective adenovirus vector genome

(b) "Gutless" adenovirus vector deleted of all viral genes

Fig. 23.3 The genomes of adenovirus gene therapy vectors, either (a) deleted for the essential E1A and E1B genes, plus nonessential E3 genes (to make more space for the transgene), or (b) deleted for all adenovirus sequences except the termini, which provide origins of replication and the signal for packaging the genome into particles. In (a), the maximum capacity for the transgene is around 7 kbp, while in (b), the maximum is around 35 kbp and, for short transgene sequences, additional "stuffer DNA" is needed to achieve the minimum genome length necessary for stable particle formation. The viral genes are shown in simplified outline in (a), with early genes colored red and late genes colored purple (see Section 9.4). Viral sequences deleted during vector construction are shown as brown boxes in the genome and the affected genes are shown striped in (a).

to cause the death of the cell, remains a very promising approach. Finally, it should be noted that, as with retroviral vectors, safety issues have emerged during clinical trials using adenovirus vectors. One patient died in 1999 following administration of a very high dose of vector particles, apparently as a result of an unusual response to the virus in that individual. Such variation at the level of the individual patient is a major problem in the development and safety testing of novel therapies.

Other viral vectors for gene therapy

The parvovirus AAV (see Sections 6.6 and 9.7) has only a small genome, so even by removing all its genes its capacity for foreign DNA is limited. However, a key advantage of AAV over other vectors seems to be that cells transduced by it elicit minimal destructive immune responses. It was also thought initially that the ability of AAV to integrate at a specific chromosomal location (see Section 6.6) might give it an advantage over the random integration of retroviral vectors, but it now appears that this property of AAV depends upon expression of the viral Rep proteins, and these genes have to be deleted when converting it to a vector. Finally, herpes simplex virus is also being considered as a vector, particularly for gene delivery to neurons where it naturally establishes latency.

Future development of viral vectors for gene therapy

A factor which limits the use of all viral vectors for gene therapy at present is the inability to target gene delivery to specific target cell populations. Ideally, vector particles should attach to, and enter, only the desired cell type, so improving effectiveness of gene delivery and limiting the scope for potential side-effects. Whilst this is not yet possible, considerable progress has been made in recent years in the re-targeting of viral infectivity onto novel receptors. This is a key area for future development. Another important area is to improve the safety of retroviral vectors, since these have been shown to be effective in the clinic but pose an unacceptable safety risk in their original form. Because of the long lead-time from basic laboratory studies to clinical trials, there are already better retroviral vectors available than those used in the trials with ADA and X-SCID patients and further developments seem likely. Finally, the problem of immune responses to adenovirus vectors needs to be addressed if these are to have a role in treating inherited disease rather than cancer. This exciting field looks set to be an important area of virology research and development for some time to come.

23.3 SUBTLE AND INSIDIOUS VIRUS–HOST INTERACTIONS

When infecting an animal, a virus is faced with a variety of differentiated cells which display a great range of structures, physiological functions, and biochemical equipment. However, an individual virus infects only certain cells and it is the specificity of this interaction which can be unexpected and interesting. The majority of virus–host interactions conform well to one or more of the models discussed in Chapter 14, but some show more subtle and surprising elements.

Endogenous retroviruses

A significant amount of the human genome is composed of endogenous retrovirus sequences. These are the residue of ancestral retroviral infections that have invaded the germ line. While most of these residual sequences are inert, some may have deleterious effects on human health while others may be beneficial or even essential to our well-being.

When retroviruses infect a cell, a DNA copy of their genome is inserted at random into the genome of the host, where it remains for the life of the cell (see Chapter 8). Whilst the vast majority of infections affect somatic cells, occasionally a retrovirus infects a germ cell. If gametes formed from that cell subsequently give rise to offspring, then every cell in the body of the new individual will contain a copy of the inserted retroviral sequences. These are known as endogenous retroviruses, as distinct from exogenous viruses that infect the individual from its environment.

Examination of the human genome reveals large numbers of endogenous retrovirus (HERV) sequences – perhaps as much as 8% of total DNA. The vast majority of these sequences are the result of ancient insertion events that predate human speciation, but some are more recent. Also, examination of the sequences of these retroviral inserts reveals that almost all are defective, i.e. unable to produce virus particles. Many have suffered gross deletion of the genome to leave just a single inserted long terminal repeat (LTR; see Section 8.4), while most of the others have deletions and/or frame-shift mutations affecting some or all of the *gag*, *pol*, and *env* genes which encode the internal structural proteins, viral particle enzymes, and envelope glycoproteins respectively. This is presumably the result of strong selective pressure to disrupt the ability of such endogenous sequences to generate virus particles.

The most recently acquired human endogenous retroviruses belong to a sequence subgroup known as HERV-K. Many of these are found only in chimpanzees and humans while a few are human-specific. Even here, most of the genes on these viruses are inactivated by mutation, but one or two of the sequences remain intact and could form particles if expressed. In addition, there is the possibility that the proteins needed for particle formation might be provided by different defective endogenous viruses (i.e. complementation). HERV-K gene activity is detected in testis tissue and particles have been seen in teratocarcinoma cells (cells from germ cell tumors).

There has been a lot of interest in the possibility that HERVs might not be inert. Either the production of virus, or of transcripts that are then reverse transcribed and inserted elsewhere on the genome, or the effects of the integrated sequences or their protein products on the host cell, might be expected to have effects on the biology of the host. Many attempts have been made to link HERV activity with complex conditions such as rheumatoid arthritis, multiple sclerosis, and schizophrenia, but it is very difficult to prove a causal connection because any activity detected is coming from multiple HERVs and it might be that only one specific HERV is relevant to a disease; these diseases are themselves heterogeneous and so probably have multiple causes, further confusing the picture. Studies of these important potential disease connections continue. One disease connection with fairly strong evidence is the activity of a subgroup of HERV-K in men with germ cell tumors. Such patients have a high frequency of antibodies to HERV-K proteins compared to those without the disease and, as already noted, cell lines from such tumors can form HERV-K particles. However, this might be a consequence of disease rather than a causal relationship.

In addition to these potential relationships to human disease, endogenous retroviruses are now recognized as one of the key problems to be overcome in advancing the science of xenotransplantation – the transplantation of organs from other species into humans. Pigs in particular have been considered as potential organ donors. However, like humans, they also harbour endogenous retroviruses (PERV) and, in this case, some of the PERV loci in the pig genome do produce virus and this can infect human cells in culture. There is therefore concern that PERV infection of tissues in the graft recipient might be damaging, or that recombination between HERV and PERV sequences might generate novel viruses that could spread to other individuals.

As well as being considered as agents of disease, one HERV at least has a positive role in human biology. A protein known as syncytin is produced from the *env* gene of a HERV-W integration in the human genome. This protein is expressed principally in syncytiotrophoblasts, multinucleated cells in the placenta that form the boundary between maternal and fetal tissue. Such giant cells form by fusion of trophoblasts to one another, and syncytin is able to fuse cells of this type in culture; it is thought to play a similar role *in vivo*. This may be viewed as an adaptation of the ancestral viral function of this protein in mediating fusion between the viral and target cell membranes.

Myalgic encephalomyelitis (ME) or postviral fatigue syndrome or chronic fatigue syndrome

ME is a distressing condition in which patients suffer extreme fatigue of voluntary muscles after moderate exertion, and this may be prolonged for a year or more, although sometimes with partial remissions. Described as a syndrome (meaning a collection of different diseases with a common expression), it is defined by exclusion of other causes of chronic fatigue.

ME has been recognized for over 30 years, although until very recently there has been considerable uncertainty about its cause – whether it was due to a persistent or latent infection or was psychosomatic or plain malingering. Now, new sensitive methods of detection have shown the presence of viral genomes in muscle biopsies of some cases of ME.

The breakthrough came through the use of cloned viral DNA. A biopsy needle was used to remove a small core of muscle, from which nucleic acid was extracted. This was then tested, as described in Table 23.1. A highly significant proportion of people with ME had enterovirus RNA in their muscle, while EBV DNA was present in a smaller proportion. None had both. Later work showed no active involvement of two other viral candidates, human herpesviruses 6 or 8. An important question was how enterovirus RNA was able to persist, when these viruses normally cause a short-lived acute infection. Part of the answer is that the viral RNA synthesis is abnormal: whereas acutely infected cells contain 99% positive sense RNA (mRNA and virion RNA), muscle biopsies from ME patients show similar amounts of positive and negative sense RNA (see Section 7.3). It will be of interest to see if full-length viral RNA is present, and if it is mutated in regions controlling RNA synthesis.

These data are in good agreement with earlier serological evidence of enterovirus infection. Here, 51% of ME patients had virion protein 1 (VP1)–antibody complexes circulating in serum, and 20% of patients excreted virus–antibody complexes in feces. Finally, electron microscopy shows the presence of abnormal mitochondria in muscle of ME patients. Obviously, any defect in energy production by the mitochondria would make muscle prone to fatigue. We now have some of the pieces of the ME jigsaw puzzle, but they and others have to be fitted together. However, the involvement of enteroviruses in ME remains contentious and other viruses may also trigger the condition.

Table 23.1

Evidence for the presence of enterovirus and Epstein–Barr virus (EBV: a herpesvirus) nucleic acid in muscle biopsy samples from patients with ME.

Technique	Primer/probe	ME	Controls
PCR	Enterovirus RNA sequence: the 5′ nontranslated region	32/60 (53%)	6/41 (15%)
Hybridization	cDNA from the enterovirus polymerase gene	34/140 (23%)*	0/152 (0%)
Hybridization	cDNA from EBV nuclear antigen 1 gene	8/89 (9%)*	0/48 (0%)

*Same set of biopsies tested; none was positive for both viruses.
cDNA, complementary DNA.

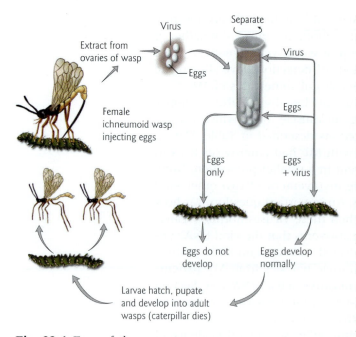

Fig. 23.4 Eggs of the ichneumonid wasp only develop in their host caterpillar when injected together with a polydnavirus.

Virus infections can give their host an evolutionary advantage

Viruses of certain parasitic wasps are telling us something that some virologists had suspected for many years – that viruses continue to exist because they can provide their host with an evolutionary edge. For the past 74 million years ichneumonid and braconid wasps have been carrying integrated copies of the genome of a bracovirus in their DNA. There is virus replication and production of free virus particles, but only in special cells of the ovary. The larvae of these wasps are parasitic on caterpillars, such as that of the tobacco hornworm moth, and the female wasp deposits eggs by injection along with bracovirus particles. The eggs hatch out and the wasp larvae feed on the living caterpillar until it dies, whereupon the wasp larvae pupate and emerge in due course as new adult wasps. However, if the eggs are first separated from the bracovirus by centrifugation and then injected into caterpillars, they fail to develop. Adding back purified virus allows normal wasp development to take place (Fig. 23.4). On examination, it was found that the virus selectively suppresses the caterpillar's immune system in favor of the wasp larva, and alters the developmental regulation of the caterpillar so that it cannot pupate and remains a caterpillar – and a suitable host for the parasite. One of the virus proteins responsible is an analog of cellular cystatins, which are inhibitors of cysteine proteases.

Thus, virus is needed to suppress (by some unknown mechanism) the immune responses of the caterpillar, and we have the situation where a virus has become essential to the successful life cycle of its host. We do not know if other viruses bestow evolutionary advantages upon their hosts, but this example stimulates us to turn the normal virus–host relationship on its head and look for positive aspects of infection. Who knows, one day it may be proved that some virus infections are good for you!

The fight between viruses and the host's interferon response

It has been known for some time that viruses do not sit back and wait to be hit by the host's immune responses but actively fight their corner, with the result that a pitched battle rages over whether or not the virus will be able to mount a successful infection. However, molecular details of how viruses accomplish this feat have only recently emerged. One of the best understood situations is the conflict between viruses and type I interferons. Much of this information has come about by the use of micro-arrays to analyze the response of the cell to infection (Section 23.1).

Interferons are cytokines produced by the innate immune system that is the first line of defence against virus infections (see Chapter 12). They induce expression of hundreds of interferon-stimulated genes, some of which have antiviral activity. These genes are responsible for the anti-viral state which is mediated through interaction of interferon with cell pathways that regulate apoptosis, inflammation, and stress responses. However, viruses have evolved to counter the antiviral response by dis-rupting the cross-talk between interferons and these cellular pathways.

Influenza A virus is just one of several viruses that have their own way of negating the interferon response. Others for which there is informa-tion are hepatitis C virus, herpes simplex virus, and vaccinia virus, and all deal with interferon with their own variations on the theme described below. Influenza A virus counters interferon in two ways. Firstly, it activ-ates P58^{IPK} that is an inhibitor of protein kinase R (PKR). PKR is norm-ally stimulated by interferon and prevents translation of all mRNAs. Secondly, influenza virus acts through its nonstructural protein, NS1. NS1 has many functions, including binding strongly to dsRNA. Deletion of the *NS1* gene makes virus so exquisitely sensitive to the effects of interferon that it can only replicate in cells in which the interferon response has been eliminated. Microarray work shows that deletion of the *NS1* gene results in an increase in the number and level of cellular genes that are induced during infection. NS1 particularly affects the transcription of the interferon response genes and NFκB-mediated gene expression. The inhibition of PKR and modulation of the interferon-regulatory factors are also involved (Fig. 23.5).

As expected, cells infected with virus that has a defective *NS1* gene produce high levels of interferon and low levels of virus and, in animals, the normally highly pathogenic virus is attenuated. The knock-on effect of these studies suggests that knowledge gained will enable us to use genetic engineering to rationally create a new generation of attenuated live virus vaccines.

23.4 EMERGING VIRUS INFECTIONS

Viruses have had a significant impact on mankind. While we are all aware of the importance of virus diseases in our lives, the most dramatic impacts are seen when a population is assaulted by a previously unknown disease, either as an introduction from one group into another, or the appear-ance of an entirely new disease.

Introduction of viruses has occurred as an unintentional consequence of invasion or exploration, such as the introduction of smallpox and measles into Mexico, South and Central America in the early sixteenth century by the Conquistadores which resulted in the death of up to half of the indigenous population, aiding the invaders. Native North American populations were devastated in the seventeenth century by smallpox introduced by European settlers. In a reversal of the situation, the US army experienced significant fatalities in Cuba and Puerto Rico through exposure to yellow fever virus during the Spanish-American war at the end of the nineteenth century, although this did not hamper their ultimate success. Improvements in our understanding of viruses as agents of disease have not diminished the risks to which we are exposed by contact with new viruses (or variants of old ones such as influenza virus). The twentieth century saw the appearance of a number of previously unknown viruses affecting humans. Some of these, such as HIV, have had an immense and devastating impact on the human population (see Chapter 19). Several other viruses have appeared which are associated with a high frequency of deaths in infected people. However, in most cases the spread of these viruses has been very restricted for one reason or another and the total number of infected individuals has remained low. Examples include the filoviruses, Marburg virus and Ebola virus in Africa and the paramyxoviruses, Nipah virus and Hendra virus in southeast Asia and Australia. Others, such as enterovirus 70 which appeared in the late 1960s and causes acute hemorrhagic conjunctivitis, are not life-threatening but can be extremely distressing.

A key factor concerning emerging viruses is that these are either natural infections of specific animal species which, due to a change in circumstances, are introduced into the human population,

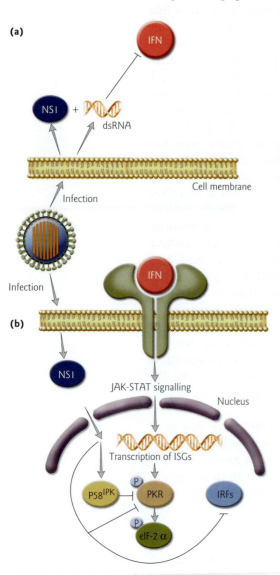

Fig. 23.5 Influenza A virus NS1 protein blocks the interferon (IFN) response on two levels: (a) virus infects cell and viral NS1 binds viral dsRNA and inhibits the production of type I interferons-α and -β; (b) virus infects cell and NS1 prevents the manifestation of the antiviral state in this cell by inhibiting protein kinase R (PKR). NS1 does this by activating the P58^{IPK}-mediated host stress response, by interacting directly with PKR, and by interfering with interferon response factors (IRFs). ISGs, interferon-stimulated genes; JAK, Janus kinases; STAT, signal transducer and activator of transcription; IF-2α, eukaryotic initiation factor-2α. Blocked lines represent inhibition. (Adapted from Katze, M. G. *et al.* 2002. *Nature Reviews Immunology* **2**, 675–687.)

or they are known viruses which are introduced into a new population; they are new only to our sphere of know-ledge and experience. In this way they emphasize our lack of knowledge about the diversity of viruses in wild, and domesticated, animals. In the natural host the viruses may cause little or no disease and it is a common feature that when a virus is transmitted from one species to a new one in a zoonotic event it frequently causes severe disease in the new host. Once a virus has invaded a new host it may, or may not, be able to be transmitted to others. It is important to appreciate that humans are not the only potential new hosts for viruses; viruses also emerge in animal populations where they may become significant threats. Examples include the appearance of canine distemper virus in the lions of the Serengeti plains and the various morbilliviruses which have appeared periodically in seals and dolphins around the world. The discovery of significant numbers of dead monkeys and gorillas, as well as other animals, in regions where Ebola virus outbreaks have occurred strongly suggests that they, like humans, are new hosts for the virus and the natural host remains to be determined.

As we extend our habitat and working environment into new territories it is inevitable that new viruses will be encountered and some of these may represent serious threats to human health. It is essential that systems are in place to identify alterations in the trends of diseases in humans and animals which may indicate the appearance of a previously unknown virus. This will offer some possibility of reacting to the new threats that will undoubtedly arise in the future.

Below we have considered three examples of emerging viruses. These have been chosen as examples of different facets of the factors which must be considered when new viruses appear.

Marburg and Ebola hemorrhagic fevers

In 1967 reports of a hemorrhagic fever in primate facilities in Marburg, Germany, and Belgrade, in what was then Yugoslavia, appeared. The disease affected staff handling primate material and spread to health workers and family members. Of 31 people affected seven died. It became clear that the cause of the disease was a previously unknown virus, named Marburg virus, which was present in monkeys captured in Uganda. The monkeys did not have any disease associated with the infection. The infection did not spread and the virus disappeared until 1975 when three people in South Africa presented with the disease, one of whom died. Two single cases of Marburg were seen in 1980 and 1987 before the appearance of a significant outbreak in the Democratic Republic of Congo where from late 1998 to late 2000, a total of 149 people were infected, with 123 deaths (82% mortality). The most recent outbreak at the time of writing (June 2006) is also the most serious recorded to date. It has affected Northern Angola and began in October 2004, and finished in late

July 2005 with a total of 374 positively confirmed cases and 329 deaths (88% of infected individuals). The numbers of deaths in this outbreak reinforces the potential seriousness of the impact of this virus. No effective treatment for Marburg virus infection has been identified.

In 1976 a hemorrhagic fever similar to that seen with Marburg virus was reported from Western Sudan and the Democratic Republic of Congo (formerly Zaire). After the first reports it became clear that the outbreak also involved Southern Sudan. Analysis of the cause quickly showed that this was due to a new virus which, though similar in structure and other features to Marburg virus, was quite distinct. The virus, named Ebola virus, is antigenically distinct from Marburg virus and has been placed in a different genus of the filovirus family to reflect the differences in detail of the molecular biology of both viruses. Since the first report, a number of outbreaks have occurred (Table 23.2) in which the human disease has been associated with a very high mortality level. All of these outbreaks, with mortality ranging from 50% to 90%, have occurred in a relatively restricted region of equatorial Africa. In 1989 a fatal disease in monkeys was reported from a primate center in Reston, Virginia. This was shown to be due to Ebola virus but fortunately this particular isolate did not cause disease in humans, though serological evidence suggested that some animal handlers may have been infected. This virus is referred to as the Reston subtype and is distinct from the three other subtypes described to date, Zaire, Sudan, and Côte d'Ivoire, all of which have caused human infection.

For both Marburg and Ebola viruses transmission between infected and susceptible individuals occurs only via close contact with body fluids, particularly blood, in which large amounts of virus particles can be detected. There is no evidence for other transmission routes. The limited route of transmission plays an important role in our ability to restrict the infection, when it appears, by taking appropriate containment measures when dealing with infected patients (and animals). A striking feature is that the Marburg and Ebola outbreaks have occurred in very restricted geographical areas with no spread beyond. This presumably reflects the range of the natural host(s) for the virus from which it is transmitted to humans. Despite a huge effort we do not yet know the natural host(s) for Marburg and Ebola viruses. While nonhuman primates such as chimpanzees have been implicated as hosts for Ebola virus they are not believed to be the reservoir from which it arises. To date the most favored potential natural host is bats.

West Nile Virus in the USA

West Nile virus was first isolated in the late 1930s. Later it was found in many parts of Europe, particularly in the Mediterranean region, and also in Africa and India. The virus is transmitted from one vertebrate host to

Table 23.2

Summary of the recorded outbreaks of Ebola virus infection in humans.

Year	Country	Cases	Deaths
1976	Democratic Republic of Congo	318	280
	Sudan	284	151
	England*	1	1
1977	Democratic Republic of Congo	34	22
1994	Gabon	49	29
	Côte d'Ivoire	1	0
1995	Democratic Republic of Congo	315	255
1996	Gabon	91	67
	South Africa†	2	1
2000–2001	Uganda	425	225
2001–2002	Gabon and Republic of Congo	122	96
Total		**1642**	**1127**

*This infection resulted from a laboratory accident.

†The first patient was a doctor who treated infected people in Gabon and the second was a nurse involved with his care.

another by mosquitoes within which the virus also multiplies. The virus is able to infect a very wide range of invertebrate and vertebrate hosts including humans and birds. However, in most mammals the virus rarely, if ever, reaches sufficient levels in the blood to be an efficient source of infection for mosquito transfer and it is believed that the natural reservoir are birds of various species, including common species such as crows and magpies. The restriction of the virus to the Eastern hemisphere of the world reflects the habitat distribution of the insect vector and natural physical barriers. The majority of infections are asymptomatic and only approximately 20% result in overt disease with a small proportion causing serious acute neurological complications which can be life-threatening.

In 1999 West Nile virus was detected during an outbreak of encephalitis in New York, the first time that it had been seen in the Americas. Since the first signs of its presence the virus has spread relentlessly across the USA and southern Canada and after 5 years was present from the east to the west coast of the continent. A particularly striking feature of West Nile virus infection in the USA is a significant mortality in horses. Infection of these animals has almost certainly aided the spread of the virus across the less densely populated parts of the continent. To date a total of more than 17,000 human cases of West Nile virus infection have been identified, with over 700 deaths (Table 23.3). At the same time the virus has been responsible for countless deaths of birds and other vertebrate species.

Table 23.3

Cases of West Nile virus (WNV) infection in humans in the USA. Figures taken from the Centers for Disease Control.

Year	Total cases	Cases of WNV fever	Cases of WNV neurological disease	WNV deaths
1999	62	3	59	7
2000	21	2	19	2
2001	66	2	64	9
2002	4156	1162	2946	284
2003	9862	6830	2860	264
2004	2539	1269	1142	100
2005	3000	1607	1294	119
Total	19,706	10,875	8464	785

The origin of the virus which initiated the introduction of West Nile virus to the USA is not yet clear. Several possibilities have been considered. From nucleotide sequence analysis of the virus genomes it is known that it is linked to virus strains circulating in birds in the Eastern Mediterranean region. The most likely explanation is that the virus was introduced from an infected bird imported (possibly illegally) into the USA. From this initial source of infection, native American mosquitoes were exposed to the virus, and from there it was spread to a wide range of vertebrates as the insects fed.

The spread of the West Nile virus infection in the USA has been extremely well documented and is an excellent case study of how a virus can enter a new territory and spread rapidly though a naïve population. The figures presented in Table 23.3 illustrate some important features of its epidemiology. At first sight the data appear to indicate that the proportions of fever, neurological disease, and deaths have altered during the course of the spread of the virus, but this is not the case and the figures merely reflect the changes in the method of detection. In the early years of the outbreak, cases of neurological disease were those which were most readily identified. As knowledge of the cause of the disease improved clinicians were more able to identify the fever associated with infection. Ultimately, testing for evidence of infection has improved and become more widespread and so the number of detected cases has risen. The rapid spread of this virus serves to emphasize the risks posed by a virus when it enters a new territory. The case illustrates that we cannot be complacent about viruses even when they appear to have established a pattern and range of infection over a long period of time.

Severe acute respiratory syndrome (SARS)

In late February 2003 a patient presented in Hanoi, Vietnam, with symptoms of a severe, acute, respiratory infection. Early symptoms were similar to those of influenza but it became rapidly clear that this was a novel infection which was considerably more severe than a typical bout of flu. A dramatic situation arose when several health workers in contact with the patient also became ill with similar symptoms. The initial patient was known to have traveled recently from Hong Kong and in March 2003 almost 50 medical staff there also presented with severe respiratory disease, which had been contracted from a patient admitted in mid February. Meanwhile, the Chinese authorities reported that since the previous November over 300 cases of atypical pneumonia resulting in five deaths had been recorded in the Guangdong Province. Over the following months as the disease spread it became clear that the Chinese outbreak was the initial stage of the spread of a previously unknown virus. For a period of some weeks the new disease made headlines around the world and led to scenes reminiscent of those seen in the 1918 influenza pandemic. A startling aspect of the disease, termed severe acute respiratory syndrome (SARS), which led to significant alarm was its rapid spread around the world. The infection appeared in other parts of Asia, Canada, and Europe and, by the end of February and within a further month or so, it had been reported in every continent of the world (Table 23.4). This rapid spread was not due to an inherent quality of the virus but instead was due to the transport of infected individuals by air. During the outbreak many countries imposed travel restrictions to reduce the risk of introduction of the virus. Between the first cases in China and the last recorded case in July 2003, in Taiwan, it is estimated that 8098 people were infected and of these 774 died, a mortality rate of 9.5%. Since July 2003 there has been no confirmed case of SARS.

The transmission of SARS requires close contact between infected and susceptible individuals such as is seen in the hospital setting. However, the virus is probably spread by respiratory transmission, in droplets shed by coughing or sneezing, and this means that the risk of spread is much higher than seen for Marburg or Ebola viruses. The rapid spread and severity of SARS led to a huge effort to identify the cause. In what is an excellent example of how the modern tools of virology can be efficiently and rapidly brought to bear on a problem, the causative agent was identified by two independent research groups in May 2003. The virus was shown to be a new member of the coronavirus family, members of which have been known to be associated with respiratory disease in some hosts, including man. The new virus, called the SARS coronavirus, has now been extensively studied in an attempt to assess its likely risk to humans in the future. Attempts to identify its origin have suggested that the natural host is the masked palm civet (often called a civet cat) which

Table 23.4

Transmission of SARS throughout the world.

First recorded case	Country (region)	Last recorded case
November 16, 2002*	China (Guangdong)	June 7, 2003
February 15, 2003	China (Hong Kong)	June 22, 2003
February 23, 2003	Vietnam (Hanoi)	April 27, 2003
February 23, 2003	Canada (Toronto)	July 2, 2003
February 25, 2003	Singapore	May 31, 2003
February 25, 2003	Taiwan	**July 5, 2003**
March 2, 2003	China (Beijing)	June 18, 2003
March 4, 2003	China (Inner Mongolia)	June 3, 2003
March 8, 2003	China (Shanxi)	June 13, 2003
March 28, 2003	Canada (Vancouver)	May 5, 2003
April 1, 2003	China (Jilin)	May 29, 2003
April 5, 2003	Mongolia (Ulaanbaatar)	May 9, 2003
April 6, 2003	Philippines (Manila)	May 19, 2003
April 12, 2003	China (Shaanxi)	May 29, 2003
April 16, 2003	China (Tianjin)	May 28, 2003
April 17, 2003	China (Hubei)	May 26, 2003
April 19, 2003	China (Jiangsu)	May 21, 2003
April 19, 2003	China (Hebei)	June 10, 2003

*The first and last recorded cases are highlighted in bold.

is sold in animal markets in China. The virus appears to require muta-tion in the spike protein genes to permit efficient replication in a human host. Serological analysis of traders in the animal markets has shown that a significant proportion of them carry SARS coronavirus-specific antibodies, suggesting that human infection is a relatively frequent event in these environments. However, it is most probable that civets are an interme-diate host of the virus rather than the natural host and the presence of a very closely related virus in Chinese horseshoe bats has been reported.

The worldwide spread of the SARS coronavirus due to air travel is the first example of its kind and raises many questions about our ability to control such infections. It also raises many unsolved questions about other viruses such as influenza A and B, whose epidemiological pattern of win-ter infection does not seem to be affected by being spread around the world by air travel.

Avian influenza A viruses

It seems inevitable that a new influenza A virus will emerge to infect the human population. One of the problems for this virus (and a blessing for us) is transmission between people. As discussed in Chapters 16 and 17, virus which is endemic in wild aquatic birds spreads to poultry, and then to humans. In 1997, H5N1 virus killed six of the 18 people (33%) in Hong Kong known to be infected, but despite there being no immunity, H5N1 virus did not spread from person to person. Since that time there have been many other examples of this situation, so it is an accepted fact. In short, although the entire influenza virus genome has been sequenced, and all its proteins identified, we know little about how the virus is transmitted. Understanding transmission is vital, as it is only the failure of viruses like H5N1 to spread that is protecting the world from a potentially catastrophic pandemic.

23.5 VIROLOGY AND SOCIETY: FOR GOOD OR ILL

As we have described in the various chapters of this book, viruses continue to pose a major threat to the health of humankind. They also threaten us indirectly, through their potential impact on agriculture and hence on our food supply and the health of our economies. Even this does not fully state the risk that viruses pose to us, for there is concern that viruses might be used as weapons by terrorists or even governments, deliberately unleashing onto vulnerable populations some of the worst effects of virus infection that we can imagine. There has never been a more important time to study viruses, to learn how they work and how they create their disease effects in a chosen host. Only with this knowledge can we seek to protect our society from the effects of both natural and deliberate infection, developing vaccines and antiviral drugs, and hope to harness viruses for our benefit, e.g. in gene therapy (Section 23.2). At the same time, we also have to consider that this knowledge could be turned against us and used to create even more potent bioterrorism agents via recombinant virus technology. Such is the climate in which science in general and virologists in particular have to operate today.

Given the crucial role that virology has to play in determining the future of humankind, and the importance of using our understanding of viruses wisely, good communication between virologists and the public is essential. It is crucial that members of the public understand what virologists are doing and why, and appreciate how that work is being sensibly regulated and its consequences considered. However, there is great ignorance about even the basics of virology. Even amongst the quality press and media sources, virologists are constantly saddened and irritated by the fundamental misconceptions that are made. If these professionals make mistakes, what hope is there for the general public? The answer, perhaps, lies in all virologists (and especially future virologists) helping to inform those around them. With greater understanding should come, for example, the willingness to accept vaccines, for, as mentioned earlier, fear and apathy make acceptance rates abysmally low. By understanding more about infectious agents, people should be able to take sensible precautions against infection, and the hysteria which has in turn accompanied increased infection with hepatitis B virus, genital herpes virus, and HIV will be replaced by more objective reactions. We are confident that the increasing numbers of students taking courses in virology will have a positive impact on the understanding of this subject by the public at large, something that will benefit us all.

KEY POINTS

- Technical advances bring new knowledge and understanding to virology.
- It may be possible one day to engineer viruses to carry therapeutic nucleic acid sequences into our cells in order to cure specific diseases.
- In addition to causing common diseases, some viruses also cause very subtle changes to our existence.
- New or emerging virus infection of humans arise continually either by viruses moving from a nonhuman host or by changes to virus genomes.
- It is important for communal health and well-being that everyone has a sound understanding of the basics of virology.

QUESTIONS

- Consider the ways in which viruses can be used as tools for the benefit of humankind.
- Why do emerging viruses continue to affect human populations, and how may such events be managed?

FURTHER READING

Burton, D. R., Barbas III, C. F. 1994. Human antibodies from combinatorial libraries. *Advances in Immunology* **57**, 191–280.

Casillas, A. M., Nyamathi, A. M., Sosa, A., Wilder, C. L., Sands, H. 2003. A current review of Ebola virus: pathogenesis, clinical presentation, and diagnostic assessment. *Biological Research for Nursing* **4**, 268–275.

Cavazzana-Calvo, M., Hacein-Bey, S., Yates, F., de Villartay, J. P., Le Deist, F., Fischer, A. 2001. Gene therapy of severe combined immunodeficiencies. *Journal of Gene Medicine* **4**, 201–206.

Espagne, E., Douris, V., Lalmanach, G. *et al.* 2005. A virus essential for insect host–parasite interactions encodes cystatins. *Journal of Virology* **79**, 9765–9776.

Fauci, A., Touchette, N. A., Folkers, G. K. 2005. Emerging infectious diseases: a 10-year perspective from the National Institute of Allergy and Infectious Diseases. *Emerging Infectious Diseases* **11**, 519–525.

Geisbert, T. W., Jahrling, P. B. 2004. Exotic emerging viral diseases: progress and challenges. *Nature Medicine* **10**, S110–S121.

Hayes, E. B. 2005. Virology, pathology, and clinical manifestations of West Nile virus disease. *Emerging Infectious Diseases* **11**, 1174–1179.

Holliger, P., Hudson, P. J. 2005. Engineered antibody fragments and the rise of single domains. *Nature Biotechnology* **23**, 1126–1135.

Imperiale, M. J., Kochanek, S. 2004. Adenovirus vectors: biology, design and production. *Current*

Topics in Microbiology and Immunology **273**, 335–357.

Jackson, N. A. C., Levi, M., Wahren, B., Dimmock, N. J. 1999. Mechanism of action of a 17 amino acid microantibody specific for the V3 loop that neutralizes free HIV-1 virions. *Journal of General Virology* **80**, 225–236.

Katze, M. G., He, Y., Gale, M. 2002. Viruses and interferon: a fight for supremacy. *Nature Reviews Immunology* **2**, 675–687.

Kay, M. A., Glorioso, J. C., Naldini, L. 2001. Viral vectors for gene therapy: the art of turning infectious agents into vehicles of therapeutics. *Nature Medicine* **7**, 33–40.

Krug, R. M., Yuan, W., Noah, D. L., Latham, A. G. 2003. Intracellular warfare between human influenza viruses and human cells: the roles of the viral NS1 protein. *Virology* **309**, 181–189.

Mayer, J., Meese, E. 2005. Human endogenous retroviruses in the primate lineage and their influence on host genomes. *Cytogenetics and Genome Research* **110**, 448–456.

McCormack, M. P., Rabbitts, T. H. 2004. Activation of the T-cell oncogene LMO2 after gene therapy for X-linked severe combined immunodeficiency. *New England Journal of Medicine* **350**, 913–922.

Melchers, W., Zoll, J., van Kuppveld, F., Swanink, C., Galama, J. 1994. There is no evidence for persistent enterovirus infections in chronic medical conditions in humans. *Reviews in Medical Virology* **4**, 235–243.

Muir, P., Archard, L. C. 1994. There is evidence for persistent enterovirus infections in chronic medical conditions in humans. *Reviews in Medical Virology* **4**, 245–250.

Pourrut, X., Kumulungui, B., Wittmann, T. *et al.* 2005. The natural history of Ebola virus in Africa. *Microbes and Infection* **7**, 1005–1014.

Rosling, M., Rosling, L. 2003. Pneumonia causes panic in Guangdong province. *British Medical Journal* **326**, 416.

Rote, N. S., Chakrabarti, S., Stetzer, B. P. 2004. The role of human endogenous retroviruses in trophoblast differentiation and placental development. *Placenta* **25**, 673–683.

Whitfield, J. B. 1990. Parasitoids, polydnaviruses, and endosymbiosis. *Parasitology Today* **6**, 381–384.

The website for the Centers for Disease Control and Prevention, Atlanta, GA. http://www.cdc.gov/

Also check Appendix 7 for references specific to each family of viruses.

Appendixes: Survey of virus properties

This chapter provides a family and thumb nail sketch for every virus referred to in this book, and gives the student an overview of viruses in the biosphere.

What follows is list of the major groups of viruses, together with a brief description and a sketch of each type of virus particle (though not drawn to scale). The classification of viruses is considered in Chapter 4. As indicated there, it is possible to classify viruses using a number of different criteria such as the nature of the host, the nature of the virus genetic material, and by using the International Committee for the Taxonomy of Viruses criteria to assign viruses to orders, families, and genera. Despite the possibilities for classification many viruses have still to be assigned to a family or genus. In the list below we have considered all of the most common classification features and for convenience have divided the viruses into those which infect animals, plants, fungi and bacteria, satellite viruses and satellites, and viroids. Some families cross these boundaries with members infecting animals or plants, etc., and the family name will be found in both categories. Here we have italicized family, subfamily, and genus names, but not those of virus name, in accordance with the conventions of the International Convention on Virus Taxonomy. The list is further ordered according to the revised Baltimore classification (see Section 4.2) and alphabetically by family. At present most is known about viruses which cause diseases of man, domestic animals, and crops. However every host species has its

Box A1

Current count of virus families and unassigned genera, according to host and the Baltimore scheme

	Animal	Plant	Algae, fungi, and protozoa	Bacteria*
Class 1	11	0	3	11
Class 2	3	2	0	2
Class 3	2	2	5	1
Class 4	16	22	3	1
Class 5	8	4	0	0
Class 6†	1	0	0	0
Class 7	1	1	0	0

*Also included are *Archaea*, *Mycoplasma* and *Spiroplasma* hosts.
†The retrotransposons of the *Metaviridae* and *Pseudoviridae* have not been included.

own broad range of viruses, and these will eventually be discovered. Thus the list is undergoing continual revision. To date 108 families/unassigned genera have been described, and are listed by host and Baltimore class in Box A1. There are over 5000 individual viruses. What is remarkable is the skewed distribution, with classes 1, 4, and 5 predominant in animals, class 4 in plants, and class 1 in bacteria. However this table does not take into account the success of individual families or even virus species. There are no class 1 viruses in plants, and few class 5 viruses in hosts other than animals. The sole family of class 6 viruses exists in animals, and there is one family of class 7 viruses in animals and another in plants.

At least one specific reference to each virus group is cited as a means of providing further information and references, and a list of general references can be found at the end of the chapter. Numerical data included with each description are approximate.

APPENDIX 1: VIRUSES THAT MULTIPLY IN VERTEBRATE AND INVERTEBRATE ANIMALS

Some vertebrate viruses have an invertebrate vector; some of these also multiply in the vector while others are carried passively on its mouthparts.

Viruses with dsDNA genomes (class 1)

Family: Adenoviridae

Nonenveloped, 80 nm icosahedral particles with a fiber protein at each vertex. Contain one molecule of linear dsDNA of 26–45,000 bp. A virus-coded protein is covalently linked to the 5′ end and forms a pseudo-circular genome by noncovalently linking to the 3′ end. Replicate and are assembled in the nucleus. Infect mainly vertebrates; one virus infects fungi (Appendix 3).

Genus: Mastadenovirus. Infects mammals, e.g. human adenovirus type 2 or type 5.
Genus: Aviadenovirus. Infect birds, e.g. fowl adenovirus type 1.
Genus: Atendovirus. Broad host range from snakes to mammals, e.g. ovine adenovirus.
Genus: Siadenovirus, e.g. frog adenovirus 1, turkey adenovirus 3.

Doefler, W., Böhm, P. 2003. Adenoviruses: model and vectors in virus–host interactions. Virion structure, viral replication, host-cell interactions. *Current Topics in Microbiology and Immunology* 272.
Doefler, W., Böhm, P. 2004. Adenoviruses: model and vectors in virus–host interactions. Immune system, oncogenesis, gene therapy. *Current Topics in Microbiology and Immunology* 273.

Family: Ascoviridae

Particle with complex symmetry, e.g. an enveloped allantoid particle, 400 × 130 nm, with one molecule of linear dsDNA of 120–180,000 bp. Replication nuclear. Infect insects, mainly *Lepidoptera,* e.g. Spodoptera ascovirus. Most are transmitted by parasitic wasps.

Cheng, X. W., Carner, G. R., Brown, T. M. 1999. Circular configuration of the genome of ascoviruses. *Journal of General Virology* **80**, 1537–1540.

Family: Asfarviridae

Genus: Asfivirus. This group shares properties with both *Iridoviridae* and *Poxviridae.* Enveloped, 200 nm icosahedral particle, containing an isometric 80 nm core. One molecule of linear dsDNA of 170–190,000 bp. Virions contain enzymes for mRNA synthesis. Cytoplasmic replication. Infect vertebrates and spread naturally by ticks. African swine fever virus.

Mebus, C. A. 1988. African swine fever. *Advances in Virus Research* **35**, 271–312.
Rojo, G., Garcia-Beato, R., Vinuela, E., Salas, M. L., Salas, J. 1999. Replication of African swine fever virus DNA in infected cells. *Virology* **257**, 524–536.

Family: Baculoviridae

Enveloped rod-shaped particles each with a nucleocapsid of 300×50 nm with one molecule of circular dsDNA of 80–180,000 bp. DNA is infectious. Virion may be occluded in a protein inclusion body, which makes it very stable – see below. A large and diverse group, common in insects – especially *Lepidoptera*, and also *Hymenoptera* and *Diptera*; also infect shrimps.

Genus: *Granulovirus*. May be occluded in a polyhedron containing one particle, e.g. Cydia pomonella granulovirus (illustrated).
Genus: *Nuclearpolyhedrosisvirus*. Virions may be occluded in a polyhedron containing many particles, e.g. Autographica californica nucleopolyhedrosis virus.

Miller, L. K. 1997. *The Baculoviruses*. Plenum, New York.
Volkman, L. E. 1997. Nuclearpolyhedrosis viruses and their insect hosts. *Advances in Virus Research* **48**, 313–348.

Family: Herpesviridae

Enveloped 200 nm particle with spikes, enclosing successively a tegument and an icosahedral nucleocapsid of 100 nm. One molecule of linear dsDNA of 125–240,000 bp. Nuclear replication. Very large family that probably infect all vertebrates; latency is a common feature of the life cycle. Some are oncogenic.

Subfamily: *Alphaherpesvirinae*
Genus: *Simplexvirus*, e.g. human herpes (simplex) virus type 1 and 2: HHV-1 and -2.
Genus: *Varicellovirus*, e.g. HHV-3 or varicella (chicken pox)-zoster virus.
Genus: *Mardivirus*, e.g. Marek's disease virus type 1; only in birds; the only oncogenic viruses in this subfamily.
Genus: *Iltovirus*, e.g. infectious laryngotracheitis virus.
Subfamily: *Betaherpesvirinae*
Genus: *Cytomegalovirus*, e.g. HHV-5 or human cytomegalovirus.
Genus: *Muromegalovirus*, e.g. mouse cytomegalovirus.
Genus: *Roseolavirus*, e.g. HHV-6, HHV-7.
Subfamily: *Gammaherpesvirinae* (lymphoproliferative viruses; oncogenic)
Genus: *Lymphocryptovirus*, e.g. HHV-4 or Epstein–Barr virus.
Genus: *Rhadinovirus*, e.g. HHV-8 (Kaposi's sarcoma-associated virus, herpesvirus saimiri.

Davison, A. J. 1991. Varicella-zoster virus. *Journal of General Virology* **72**, 475–486.
Davison, A. J. 2002. Evolution of the herpesviruses. *Veterinary Microbiology* **86**, 69–88.
Jones, C. 1999. Alphaherpes virus latency: its role in disease and survival of the virus in nature. *Advances in Virus Research* **51**, 81–133.
Rixon, F. J., Chiu, W. 2003. Studying large viruses. *Advances in Protein Chemistry* **64**, 379–408.

Family: Iridoviridae

Nonenveloped, icosahedral particles of 120–200 nm that contain lipid. Purified virus pellets are iridescent blue or green. One molecule of linear dsDNA of 140–300,000 bp. Contain several enzymes. Initial replication is nuclear. Includes viruses of insects, fish (flounder, dab and goldfish), and of many frog species, e.g. frog virus 3.

Chinchar, V. G. 2002. Ranaviruses (family *Iridoviridae*): emerging cold-blooded killers. *Archives of Virology* **147**, 447–470.
Williams, T. 1996. The iridoviruses. *Advances in Virus Research* **46**, 345–412.

Family: Nimaviridae

Genus: *Whispovirus*. Enveloped ovoid particles (130 × 280 nm) with a thread-like extension at one end. Have an inner nucleocapsid with one molecule of circular dsDNA of 300,000 bp. Nuclear. Infect marine and fresh-water crustaceans, e.g. white spot syndrome virus 1.

Yang, F., He, J., Lin, X., Li, Q., Pan, D., Zhang, X., Xu, X. 2001. Complete genome sequence of the shrimp white spot bacilliform virus. *Journal of Virology* **75**, 11811–11820.

Family: Papillomaviridae

Nonenveloped, 55 nm icosahedral particle with 72 capsomers in a skewed (T = 7) arrangement. One 8000 bp molecule of covalently closed, circular dsDNA. Replicates and assembles in the nucleus. Classification based on sequence. Over 90 human papillomaviruses (HPVs). Some are oncogenic. Infect many different vertebrate species. 16 genera. No cell culture system for replication.

Saveria Campo, M., ed. 2006. *Papillomavirus Research: from natural history to vaccines and beyond*. Caister Academic Press, Portland, OR.
zur Hausen, H. 2002. Papillomaviruses and cancer: from basic studies to clinical application. *Nature Reviews Cancer* **2**, 342–350.

Family: Polydnaviridae

Enveloped particles, but few other structural features in common. Multiple copies of circular dsDNA of 2000–30,000 bp, but the entire genome is about 250,000 bp. Replication is nuclear. Infect insects. Genera have a similar natural history.

Genus: *Ichnovirus*. 330 nm long × 85 nm diameter particle with 1 nucleo-capsid (left-hand figure). Infect parasitic ichneumoid wasps.

Genus: *Bracovirus*. Drop-shaped virions 8–150 nm long × 40 nm diameter, containing 1–12 nucleocapsids inside two membranes (right-hand figure above). Infect parasitic braconid wasps.

Gruber, A., Stettler, P., Heiniger, P., Schumperli, D., Lanzrein, B. 1996. Polydnavirus DNA of the braconid wasp *Chelonis inanitus* is integrated in the wasp's genome and excised only in later pupal and adult stages of the female. *Journal of General Virology* **77**, 2873–2879.
Whitfield, J. B. 1990. Parasitoids, polydnaviruses, and endosymbiosis. *Parasitology Today* **6**, 381–384.

Family: Polyomaviridae

Nonenveloped 40–45 nm icosahedral particle with 72 capsomers in a skewed arrangement. One 5000 bp molecule of covalently closed, circular dsDNA. Replicate and assemble in the nucleus. Include murine polyomavirus, simian virus 40 (SV40), and the human JC and BK viruses. Oncogenic.

McCance, D. J. 1998. *Human Tumor Viruses*. ASM Press, Herndon,VA.

Family: Poxviridae

Either brick-shaped or ovoid virions of 220–450 nm long × 140–260 nm wide. Some are enveloped; all have lipid. Complex structure enclosing two lateral bodies and a biconcave core; have all the enzymes required for mRNA synthesis. One molecule of linear dsDNA of 130–375,000 bp. Cytoplasmic replication. Infect mostly vertebrates but also insects.

Subfamily: *Chordopoxvirinae* (viruses of vertebrates).
Genus: *Orthopoxvirus*, e.g. vaccinia, variola of humans and related viruses.
Genus: *Avipoxvirus*, e.g. fowlpox and related viruses.
Genus: *Capripoxvirus*, e.g. sheeppox and related viruses.
Genus: *Leporipoxvirus*, e.g. myxoma of rabbits and related viruses; spread passively by arthropods.
Genus: *Molluscipoxvirus*, e.g. molluscum contagiosum virus of humans.
Genus: *Parapoxvirus*, e.g. orf virus/milker's node virus and related viruses.
Genus: *Suipoxvirus*, e.g. swinepox virus.
Genus: *Yatapoxvirus*, e.g. yabapox and tanapox and related viruses of monkeys.
Subfamily: *Entomopoxvirinae* (viruses of insects); three genera.

Moyer, R. W., Turner, P. C. 1990. Poxviruses. *Current Topics in Microbiology and Immunology* **163**, 1–211.
Seet, B. T., Johnston, J. B., Brunetti, C. R. *et al.* 2003. Poxviruses and immune evasion. *Annual Review of Immunology* **21**, 377–423.

Viruses with ssDNA genomes (class 2)

Family: Circoviridae

Nonenveloped small icosahedral particles with one molecule of circular ssDNA of 1800–2300 nt. Common worldwide.

Genus: *Circovirus.* 12–20 nm virion, with an ambisense genome, e.g. porcine circovirus-1.
Genus: *Gyrovirus.* 19–26 nm virion with a negative sense genome, e.g. chicken anaemia virus.

Todd, B. J., McNulty, M. S., Adair, B. M., Allan, G. M. 2001. Animal circoviruses. *Advances in Virus Research* **57**, 1–70.

Family: Unassigned

Genus: *Anellovirus.* 30 nm nonenveloped isometric particles with one molecule of negative sense, circular ssDNA of 3500 nt; lack of a cell system limits investigation, e.g. TT (torque teno) virus of humans, nonhuman primates, and other mammals.

Okamoto, H., Mayumi, M. 2001. TT virus: virological and genomic characteristics and disease associations. *Journal of Gastroenterology* **36**, 519–529.

Family: Parvoviridae

Nonenveloped icosahedral particles of 18–26 nm which contain no enzymes. One 5000 nt. linear molecule of *either* negative *or* positive sense ssDNA per particle, although packaging of the negative strand may be favored. On extraction these form an artefactual double strand. Replication is nuclear.

Subfamily: *Parvovirinae* (viruses of vertebrates).
Genus: *Parvovirus*, e.g. canine parvovirus, minute virus of mice.
Genus: *Erythrovirus.* B19 virus, which causes fifth disease in children.
Genus: *Dependovirus.* Adeno-associated virus. These are satellite viruses (Appendix 5).
Genus: *Amdovirus.* Aleutian disease of mink. Causes a disorder of the immune system with high levels of virus-specific antibody.
Genus: *Bocavirus.* Bovine parvovirus.
Subfamily: *Densovirinae* (viruses of mainly insects).

Heegaard, E. D., Brown, K. E. 2002. Human parvovirus B19. Clinical *Microbiology Reviews* **15**, 485–505.
Parrish, C. R., 1995. Autonomous animal parvoviruses. *Seminars in Virology* **6**, 269–355.

Viruses with dsRNA genomes and a virion-associated RNA-dependent RNA polymerase (class 3)

Family: Birnaviridae

Nonenveloped icosahedral 60 nm particles with a 45 nm core containing two segments of linear dsRNA of 2800 and 3100 bp. Has a VPg covalently linked to 5′ end of each segment. Cytoplasmic replication.

Genus: *Aquabirnavirus*, e.g. pancreatic necrosis virus. Infect fish, molluscs, and crustacea.
Genus: *Avibirnavirus*, e.g. infectious bursal disease. Bird hosts only.
Genus: *Entomobirnavirus*, e.g. *Drosophila* X virus. Insect hosts only.

Dobos, P. 1995. The molecular biology of infectious pancreatic necrosis virus. *Annual Review of Fish Diseases* **5**, 25–54.

Family: Reoviridae

Large family with members found in vertebrates, insects, fungi (Appendix 3) and plants (Appendix 2). Nonenveloped 60–80 nm icosahedral particle containing an isometric nucleocapsid with 10, 11, or 12 segments of linear dsRNA each of 1–4000 bp. Cytoplasmic replication. Within a genus RNA segments in a mixed infection readily assort to form genetically stable hybrid virus.

Genus: *Orthoreovirus*. Infect only vertebrates (man, dogs, cattle, birds). 10 dsRNA segments, e.g. reoviruses of man.
Genus: *Orbivirus*. Viruses of vertebrates and their insect vectors. 10 dsRNA segments, e.g. bluetongue virus of sheep.
Genus: *Rotavirus*. The name comes from the appearance of its wheel-like particle. Causes life-threatening diarrhea in very young vertebrates of many species, including man. 11 dsRNA segments.
Genus: *Aquaerovirus*. Viruses of fish and crustacea. 11 dsRNA segments.
Genus: *Coltivirus*. Colorado tick fever virus of vertebrates. 12 dsRNA segments. Insect vector.
Genus: *Cypovirus*. Cytoplasmic polyhedrosis viruses of insects (including *Lepidoptera*, *Hymenoptera*, and *Diptera*). 10 RNA segments.
Genus: *Seadornavirus*. Infects humans. Mosquito vector. 12 dsRNA segments.
Genus: *Idnoreovirus*. Infects insects. 10 dsRNA segments.

Tyler, K. L., Oldstone, M. B. A. 1998a. Reoviruses I. Structure, proteins and genetics. *Current Topics in Microbiology and Immunology* **233**, 1–213.
Tyler, K. L., Oldstone, M. B. A. 1998b. Reoviruses II. Cytopathogenicity and pathogenesis. *Current Topics in Microbiology and Immunology* **233**, 1–183.

Viruses with positive sense ssRNA genomes (class 4)

Family: Arteriviridae (Nidovirales)

With the *Coronaviridae* and *Roniviridae* form the order *Nidovirales*. A 45–60 nm enveloped particle containing an iscosahedral nucleocapsid with one molecule of linear positive sense ssRNA of 13,000–16,000 nt. Cytoplasmic replication; has a nested set of subgenomic mRNAs with a common leader sequence.

Genus: *Arterivirus*, e.g. equine arteritis virus, lactate dehydrogenase-elevating virus.

Snijder, E. J. & Meulenberg, J. J. M. (1998). The molecular biology of arteriviruses. *Journal of General Virology* **79**, 961–979.

Family: Astroviridae

Nonenveloped icosahedral 30 nm star-like particle with five or six points. Contain one molecule of linear positive sense ssRNA of 6000–7000 nt. Cytoplasmic replication.

Genus: *Avastrovirus*. Infect chickens, ducks and turkeys.
Genus: *Mamastrovirus*. Gastroenteritis in humans and other mammals.

Koci, M. D., Schultz-Cherry, S. 2002. Avian astroviruses. *Avian Pathology* **31**, 213–227.
Willocks, M. M., Carter, M. J., Madeley, C. R. 1992. Astroviruses. *Reviews in Medical Virology* **2**, 97–106.

Family: Caliciviridae

Nonenveloped icosahedral 27–40 nm particle with calix-like (cup-shaped) depressions. Contain one molecule of linear positive sense ssRNA of 7500 nt. 5′ end has a VPg. Cytoplasmic replication with subgenomic mRNAs. Includes viruses of many vertebrate species.

Genus: *Lagovirus*. Rabbit hemorrhagic disease virus. Epidemics with high mortality.
Genus: *Norovirus*. Norwalk virus, associated with epidemic gastroenteritis in humans.
Genus: *Vesivirus*. Vesicular exanthema of swine virus.
Genus: *Sapovirus*. Sapporo virus. Gastroenteritis in humans.

Clarke, I. N., Lambden, P. R. 1997. The molecular biology of caliciviruses. *Journal of General Virology* **78**, 291–301.

Family: Unassigned

Genus: *Hepevirus*. Nonenveloped icosahedral 30 nm particle, with one molecule of linear positive sense ssRNA of 7200 nt. that has a 5′ cap. Hepatitis E virus of humans. May be zoonotic.

Emerson, S. U., Purcell, R. H. 2003. Hepatitis E virus. *Reviews in Medical Virology*
13, 145–154.

Family: Coronaviridae (Nidovirales)

Enveloped particles of 120–160 nm with club-shaped sparse protein
spikes. Contains a helical nucleocapsid with one molecule of positive sense
ssRNA of 30,000 nt. One of the largest RNA genomes. Cytoplasmic replication with a nested set of subgenomic mRNAs with a common leader
sequence. Virions bud from the endoplasmic reticulum. Includes viruses
of mammals and birds.

Genus: *Coronavirus.* Spherical particles. Avian infectious bronchitis virus
and human coronavirus 229E.
Genus: *Torovirus.* Biconcave or toroidal or donut-shaped particles. Equine
torovirus.

Koopmans, M. M., Horzinek, M. C. 1994. Toroviruses of animals and humans.
Advances in Virus Research **43**, 233–273.
Lai, M. M. C., Cavanagh, D. 1997. The molecular biology of coronaviruses. *Advances
in Virus Research* **48**, 1–100.

Family: Dicistroviridae

Genus: Cripavirus: Nonenveloped 30 nm isometric particles. One molecule of linear positive sense ssRNA of 10,000 nt. Has a 5' VPg. Cytoplasmic replication. Mostly infect insects (e.g. cricket paralysis virus) but one
virus isolated from shrimps.

Mari, J., Poulos, B. T., Lightner, D. V., Bonami, J.-R. 2002. Shrimp Taura syndrome virus: genomic characterization and similarity with members of the genus
cricket paralysis viruses. *Journal of General Virology* **83**, 915–926.

Family: Flaviviridae

Enveloped 40–60 nm particles with an isometric nucleocapsid of 25–30
nm and one molecule of linear positive sense ssRNA of 9500–12,500 nt.
Differ from *Togaviridae* by the presence of a matrix protein, the lack of
subgenomic mRNAs, and budding from the endoplasmic reticulum.
Cytoplasmic replication. Spread by arthropods unless stated.

Genus: *Flavivirus.* Large group of viruses that multiply in the vertebrate host
and insect or tick vector, e.g. yellow fever virus, tick-borne encephalitis
virus group, dengue virus group, Japanese encephalitis virus group.
Genus: *Pestivirus.* Includes bovine diarrhea virus, border disease virus
(sheep), classical swine fever (hog cholera) virus. No vector.
Genus: *Hepacivirus.* Hepatitis C virus of man. No vector.

Lindenbach, B. D., Rice, C. M. 2001. *Flaviviridae*: the viruses and their replication. In *Fields Virology*, 4th edn (D. M. Knipe, and P. M. Howley, eds.), pp. 991–1041. Lippincott Williams and Wilkins, Philadelphia.

Meyers, G., Thiel, H.-J. 1996. Molecular characterization of pestiviruses. *Advances in Virus Research* **47**, 53–118.

Wieland, S. F., Chisari, F. V. 2005. Stealth and cunning: hepatitis B and hepatitis C viruses. *Journal of Virology* **79**, 9369–9380.

Family: Unassigned

Genus: Iflavirus. One linear positive sense ssRNA of 9000 nt. in a nonenveloped isometric 30 nm particle. Has a 5′ VPg protein. Infect insects, including moths and the honeybee, e.g. infectious flacherie virus.

Wu, C.-Y., Lo, C.-F., Huang, C.-J., Yu, H.-T., Wang, C.-H. 2002. The complete genome sequence of *Perina nuda* picorna-like virus, an insect-infecting RNA virus with a genome organization similar to that of the mammalian picornaviruses. *Virology* **294**, 312–323.

Family: Tymovirus

Genus: Marafivirus. Nonenveloped isometric 30 nm particle with one molecule of linear ssRNA of 6000–7000 nt. Multiply in leafhoppers and transmitted to plants. Cytoplasmic replication, e.g. maize rayado fino virus. Also multiply in plants – Appendix 2.

Hammond, R. W., Ramirez, P. 2001. Molecular characterization of the genome of *Maize rayado fino virus*, the type member of the genus *Marafivirus*. *Virology* **282**, 338–347.

Family: Nodaviridae

Two molecules of linear positive sense ssRNA of 1400 and 3000 nt in one nonenveloped 30 nm particle. Cytoplasmic replication with subgenomic mRNAs.

Genus: Alphanodavirus. Insect viruses (*Lepidoptera* and *Coleoptera*) but some grow unnaturally in suckling mice or vertebrate cells, e.g. Nodamura virus and black beetle virus.

Genus: Betanodavirus. Infect vertebrates, e.g. striped jack nervous necrosis virus.

Schneemann, A., Reddy, A., Johnson, J. E. 1998. The structure and function of nodavirus particles: a paradigm for understanding chemical biology. *Advances in Virus Research* **50**, 381–446.

Family: Picornaviridae

Nonenveloped viruses of mainly vertebrates. One molecule of linear positive sense ssRNA of 7000–8500 nt in 30 nm icosahedral particles. Has a 5′ VPg. Cytoplasmic replication. Most restricted to one host species.

Genus: *Enterovirus*. Acid-resistant, primarily of the intestinal tract, e.g. polioviruses, most echoviruses, coxsackie viruses of humans, and various nonhuman enteroviruses.

Genus: *Aphthovirus*. The economically important foot-and-mouth disease viruses of cattle and other ruminants.

Genus: *Cardiovirus*, e.g. encephalomyocarditis (EMC) virus of mice.

Genus: *Erbovirus*, e.g. equine rhinitis B virus.

Genus: *Hepatovirus*, e.g. human hepatitis A virus.

Genus: *Kobuvirus*, e.g. Aichi virus; causes gastroenteritis of humans associated with eating shellfish.

Genus: *Parechovirus*, e.g. human parechoviruses (formerly echoviruses) 22 and 23.

Genus: *Rhinovirus*. Acid-labile; infect the upper respiratory tract. Include about 100 common cold viruses.

Genus: *Teschovirus*. Porcine teschoviruses.

Stanway, G., Hovi, M. E., Knowles, J. R., Hyypia, T. 2002. Molecular and biological basis of picornavirus taxonomy. In *Molecular Biology of Picornaviruses* (B. L. Semler, and E. Wimmer, eds), pp. 17–24. ASM Press, Washington, DC.

Family: Roniviridae (Nidovirales)

Genus: *Okavirus*. Enveloped bacilliform virions 170 × 50 nm with protein spikes; nucleocapsid with one molecule of linear, positive sense ssRNA of 26,000 nt. Produces a nested set of 3′ coterminal mRNAs, like other nidoviruses. Infect crustacea, e.g. gill-associated virus.

Cowley, J. A., Walker, P. J. 2002. The complete genome sequence of gill-associated virus of *Penaeus monodon* prawns indicates a gene organisation unique among nidoviruses. *Archives of Virology* **147**, 1977–1987.

Family: Tetraviridae

Nonenveloped 40 nm icosahedral particle with one molecule (6500 nt) or two molecules (2500 and 5500 nt) of linear, positive sense ssRNA. No infection of cultured cells. All isolated from *Lepidoptera*, e.g. Naudaurelia capensis β virus.

Hanzlik, T. N., Gordon, K. H. J. 1997. The *Tetraviridae. Advances in Virus Research* **48**, 101–168.

Family: Togaviridae

Enveloped 70 nm particles containing an icosahedral nucleocapsid and one molecule of linear, positive sense ssRNA of 9500–12,000 nt. Cytoplasmic replication; bud from the plasma membrane. Have an intracellular subgenomic mRNA.

Genus: *Alphavirus*. Multiply in vertebrate host and insect vector, e.g. Sindbis and Semliki Forest viruses.

Genus: *Rubivirus*. Rubella virus of man; no vector.

Frey, T. K. 1994. Molecular biology of rubella virus. *Advances in Virus Research* **44**, 69–160.

Strauss, J. H., Strauss, E. G. 1994. The alphaviruses: gene expression, replication, and evolution. *Microbiological Reviews* **58**, 491–562.

Viruses with negative sense/ambisense ssRNA genomes and a virion-associated RNA-dependent RNA polymerase (class 5)

Family: Arenaviridae

Genus: *Arenavirus*. Enveloped isometric usually 120 nm particles with club-shaped spikes. Genome is contained in two helical nucleocapsids, the larger with one molecule of negative sense ssRNA of 75,000 nt and the smaller of 3500 nt. The latter is ambisense. May package more than two nucleocapsids per virion. Virions contain host cell ribosomes with no known function. Cytoplasmic replication; buds from the plasma membrane. Divided into the Old World Arenaviruses (e.g. lymphocytic choriomeningitis virus, lassa virus) and the New World Arenaviruses (e.g. Tacaribe, Junín, Pichinde).

Salvato, M. S., Ed. 1993. *The Arenaviridae*. Plenum Press, New York.

Family: Bornaviridae

Genus: *Bornavirus*.
Enveloped isometric 90 nm virions with a helical nucleocapsid and one molecule of linear negative sense ssRNA of 9000 nt. The only non-segmented negative strand virus in which RNA synthesis is nuclear. Has various spliced transcripts. Recently isolated from people with behavioral/neuropsychiatric problems but not known if causal. Known to infect horses since the eighteenth century and several other domestic vertebrate species. Highly neurotropic. Natural host unknown. Borna disease virus.

Pringle, C. R., Easton, A. J. 1997. Monopartite negative strand RNA genomes. *Seminars in Virology* **8**, 49–57.

Family: Bunyaviridae

Enveloped isometric 100 nm particle with 10 nm spikes. Contains three helical nucleocapsids, each with one molecule of linear, negative sense ssRNA: large, 7000–12,000 nt; medium, 4000–5000 nt; small, 1–3000 nt. Cytoplasmic replication; buds from the Golgi. Infect vertebrates and spread by arthropods unless stated; some infect plants (Appendix 2).

Genus: *Orthobunyavirus*. Mosquito and gnat vectors, e.g. Bunyamwera and 150 or so other viruses.

Genus: *Hantavirus*, e.g. Hantaan virus. No arthropod vector.

Genus: *Nairovirus*, e.g. Nairobi sheep disease virus. Tick vector.

Genus: *Phlebovirus*. The S RNA is ambisense. Sandfly, gnat, and tick vectors, e.g. rift valley fever virus.

Elliott, R. M., ed. 1996. *The Bunyaviridae*. Plenum Press, New York.

Kolakofsky, D. 1991. Bunyaviridae. *Current Topics in Microbiology and Immunology* **169**, 1–256.

Family: Filoviridae

Enveloped bacilliform or filamentous, and sometimes branched particles 800–900 (sometimes 14,000) × 80 nm with helical nucleocapsid of 50 nm diameter. One molecule of negative sense ssRNA of 19,000 nt. Buds from plasma membrane. Zoonotic but the natural reservoir is not known. Highly pathogenic for man, with a hemorrhagic fever – impaired clotting that leads to bleeding in many tissues.

Genus: *Marburgvirus*. Marburgviruses. 23–88% lethality in humans.

Genus: *Ebolavirus*. Ebolaviruses. 50–90% lethality in humans.

Klenk, H.-D., ed. 1999. Marburg and Ebola viruses. *Current Topics in Microbiology and Immunology* **235**, 1–225.

Family: Orthomyxoviridae

Enveloped pleomorphic 120 nm (sometimes filamentous) particles with a dense layer of protein spikes. Between six and eight helical nucleocapsids, 9 nm in diameter with transcriptase activity. Each contains one molecule of linear, negative sense ssRNA, totalling 10–15,000 nt. All RNA synthesis is nuclear. Within a genus, RNA segments in a mixed infection readily assort to form genetically stable hybrid virus. Plasma membrane budding.

Genus: *Influenzavirus A*. Genome comprises eight molecules of RNA ranging from 900 to 2350 nt. Virions have separate hemagglutinin and neuraminidase spike proteins. Undergo antigenic shift and drift. Natural reservoir is seashore birds; infect several other vertebrate species including man.

Genus: *Influenzavirus B*. Genome comprises eight molecules of RNA ranging from 900 to 2350 nt. Virions have separate hemagglutinin and neuraminidase spike proteins. Undergo antigenic drift only. Infect man.

Genus: *Influenzavirus C*. Genome comprises seven molecules of RNA ranging from 1000 to 2350 nt. Virions have no neuraminidase but have a hemagglutinin-esterase-fusion spike protein. The esterase is a receptor-destroying enzyme. Undergo minor antigenic variation. Infect man.

Genus: *Thogotovirus*. Genome comprises six (Thogoto virus) or seven (Dhori virus) RNA segments. Has one glycoprotein. Carried by ticks and can infect man. Genome comprises six molecules of RNA. Found in Asia, Africa, and Europe.

Genus: *Isavirus*. Genome comprises eight RNA segments. Infectious salmon anemia virus.

Nicholson, K. G., Wood, J. N., Zambon, M. 2003. Influenza. *Lancet* **362**, 1733–1745.

Family: Paramyxoviridae

Enveloped pleomorphic particles usually 150–200 nm in diameter. Have a dense layer of fusion protein spikes and attachment protein spikes. Contain one helical nucleocapsid 12–17 nm in diameter with one molecule of linear, negative sense ssRNA of 13–18,000 nt. May form filaments of 10–10,000 nm. Cytoplasmic replication; buds from plasma membrane. Infect vertebrates. Most, but not all, are respiratory viruses.

Subfamily: *Paramyxovirinae*.
Genus: *Respirovirus*, e.g. Human parainfluenzavirus type 1.
Genus: *Morbillivirus*. Measles virus, rinderpest virus, and canine distemper virus.
Genus: *Avulavirus*, e.g. Newcastle disease virus of poultry.
Genus: *Rubulavirus*, e.g. mumps virus or avian parainfluenzavirus type 1.
Genus: *Henipavirus*. Hendra virus and Nipah virus. Newly emerged infections of humans and domesticated animals. Natural hosts include bats.
Subfamily: *Pneumovirinae*.
Genus: *Pneumovirus*, e.g. human and bovine respiratory syncytial viruses, pneumonia virus of mice.
Genus: *Metapneumovirus*, e.g. turkey rhinotracheitis virus; human metapneumovirus, a newly discovered infection.

Curran, J., Kolakofsky, D. 1999. Replication of paramyxoviruses. *Advances in Virus Research* **54**, 403–422.

Pringle, C. R., Easton, A. J. 1997. Monopartite negative strand RNA genomes. *Seminars in Virology* **8**, 49–57.

Family: Rhabdoviridae

Enveloped, bullet-shaped particles 100–430 nm × 45–100 nm. Have 5–10 nm spikes. Inside is a helical nucleocapsid with one molecule of linear, negative sense ssRNA of 11–15,000 nt. Cytoplasmic replication and buds from plasma membrane. Infect vertebrates and plants (Appendix 2).

Genus: *Vesiculovirus*, e.g. vesicular stomatitis virus.
Genus: *Lyssavirus*, e.g. rabies virus; neurotropic; unusual as infects all mammals. Maintained naturally in bats and small carnivores; humans are dead-end hosts.
Genus: *Ephemerovirus*, e.g. bovine ephemeral fever virus.

Genus: *Novirhabdovirus*, e.g. infectious hematopoietic virus of fish.

Pringle, C. R., Easton, A. J. 1997. Monopartite negative strand RNA genomes. *Seminars in Virology* **8**, 49–57.

Viruses with RNA genomes that replicate through a DNA intermediate (class 6)

Family: Retroviridae

Enveloped 80–100 nm particles with spikes. Nucleocapsid can be isometric or a truncated cone (see genera); contains two identical copies of a linear, positive sense ssRNA of 7000–11,000 nt. Virions contain a reverse transcriptase and integrase enzymes. The DNA provirus is nuclear and integrated with host DNA. Transmission is horizontal or vertical. Associated with many different diseases. Not all viruses are oncogenic.

Subfamily: *Orthoretrovirinae*.
Genus: *Alpharetrovirus*, e.g. avian leukosis virus, Rous sarcoma virus. Concentric nucleocapsid. Oncogenic.
Genus: *Betaretrovirus*, e.g. mouse mammary tumor virus. Eccentric nucleocapsid. Oncogenic.
Genus: *Gammaretrovirus*, e.g. murine leukaemia virus. Concentric nucleocapsid. Oncogenic.
Genus: *Deltaretrovirus*, e.g. bovine leukaemia virus, human T-cell lymphotropic virus types 1–3. Concentric nucleocapsid. Oncogenic.
Genus: *Epsilonretrovirus*, e.g. walleye dermal sarcoma virus of fish.
Genus: *Lentivirus*. Infect primates (human immunodeficiency virus types 1 and 2, simian immunodeficiency viruses), horses, sheep and goats, cattle, and cats. Variously associated with immunodeficiencies, arthritis, and neurological disorders. Cone-shaped nucleocapsid.
Subfamily: *Spumavirinae*.
Genus: *Spumavirus*, e.g. simian foamy virus; named after their foamy cytopathogenic effect. Concentric nucleocapsid. Also infect cats, horses, and cattle. No disease known. No natural human infection.

Coffin, J., Hughes, S., Varmus, H. 1997. *Retroviruses*. Cold Spring Harbor Laboratory Press, Cold Spring Harbor, NY.
Levy, J. A. 1998. *HIV and the Pathogenesis of AIDS*, 2nd edn. ASM Press, Herndon, VA.

Viruses with a DNA genome that replicate through an RNA intermediate (class 7)

Family: Hepadnaviridae

A 42–50 nm enveloped particle, with projections, that contains an isometric nucleocapsid with DNA polymerase and protein kinase activities. One partially ds, circular DNA molecule that is not covalently closed. This has a complete negative sense strand of 3000 nt with a 5' terminal protein, and a variable length positive sense strand of 1700–2800 nt. The circularization overlaps the 3′ and 5′ termini of the negative sense DNA. Has a reverse transcriptase that converts an RNA intermediate into genomic DNA.

Genus: *Orthohepadnavirus*, e.g. hepatitis B (HBV) of humans, which is also associated with liver cancer, and woodchuck hepatitis virus.
Genus: *Avihepadnavirus*, e.g. duck hepatitis B virus.

Wieland, S. F., Chisari, F. V. 2005. Stealth and cunning: hepatitis B and hepatitis C viruses. *Journal of Virology* **79**, 9369–9380.

APPENDIX 2: VIRUSES THAT MULTIPLY IN PLANTS

Knowledge of plant virus multiplication is harder to obtain than that of animal viruses as plant cell culture systems are less manageable. Work tends to concentrate on physical properties and disease characteristics, especially as many are important in agriculture. Differences in virus proteins, translation strategy (which may not be mentioned below), and vector are important criteria in plant virus classification. Some plant viruses also multiply in their animal vector (*Bunyaviridae*, *Marafivirus*, *Reoviridae*, *Rhabdoviridae*), so the plant/animal virus distinction becomes uncertain.

Viruses with ssDNA genomes (class 2)

Family: Nanoviridae

Five nonenveloped small isometric particles with positive sense, circular ssDNA. 18 nm virions; genome comprises between six and eight molecules of DNA of approximately 1000 nt, with one DNA molecule per particle. Persists but does not multiply in its aphid vector, e.g. subterranean clover stunt virus, banana bunchy top virus.

Katul, L., Timchenko, T., Gronenborn, B., Vetten, H. J. 1998. Ten distinct circular ssDNA components, four of which encode putative replication-associated proteins, are associated with the faba bean necrotic yellows virus genome. *Journal of General Virology* **79**, 3101–3109.

Family: Geminiviridae

Unique virions composed of two geminate (twinned or joined) incomplete icosahedra, 30 nm long × 18 nm. Nonenveloped. Most have one molecule of closed circular ssDNA of 2500–3000 nt, but most Begamoviruses have two DNAs. Nuclear replication. Persistent or not in an insect vector, but does not multiply in the vector, e.g. maize streak virus, bean golden yellow mosaic virus.

Palmer, K. E., Rybicki, E. P. 1998. The molecular biology of the mastreviruses. *Advances in Virus Research* **50**, 183–234.

Viruses with dsRNA genomes and a virion-associated RNA-dependent RNA polymerase (class 3)

Family: Partitiviridae

Nonenveloped isometric particles of 30–40 nm, each containing two molecules of linear dsRNA of 1400–3000 bp. Have a transcriptase. Cytoplasmic replication. Spread by seed, pollen or mechanically, e.g. white clover cryptic virus 1. Also infect fungi (Appendix 3).

Rong, R., Rao, S., Scott, S. W., Carner, G. R., Tainter, F. H. 2002. Complete sequence of the genome of two dsRNA viruses from *Discula destructiva*. *Virus Research* **90**, 217–224.

Family: Reoviridae

Also found in vertebrates, insects and fungi (Appendixes 1 and 3).

Genus: *Phytoreovirus*. 12 dsRNA segments. Multiply in leafhopper insect vector, e.g. wound tumor virus of clover.
Genus: *Fijivirus*. 10 dsRNA segments. Multiply in the planthopper insect vector, e.g. Fiji disease virus.
Genus: *Oryzavirus*. 10 dsRNA segments. Multiply in planthopper insect vector, e.g. rice ragged stunt virus.

Uyeda, I., Milne, R. G. 1995. Genomic organisation, diversity and evolution of plant reoviruses. *Seminars in Virology* **6**, 85–139.

Family: Unassigned

Genus: *Endornavirus*. No virus particle; dsRNA of 14–18,000 bp with a site-specific break in the coding strand. Encodes an RNA-dependent RNA polymerase and a helicase. Associated with lipid membrane in the cytoplasm. Seed transmitted, e.g. vicia faba endornavirus.

Horiuchi, H., Udagawa, T., Koga, R., Moriyama, H., Fukuhara, T. 2001. RNA-dependent RNA polymerase activity with endogenous double-stranded RNA in rice. *Plant Cell Physiology* **42**, 197–203.

Viruses with positive sense ssRNA genomes (class 4)

(a) Isometric virions

Family: Comoviridae

Two molecules of linear, positive sense ssRNA each encapsidated separately in a nonenveloped 30 nm icosahedral particle. The small RNA is about 4000 nt and the larger about 7000 nt. Both RNAs are needed for infectivity. Cytoplasmic replication. Different genera are transmitted by beetles, aphids, or nematodes, e.g. cowpea mosaic virus, broad-bean wilt virus 1, tobacco ringspot virus.

Harrison, B. D., Murant, A. F. 1996. *The Plant Viruses*, 5th edn. Plenum Press, New York.

Family: Luteoviridae

Nonenveloped isometric 25–30 nm particles with one molecule of linear, positive sense ssRNA of 5500 nt. Some have a 5′ VPg. Does not multiply in aphid vector, but persists, e.g. barley yellow dwarf virus, potato leafroll virus, pea enation mosaic virus-1.

Mayo, M. A., Ziegler-Graff, V. 1996. Molecular biology of luteoviruses. *Advances in Virus Research* **46**, 413–460.

Family: Unassigned

Genus: *Sadwavirus*. Nonenveloped isometric 25–30 nm particles with two molecules of linear, positive sense ssRNA of 7000 and 5000 nt, probably in separate particles. Have a 5′ VPg, e.g. Satsuma dwarf virus.

Karasev, A. V., Han, S. S., Iwanami, T. 2001. Satsuma dwarf and related viruses belong to a new lineage of plant picorna-like viruses. *Virus Genes* **23**, 45–52.

Family: Sequiviridae

Nonenveloped icosahedral 25–30 nm particle with one molecule of linear, positive sense ssRNA of 10,000–12,000 nt. with a 5′ VPg. Cytoplasmic replication. Nonpersistent in aphid vector but sequiviruses need a helper waikavirus virus for transmission. Resemble picornaviruses, e.g. parsnip yellow fleck virus, rice tungro spherical virus.

Hull, R. 1996. Molecular biology of rice tungro virus. *Annual Review of Phytopathology* **34**, 275–297.

Family: Unassigned

Genus: *Sobemovirus*. Nonenveloped icosahedral 30 nm particle with one molecule of linear, positive sense ssRNA of 4000 nt. with a 5′ VPg. Virions found in both cytoplasm and nucleus, e.g. southern bean mosaic virus. Have small circular satellite RNAs.

Tamm, T., Truve, E. 2000. Sobemoviruses. *Journal of Virology* **74**, 6231–6241.

Family: Tombusviridae

Nonenveloped icosahedral 35 nm particle most (except the *Dianthovirus* genus) with one molecule of linear, positive sense ssRNA of 4000–5000 nt.

Dianthoviruses have two RNAs of 1500 and 4000 nt each in a different particle, both of which are needed for infectivity. Cytoplasmic replication. Transmitted by soil, fungi, mechanically, or by seed, e.g. *Pothos* latent virus, oat chlorotic stunt virus, *Panicum* mosaic virus, tomato bushy stunt virus, carnation mottle virus, tobacco necrosis virus A, maize chlorotic mottle virus, carnation ringspot virus.

Russo, M., Burgyan, J., Martelli, G. P. 1994. The molecular biology of Tombusviridae. *Advances in Virus Research* **44**, 321–428.

Family: Unassigned

Genus: *Idaeovirus*. Two nonenveloped isometric 33 nm particles with two molecules of linear ssRNA of 2200 and 5500 nt. Particles also contain an mRNA encoded by RNA-2 of 1000 nt. Cytoplasmic replication. Transmitted in pollen and seed, e.g. raspberry bushy dwarf virus.

Jones, A. T., McGavin, W. J., Mayo, M. A., Angel-Diaz, J. E., Karenlampi, S. O., Kokko, H. 2000. Comparisons of some properties of two laboratory strains of raspberry bushy dwarf virus (RBDV) with those of three previously published RBDV isolates. *European Journal of Plant Pathology* **106**, 623–632.

Family: Unassigned

Genus: *Cheravirus*. Two nonenveloped isometric 30 nm particles with two molecules of linear ssRNA of 3300 and 7000 nt. Nematode and seed transmitted, e.g. cherry rasp leaf virus.

James, D., Upton, C. 2002. Nucleotide sequence of RNA-2 of flat apple isolate of cherry rasp leaf virus with regions showing greater identity to animal picornaviruses than to related plant viruses. *Archives of Virology* **147**, 1631–1641.

Family: Tymovirus

Nonenveloped isometric 30 nm particle with one molecule of linear ssRNA of 6000–7000 nt. Transmitted semipersistently in beetle vectors, mechanically, by leafhoppers or seed. Cytoplasmic replication. Tymovirus, Marafivirus, and Maculavirus genera, e.g. turnip yellow mosaic virus, maize rayado fino virus, southern bean mosaic virus. Marafiviruses multiply in their leafhopper vector (Appendix 1).

Hammond, R. W., Ramirez, P. 2001. Molecular characterization of the genome of *Maize rayado fino virus*, the type member of the genus *Marafivirus*. *Virology* **282**, 338–347.

Family: Unassigned

Genus: *Umbravirus*. Uncharacterized 50 nm particles with one molecule of linear, positive sense ssRNA of 4000 nt. Infectivity is sensitive to lipid solvents, so particles may have an associated membrane. Virions are formed only with the coat protein of a coinfecting virus, often from the *Luteoviridae*. A third component (a satellite RNA: Appendix 5) is needed for disease and for successful transmission by aphids. Cytoplasmic replication. Persistent but does not multiply in aphids, e.g. groundnut rosette virus.

Naidu, R. A., Kimmina, F. M., Deom, C. M., Subramanyam, P., Chiyembebekeza, A. J., van der Merwe, P. J. A. 1999. Groundnut rosette – a virus disease affecting groundnut production in sub-Saharan Africa. *Plant Disease* **83**, 700–709.

(b) Isometric virions and virions that are short rods

Family: Bromoviridae

Nonenveloped particles with different morphologies (see below). Genome comprises three linear, positive sense ssRNAs of approximately 2000–3000 nt, totalling about 8000 nt. One RNA per particle. A fourth RNA (1000 nt) is also encapsidated; this is a coat protein mRNA encoded by RNA 3. Infectivity requires RNAs 1–3 + coat protein or RNA 4. Cytoplasmic replication.

Genera: *Ilarvirus, Bromovirus, Cucumovirus*. Three isometric particles of 30 nm, e.g. tobacco streak virus, brome mosaic virus, cucumber mosaic virus. Genus: *Alfamovirus*. Four RNAs in four bacilliform particles (short rods) 56, 43, 35 and 30 × 18 nm. Nonpersistent in aphid vector, e.g. alfalfa mosaic virus.

Family: Unassigned

Genus *Oleavirus*. Quasi-spherical to bacilliform particles of 55, 48, 43 and 37 × 18 nm with one RNA per virion, e.g. olive latent virus 2.

Palukaitis, P., Roossinck, M. J., Dietzgen, R. G., Francki, R. I. B. 1992. Cucumber mosaic virus. *Advances in Virus Research* **41**, 282–348.

(c) Virions that are rigid rods

Family: Unassigned

Genus: *Benyvirus*. Four nonenveloped rods of 390, 265, 100 and 85 × 20 nm with helical symmetry, each encapsidating one molecule of linear ssRNA of 6700, 4600, 1800, or 1300 nt. All required for infectivity, e.g. beet necrotic yellow vein virus. Some viruses may have another small RNA. Fungal vector.

Morales, F. J., Ward, E., Castano, M., Arroyave, J. A., Lozano, I., Adams, M. J. 1999. Emergence and partial characterization of rice stripe necrosis virus and its fungal vector in South America. *European Journal of Plant Pathology* **105**, 643–650.

Family: Unassigned

Genus: *Furovirus.* Two nonenveloped rods, 300 × 20 nm and 150 × 20 nm, with helical symmetry. Contain respectively one molecule of linear, positive sense ssRNA of 7000 and 3500 nt. Both required for infectivity, e.g. soil-borne wheat mosaic virus. Cytoplasmic replication. Fungal vector.

Shirako, Y., Suzuki, N., French, R. C. 2000. Similarity and divergence among viruses in the genus *Furovirus. Virology* **270**, 201–207.

Family: Unassigned

Genus: *Hordeivirus.* Three nonenveloped rods, 148, 126, 109 × 20 nm, with helical symmetry. Contain respectively one molecule of linear, positive sense ssRNA of 3800, 3200 or 2800 nt. All required for infectivity. Cytoplasmic replication. Transmitted mechanically and by seed, e.g. barley stripe mosaic virus.

Jackson, A. O., Hunter, B. G., Gustafson, G. D. 1989. Hordeivirus relationships and genome organization. *Annual Review of Phytopathology* **27**, 95–121.

Family: Unassigned

Genus: *Ourmiavirus.* Four nonenveloped bacilliform particles (short rods), mostly 30 or 40 × 18 nm, but also 45 and 60 × 18 nm, with three molecules of linear, positive sense ssRNA of (from largest to smallest) 3000, 1100, and 1000 nt. Mechanical transmission, e.g. ourmia melon virus.

Accotto, G. P., Riccioni, L., Barba, M., Boccardo, G. 1997. Comparison of some molecular properties of Ourmia melon and Epirus cherry viruses, two representatives of a proposed new group. *Journal of Plant Pathology* **78**, 87–91.

Family: Unassigned

Genus: *Pecluvirus.* Two nonenveloped rods of 245 and 190 × 21 nm with helical symmetry and each with a linear, positive sense ssRNA of 6000 and 4500 nt. Both required for infectivity, e.g. peanut clump virus. Fungal vector.

Herzog, E., Guilley, H., Manohar, S. K., Dollet, M., Richards, K. E., Fritsch, C., Jonard, G. 1994. Complete nucleotide sequence of peanut clump virus RNA 1 and relationships with other fungus-transmitted rod-shaped viruses. *Journal of General Virology* **75**, 3147–3155.

Family: Unassigned

Genus: Pomovirus. Three nonenveloped rods of 300, 150 and 70 × 18 nm with helical symmetry and a genome of three linear, positive sense ssRNAs of 6000, 3500 and 2700 nt, e.g. potato mop-top virus. Fungal vector.

Koenig, R., Pleij, C. W. A., Buttner, G. 2000. Structure and variability of the 3′ end of RNA 3 of beet soil-borne pomovirus – a virus with uncertain pathogenic effects. *Archives of Virology* **145**, 1181–2000.

Family: Unassigned

Genus: Tobamovirus. Nonenveloped rod, 300 × 18 nm, with helical symmetry. Contains one molecule of linear, positive sense ssRNA of 6500 nt. Transmitted mechanically or by seed, e.g. tobacco mosaic virus.

Buck, K. 1999. Replication of tobacco mosaic virus RNA. *Philosophical Transactions of the Royal Society of London, Series B* **354**, 613–627.

Family: Unassigned

Genus: Tobravirus. Two nonenveloped rods, 200 × 22 and 46–115 × 22 nm, with helical symmetry. Contain respectively one molecule of linear, positive sense ssRNA of 7000 or 2000–4000 nt. The larger RNA alone is infectious and the smaller specifies the coat protein. Both are needed for synthesis of new virions. Cytoplasmic replication. Nematode vector, e.g. tobacco rattle virus.

MacFarlane, S. A. 1999. Molecular biology of the tobraviruses. *Journal of General Virology* **80**, 2799–2807.

(d) Virions that are flexuous rods (i.e. with bends but not necessarily flexible)

Family: Closteroviridae

One or two nonenveloped long, 12 nm diameter, flexuous rods containing one or two molecules of linear, positive sense ssRNA totalling 7500–19,500 nt. Helical symmetry. Nonpersistent in insect vector.

Genera: Closterovirus and Ampelovirus. One variable length rod (1250–2200 nm) containing one molecule of proportionally sized RNA (15,000–19,000 nt). Aphid, mealy bug, or whitefly vectors, e.g. beet yellows virus, grapevine leafroll-associated virus 3.

Genus: Crinivirus. Large genome comprising 2 RNAs of 16,000 and 18,000 nt in separate 700 and 800 nm particles. Both are needed for infectivity, e.g. lettuce infectious yellows virus.

Agranovsky, A. A. 1996. Principles of molecular organization, expression, and evolution of closteroviruses: over the barriers. *Advances in Virus Research* **47**, 119–158.

Family: Potyviridae

Nonenveloped flexuous rod/s, 11–15 nm diameter, with helical sym-metry. One or two molecules of linear, positive sense ssRNA. Have a VPg. Cytoplasmic replication. Genera have different vectors. Many members.

Genera: *Potyvirus, Macluravirus, Ipomovirus, Rymovirus, Tritimovirus*. One par-ticle of approximately 700 nm with one RNA molecule of 10,000 nt. Cytoplasmic replication. Aphid, whitefly, mite vectors, e.g. potato virus Y, maclura mosaic virus, sweet potato mild mottle virus, ryegrass mosaic virus, wheat streak mosaic virus.
Genus: *Bymovirus*. Two particles of 300 and 600 nm with two RNAs of 8000 and 3500 nt. Both needed for infectivity e.g. barley yellow mosaic virus. Fungal vector.

Stengar, D. C., Hall, J. S., Choi, I. R., French, R. 1998. Phylogenetic relationships within the family *Potyviridae*: wheat streak mosaic virus and brome streak mosaic virus are not members of the genus *Rymovirus*. *Phytopathology* **88**, 782–787.

Family: Flexiviridae

Nonenveloped flexuous rod, 470–1000 length × 12 nm diameter, with hel-ical symmetry. Contains one molecule of linear, positive sense ssRNA of 6000–9000 nt. Cytoplasmic replication. Mite, aphid, seed and mechanical transmission.

Genera: *Allexivirus, Capillovirus, Carlavirus, Foveavirus, Mandarivirus, Potex-virus, Trichovirus, Vitivirus*, e.g. shallot virus X, apple stem grooving virus, carnation latent virus, apple stem pitting virus, Indian citrus ringspot virus, potato virus X, apple chlorotic leaf spot virus, grapevine virus A.

Gambley, C. F., Thomas, J. E. 2001. Molecular characterization of banana mild mottle virus, a new filamentous virus in *Musa* spp. *Archives of Virology* **146**, 1369–1379.

Viruses with negative sense/ambisense ssRNA genomes and a virion-associated RNA-dependent RNA polymerase (class 5)

Family: Bunyaviridae

Other genera infect animals (Appendix 1)
Genus: *Tospovirus*. The M RNA is ambisense. Transmitted by (and mul-tiply in) insect (thrips) vector, e.g. tomato spotted wilt virus.

Goldbach, R., Kuo, G. 1996. Introduction (Tospoviruses and thrips). *Acta Hor-ticultura* **431**, 21–26.

Family: Rhabdoviridae

Other genera infect animals (Appendix 1)

Genus: *Cytorhabdovirus*. Cytoplasmic replication. Spread by (and multiply in) insect vector, grafting, or mechanically, e.g. lettuce necrotic yellows virus.

Genus: *Nucleorhabdovirus*. Nuclear. Spread by (and multiply in) insect vector, grafting, or mechanically, e.g. potato yellow dwarf virus.

Tanno, F., Nakatsu, A., Toriyama, S., Kojima, M. 2000. Complete nucleotide sequence of Northern cereal mosaic virus and its genome organization. *Archives of Virology* **145**, 1383–1384.

Family: Unassigned

Genus: *Varicosavirus*. Nonenveloped helical straight rods of 320–360 × 18 nm. Contain two linear ssRNA molecules of 6000 and 7000 bp. Cytoplasmic replication, e.g. lettuce big-vein virus. Fungal vector.

Huijberts, N., Blystad, D.-R., Bos, L. 1990. Lettuce big-vein virus – mechanical transmission and relationship to tobacco stunt virus. *Annals of Applied Biology* **116**, 463–475.

Family: Unassigned

Genus: *Ophiovirus*. Virions are nonenveloped variable-length flexuous, filamentous, naked nucleocapsids, > 760 × 3 nm in diameter, with three or four molecules of linear, negative sense ssRNA of 9000, 1700, 1500 nt, e.g. citrus psorosis virus. Differ from tenuiviruses as do not infect grasses. Related to animal negative strand viruses.

Van der Wilk, F., Dullemans, A. M., Verbeek, M., van den Heuvel, J. F. J. M. 2002. Nucleotide sequence and genomic organization of an ophiovirus associated with lettuce big-vein disease. *Journal of General Virology* **83**, 2869–2877.

Family: Unassigned

Genus: *Tenuivirus*. Virions are nonenveloped variable-length flexuous, filamentous, nucleocapsid, 900–1300 × 3–10 nm in diameter. Four or more molecules of linear, negative sense ssRNA of 9000, 3500, 2500, and 2000 in rice stripe virus. Segments 2, 3, and 4 are ambisense. Transmission by leafhoppers. Differ from ophioviruses as only infect grasses, e.g. rice stripe virus. Related to bunyaviruses (phleboviruses).

de Miranda, J. R., Munoz, M., Wu, R., Ezpinoza, A. M. 2001. Phylogenetic position of a novel tenuivirus from the grass *Urochloa plantaginea*. *Virus Genes* **22**, 329–333.

Viruses with DNA genomes that replicate through a RNA intermediate (class 7)

Family: Caulimoviridae

Nonenveloped particles that are isometric or bacilliform. Contain one partially ds circular DNA molecule that is not covalently closed (like that of hepadnaviruses), with a complete negative sense strand of 7000–8000 nt and an incomplete positive strand. The circularization overlaps the 3′ and 5′ termini of the negative sense DNA. Have a reverse transcriptase. Replication is nuclear.

Genera: Caulimovirus, Cavemovirus, Petuvirus, Soymovirus . Isometric 50 nm particles, e.g. cauliflower mosaic virus, cassava vein mosaic virus, petunia vein clearing virus, Soybean chlorotic mottle virus.
Genera: Badnavirus, Tungrovirus. Rod-shaped particles 130 (but widely variable) × 30 nm. Nonpersistent in beetle larvae or leafhoppers, e.g. commelina yellow mottle virus, rice tungro bacilliform virus.

Yang, I., Haffner, G. J., Dale, J. L., Harding, R. M. 2003. Genomic characterization of *Taro bacilliform virus*. *Archives of Virology* **148**, 937–949.

APPENDIX 3: VIRUSES THAT MULTIPLY IN ALGAE, FUNGI, AND PROTOZOA

Viruses with dsDNA genomes (class I)

Family: Adenoviridae

Other genera infect animals (Appendix 1).
Genus: *Rhizidiovirus*. Infects *Rhizidiomyces* fungi.

Family: Phycodnaviridae

Infect algae. Polyhedral, nonenveloped isometric 100–200 nm particles. Members have either one molecule of linear dsDNA or dsDNA that is circularized by annealing of complementary ends.100–550,000 bp. Particles contain internal lipid. Infect *Paramoecium*, *Chlorella* and *Hydra* spp. May have a use in controlling algal blooms.

Genera: Chlorovirus, Coccolithovirus, Prasinovirus, Prymnesiovirus, Phaeovirus, and Raphidovirus.

Van Etten, J. L., Graves, M. V., Muller, D. G., Boland, W., Delaroque, N. 2002. *Phycodnaviridae* – large DNA algal viruses. *Archives of Virology* **147**, 1479–1516.

Family: Unassigned

Genus: *Mimivirus.* Infect protozoa – recently discovered. Nonenveloped, spherical particles of 400 nm surrounded by an icosahedral capsid and fibrils. Circular dsDNA of 1,180,000 bp. The largest known virus: Acanthamoeba polyphaga mimivirus.

La Scola, B., Audic, S., Robert, C. *et al.* 2003. A giant virus in amoebae. *Science* **299**, 2033.

Viruses with dsRNA genomes and a virion-associated RNA-dependent RNA polymerase (class 3)

Family: Chrysoviridae

Genus: *Chrysovirus.* Infect fungi. Nonenveloped, isometric 35–40 nm particles with four linear, dsRNAs of 3000–3500 bp., e.g. penicillium chrysogenum virus.

Jiang, D., ghabrial, S. A. 2004. Molecular characterization of Penicillium Chrysogenum virus: reconsideration of the taxonomy of the genus *Chrysovirus. Journal of General Virology* **85**, 2111–2121.

Family: Hypoviridae

Genus: *Hypovirus.* Infect fungi. No real virions, but intracellular vesicles of 50–80 nm containing one molecule of linear dsRNA of 9000–13,000 bp. There is polymerase activity but no structural proteins. Surrounded by rough endoplasmic reticulum in infected cells. Reduces the virulence of chestnut blight fungus, e.g. *Cryphonectria* hypovirus 1.

Choi, G. H., Nuss, D. L. 1992. Hypovirulence of chestnut blight fungus conferred by an infectious viral cDNA. *Science* **257**, 800–803.

Family: Partitiviridae

Other genera infect plants (Appendix 2)
Genera: *Partitivirus.* Infect fungi, e.g. Atkinsonella hypoxylon virus.

Strauss, E. E., Lakshman, D. K., Tavantzis, S. M. 2000. Molecular characterization of the genome of a partitivirus from the basidomycete *Rhizoctonia saloni. Journal of General Virology* **81**, 549–555.

Family: Reoviridae

Other genera infect animals and plants (Appendixes I and 2)
Genus: *Mycoreovirus.* 11 or 12 RNA segments. Infect parasitic fungi and make them less virulent, e.g. mycoreovirus 1.

Hillman, B. I., Supyani, S., Kondo, H., Suzuki, N. 2004. A reovirus of the fungus *Cryphonectria parasitica* that is infectious as particles and related to the *Coltivirus* genus of animal pathogens. *Journal of Virology* **78**, 892–898.

Family: Totiviridae

Infect fungi and protozoa. Nonenveloped, isometric particles of 30–40 nm with one molecule of linear dsRNA of 5000–7000 bp. Cytoplasmic replication.

Genera: *Totivirus*, infect the fungus *Saccharomyces cerevisiae*, and can have satellite RNAs that encode a toxin; *Giardiavirus*, infect the protozoan *Giardia lamblia; Leishmaniavirus*, infect protozoa *Leishmania* spp.

Ghabrial, S. A. 1994. New developments in fungal virology. *Advances in Virus Research* **43**, 303–388.

Viruses with positive sense ssRNA genomes (class 4)

Family: Barnaviridae

Genus: *Barnavirus*. Nonenveloped, short rod, 50 × 19 nm, with one molecule of linear positive sense ssRNA of 4000 nt. Infect fungi, e.g. mushroom bacilliform virus.

Ghabrial, S. A. 1994. New developments in fungal virology. *Advances in Virus Research* **43**, 303–388.

Family: Marnaviridae

Genus: *Marnavirus*. One linear positive sense ssRNA of 9000 nt in a non-enveloped isometric 25 nm particle. No VPg. One member, heterosigma akashiwo RNA virus, infects a marine photosynthetic alga.

Lang, A. S., Culley, A. I., Suttle, C. A. 2003. Nucleotide sequence and characterization of HaRNAV: a marine virus related to picorna-like viruses infecting the photosynthetic alga *Heterosigna akashiwo. Virology* **310**, 359–371.

Family: Narnaviridae

Infect fungi. No real virions, but present in cytoplasm as a ribonucleoprotein complex containing one molecule of linear, positive sense ssRNA of 2500 nt.

Genera: *Narnavirus*, e.g. saccharomyces cerevisiae 20S narnavirus; *Mitovirus*, e.g. *Cryphonectria* parasitica virus NB631.

Ghabrial, S. A. 1994. New developments in fungal virology. *Advances in Virus Research* **43**, 303–388.

APPENDIX 4: VIRUSES (PHAGES) THAT MULTIPLY IN *ARCHAEA*, BACTERIA, *MYCOPLASMA*, AND *SPIROPLASMA*

The well-known molecular biology of phages is based upon a detailed study of just a few representatives, and surprisingly little is known of their comparative biology.

Viruses with dsDNA genomes (class I)

(a) Viruses that have a head–tail structure (*Caudovirales*): families in order of decreasing tail length

Family: Siphoviridae

Nonenveloped particle with a long, noncontractile, often flexible tail up to 570×8 nm. Icosahedral head of 60 nm. Contain one molecule of linear dsDNA of 48,500 bp. Cause no host DNA breakdown. Infect bacteria including *Enterobacteria*, *Mycobacteria*, *Lactococcus*, *Methanobacteria*, *Streptomyces*, and *Vibrio*. Include phage lambda (λ), T5, chi (χ), and phi (ϕ) 80.

Ackermann, H.-W. 1999. Tailed bacteriophages. *Advances in Virus Research* **51**, 135–201.

Family: Myoviridae

Nonenveloped particle with a complex, rigid, contractile tail 80–455×16 nm. Contraction requires ATP. Head separated from tail by a neck; isometric (as shown) or elongated, 110×80 nm. Have one molecule of linear dsDNA of 336,000 bp. Infect *Enterobacteria*, *Bacillus*, *Vibrio*, and *Halobacteria (Archaea)*. Include the "T-even" enterobacteria phages T2, T4, T6, and PBS1, SP8, SP50, P1, P2, 21, 34, and Mu.

Miller, E., Kutter, E., Mosig, G., Arisaka, F., Kunisawa, T., Ruger, W. 2003. Bacteriophage T4. *Microbiology and Molecular Biology Reviews* **67**, 86–156.

Family: Podoviridae

Nonenveloped particle with a short, noncontractile tail of 20×8 nm. Icosahedral head of 60 nm. One molecule of linear dsDNA of 40,000 bp. Cause host DNA break down. Include the "T-odd" phages T3 and T7, enterobacteria phage P22, and bacillus phage phi (ϕ) 29.

Ackermann, H.-W. 1999. Tailed bacteriophages. *Advances in Virus Research* **51**, 135–201.

(b) Viruses that do not have a head–tail structure

Family: Tectiviridae

Nonenveloped 63 nm icosahedron. Have unusual internal lipid envelope around the nucleoprotein. One molecule of linear dsDNA of 150,000 bp. A tail of 60 × 10 nm is only apparent after DNA release. Infect Gram-negative bacteria carrying drug-resistance plasmids, e.g. enterobacteria phage PRD1.

Bamford, D. H., Caldentey, J., Bamford, J. K. H. 1995. Bacteriophage PRD1: a broad host range dsDNA tectivirus with an internal membrane. *Advances in Virus Research* **45**, 281–319.

Family: Corticoviridae

Nonenveloped, nontailed icosahedral 56 nm particle formed of several layers including one of internal lipid. Spikes at vertices. One molecule of circular dsDNA of 10,000 bp. Infect *Pseudomonas*, e.g. pseudoalteromonas phage PM2.

Kivela, H. M., Kalkkinen, N., Bamford, D. H. 2002. Bacteriophage PM2 has a protein capsid surrounding a spherical lipid-protein core. *Journal of Virology* **76**, 8169–8178.

Family: Fuselloviridae

Enveloped lemon-shaped particle with a noncontractile tail. Head is 100 × 60 nm and overall is 300 nm. One molecule of circular dsDNA of 17,000 bp. Infect thermophilic *Archaea* such as *Sulfolobus*, e.g. sulfolobus spindle-shaped virus 1.

Prangishvili, D., Stedman, K., Zillig, W. 2001. Viruses of the extremely thermophilic archeon *Sulfolobus*. *Trends in Microbiology* **9**, 39–43.

Family: Unassigned

Genus: *Salterprovirus*. Lemon-shaped, probably enveloped, particle of 74 × 44 nm, with a short 7 nm tail. Requires high salt for stability. Linear dsDNA of 14,500 bp. Lytic. Infect the halophilic archeon, *Haloarcula hispanica*, e.g. His1 virus.

Bath, C., Dyall-Smith, M. L. 1998. His1, an archeal virus of the *Fuselloviridae* family that infects *Haloarcula hispanica*. *Journal of Virology* **72**, 9392–9395.

Family: Guttaviridae

Droplet-shaped virions 180 × 80 nm with a beard of filaments at one end. Contain covalently closed circular dsDNA of 20,000 bp that resists digestion by many restriction endonucleases; probably extensively methylated.

Infect thermophilic *Archaea*, e.g. sulfolobus newzealandicus droplet-shaped virus (SNDV).

Arnold, H. P., Ziese, U., Zillig, W. 2000. SNDV, a novel virus of the extremely thermophilic and acidophilic archeon, *Sulfolobus*. *Virology* **272**, 409–416.

Family: Plasmaviridae

Enveloped pleomorphic 80 nm particle with small dense core. 50 and 125 nm particles also produced. Lacks a regular capsid structure. One molecule of circular dsDNA of 12,000 bp. Infect *Mycoplasma*, e.g. acholeplasma phage L2.

Maniloff, J., Kampo, G. K., Dascher, C. C. 1994. Sequence analysis of a unique temperate phage: mycoplasma virus L2. *Gene* **141**, 1–8.

Family: Rudiviridae

Nonenveloped rigid rod of 830–900 × 23 nm with helical symmetry, with a plug and three tail fibers at each end. One molecule of linear dsDNA of 33,000 bp. Nonlytic. Infect the thermophilic archeon *Sulfolobus*, e.g. sulfolobus islandicus rod-shaped virus 2.

Prangishvili, D., Arnold, H. P., Gotz, D. *et al.* 1999. A novel virus family, the *Rudiviridae*: structure, virus–host interactions and genome variability of the sulfolobus viruses SIRV-1 and SIRV-2. *Genetics* **152**, 1387–1396.

Family: Lipothrixviridae

Enveloped, rods varying from rigid (left) to flexible (right), 400–2000 × 30 nm, with protrusions at both ends that participate in cell attachment. One molecule of linear dsDNA of 16,000–42,000 bp. Infect thermophilic *Archaea*, e.g. thermoproteus tenax virus 1.

Prangishvili, D., Stedman, K., Zillig, W. 2001. Viruses of the extremely thermophilic archeon *Sulfolobus*. *Trends in Microbiology* **9**, 39–43.

Viruses with ssDNA genomes (class 2)

Family: Inoviridae

Genera differ somewhat: both are nonenveloped, rods with one molecule of circular, positive sense ssDNA.
Genera: *Inovirus*. Filamentous rod of 700–2000 × 7 nm with ssDNA of 4500–8500 nt. Host bacteria not lysed. Infect *Enterobacteria*, e.g. enterobacteria phage M13 and enterobacteria phage fd; and *Vibrio cholera* (left illustration).

Genus: *Plectrovirus*. Rigid rods infecting *Acholeplasma* are 80 × 15 nm with ssDNA of 4500 nt, e.g. acholeplasma phage virus type L51. Particles infecting *Spiroplasma* are 260 × 15 nm with DNA of 8000 nt.

Marvin, D. A. 1998. Filamentous phage structure, infection and assembly. *Current Opinion in Structural Biology* **8**, 150–158.

Family: Microviridae

Nonenveloped, icosahedral 25 nm particle with knobs on vertices. One molecule of circular ssDNA of 6000 nt (*Microvirus*) or 4400 nt (others). The *Microvirus* genus differs significantly from the others.

Genus: *Microvirus*. Infect bacteria, e.g. enterobacteria phage phi (φ) X174.
Genus: *Spiromicrovirus*. Infect *Spiroplasma*, e.g. spiroplasma phage 4.
Genus: *Chlamydiamicrovirus*, e.g. chlamydia phage 1.
Genus: *Bdellomicrovirus*, e.g. bdellovibrio phage MAC 1.

Brentlinger, K., Hafenstein, S., Novak, C. R. *et al.* 2002. *Microviridae*, a family divided. Isolation, characterization and genome sequence of phiMH2K, a bacteriophage of the obligate intracellular parasitic bacterium *Bdellovibrio bacteriovirus*. *Journal of Bacteriology* **184**, 1089–1094.

Viruses with dsRNA and a virion-associated RNA-dependent RNA polymerase (class 3)

Family: Cystoviridae

Enveloped, icosahedral 85 nm particle with 8 nm spikes. Has a 58 nm nucleocapsid and a 43 nm core. Each particle contains three molecules of linear dsRNA of 6374, 4057, and 2948 bp. Infect *Pseudomonas*, e.g. pseudomonas phage phi (φ) 6.

Mindich, L., Qiao, X., Qiao, J., Onodera, S., Romantschuk, M., Hoogstraten, D. 1999. Isolation of additional bacteriophages with genomes of segmented double-stranded RNA. *Journal of Bacteriology* **181**, 4505–4508.

Viruses with positive sense ssRNA (class 4)

Family: Leviviridae

Nonenveloped 26 nm icosahedral particles with one molecule of linear positive sense ssRNA of 3500–4300 nt. Include enterobacteria phages R17, MS2 and Qβ.

Bollback, J. P., Huelsenbeck, J. P. 2001. Phylogeny, genome evolution and host specificity of single-stranded RNA bacteriophage (Family *Leviviridae*). *Journal of Molecular Evolution* **52**, 117–128.

APPENDIX 5: SATELLITE VIRUSES AND SATELLITE NUCLEIC ACIDS OF ANIMALS, PLANTS, AND BACTERIA

The replication of satellite viruses and satellite nucleic acids depends upon coinfection of a host cell with a helper virus. Here we use a broad definition of a *satellite virus/satellite nucleic acid* as being incapable of independent production of progeny particles, yet may be able to replicate itself. *Satellite viruses* encode their own coat protein, while *satellite nucleic acids* encode nonstructural proteins or no proteins at all, and use the coat protein of their helper virus. Exactly what help satellite viruses need from their helper virus is not always clear. The *satellite virus/satellite nucleic acid genome* has no significant homology with that of the helper virus and hence differs from other types of dependent nucleic acid molecules, like defective interfering virus genomes. Some *satellite viruses/satellite nucleic acids* modulate the replication of the helper virus and exacerbate or diminish disease.

Vogt, P. K., Jackson, A. O. 1999. Satellites and defective viral RNAs. *Current Topics in Microbiology and Immunology* **239**, 1–180.

Satellite dsDNA (class 1)

Enterobacteria phage P4 satellite has one molecule of linear dsDNA of 11,627 bp. Its helper, enterobacteria phage P2 (*Myoviridae*), provides head and tail proteins to form the P4 virion.

Christie, G. E., Calender, R. 1988. Interactions between satellite P2 and its helpers. *Annual Review of Genetics* **24**, 465–490.

Satellite viruses and satellite nucleic acids with ssDNA genomes (class 2)

(a) Satellite virus

Family: Parvoviridae

See Appendix 1 for the independently replicating (autonomous) parvoviruses.

Subfamily: *Parvovirinae*.
Genus: *Dependovirus*. Have ssDNA of 4700 nt. Dependent on helper adenoviruses or herpes viruses for efficient replication, but in some cell cultures replicate independently. Particle populations contain equal numbers of positive and negative DNAs. Either type of particle is infectious alone. Adeno-associated viruses. Infect vertebrates.

Berns, K. I., Giraud, C. 1996. Biology of adeno-associated virus. *Current Topics in Microbiology and Immunology* **218**, 1–23.

(b) Satellite

682 nt genome with no ORF. Geminiviruses (Begomovirus) helpers, e.g. tomato leaf curl virus satellite DNA.

Mansoor, S., Briddon, R. W., Zafar, Y., Stanley, J. 2003. Geminivirus disease complexes: an emerging threat. *Trends in Plant Science* **8**, 128–134.

Satellite dsRNA (class 3)

Nonenveloped isometric virions with a *Totiviridae* helper virus. Linear dsRNA of 500–1800 bp inside a virion made of helper virus coat protein. Infect fungi, e.g. M satellite of Saccharomyces cerevisiae L-A virus.

Wickner, R. B. 1996. Double-stranded viruses of *Saccharomyces cerevisiae*. *Microbiological Reviews* **60**, 250–265.

Satellite viruses and satellites with positive sense ssRNA genomes (class 4)

(a) Satellite viruses

Chronic bee-paralysis virus-associated satellite virus group. Populations of 12 nm particles yield three molecules of satellite RNA each of 1000 nt. Particles are unrelated to those of the chronic bee-paralysis virus helper. The satellite interferes with the helper replication. Infect animals.

Overton, H. A., Buck, K. W., Bailey, L., Ball, B. V. 1982. Relationships between the RNA components of chronic bee-paralysis virus and those of chronic bee-paralysis virus associate. *Journal of General Virology* **63**, 171–179.

Tobacco necrosis virus satellite virus group. Isometric particles of 17 nm with one RNA of e.g. 1239 nt. Have one ORF that encodes the coat protein and some have a second ORF. Helpers from diverse groups. Fungal vector. Infect plants.

Dodds, J. A. 1999. Satellite tobacco mosaic virus. *Current Topics in Microbiology and Immunology* **239**, 145–147.

(b) Satellites

Large ssRNA satellites. Encode a nonstructural protein. Rarely modify disease. Linear RNA of 800–1500 nt. Infect plants, e.g. arabis mosaic virus large satellite RNA.

Mayo, M. A., Taliansky, M. E., Fritsch, C. 1999. Large satellite RNA: molecular parasitism or molecular symbiosis. *Current Topics in Microbiology and Immunology* **239**, 65–79.

Small ssRNA satellites. Linear RNA of <700 nt. Have no functional open reading frame. Infect plants and can exacerbate disease, e.g. cucumber mosaic virus satellite RNA.

Garcia-Arenal, F., Palukaitis, P. 1999. Structure and functional relationships of satellite RNAs of Cucumber mosaic virus. *Current Topics in Microbiology and Immunology* **239**, 37–63.

Circular ssRNA satellites. Covalently closed circular RNA of usually 350 nt. Smallest is 220 nt. All form a ds rod like viroid RNA. This is encapsidated. Also called virusoids. Have self-cleaving (ribozyme) activity. Several have sobemovirus helpers. Infect plants, e.g. velvet tobacco mottle virus satellite RNA.

Symons, R. H., Randles, W. J. 1999. Encapsidated circular viroid-like satellite RNAs (virusoids) of plants. *Current Topics in Microbiology and Immunology* **239**, 81–105.

Satellite negative sense ssRNA (class 5)

Deltavirus. Hepatitis delta virus. One circular negative sense ssRNA molecule of 1700 nt in an isometric particle of 36 nm, with lipid and all three surface proteins of the hepatitis B virus helper. Two delta virus encoded proteins form the core. Despite its RNA genome is apparently replicated by host DNA-dependent RNA polymerase II. RNA has self-cleaving (ribozyme) activity. Exacerbates hepatitis B virus infection. Infects man.

Taylor, J. M. 1999. Human hepatitis delta virus: an agent with similarities to certain satellite RNAs of plants. *Current Topics in Microbiology and Immunology* **239**, 107–122.

APPENDIX 6: VIROIDS (GENOME UNCLASSIFIED AS THEY SYNTHESIZE NO mRNA)

Viroids are small, circular, infectious ssRNAs which are never encapsidated, have no helper virus (and so are distinguished from satellite circular ssRNA), and many are serious plant pathogens. Some are silent. RNAs have extensive internal base-pairing, and resemble double-stranded rods rather than circles. There is no open reading frame in either sense strand, and hence viroids encode no protein and are not classified under the Baltimore scheme. Transmitted through vegetative propagation of the host, by seed, by aphids or through mechanical damage, and so overcome the problem of there being no receptor for naked RNA.

Family: Pospiviroidae

RNA of 350 nt with a central conserved sequence and no ribozyme self-cleavage. Replicated in the nucleus by the host's DNA-dependent RNA polymerase II. Many members, e.g. potato spindle tuber viroid, coconut cadang cadang viroid. The latter is a major economic problem.

Family: Avsunviroidae

RNA from 246 nt. No central conserved sequence. Replicate in chloroplasts. Undergo self-cleavage via their hammerhead ribozyme, e.g. avocado sunblotch viroid, potato spindle tuber viroid.

Flores, R., Daros, J. A., Hernandez, C. 1997. Viroids: the noncoding genomes. *Seminars in Virology* **8**, 65–73.

APPENDIX 7: FURTHER READING

Fauquet, C. M., Mayo, M. A., Maniloff, J., Desselberger, U., Ball, L. A., eds. 2005. *Virus Taxonomy: VIIIth report of the International Committee on Taxonomy of Viruses*, 8th edn. Elsevier, Amsterdam.

Knight, D. M., Howley, P. M. 2001. *Fundamental Virology*, 4th edn. Lipincott Williams & Wilkins, Philadelphia, PA.

The website of the International Committee on Taxonomy of Viruses (ICTV): http://www.ncbi.nlm.nih.gov/ICTVdb/index.htm

Virus particle graphics on the VirusWorld web site: http://www.rhino.bocklabs.wisc.edu/virusworld

Another ICTV database and links to other mirror sites around the world: http://www.ictvdb.rothamstead.ac.uk/index.htm

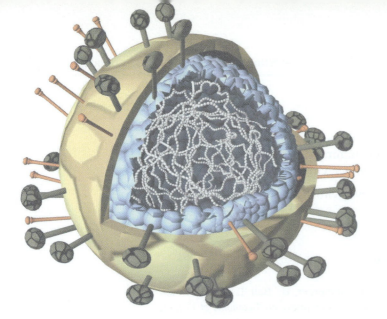

Index

Page numbers in *italics* refer to figures; those in **bold** refer to tables or boxes. The index is organized in letter-by-letter order, i.e. spaces and hyphens are ignored in the alphabetical sequence. Since the major subject of this book is viruses, few index entries are listed under headings beginning with 'virus' or 'viral'. Readers are advised to seek more specific references.

transmission 320–1
  in infected blood 266–7, 320–1
  sexual 264–5, 320
  vertical 268, **268**, 321
unresolved issues 338
vaccines *see* HIV vaccines
Vif protein 316, 318
virion proteins 318, *319*
Vpr protein 316, 318
Vpu protein 316, 318
HIV-2 114, 310, 313, **314**
  origins 313
  transmission 320–1
HIV infection 310
  animal models 331–2, **331**
  biology 311–21
  chemotherapy 329–30, **331**
    see also antiretroviral drugs
  control measures 327–37, **328**
  cost of pandemic 337–8
  course **321**, 321–3, *327*
  death 323–4
  diagnosis *328*, 328–9
  genetics of susceptibility 298
  global spread 311–12, **312**, *315*
  history 311, 312–13
  immunological abnormalities 323,
    324, 324–5
  incubation period **321**, 325, **326**
  prevention 327–8
  primary **321**, 322
  slow progression 238, 325, **326**
  types of **229**
  see also AIDS
HIV vaccines 233, 327, 330–7
  animal model systems 330–2
  clinical trials 335, **335**
  failure of experimental 333–6, **337**
  future 337
  immunity to HIV and 332–3
  types investigated 333, **334**
  vaccinia virus-based 379–80
HLA **195**
Holmes, F.O. 7
Hordeivirus 465
horizontal transmission 260, 261–7
host
  cells *see* cell(s)
  dependency of DNA replication 95

evolutionary advantage to 432, *432*
impact of viruses on **4**
species range of viruses 3, 49, 272
survival of viruses outside 260,
    263, 296
virus classification based on 50–51
virus interactions *see* virus–host
    interactions
HPV *see* human papillomavirus
HSV *see* herpes simplex virus
HTLV-1 *see* human T-cell
    lymphotropic virus type 1
human disease *see* disease, viral
human endogenous retrovirus
    (HERV) sequences 429–30
human foamy virus 122
human herpes virus type 8 (HHV-8)
    311
  oncogenicity **346**, 355–6
human immunodeficiency virus *see*
    HIV
humanized mouse monoclonal
    antibodies 418–19
human leukocyte antigen (HLA) **195**
human papillomavirus (HPV)
  carcinogenesis 352–3
  E6 and E7 proteins *351*, 353, **353**
  gene expression 132, 133
  interferon therapy 397
  oncogenicity **241**, **346**
  transformation by **348**
  transmission 265, **268**
  vaccine 353, 361
human respiratory syncytial virus
    (HRSV) *see* respiratory syncytial
    virus
human T-cell lymphotropic virus type
    1 (HTLV-1) 114
  gene expression 145
  oncogenicity **346**, 358–9
  Tax protein 359
humoral immunity **194**, *196*, 200,
    205–8
  in primary infections 232–3, **233**
  see also antibodies
hybridoma 29
hydroxymethylcytosine (HMC) 12
Hypoviridae 470
Hypovirus 470